The Kachere Tree Body of Christ (Photo: MISSIO, Munich)

The Kachere Tree is propagated by birds who deposit the pits of its fruit on the branches of other trees. Airborne roots float down and take root in the soil. As the Kachere tree grows it incorporates and changes its host. Malawian sculptors from the Kun'goni Art Centre in Mua saw this descent and transformation as a profound symbol of God's grace. From a towering Kachere tree they created this 3m high figure in order to concretize Paul's vision of the Body of Christ. The torso and limbs are made up of tiny carvings of those who have been transformed through incorporation into the Christian community. Together they are the fruit-bearing physical presence of Christ in the world.

Paul

A Critical Life

Jerome Murphy-O'Connor, OP

CLARENDON PRESS · OXFORD

1996

Oxford University Press, Walton Street, Oxford OX2 6DP
Oxford New York
Athens Auckland Bangkok Bombay
Calcutta Cape Town Dar es Salaam Delhi
Florence Hong Kong Istanbul Karachi
Kuala Lumpur Madras Madrid Melbourne
Mexico City Nairobi Paris Singapore
Taipei Tokyo Toronto
and associated companies in
Berlin Ibadan

Oxford is a trade mark of Oxford University Press

Published in the United States
by Oxford University Press Inc., New York

© *Jerome Murphy-O'Connor OP, 1996*

British Library Cataloguing in Publication Data
Data available

Library of Congress Cataloging in Publication Data
Paul: a critical life / Jerome Murphy-O'Connor.
Includes bibliographical references and index.
1. Paul, the Apostle, Saint. 2. Christian saints—Turkey—Tarsus—
Biography. 3. Apostles—Biography. I. Title.
BS2506.M855 1996 225.9'2—dc20 [B] 95–49173
ISBN 0–19–826749–5

1 3 5 7 9 10 8 6 4 2

Typeset by Joshua Associates Ltd., Oxford
Printed in Great Britain
on acid-free paper by
Bookcraft (Bath) Ltd, Midsomer Norton

Preface

NOT the least of problems faced by the author of a biography of the Apostle, Paul of Tarsus, is to find a title that will distinguish it from its many eminent predecessors. My choice of *Paul: A Critical Life* was dictated by the polyvalence of the adjective, whose range of meanings may serve both to explain my purpose and to highlight the specific contributions of this volume.

One sense of 'critical' is 'involving suspense as to the issue', but it can also mean 'decisive, crucial'. Both are applicable to Paul, and to my own life in relation to his. The church of Antioch was responsible for the missionary outreach, which demanded of pagan converts only faith in Jesus. It was in this spirit of freedom that Paul laboured in Asia Minor and Greece. When Antioch later changed its stance and aligned itself with Jerusalem, which insisted on observance of the Law, the the status of its churches to the north and west came under attack. The very nature of Gentile Christianity was put at risk. Paul was its main defender. For five or six years in the middle of the first century AD he invested every ounce of his energy, and every scintilla of his intelligence, in devising a response which was ultimately to prevail. Even if his writings were not part of the canon, the incalculable debt we owe him is adequate justification for yet another attempt to understand how and why he achieved what he did.

On a more personal level, I wrote my doctoral dissertation at the University of Fribourg, Switzerland, on Paul's understanding of the function of preaching, and it was to prove crucial to my future career. Not only did it lead to my nomination to the École Biblique in 1965, which has been my academic home ever since, but it stimulated a life-time interest in the Pauline writings. At first my concern was with the exact interpretation of points of detail, with a view to a better understanding of his theology, but almost insensibly my focus gradually shifted to the historical dimension of his life and work. The more conscious I became of the way theological thought actually develops—by historically conditioned insights rather than by logical deduction from a deposit of faith—the more I wanted to encounter the personality behind the letters, and to determine the factors which led him to think in a particular way. This book contains the fruits of that quest, which are displayed with a certitude that all historians will recognize as spurious. Only definiteness, however, can provoke the reactions that in dialogue lead to progress. I make my own what J. A. T. Robinson said in the conclusion to a much more challenging work, 'all the statements of this book should be taken as questions' (1976: 357).

I try to be as 'critical' as possible in the sense of 'exercising careful judgement', above all with respect to the use of material from the Acts of the Apostles. The tradition of lives of Paul has been to accept the framework provided by Luke, and into it to integrate material from the letters. The appropriateness of this approach, which subordinated the testimony of the individual concerned to that of a tendentious theologian, was questioned by J. Knox, who, in consequence, laid down the methodological principle, 'A fact only suggested in the letters has a status which even the most unequivocal statement of Acts, if not otherwise supported, cannot confer. We may, with proper caution, use Acts to supplement the autobiographical data of the letters, but never to correct them' (1950: 32). Recent lives of Paul (e.g. Fitzmyer, Baslez, Légasse) all pay lip-service to this principle, but in practice they not only permit Luke to exercise decisive control over the presentation of Paul's career, but fail to recognize the problems of extracting historical data from the Acts of the Apostles. The sporadic criticisms of Luke's portrait of Paul, which are scattered through many commentaries on Acts, has been recently competently synthesized by J. C. Lentz, Jr., in his *Luke's Portrait of Paul* (1993).

I may have gone to the other extreme in the way I use the letters as the principal source of Paul's biography, but the publications of three of my colleagues have made it impossible for me to continue to read the Acts of the Apostles with the naïvety that characterized some of my earlier work. Advances in textual criticism mean that it is no longer acceptable to move without comment or justification from the Western text to the Alexandrian text and back again. Moreover, in their present form both are corrupt. To date the only fully documented critical text is that provided by M.-E. Boismard and A. Lamouille in their monumental *Texte Occidental des Actes des Apôtres* (1984). A by-product of their work is a heightened awareness of the complexity of the literary development of the Acts of the Apostles.

They have attempted to determine its various stages in their *Les Actes des deux Apôtres*, i–iii (1990), whose implications for the history of the early church are being worked out by Justin Taylor, SM His first volume, *Les Actes des deux Apôtres*, v. *Commentaire historique (Act 9. 1–18, 22)*, appeared in 1994. Only those who have attempted to reconstruct history will recognize the inestimable advantage of working beside a colleague who approaches the same situations from a different perspective. The interaction enlightened me in ways that I would not have thought possible, and his publications dispense me from dealing with the data of Acts in greater detail. Our books should be considered complementary.

The decision to use the letters as the controlling source in the reconstruction of Paul's life has had important consequences for the organization of the book. The authenticity, integrity, and order of the letters had to be addressed more thoroughly than in comparable biographies. A strictly chronological order proved incapable of integrating such discussions, which often provided valuable insights into Paul's attitude at a given moment. Thus I deal with all his contacts with a given church in the same chapter, even though events concerning other churches may have intervened. A life, however, moves forward through time. In order to keep this dimension in focus, I begin by establishing a general chronological framework of Paul's career (Ch. 1), which subsequently is made more precise for the crucial two years and three months that he spent in Ephesus, whence he wrote half of his letters (Ch. 7).

This two-pronged approach has the advantage of throwing into relief the essentially dialogical character of Paul's thought. Each community generated questions to which he had no ready-made answers. His response in each case is tailored to the particular situation, but rooted in a consistent core, which is his vision of Christ. I have devoted particular attention to isolating the new ideas, and the improved or modified formulations, that changing circumstances forced him to develop. Only thus can one come to a proper appreciation of the quality of his intellectual training and the extraordinary flexibility of his mind.

The bibliography reveals my indebtedness to the generations of Pauline scholars on whose shoulders I stand. Three deserve to be singled out—Ceslaus Spicq, OP, of the University of Fribourg, Ernst Käsemann of the University of Tübingen, and Charles Kingsley Barrett of the University of Durham—because, in addition to the illumination of their writings, their friendship and personal example as teachers and ministers have greatly influenced the way I see the role of scholarship in the church.

I must also express my gratitude to Santiago Martinez Caro of Ankara, Turkey, who surveyed Galatia for me, and to Anthony Ward. S.M., erstwhile Librarian of the École Biblique and now Chief Archivist of the Basilica of Saint Peter in Rome, my colleague Justin Taylor, SM, and Terence Prendergast, SJ, whom the Catholic Biblical Association of America sent to the École Biblique as a visiting professor at just the opportune moment. The three last named read the manuscript and saved me from many mistakes. Those that remain are my own.

J.M.-O'C.

1 September 1995

Contents

List of Figures

Abbreviations

AARAS	*American Academy of Religion Academy Series*
AB	*Anchor Bible*
ABD	*Anchor Bible Dictionary*
ABRL	Anchor Bible Reference Library
AJ	Josephus, *Antiquities of the Jews*
AJA	*American Journal of Archaeology*
AnBib	Analecta Biblica
ANRW	*Aufstieg und Niedergang der römischen Welt*
AusBR	*Australian Biblical Review*
BA	*Biblical Archaeologist*
BAGD	W. Bauer, W. F. Arndt, F. W. Gingrich, F. W. Danker, *A Greek-English Lexicon of the New Testament and Other Early Christian Literature*, 2nd edn. (Chicago: University of Chicago Press, 1979)
BASOR	Bulletin of the American Schools of Oriental Research
BBB	Bonner biblische Beiträge
BCH	*Bulletin de correspondance Héllenistique*
BdeJ	*Bible de Jérusalem*
BDF	F. Blass, A. Debrunner, R. Funk, *A Greek Grammar of the New Testament and Other Early Christian Literature* (Cambridge: Cambridge University Press, 1961)
BETL	Bibliotheca ephemeridum theologicarum lovaniensium
BHT	Beiträge zur historischen Theologie
BJRL	*Bulletin of the John Rylands Library*
BNTC	Blacks New Testament Commentary
BR	*Bible Review*
BZ	*Biblische Zeitschrift*
BZAW	Beihefte zur Zeitschrift für die altestamentliche Wissenschaft
BZNW	Beihefte zur Zeitschrift für die neutestamentliche Wissenschaft
CAH	*Cambridge Ancient History*
CBQ	*Catholic Biblical Quarterly*
CCSL	Corpus Christianorum Series Latina
CD	*Damascus Document* (Qumran)
CGTC	Cambridge Greek Testament Commentary
CIG	*Corpus Inscriptionum Graecarum*
CIL	*Corpus Inscriptionum Latinarum*

CNT	Commentaire du Nouveau Testament
ConBNT	Coniectanea biblica. New Testament
CRB	Cahiers de la Revue biblique
DBSup	*Dictionnaire de la Bible. Supplément*
Digest	*The Digest of Justinian*, ed. T. Mommsen; trans. A. Watson (Philadelphia: Pennsylvania State University Press, 1985)
ÉBib	Études bibliques
ETL	*Ephemerides theologicae lovanienses*
ExpTim	*Expository Times*
FRLANT	Forschungen zur Religion und Literatur des Alten und Neuen Testaments
FS	Festschrift
GNS	Good News Studies
HDB	J. Hastings, *Dictionary of the Bible*, 2nd edn., revised by F. C. Grant and H. H. Rowley (Edinburgh: Clark, 1963)
HNTC	Harpers New Testament Commentary
HSCP	*Harvard Studies in Classical Philology*
HTKNT	Herders theologischer Kommentar zum Neuen Testament
HTR	*Harvard Theological Review*
HUZT	Hermeneutische Untersuchungen zur Theologie
ICC	International Critical Commentary
Int	*Interpretation*
IvEph	*Die Inschriften von Ephesos*, ed. C. Borker *et al.* (Bonn: Habelt, 1979–81)
JB	Jerusalem Bible
JBC	Jerome Biblical Commentary
JBL	Journal of Biblical Literature
JCS	Journal of Classical Studies
JHS	Journal of Hellenic Studies
JQR	Jewish Quarterly Review
JRS	Journal of Roman Studies
JSJ	Journal for the Study of Judaism
JSNT	Journal for the Study of the New Testament
JSNTSup	Journal for the Study of the New Testament Supplements
JSOTSup	Journal for the Study of the Old Testament Supplements
JSP	*Journal for the Study of the Pseudepigrapha*
JTS	*Journal of Theological Studies*
JW	Josephus, *Jewish War*
LCL	Loeb Classical Library
LD	Lectio Divina
LSJ	H. G. Liddell, R. Scott, *A Greek-English Lexicon*, revised by H. S. Jones, with supplement (Oxford: Clarendon Press, 1968)

LXX	The Septuagint
MAMA	*Monumenta Asiae Minoris Antiqua*, ed. W. M. Calder (London: Manchester University Press, 1928).
MeyerK	H. A. W. Meyer, *Kritisch-exegetischer Kommentar über das Neue Testament*
MQR	*Michigan Quarterly Review*
NAB	*New American Bible*
NCB	New Clarendon Bible
NEB	*New English Bible*
NH	Pliny, *Natural History*
NICNT	New International Commentary on the New Testament
NIGTC	New International Greek Testament Commentary
NJB	*New Jerusalem Bible*
NJBC	*New Jerome Biblical Commentary*
NovT	*Novum Testamentum*
NovTSup	Novum Testamentament Supplements
NRSV	*New Revised Standard Version*
NRT	*Nouvelle Revue Théologique*
NTAbh	Neutestamentliche Abhandlungen
NTOA	Novum Testamentum et Orbis Antiquus
NTS	*New Testament Studies*
OBO	Orbis Biblicus et Orientalis
OCD	*Oxford Classical Dictionary* (1949)
OCCL	*Oxford Companion to Classical Literature* (1989)
PG	*Patrologia Graeca*, ed. J. P. Migne
PIBA	*Proceedings of the Irish Biblical Association*
PL	*Patrologia Latina*, ed. J. P. Migne
PW	Pauly-Wissowa-Kroll, *Real-Encyclopädie der classischen Altertumswissenschaft* (Stuttgart, 1894–1980)
PWSup	Supplement to PW
RB	*Revue biblique*
RechBib	Recherches bibliques
REG	*Revue des Études Grecques*
PL	Patrologia Latina
RHPR	*Revue d'Histoire et des philosophie religieuses*
RNT	Regensburger Neues Testament
RQ	*Revue de Qumran*
RSPT	*Revue des sciences philosophiques et théologiques*
RSR	*Recherches de science religieuse*
RSV	*Revised Standard Version*
SBLDS	Society of Biblical Literature Dissertation Series
SBS	Stuttgarter Bibelstudien

SBT	Studies in Biblical Theology
SC	Sources Chrétiennes
SD	Studies and Documents
SJLA	Studies in Judaism in Late Antiquity
SJT	*Scottish Journal of Theology*
SNTSMS	Society for New Testament Studies Monograph Series
SPB	Studia Postbiblica
StNT	Studia Neotestamentica
STRT	Studia Theologica Rheno-Traiectina
SUNT	Studien zur Umwelt des Neuen Testaments
TDNT	*Theological Dictionary of the New Testament*
THKNT	Theologischer Handkommentar zum Neuen Testament
TPAPA	*Transactions and Proceedings of the American Philological Association*
TS	*Theological Studies*
TSAJ	Texte und Studien zum antiken Judentum
TSK	*Theologische Studien und Kritiken*
TU	Texte und Untersuchungen
TynB	*Tyndale Bulletin*
TynNTC	Tyndale New Testament Commentary
TZ	*Theologische Zeitschrift*
UNT	Untersuchungen zum Neuen Testament
VC	*Vigiliae Christianae*
WBC	Word Biblical Commentary
WUNT	Wissenschaftliche Untersuchungen zum Neuen Testament
ZNW	*Zeitschrift für die neutestamentliche Wissenschaft*
ZPE	*Zeitschrift für Papyrologie und Epigraphik*
ZTK	*Zeitschrift für Theologie und Kirche*

— 1 —

The Chronological Framework

THE chronological reference points which situate Paul's life and ministry within the history of the first century are not numerous. Paul himself provides two, from which other episodes can be dated. Others are given by Luke in the Acts of the Apostles.

THE EVIDENCE FROM THE PAULINE LETTERS

Date of Birth

In the letter to Philemon 9 Paul calls himself a *presbytês*. Both the *NAB* and *RSV* have the translation 'ambassador', which means they accept the variant *presbeutês*. This reading, however, is without manuscript support, and is no more than a conjecture based on a misunderstanding of the letter. An evocation of authority does not suit the context where Paul has just refused to order Philemon to do something. A mention of his age fits better with the appeal he is making. Hence, we must translate 'old man' (*NRSV*, *NJB*) or more accurately 'elderly'.[1] The meaning of the adjective is vague, but some precision as regards the way it was used in the first century can be derived from ancient versions of Shakespeare's seven ages:

> All the world's a stage,
> And all the men and women merely players:
> They have their exits and their entrances;
> And one man in his time plays many parts,
> His acts being seven ages.
>
> (*As You Like It* II. 7)

All unattributed translations are my own

[1] Lohse (1968), 277; Bornkamm, *TDNT* 6. 683; against Knox (1950), 74.

The Greek Tradition

Shakespeare's use of the figure 7 reflects a venerable tradition, but he substitutes vivid descriptions for the precise figures of the ancients. A contemporary of Paul, Philo, the great Jewish philosopher from Alexandria, wrote in his *De Opifico Mundi* (103–5; trans. Yonge):

> The growth of men from infancy to old age, when measured by the number seven, displays in a most evident manner its perfecting power; for in the first period of seven years, the putting forth of the teeth takes place. And at the end of the second period of the same length, he arrives at the age of puberty; at the end of the third period, the growth of the beard takes place. The fourth period sees him arrive at the fulness of his manly strength. The fifth seven years is the season for marriage. In the sixth period he arrives at the maturity of his understanding. The seventh period is that of the most rapid improvement and growth of both his intellectual and reasoning powers. The eight is the sum of the perfection of both. In the ninth, his passions assume a mildness and gentleness, from being to a great degree tamed. In the tenth, the desirable end of live comes upon him, while his limbs and organic senses are still unimpaired; for excessive old age is apt to weaken and enfeeble them all.

> And Solon, the Athenian lawgiver [635–560 BC], described these different ages in the following elegiac verses:

> > In seven years from th'earliest breath,
> >   The child puts forth his hedge of teeth;
> > When strengthened by a similar span,
> >   He first displays some signs of man.
> > As in a third, his limbs increase,
> >   A beard buds o'er his changing face.
> > When he has passed a fourth such time,
> >   His strength and vigour's in its prime.
> > When five times seven years o'er his head
> >   Have passed, the man should think to wed;
> > At forty-two, the wisdom's clear
> >   To shun vile deed or folly or fear;
> > When seven times seven years, to sense
> >   Add ready wit and eloquence.
> > And seven years further skill admit
> >   To raise them to their perfect height.
> > When nine such periods have passed,
> >   His powers, though milder grown, still last;
> > When God has granted ten times seven,
> >   The aged man prepares for heaven.

> Solon therefore thus computes the life of man by the aforesaid ten periods of seven years. But [Pseudo] Hypocrates the physician says that there are seven ages of man—infancy, childhood, boyhood, youth, manhood, middle age, old

age—and that these too are measured by periods of seven, though not in the same order. And he speaks thus:

> In the nature of man there are seven seasons, which men call ages: infancy, childhood, boyhood and the rest:
> He is an infant [*paidion*] till he reaches his seventh year, the age of the shedding of his teeth.
> He is a child [*pais*] until he arrives at the age of puberty, which takes place in fourteen years.
> He is a boy [*meirakion*] till his beard begins to grow, and that time is the end of a third period of seven years.
> He is a youth [*neaniskos*] till the completion of the growth of his whole body, which coincides with the fourth seven years.
> Then he is a man [*anêr*] until he reaches his forty-ninth year, or seven times seven periods.
> He is an elderly man [*presbytês*] till he is fifty-six, eight times seven years old.
> And after that he is an old man [*gerôn*].

The importance both Solon and Hippocrates attached to the figure 7 is manifest, but Solon's use is much more realistic than that of Hippocrates. For the latter, perfection is achieved at 49, because this is the square of 7; the artificiality is blatant.[2] After 50 it is all downhill, and the author's lack of interest is underlined by his failure to complete the schema. Solon's estimate of the average lifespan[3] corresponds to that of the Bible, 'The days of our life are seventy years, or perhaps eighty, if we are strong' (Ps. 90: 10),[4] and he extends the vigour of the prime of life to 56. The characteristics of his ninth stage (gentle but not marked decline) are those we associate with the 'elderly' as opposed to the 'old'.

The Jewish Tradition

Jewish tradition confirms this interpretation. The redemption price of 50 shekels was valid only for a man in the prime of life, i.e. between the ages of 20 and 60. Once past 60 his value dropped to 15 shekels (Lev. 27: 2–7).[5] The suggestion that decline begins at 60 is explained by the *Damascus Document* which legislates,

> Let no man be in office from 60 years and over as judge of the congregation; for because of the unfaithfulness of men their days are diminished, and in the heat of his anger against the inhabitants of the earth God has ordained that their understanding should decline before their days are fulfilled.
>
> (*CD* 10: 7–10)

---

[2] None the less it is accepted by Fitzmyer (1990), 14.

[3] Herodotus quotes Solon as saying, 'I set the limits of man's life at 70 years' (*Histories* 1. 32).

[4] 'Then they shall say, "The days of the forefathers were many, even unto a thousand years, and were good, but, behold, the days of our life, if a man has lived many, are three score years and ten, and, if he is strong, four score years and those evil"' (*Jubilees* 23: 15).

[5] See Philo, *Spec. Leg.* 2. 33.

A slightly different perspective is furnished by a rabbinic list which specifies 60 as the age when one is classed as an elder,

> At 5 years old one is fit for the Scripture,
> At 10 for the Mishnah,
> At 13 for the fulfilling of the commandments,
> At 15 for the Talmud,
> At 18 for the bride-chamber,
> At 20 for pursuing a calling,
> At 30 for authority,
> At 40 for discernment,
> At 50 for counsel,
> At 60 for to be an elder,
> At 70 for grey hairs,
> At 80 for special strength,
> At 90 for bowed back,
> At 100 a man is as one that has already died.
>
> (*m. Aboth* 5. 21; trans. Danby)

This list is attributed variously to Rabbi Samuel the Younger (end of the first century AD) or R. Judah ben Tema (end of the second century AD) but, if we excise the references to Mishnah and Talmud, there is no reason why the remainder should not have been in circulation at least by the first century. There are manifest links with the legislation of the Essenes,[6] according to which education began in early childhood, the age of adulthood (including the right to marry) was 20, initial responsibility began at 25 (cf. *CD* 10: 6–7), and a man could become a judge at 30 (1QSa 1. 6–16).

To sum up: for Paul's contemporaries, any male in his late fifties or early sixties would have been considered 'elderly'. We must presume that Paul shared this assessment. His concern, it will be recalled, was to present himself to Philemon as a pathetic figure. Thus, if, as I shall argue later,[7] the letter to Philemon was written in AD 53, Paul would then have been about 60, which would put his birth in the last years of the pre-Christian era. In other words, he would have been born about the same time as Christ.[8]

Date of Departure from Damascus

'At Damascus, the governor [*ho ethnarchês*] of King Aretas guarded the city of Damascus in order to seize me, but I was let down in a basket through a window

---

[6] On the underlying common tradition, see Borgen (1961).

[7] See Ch. 7, 'Imprisonment'.

[8] For Légasse (1991), 31, Paul would have been ten years younger. Baslez (1991), 22, dates his birth to AD 15 on the grounds that he was 'young' (Acts 7: 58) at the time of his conversion in AD 34. *Neanis*, however, is applicable to anyone between 24 and 40 (BAGD s.v.).

in the wall, and escaped his hands' (2 Cor. 11: 32–3). This ironic 'accomplishment' must be integrated into data supplied by Paul in Galatians: 'I went away into Arabia, and again I returned to Damascus. Then after three years I went up to Jerusalem to visit Cephas and remained with him fifteen days' (1: 17–18).

Arabia was the territory controlled by the Nabataeans in what is today the Kingdom of Jordan.[9] The Aretas mentioned here can only be Aretas IV the Arabian.[10] According to inscriptions and coins,[11] he ruled the Nabataeans for 48 years. Since his reign began in 9 BC,[12] his death (given the flexibility of ancient counting systems) must be placed between AD 38 and 40. AD 39 appears to be the most probable.[13]

The simplest way of integrating the data of Paul's two allusions is to recognize that the political situation of Damascus differs in the references in Galatians and 2 Corinthians,[14] and to postulate that he got into trouble among the Nabataeans[15] in Arabia and was forced to return to Damascus in some haste. He there enjoyed three years of peace until the Nabataeans obtained control of the city which obliged him to flee again, this time to Jerusalem.

This reconstruction highlights the importance of 2 Corinthians 11: 32–3. If a date can be extracted from it, we can then date Paul's conversion, his visit to Arabia, and his first visit to Jerusalem as a Christian. It would be a key chronological element in the early part of the Apostle's life.

Nabataean Control of Damascus

From 65 BC when Pompey marched the legions into the East, the great trading city of Damascus was an integral part of the Roman province of Syria.[16] When did it pass into Nabataean hands? Some deny that it ever did,[17] because there is no independent confirmation of Paul's information, and because when Luke recounts the Apostle's flight he blames the hostility of the Jews; the Nabataeans are not mentioned, 'When many days had passed the Jews plotted to kill him, but their plot became known to Saul. They were watching the gates day and night to kill him, but the disciples took him by night and let him down over the wall in a basket' (Acts 9: 23–5).

---

[9] For details see Ch. 4 'Arabia'.
[10] Josephus, *JW* 2. 68; 2 Macc. 5: 8.
[11] Meshorer (1975), 41–63.
[12] Josephus, *AJ* 16. 294.
[13] So rightly Jewett (1979), 30, who follows Gutschmid (1885), 84–9; see also PW 2. 674.
[14] Burton (1921), 57.
[15] What happened will be discussed later; see Ch. 4, 'Arabia'.
[16] Josephus, *AJ* 14. 29; *JW* 1. 127. The history of the province is given in Schürer (1973–87), 1. 242–66.
[17] e.g. Lüdemann (1984), 31 n. 10, who, in consequence, attaches no chronological value to 2 Cor. 11: 32–3, and thus is of necessity rather speculative in the dates he assigns to the early part of Paul's career, namely conversion in AD 30, or less probably 33, and first visit to Jerusalem in 33, or less probably 36 (p. 262).

On methodological grounds alone, Paul's first-hand account is certainly to be preferred to Luke's second-hand version, which moreover is a tissue of implausibilities.[18] Why should the Jews take the risk of killing Paul? Their purpose would have been fulfilled by declaring him *persona non grata* in the Jewish *politeuma*, and asking the authorities to expel him from the city. Why should the Jews watch the gates, when it would have been perfectly easy to find out where Paul was living and arrange an 'accident' there? The seven gates of Damascus were manned by official guards, whose presence would have made an assassination attempt unwise. Manifestly, Luke knew only the means of Paul's escape from Damascus, and to explain it invented the motive of Jewish hostility, which occurs as a refrain in Acts (20: 3, 19; 23: 30).

Others ascribe the danger to Nabataeans, but deny that they controlled Damascus.[19] One version claims that the troops of Aretas were lying in ambush for Paul outside the city. This, of course, is sheer nonsense. 2 Corinthians 11: 32–3 clearly indicates that the danger to Paul was inside the city; the open country outside promised safety.[20] Another version suggests, on the basis of a parallel with the Jewish colony of Alexandria, whose head had the title ethnarch,[21] that the ethnarch at Damascus was merely a sort of 'consul' who headed the Nabataean trading colony there.[22] The existence of such an institution is very probable, even though specific evidence is lacking.[23] The authority of such a 'consul', however, was limited to the members of his own colony. He had none over outsiders such as Paul, and certainly could not 'guard the city'. The most satisfactory explanation of the use of *ethnarchês* in place of the *stratêgos* normally used of Nabataean governors is that it reflected the governor's rank as a prince of his own tribe.[24]

In default of any reliable evidence to the contrary, therefore, Paul's assertion of a Nabataean presence in Damascus prior to the death of Aretas IV must be considered trustworthy.[25] Jewett convincingly argues that this could not have happened prior to 16 March AD 37, the day on which the emperor Tiberius died. The latter's policy for the eastern frontier of the empire demanded regularly organized provinces rather than client kingdoms. In reaction, his successor, Gaius (Caligula), reinstated a number of client kings, e.g. Antiochus and Agrippa, who got Commagene and part of Transjordan respectively in AD 37,

[18] Harding (1993).
[19] e.g. Haenchen (1966), 268–9; Suhl (1975), 314–15.
[20] Jewett (1979), 31, adds other obvious criticisms; the Romans would not have permitted a hostile force to control the roads to Damascus; the lowering of a basket would have been a certain way to attract the attention of observers outside.
[21] Josephus, *AJ* 14. 117.
[22] Knauf (1983), 145–7; Légasse (1991), 75; Klauck (1986), 90.
[23] Smallwood (1981), 225–6.
[24] J. J. Taylor (1992), 719–28.
[25] Slight confirmation is furnished by a break in the Roman coin series. 'There are no coins of Caligula and Claudius, though some exist from Nero onwards.' (Schürer (1973–87) 2. 129).

and Sohaemus who got Iturea in AD 38. This attitude did not survive long into the four-year reign of Gaius, who revived Tiberius' policies to spite the Senate.[26]

No mention is made of any grant of territory to the Nabataeans, but, if there was any city they wanted, it was the great crossroads of Damascus, which they had controlled in the early first century BC,[27] and Gaius had a reason to be grateful to them. When he was with his father Germanicus in the east in AD 18–19, he was witness to the support the Nabataeans gave Germanicus against Gnaeus Calpurnius Piso, the governor of Syria, despite the damage that the latter could do to their northern trade, which had to pass through Syria.[28]

The Implications for Pauline Chronology

It is very probable, therefore, that the Nabataeans acquired control of Damascus in the latter half of AD 37. This, then, must date Paul's departure from the city, and his journey to Jerusalem. From it we can work both backwards and forwards.

Looking backwards, it establishes that Paul's three-year stay in Damascus was from AD 34–37, and demands that his visit to Arabia took place in AD 33 or 34. Moreover, since Paul gives the impression that his conversion immediately preceded his time in Arabia (Gal. 1: 16–17), his encounter with Christ should be assigned to AD 33. At this stage, as we have seen, Paul would have been in his late thirties.

Looking forwards, it permits us to date Paul's next visit to Jerusalem to the year 51, because he tells us, 'Then after 14 years I went up again to Jerusalem' (Gal. 2: 1).[29] Earlier pedantic speculation as to whether or not these fourteen years included the three years of Galatians 1: 18 has now given way to the commonsense recognition that the 'then ... then ... then' of Galatians 1: 18, 21 and 2: 1 mark the successive steps of a chronological series.[30] A certain mechanical

[26] Jewett (1979), 32–3. The suggestion of Baslez (1991), 74, that the Nabataeans acquired control of Damascus without Roman consent mistakes both Roman frontier policy and Nabataean military might.

[27] Josephus, *AJ* 13. 387–92; *JW* 1. 99–102. According to Schwank (1983), 434–5, Paul's assertion of Nabataean control over Damascus was a mistake based on his discovery there of coins of Aretas III (85–62 BC)!

[28] According to Tacitus, 'Piso was heard to remark at a banquet at the Nabataean court, when massive golden crowns were offered to Germanicus and [his wife] Agrippina, and lighter specimens to Piso and the rest, that this was a dinner given to the son, not of a Parthian king, but of a Roman prince. At the same time, he tossed his crown aside, and added a diatribe on luxury which Germanicus, in spite of its bitterness, contrived to tolerate' (*Annals* 2. 57; trans. Jackson).

[29] Betz (1992), 191 dates this event to AD 43–44 on the grounds that the James mentioned in Gal. 2: 9 was James the brother of John executed by Herod in 44. In fact, of course, he was the brother of the Lord who was executed in 66 (Josephus, *AJ* 20. 200).

[30] Burton (1921), 68. Similarly but in more detail in Jewett (1979), 52–3. Without adequate justification, Longenecker (1990), 45, while conceding that Paul's intention is to lay out in successive fashion his contacts with Jerusalem, none the less considers it more probable that the three years of 1: 18 and the fourteen years of 2: 1 are to be understood concurrently and not consecutively.

literalism, however, has led some scholars to find the referent of 'then' in Galatians 2: 1, not in the Apostle's first visit to Jerusalem (Galatians 1: 18), but in Galatians 1: 21 'Then I went into the regions of Syria and Cilicia.'[31] Were this in fact the case, it would make Galatians 2: 1 worthless for chronological purposes, because the duration of the mission in Syria and Cilicia cannot be determined. This interpretation, however, cannot be maintained. Paul's concern in Galatians 1 and 2 is to prove his independence of Jerusalem. Syria and Cilicia are mentioned only to prove that he left the Holy City, and fourteen years is manifestly the time he stayed away.[32] This view is confirmed, not only by the 'again' prefixed to 'to Jerusalem', but by the mention of the churches of Judaea in Galatians 1: 22, which is the proximate point of reference influencing Paul's formulation.

What has been learnt from the letters can now be summarized as follows:

Birth	*c.*6 BC
Conversion	AD 33
Arabia	34
Damascus	34–37
Jerusalem (1st visit)	37
Syria and Cilicia	37–?
Jerusalem (2nd visit)	51

These conclusions, however, must be confronted with the evidence of Acts.

THE EVIDENCE FROM THE ACTS
OF THE APOSTLES

It is not my intention here to discuss the dating of all Luke's allusions to secular history. Such material is handled in every commentary on Acts with generally inconclusive results. The best presentation in terms of both prudence and ingenuity remains that of R. Jewett,[33] but a new approach and new data demand that two crucial points be considered.

---

[31] Betz (1979), 83. Similarly Lüdemann (1984), 63–4, who manages to have his cake and eat it by deducting one year for the mission in Syria and Cilicia from the fourteen! His method for determining the length of the mission in Syria and Cilicia is not explained.

[32] This is the reason why Gal. 2: 1 cannot be translated 'once again in 14 years' despite the evidence for *dia* the sense of 'during' (Rom. 11: 10; 2 Thess. 3: 16) and the logic of Giet (1953), 323–8. The grammatical and contextual point is made with perfect clarity by Burton (1921), 68, who considers that Paul's argument for his independence of Jerusalem 'is somewhat strengthened by the use of the preposition *dia*, which, meaning properly "through," and coming to signify "after" only through the thought of a period passed through, also suggests that the period of fourteen years constitutes a unit in the apostle's mind—an unbroken period of non-communication with the apostles'.

[33] (1979), 33–62.

The Edict of Claudius

Luke opens his account of Paul's visit to Corinth with the words, 'After this he left Athens and went to Corinth. And he found a Jew named Aquila, a native of Pontus, lately come from Italy with his wife Priscilla, because Claudius had commanded all the Jews to leave Rome' (Acts 18: 1–2). One very important witness, the Codex Vaticanus, omits the name of the emperor Claudius (AD 41–54), but this is certainly due to a scribal accident, because the grammatical structure of the phrase demands a proper name. Luke's intention is certainly to date Paul's arrival in Corinth, but his roundabout way of doing it betrays a certain hesitation, which contrasts vividly with the certitude of Acts 18: 12 (see below).[34] He insinuates that the Apostle's arrival followed very quickly on that of Aquila and Priscilla, whose presence in the city is explained by a decree of Claudius, an event whose date Luke presumes is clear in his readers' minds. We are less fortunate.

The imperial edict evoked by Luke is also mentioned by Suetonius, 'He expelled from Rome the Jews constantly making disturbances at the instigation of Chrestus' (*Claudius* 25. 4).

The Evidence of Orosius

The modern consensus supplies the date unfortunately omitted by Suetonius on the basis of the fifth century Christian historian Paul Orosius, who cites his pagan predecessor:

> Josephus refers to the expulsion of Jews by Claudius in his ninth year. But Suetonius touches me more in saying, 'Claudius expelled from Rome the Jews constantly making disturbances at the instigation of Christus.' It cannot be determined whether he ordered only the Jews agitating against Christ to be restrained and suppressed, or whether he also wanted to expel Christians as being men of a related faith. (*Historiae adversus paganos* 7. 6. 15–16)

According to Roman custom the regnal years of an emperor were counted from the moment he assumed power. Since Claudius was acclaimed by the Praetorian Guard on 25 January AD 41, his ninth year ran from 25 January AD 49 to 24 January AD 50. In consequence, the majority of scholars date Paul's arrival in Corinth to AD 49.[35]

Such credence, however, is regularly qualified by expressions of doubt about Orosius' reliability, because no such reference appears in Josephus.[36] Why then is Orosius believed? Jewett gives the only possible answer, which robs Orosius

---

[34] On the source-critical position, see Boismard and Lamouille (1990), 2. 247, 301.
[35] Most recently Jewett (1979), 37–8; Fitzmyer (1990), 10; Klauck (1984) 6.
[36] The attempt of Smallwood (1981), 211–13, to find indirect evidence in support of Orosius evaporates on close analysis; see my (1992e), 140–2.

of all independent value, 'Despite the sometimes questionable accuracy of Orosius, this conclusion [the expulsion of Jews from Rome in AD 49] is corroborated by the striking correlation with the arrival of Paul as reckoned on the basis of the Gallio inscription.'[37] In other words, Orosius is accepted because his date harmonizes perfectly with the date acquired by subtracting eighteen months (Acts 18: 11) from the time of Paul's encounter with the proconsul Gallio (Acts 18: 12). It is entirely possible, however, that Orosius made precisely the same deduction in order to date Suetonius' information! Following Harnack's lead,[38] Lüdemann has shown that Orosius had no knowledge of Josephus, that his sources in Books 6 and 7 were Eusebius and Suetonius, and that the awkwardness of his formulation betrays the consciousness of invention.[39]

The silence of Tacitus tends to confirm this interpretation. His *Annals* are complete for AD 49, but there is not a single allusion to any action, taken or contemplated, against the Jews of Rome in that year.

The Evidence of Dio Cassius

If we extend our investigation to include the whole of the reign of Claudius, Dio Cassius furnishes an important item of information regarding the relationship of the emperor and the Jewish community:

> As for the Jews, who had again increased so greatly by reason of their multitude that it would have been hard without raising a tumult to expel them from the city, he did not drive them out, but ordered them, while continuing their traditional mode of life, not to hold meetings. (*History* 60. 6. 6; trans. Cary)

Given Dio Cassius' annalistic arrangement, there is no doubt that this episode must be dated to the first year of the reign of Claudius, AD 41.[40] Is it the event to which Suetonius refers? The similarities suggest an affirmative answer, but the differences demand a more complex answer.

Suetonius and Dio Cassius have in common: (1) an action of Claudius, (2) evoking expulsion (3) of Jews (4) from Rome, and (5) the idea of tumult. The accounts differ in two respects: (1) the former affirms that Jews were expelled; the latter denies it; (2) the feared tumult in the latter has taken place in the former.

---

[37] (1979), 38. Jewett also makes explicit what all followers of Orosius presume, 'Reckoning several months for travel from Rome and resettlement in Corinth, the encounter [of Prisca and Aquila with Paul] must have taken place sometime between spring of AD 49 and the spring of AD 50' (p. 38). Such precision is impossible.

[38] (1912), 675–6.

[39] (1991), 289–98. Schwartz (1990), 94–5, speculates that Orosius had a source which correlated the explusion with the advent to power of an Agrippa, and confused Agrippa I, who became king of Judea in AD 41, with Agrippa II, who was crowned in AD 48–49.

[40] A clear summary of the argument is given by Slingerland (1988), 307–8.

A number of authors are more impressed by the differences than the similarities, and postulate two imperial actions against Roman Jews, the first in AD 41 (Dio Cassius) and the second in AD 49 (Suetonius dated by Orosius).[41] Slingerland, the most recent advocate of this view, attaches particular importance to the 'again' in Dio Cassius' account, and relates it to the latter's earlier mention of the expulsion of Jews from Rome which took place in AD 19 under Tiberius.[42] In consequence, he maintains that Dio Cassius' narrative is completely intelligible in itself and does not need to be supplemented by that of Suetonius.[43]

Slingerland's preoccupation with methodology unfortunately has blinded him to the fact that the account of Dio Cassius is both incomplete and implausible. Incomplete in so far as Dio assigns no meaningful justification for the action of Claudius. Tiberius, on the contrary, feared the proselytization of the upper classes.[44] Implausible in so far as the refusal to permit assemblies of Jews would have created more problems than it solved. From a purely administrative point of view it would have been preferable to expel them. Jews were accustomed to meet every Saturday to study the Law prayerfully,[45] and the effect of the refusal of this right, which had been guaranteed by Roman law,[46] would have been to create a perpetual grievance that could not but lead to periodic eruptions, which is precisely what the emperor was trying to avoid. In other words, the measure reported by Dio Cassius would have exacerbated the situation. His report, therefore, cannot be taken at face value, and we are forced to delve beneath the surface.[47]

Why did Dio Cassius write as he did? To answer this we must remember that he wrote in the first decades of the third century AD and necessarily had to rely on sources. What exactly these were, we do not know, but they must have belonged to the same annalistic tradition that served Suetonius and Tacitus. Thus, there is the distinct possibility that Dio Cassius had to rely on inaccurate information. Alternatively, he may have misunderstood what he read.

Light begins to dawn when we compare Dio Cassius and Suetonius. I have translated the latter's text (cf. above) in such a way as to bring out its most natural interpretation, i.e. Claudius expelled only the trouble-makers among the Jews. However, it could be read in another way, namely, 'Since the Jews constantly made disturbances at the instigation of Chrestus, he expelled them from Rome' (LCL translation). This would mean that all Jews were expelled.

---

[41] Smallwood (1981), 215; Bruce (1985), 281.
[42] *History* 57. 18. 5.
[43] (1990), 686–90.
[44] For details see Smallwood (1981), 201–10.
[45] Josephus, *Against Apion*, 2. 175; Philo, *Vita Mosis* 2. 216.
[46] See Smallwood (1981), 558–60; Saulnier (1981), 161–98.
[47] The pioneer of this indispensable form and source-critical approach has been Lüdemann (1984), 164–70; (1991), 289–98.

With great perception, Lüdemann noted that Dio Cassius' text looks very like a conscious correction of this reading.[48] The Roman historian was not aware of any punishment of Jews on such a massive scale, and so substituted what he considered a lesser penalty. He did not know, however, that, in opposition to the central organization of Alexandria, the Roman community was divided into a number of distinct synagogues,[49] and so he thought of the punitive action as affecting the entire Jewish population of the city. In reality it would have been directed only against a single synagogue, which would have been closed until there was a guarantee that there would be no further disturbances.

This, of course, does not at all square with what Suetonius says, for he speaks of expulsion and not of closure. But there are two difficulties in his account which makes its details suspect. He names Chrestus as the chief agitator but, as all have recognized,[50] this is a misunderstanding of the role of Christ, whose person was the subject of the dispute. In addition it was not legally possible to simply expel Jews who were Roman citizens; such a measure could be applied only to those who had no right of residence.[51]

If neither Suetonius nor Dio Cassius can be taken at face value, we cannot conclude that they are referring to two distinct events. It is preferable, according to the rules of normal literary criticism, to see them as partial accounts, confused and inaccurate, of the same episode. The historical kernel underlying both accounts can be reconstructed as follows: as the result of a disturbance concerning Christ in a Roman synagogue, Claudius in AD 41 expelled the missionaries who were not Roman citizens, and temporarily withdrew from that specific Jewish community the right of assembly.[52]

This reconciliation of the accounts of Suetonius and Dio Cassius is admittedly tenuous,[53] but it is confirmed by Philo. He completed his *Legatio ad Gaium* in AD 41 at Rome while waiting for an audience with Claudius.[54] In it we find the following passage:

> Augustus knew that they had synagogues, and they were in the habit of visiting them, and most especially on the sacred sabbath days, when they publicly cultivate their national philosophy. He knew also that they were in the habit of contributing sacred sums of money from their first fruits and sending them to Jerusalem by the hands of those who were to conduct the sacrifices. But *he*

---

[48] (1984), 165.   [49] Smallwood (1981), 138.
[50] Slingerland (1989) unconvincingly argues against the identification.
[51] Smallwood (1981), 216.
[52] Lüdemann (1984), 166. Similarly Schwartz (1990), 94–6.
[53] Slingerland (1988), 321, lists nine other hypothetical reconciliations of Suetonius and Dio Cassius as if this was in itself an objection. The opposite is in fact the case; the number of suggestions is evidence of a widespread conviction that the two accounts concern the same event. Lüdemann's hypothesis is the simplest; it has been accepted by Penna (1982), 331.
[54] Smallwood (1981), 30, 151.

never removed them from Rome, nor did he ever deprive them of their rights as Roman citizens, because he had a regard for Judaea, nor did he ever meditate any new steps of innovation or rigour with respect to their synagogues, *nor did he forbid their assembling for the interpretation of the Law*, nor did he make any opposition to their offerings of first fruits. (156–7; trans. Yonge)

The relevance of the two phrases, which I have italicized, to the point at issue is immediately obvious, and they become highly significant when it is recognized that in no source is there any hint that Augustus even contemplated such measures. Their intention becomes clear if we assume with Smallwood[55] that in AD 41 Philo had heard rumours of such actions on the part of Claudius. The latter revered Augustus as the model Roman ruler, and by citing Augustus as a counter-precedent Philo was in fact making a protest to which the touchy emperor could not take exception.

In sum, therefore, a very high degree of probability can be accorded to the hypothesis that as the result of an imperial action in AD 41 some Jews were expelled from Rome.

The Attitude of Claudius to the Jews

It has been objected, however, that this conclusion conflicts with what is known of Claudius' attitude towards the Jews at the beginning of his reign. He restored to them all the privileges which had been abrogated by his predecessor Gaius (Caligula). The letter has been preserved by Josephus in his *Antiquities of the Jews* (19. 287–91). The tenor of this letter leaves no doubt about Claudius' positive attitude towards Jews throughout the empire; all their traditional rights are reaffirmed in the most explicit and public way possible. The third paragraph of the letter, however, contains a warning which should not be overlooked,

> It will, therefore, be fit to permit the Jews, who are in all the world under us, to keep their ancient customs without being hindered in so doing. But I charge them also to use this my kindness to them with moderation, and not to show contempt of the superstitious observances of other nations, but to keep their own laws only. (290; trans. Whiston and Margoliouth)

Claudius does not give them *carte blanche;* there are limits to what he is prepared to tolerate. This point is made formally in a letter addressed by Claudius to the city of Alexandria on 10 November AD 41, which complements the fulsome letter restoring Jewish privileges preserved by Josephus in *Antiquities of the Jews* (19. 281–5). In the translation of C. K. Barrett the most important paragraph reads as follows:

> on the other hand I explicitly order the Jews not to agitate for more privileges than they formerly possessed, and not in future to send out a separate embassy

[55] (1981), 214.

as if they lived in a separate city, a thing unprecedented, and not to force their way into gymnasiarchic or cosmetic games, while enjoying their own privileges and sharing a great abundance of advantages in a city not their own, and not to bring in or admit Jews who come down the river from Syria or Egypt, a proceeding which will compel me to conceive serious suspicions; otherwise I will by all means take vengeance upon them as fomentors of what is a general plague infecting the whole world.[56]

This document makes it perfectly clear that, from the outset of his reign, Claudius was prepared to react vigorously against anything that could be interpreted as a threat to public order. In AD 41 if certain Jews (from whom Christians were not then distinguished) at Rome had been seen as agitators, the emperor would definitely have moved against them in the way suggested by Suetonius and Dio Cassius. This year, therefore, certainly enjoys greater probability as the date of the edict of Claudius than the alternative of AD 49, which is based exclusively on the unreliable testimony of Orosius.

We are now in a position to try to fix the beginning of Paul's ministry in Corinth. Luke gives the impression that he arrived more or less on the heels of Aquila and Priscilla, who had 'recently' come from Italy as a result of the edict of Claudius (Acts 18: 1–2). Thus, Paul would have arrived in Corinth by AD 42 at the latest.

Lüdemann distinguishes this founding visit from a subsequent visit when Paul encountered the proconsul Gallio (Acts 18: 12).[57] I doubt that this can be maintained. In the first place, Luke is much less precise than appears at first sight. The edict of Claudius, as we have just seen, concerned only a single synagogue in Rome, and Luke does not say that Aquila and Priscilla came from that city; they came from Italy. Moreover, the edict involved only banishment from the city, not exile from the country. We cannot assume that the expelled Jews immediately took to the boats. It is more reasonable to assume that they took up residence somewhere outside the city in order to see how the situation would develop. How long they might have stayed, no one can say. Nor can we determine with any exactitude what Luke might have meant by 'recently'. Secondly, the datum from Acts must be confronted with the evidence from Paul's letters. We have already seen that Paul's missionary activity began in late AD 37—a date that Lüdemann does not accept—and as we analyse the pre-Corinth phase of this activity, it will become clear that it cannot be fitted into the four-year period postulated by Lüdemann.

Unless we are prepared to ignore the chronological data that Paul himself gives us in his letters, we must question whether there was any real relationship between the edict of Claudius and the move to Corinth of Aquila and Priscilla. Luke is unlikely to have had solid information on such a minor point, and in all

---

[56] (1958), 46.
[57] (1984), 171.

probability combined vague memories to produce a scenario which, on other grounds, we know cannot be factual. If this hypothesis is correct, Acts 18: 1–2 is but another example of the phenomenon manifested in Luke 2: 2. In this latter passage Luke claims that the motive for the journey of the Holy Family from Nazareth to Bethlehem was the census of Quirinius. This census, however, took place in AD 6–7,[58] some ten years after the birth of Jesus in the days of Herod the Great, who died in 4 BC. In both instances Luke attempted to link sacred and secular history and got it slightly wrong.

Paul's Encounter with Gallio

'When Gallio was proconsul of Achaia, the Jews made a united attack on Paul and brought him before the tribunal' (Acts 18: 12). This assertion that Paul's ministry in Corinth overlapped, at least in part, with the term of office of the Roman governor Gallio is the linchpin of Pauline chronology.[59] It is the one link between the Apostle's career and general history that is accepted by all scholars. Our only means of dating the presence of this official in Corinth is a badly broken inscription containing a letter of the emperor Claudius.

The Delphi Inscription

Four fragments, discovered during the French excavations at Delphi, were first joined and published by Emile Bourguet in 1905. In 1910 he found three more fragments belonging to the same inscription; these were published by A. Brassac in 1913. Further fragments were discovered subsequently, but the fact that the new fragments were not included in the third edition of W. Dittenberger's authoritative *Sylloge Inscriptionum Graecarum* (1915–24) meant that they were ignored in all subsequent discussions of the inscription. In 1967 A. Plassart succeeded in joining the two groups of fragments, and added two more. His official publication of the nine fragments appeared in 1970. That same year a number of his readings were improved by J. H. Oliver. The one remaining doubt—whether line seven mentions an individual (Plassart) or a group (Oliver)—appears to be resolved in the latter's favour by the observation that two of the original fragments fit together at the back.[60]

> Tiber[ius Claudius Caes]ar Au[gus]tus Ge[rmanicus, invested with tribunician po]wer [for the 12th time, acclaimed Imperator for t]he 26th time, f[ather of the fa]ther[land ... sends greetings to ...]. For a l[ong time I have been not onl]y [well disposed toward t]he ci[ty] of Delph[i, but also solicitous for its pro]sperity, and I have always guar[ded th]e cul[t of t]he [Pythian] Apol[lo.

[58] Schürer (1973–87), 2. 399–427.
[59] Acts 18: 12 belongs to one of the most primitive levels of Acts; see Boismard and Lamouille (1990), 2. 249.
[60] Hemer (1989), 252 n. 19.

But] now [since] it is said to be desti[tu]te of [citi]zens, as [L. Jun]ius Gallio, my fri[end] an[d procon]sul, [recently reported to me, and being desirous that Delphi] should continue to retain [inta]ct its for[mer rank, I] ord[er you (pl.) to in]vite [well born people also from ot]her cities [to Delphi as new inhabitants and to] all[ow] them [and their children to have all the] privi[leges of Del]phi as being citi[zens on equal and like (basis)].[61]

Dating the Letter of Claudius

The letter was written after Claudius had been acclaimed emperor for the twenty-sixth time. Such acclamations were ritualized public applause that sanctioned a triumph of the emperor, e.g. the conclusion of a successful military campaign or a specially significant victory. Unfortunately, we have no text which dates the twenty-sixth acclamation precisely. The problem, then, is to delimit as tightly as possible the time-span within which it must have occurred.

The upper limit is fixed by the twenty-seventh acclamation which took place before 1 August 52. Frontinus (AD 30–104), speaking of two aqueducts begun by the emperor Gaius, says, 'These works Claudius completed on the most magnificent scale, and dedicated in the consulship of Sulla and Titianus, on the 1st of August in the year 803 after the founding of the City'. (*Aqueducts*, 1. 13; trans. Bennett and McElwain). The dedicatory inscription on one of these aqueducts, the Aqua Claudia, reads in part, 'Tiberius Claudius son of Drusus Caesar Augustus Germanicus Pontifex Maximus, 12th year of tribunician power, consul for the 5th time, acclaimed emperor for the 27th time, father of the country.'[62]

Tribunician power was accorded to an emperor at the moment of his accession to the purple, and for each year of his reign he added one unit. Since the first tribunician year of Claudius was 25 January 41 to 24 January 42, his twelfth year was 25 January 52 to 24 January 53. Thus, the year mentioned by Frontinus must be wrong, because 803 AUC = AD 50; that this is due to scribal error is confirmed by the names of the consuls, because Sulla and Titianus held office in AD 52.

The twenty-seventh acclamation, therefore, took place between 25 January 52 and 1 August 52. This period, however, can be narrowed significantly. Acclamations were related to military prowess, and normally no major campaigns were undertaken in winter; the battle season was from late March to early November. Thus, the twenty-seventh acclamation must be dated between April and July 52. In consequence, the twenty-sixth acclamation must have taken place in the same period or in or before the previous November.

This ambiguity remains despite the perfectly preserved inscription of Kys in Caria which reads, 'Tiberius Claudius Caesar Germanicus Emperor God

[61] The translation and restoration brackets are those of Fitzmyer (1990), 9. The Greek text is given in my (1992e), 179.   [62] *CIL* 6. 1256; quoted in Brassac (1913), 42.

Augustus, Pontifex Maximus, 12th year of tribunician power, consul for the 5th time, acclaimed emperor for the 26th time, father of the country'.[63]

While this establishes a correlation between the twelfth year and the twenty-sixth acclamation, it does not exclude the possibility that the twenty-sixth acclamation had been accorded prior to the beginning of the twelfth year. The latter changed automatically on 25 January for Claudius, whereas acclamations followed no set calendar. It should be kept in mind that the reference to the twelfth year in the letter to Delphi is a restoration, and cannot be used as an argument.

The lower limit is fixed by a series of inscriptions (whose texts are substantially identical with that on the Aqua Claudia save for changes in the numbers),[64] which affirm that the twenty-second, twenty-third, and twenty-fourth acclamations took place in the eleventh tribunician year of Claudius, i.e. 25 January 51 to 24 January 52. No inscription correlates the twenty-fifth acclamation with a tribunician year.

At this stage we are forced into speculation on probabilities concerning the relation of six acclamations (the twenty-second to the twenty-seventh inclusive) with two time-spans, namely, the battle seasons April to November 51 and April to July 52. Since we have no dates for the twenty-fifth and twenth-sixth acclamations, we have to assign them to one period or the other. The possibilities are:

(1)	April–November 51	April–July 52
	22nd 23rd 24th	25th 26th 27th
(2)	April–November 51	April–July 52
	22nd 23rd 24th 25th	26th 27th
(3)	April–November 51	April–July 52
	22nd 23rd 24th 25th 26th	27th

The criterion for a decision between the three options can only be the assumption that the symbolic value of the acclamations would diminish in direct proportion to their frequency. On this basis, possibilities (1) and (3) appear less probable than possibility (2). In the first there are too many acclamations in 52, while in the third there are too many in 51, but in the second we get a much better balance, i.e. the acclamations average one every two months. The argument is tenuous but, given the present dearth of evidence, it is the only one possible, and this must be kept clearly in mind when assessing the value of the conclusion.

In the above hypothesis, the twenty-sixth acclamation would have taken place after the first significant victory in the spring campaign of AD 52, i.e. in

---

[63] *BCH* 11 (1887), 306–7; quoted in Brassac (1913), 44.
[64] They are cited in Brassac (1913), 43–4.

April at the earliest. Thus, the letter of Claudius was probably written in the late spring or very early summer of that year.[65]

The purpose of the letter is to deal with a social problem, the depopulation of Delphi, to which the emperor's attention had been drawn by a report from Gallio. The formulation makes it certain that the latter was not the recipient of the letter, as older studies would make it appear. The letter was addressed either to a proconsul of Achaia or to a group such as the city council of Delphi. There is no justification for supposing that the city was empty, as in an evacuation which might have followed a plague. We have to do with a significant drop in the number of citizens, which is most probably to be accounted for by a change in the economic situation. Delphi was not a trade centre; it was located off the main routes and in very difficult terrain, particularly in winter. It was the sanctuary of the Pythian Apollo, the oldest and most venerated shrine of Greece, and any decrease in the numbers of pilgrims would have affected the revenues of the city.

Delphi had been in decline for well over a century by the time this letter was written. It had been pillaged many times, but more importantly its prestige had evaporated. It survived on the memories of a glorious past, but it was no longer a vital spiritual centre. As fewer visitors came, more families, unable to procure a livelihood, would have drifted away, with the inevitable consequence that even fewer services were available for those who did come on pilgrimage.

This vicious circle appears to have engendered a sense of hopelessness, for the council which ran the city did not take the initiative in seeking imperial aid. The formulation of the letter clearly indicates that they did not petition the emperor either directly, or indirectly through the good offices of Gallio. The fact that the reaction of Claudius is expressed in the form of a command hints at a certain apathy, if not reluctance, on the part of the Delphians. He is certainly not acceding to a request for more citizens.

Dating the Proconsulship of Gallio

The view that provincial office-holders began their functions on 1 July is a deduction from an ordinance made in AD 15 by the emperor Tiberius that they 'should take their departure [from Rome] by the first day of June.'[66] It is reasonable that adequate time should be allotted for the journey. The departure date was advanced by Claudius in AD 42: 'The governors who were chosen by lot were to set out before the first day of April, for they had been in the habit of tarrying a long time in the city.'[67] Since this had little effect, presumably because travel was still difficult in the early spring, the emperor had to repeat the ordinance in a slightly mitigated form the following year, insisting that

---

[65] Similarly Brassac (1913), 45; Hemer (1989), 253.
[66] Dio Cassius, *History*, 57. 14. 5; trans. Cary.
[67] Ibid. 60. 11. 6.

governors had to be out of Rome by mid-April at the latest.[68] This has led some to consider that the year of office began on 1 June.[69] The purpose of the regulation, however, was to ensure that governors got to their destinations in time, no matter how slowly they travelled.

Assuming that the average journey to a post would have taken a month, this would have given the newcomer a further month in which to familiarize himself with the local situation before assuming full responsibility. Some such arrangement was imperative, because proconsuls held office for only one year, and one could not have a significant period of each year wasted while the new appointee found his way around. We can safely assume, therefore, that a proconsul's term of office ran from the beginning of July to the end of the following June.

Since the letter of Claudius in which Gallio is mentioned was probably written in April or May AD 52, the proconsular year of Gallio cannot have been 52–53, since the emperor's response cannot antedate the report.[70] The last year in which Gallio could possibly have been proconsul is 1 July 51 to 30 June 52. However, we cannot simply assume, from the title given to Gallio in the letter, that this was in fact the year.[71] Even if he were already out of office, he would naturally have been given the title which authorized his report to the emperor.

It is considerably more difficult to determine the earliest year that Gallio could have been proconsul. It does seem very unlikely, however, that he could have been appointed before AD 49. Only in that year did his brother, the philosopher Seneca, return from Corsica, whither he had been exiled by Claudius in AD 41. In the Roman system, the disgrace of one member touched the whole family. Since Gallio does not appear to have been a man of exceptional character or ability, it is very probable that his nomination as proconsul of Achaia was materially assisted by the influence of his brother, who, on his rehabilitation, had been named to the imperial court as the tutor of Nero.

Theoretically, it is not impossible that Gallio served the two years, AD 50–52, because he may have been *extra sortem*, and officials of this type, like legates in the imperial provinces, served at the emperor's pleasure; their tenure was not limited to one year. In reality, however, the question of Gallio's precise status is irrelevant because he did not complete his term of office. Thus, he only served part of 50–51 or of 51–52. We know this from a note by his brother Seneca: 'When, in Achaia, he began to feel feverish, he immediately took ship, claiming that it was not a malady of the body but of the place' (*Epistulae morales* 104. 1; trans. Gummere). The impression of a fussy hypochondriac is confirmed by Pliny's report that Gallio felt the need for a long sea-voyage to recuperate after his consulship (*NH* 31. 62). If we accept what Seneca says of a 'malady of the

---

[68] Ibid. 60 17. 3.
[69] Fitzmyer (1990), 9.
[70] Ibid.
[71] As does Lüdemann (1984), 163.

place', it is natural to assume that Gallio took a dislike to Achaia and used a minor illness as an excuse to leave his post. Such an unreasoning aversion to a place is normally the result of a first impression; it may intensify with the passage of time, but it does not usually begin late. If this assessment is correct, it is unlikely that Gallio remained in Achaia more than four months, i.e. from June to September.

Otherwise, he would have been stuck there for the winter. The danger of winter travel in the eastern Mediterranean is underlined by Luke, 'The voyage was already dangerous because the Fast [i.e. Yom Kippur, celebrated near the autumnal equinox] was already over' (Acts 27: 9; cf. 28: 11). Pliny makes the same point more succinctly, 'Spring opens the sea to voyagers' (*NH* 2. 122).[72] The note of Dio Cassius, 'If anyone ever risked a voyage at that season [winter] he was sure to meet with disaster',[73] is unconsciously confirmed by Suetonius who, after recounting how Claudius had been mobbed by the Roman crowd because of the lack of grain, continues, 'After this experience he resorted to every possible means to bring grain to Rome, even in the winter season' (*Claudius*, 18; trans. Rolfe). In other words, ships put to sea from November to March only for the most serious reasons, because one could be tossed by storms for three continuous months.[74] In winter and early spring no one made a trip that could be deferred.[75]

Even if Gallio did not depart before the winter, the 'closed sea' has obvious implications for the time of the transmission of his report concerning Delphi to Rome. It was not a matter of high priority, and since we have no reports of any serious disturbances in Achaia in AD 50–52 which would demand immediate communication with the capital, it must be assumed that the report went with the normal courier traffic outside the winter season.

Nothing so far has permitted us to choose between AD 50– 51 and 51–52 as the year of Gallio's proconsulship. The only basis on which a decision can be made is Claudius' administrative ability. Did he deal with problems quickly or did he let them drag on? As usual, it is difficult to answer this question with any certitude. His life, as recounted by Suetonius, produces a very mixed impression. Claudius appears as extremely erratic: 'He showed strange inconsistency of temper, for he was now careful and shrewd, sometimes hasty and inconsiderate, occasionally silly and like a crazy man' (*Claudius*, 15; trans. Rolfe).

---

[72] In the same context, Pliny mentions 11 November and 8 February as the traditional dates for the beginning of winter and spring. Vegetius (*Epitoma rei militaris* 4. 39) extends winter to 10 March, but considers the safe sailing season to run only from 27 May to 14 September (5. 9).

[73] *History* 60. 11. 2. Instead of waiting for spring, Julius Caesar took advantage of a spell of fine weather to cross the Adriatic. Not all of his army were able to follow, forcing him to recognize that his voyage was 'more fortunate than prudent' (41. 44. 1–4). When Caesar tried to return to Italy, a storm forced him back (41. 46. 2–4).

[74] Josephus, *JW* 2. 200–3.

[75] For the stories of those obliged to go to sea in winter, see Spicq (1969), 145 n. 2.

This is certainly true, if one looks at the emperor's life as a whole. A strict chronological order, however, imposes a different assessment. From the period AD 41–50 'a large number of imperial enactments [survive and show] ... profound administrative common-sense', but in the last four years his powers began to fail (*OCD* 197). Just at the point that interests us, the weak side of his character began to predominate and we have no way of knowing how quickly he disposed of business. However, one factor disposes me to think that he reacted quickly to Gallio's information, namely, his known fondness for Achaia. According to Suetonius,

> He gave no less attention to Greek studies, taking every occasion to declare his regard for that language and its superiority ... and in commending Achaia to the senators he declared that it was a province dear to him through the association of kindred studies. (*Claudius*, 42; trans. Rolfe)

It would be a mistake to imagine that Gallio had any particular interest in Delphi, or that his report was motivated by a high sense of duty. Delphi was the traditional centre of the culture that the emperor admired so extravagantly, and Gallio was astute enough to recognize that evidence of his concern for that city would place him in the good graces of Claudius. Perhaps he even brought the report with him in an effort to assuage the imperial anger at his abandonment of his post. Be that as it may, it seems probable that Claudius would have reacted rather quickly to anything concerning Delphi. In this case, Gallio's report would have reached him at the earliest by late autumn AD 51 or at the latest when the sea was again opened for normal shipping in the spring of AD 52. Gallio's term of office, therefore, is more likely to have been AD 51–52 than the previous year.[76]

The line of argument developed to support this conclusion is admittedly tenuous, but some such approach is necessary in order to justify the current consensus. It should also be noted that this consensus is no more than a lucky accident, because it depends on a misunderstanding of what the letter of Claudius was all about since only four of the nine available fragments were used to reconstitute it.

In the light of Seneca's statement that his brother did not finish his term of office, it is impossible to place Gallio's encounter with Paul (Acts 18: 12–17) in the latter part of the proconsular year AD 51–52. The encounter must have taken place between July, when Gallio arrived in Corinth, and September AD 51, the last date when he could have sailed to Rome.

This conclusion finds positive confirmation in Galatians 2: 1, which places Paul in Jerusalem in AD 51. Luke gives the impression that Paul left Corinth for Jerusalem by ship shortly after his encounter with the proconsul (Acts 18: 18–22). This cannot have been later than mid-September, because of the 'closed

---

[76] Similarly Hemer (1989), 252.

sea'. The run to the south-east was the best and fastest point of sailing for ships of the period. Paul could easily have been in Jerusalem by the autumn of AD 51, precisely as we have already deduced from Galatians 2: 1. This correlation has two important corollaries. First, the figure 14 in Galatians 2: 1 is not a round number.[77] Secondly, the visit to Jerusalem of Acts 18: 22 is a doublet of that in Acts 15: 2– 3.

Negative confirmation is furnished by the fact that Paul did not preach the collection in Corinth during his founding visit.[78] The only explanation is that it had not yet been decided at the Jerusalem Conference. Hence, the founding visit to Corinth must be placed before AD 51.

There is no independent support for Luke's assertion that Paul stayed in Corinth for eighteen months (Acts 18: 11), but the figure is plausible when one considers the time necessary to make conversions and establish a community. Moreover, it means that, if Paul left Corinth in September AD 51, he would have arrived there in April AD 50. In other words, it accords with the fact that the ancients did not travel in winter (cf. 1 Cor. 16: 6) except in abnormal circumstances; conditions were too difficult.[79] Paul would have set out from his previous location only with the advent of spring.

Hearings before Felix and Festus

After Paul's arrest in Jerusalem, Luke tells us that he was sent to the governor Antonius Felix in Caesarea (Acts 23: 24). Two years later (24: 27),[80] Felix was replaced by Porcius Festus (Acts 25: 1). The dating of the inception of Felix's term of office is determined by two rather vague references. According to Josephus, 'Claudius sent Felix, the brother of Pallas, to take care of the affairs of Judea, and, when he had already completed the twelfth year of his reign, he had bestowed upon Agrippa the tetrarchy of Philip and Batanea' (*AJ* 20. 137– 8; trans. Whiston and Margoliouth).

Since Claudius came to power on 25 January 41, his twelfth year ran from 25 January 52 to 24 January 53. The gift of territory to Agrippa II, therefore took

---

[77] This in turn makes it probable that three years (Gal. 1: 18) is also accurate. The fact that some ancients found it at times convenient to think in round numbers, e.g. two years and three months (Acts 19: 8–10) becomes three years (Acts 20: 31), does not create an assumption that any given year might have been rounded upward from as little as one month, i.e. in this case that the three years and fourteen years might be as little as one year and a couple of months and twelve years and a couple of months, respectively. The sensible remarks of Jewett (1979), 53–4, are a necessary antidote to the excessive prudence of Lüdemann's observation apropos of 2 Cor. 9: 2, 'Theoretically, the expression "a year ago" can cover a period of one to twenty-three months' (1984), 134 n. 178.

[78] So rightly Lüdemann (1984), 81–3.

[79] The most graphic description is that of Aelius Aristides (*Discourse*, 48), who took a hundred days to slog from Smyrna to Rome in January–March AD 143.

[80] Given Luke's interests and the focus of his narrative, this period most probably refers to the length of Paul's imprisonment, and not to the duration of Felix's term of office.

place in AD 53, probably in the second half of the year.[81] The juxtaposition to the mention of Antonius Felix gives rise to the assumption that the latter's nomination should be dated not long before, i.e. towards the end of AD 52 or in the early part of AD 53. The formulation of the *Jewish War* makes a temporal link between the two events even stronger, although neither is dated. 'After this Caesar sent Felix, the brother of Pallas, to be procurator of Samaria and Galilee and Perea, and removed Agrippa from Chalcis unto a greater kingdom' (2. 247; trans. Whiston and Margoliouth).

For his part, Tacitus records the dismissal of Felix's predecessor as procurator of Judaea, Cumanus, among the events of AD 52 (*Annals* 12. 54). Unfortunately, his account of the circumstances leading up to the disgrace of Cumanus is totally implausible,[82] and one would have no reason for retaining the date unless it harmonized roughly with what might be deduced from Josephus.

Festus died in office in AD 62.[83] The date is calculated from the fact that his successor Albinus was in Judaea by the feast of Sukkot four years before the outbreak of the First Revolt in AD 66.[84]

The crucial question for Pauline chronology is: at what moment in the ten-year period, AD 52– 62, did power pass from Felix to Festus? The one thing that can be said with certitude is that it was after 13 October 54 (the death of Claudius) because Felix was deposed by Nero, and was only saved from punishment by the entreaties of his brother Pallas.[85] Eusebius dates it to the second year of Nero, but it has been well argued that Eusebius was four years out in his estimate of Nero's regnal years; in other words, the 'second' was really the 'sixth', i.e. between October 59 and October 60.[86] Confirmation of this date may be drawn from a change of the provincial coinage of Judaea in the fifth year of Nero, i.e. AD 58–59. As Smallwood points out, this is 'more likely to be the work of a new procurator than of an outgoing one who had already minted a large issue'.[87] Hence, the change of procurators is dated to AD 59,[88] or to AD 60.[89] In order to underline how tenuous this conclusion is, it is imperative to remember that Josephus was much less interested in good procurators than in those whose bad administration contributed to the outbreak of the First Revolt. Luke, however, does say that Felix was procurator 'for many years' (Acts 24: 10).

[81] Schürer (1973–87), I. 472.
[82] See Haenchen (1971), 68–9; Bruce (1985), 285–6.
[83] Josephus, *AJ* 20. 197.
[84] Josephus, *JW* 6. 300–9.
[85] Josephus, *AJ* 20. 182.
[86] Bruce (1985), 287.
[87] (1981), 269 n. 40.
[88] Jewett (1979), 43.
[89] Schürer (1973–87), I. 465 n. 42.

DEVELOPING A CHRONOLOGY FOR
PAUL'S LIFE

Subsequent to Paul's departure from Jerusalem in the autumn of AD 37, the securely datable facts of his ministry are, as we have seen, his encounter with Gallio in the summer of AD 51, and his presence in Jerusalem later that same year. What do we know about Paul's movements in the periods preceding and following this year, and is it possible to date them?

Prior to AD 51

In Galatians Paul provides a formal statement and a clear hint as to where he was during the fourteen years which separated his two visits to Jerusalem. On leaving the Holy City, he went into 'the regions of Syria and Cilicia' (Gal. 1: 21). His activity was not limited to these areas, for in Galatians 2: 5 he says, 'to them we did not yield submission even for a moment in order that the truth of the gospel might continue among you'. Older translations which give the purpose clause an exclusively future meaning (e.g. *RSV, BdeJ, NAB*) were influenced by Luke's presentation of the evangelization of Galatia (Acts 16: 6) as subsequent to the Jerusalem Conference (Acts 15). The only attested sense of *diamenein*, however, is 'to remain, continue',[90] which necessarily implies that at the time of the Jerusalem Conference the truth of the gospel had already been preached to the Galatians.[91] Thus we can deduce that Paul went west from Cilicia. He did not have to spell out to the Galatians that it was to them he travelled.[92]

Galatia and a Journey into Europe

At some point prior to AD 51, therefore, Paul had been in central Asia Minor. He further reveals that his presence among the Galatians was the result of an accident; 'you know that it was because of a bodily ailment that I preached the gospel to you' (Gal. 4: 13). If his visit was unplanned, he must have been going somewhere else, and we must assume that, once recovered, he set out in pursuit of his original goal. Where was he going?

Luke in Acts 16–18 provides an answer, but it would be illegitimate to assume that his information is correct. The epistles must be approached first,

[90] LSJ 403b; BAGD 186.
[91] First emphasized by Lüdemann (1984), 71, and approved by Jewett (1979), 84, this is now the common interpretation of translations (e.g. *JB, NRSV*) and commentaries, e.g. Longenecker (1990), 53. The attempt of Rolland (1992), 879, to retain a future aspect by interpreting *pros hyman* to mean 'for your profit' is unconvincing; see Burton (1921), 86.
[92] The insistence of Dunn (1993), 80, that Paul did not move outside Syria and Cilicia is misplaced.

and fortunately they can be seduced into yielding an answer, if one takes the trouble of correlating a series of rather clear hints.

According to 1 Thessalonians, Paul's visit to Thessalonica, had been preceded by a visit to Philippi, where he had been badly treated (2: 2). Subsequently, he was in Athens (3: 1), and then, it would appear, in a different and unnamed location whence the letter was written (3: 6). The anxiety Paul demonstrates for the perseverance of the Thessalonians, and the warmly affectionate tone in which he speaks of his longing for them, suggest (1) that the church at Thessalonica had been but recently established, and (2) that Paul had not yet achieved anything significant in his new mission field. Combining these two, we can infer (3) that it is question of Paul's first missionary venture into Europe.

In writing to the Philippians, Paul notes that 'in the beginning of the gospel,[93] when I left Macedonia, no church entered into partnership with me in giving and receiving except you only; for even in Thessalonica you sent me help once and again' (4: 15). The implications of this text are double: (1) since Thessalonica is in Macedonia, the money must have been sent to Paul somewhere in Achaia; (2) prior to establishing the church in Philippi, Paul had founded other communities capable of aiding him financially. But (3), since Paul's route in Greece was from north to south, these other churches must lie further east, i.e. in Asia Minor.

Paul evokes financial aid from Macedonia in 2 Corinthians, 'I robbed other churches by accepting support from them in order to serve you [Corinthians]. And when I was with you and in need, I did not burden any one, for my needs were supplied by the brethren who came from Macedonia' (11: 8–9). The possibility that the allusion is to his second visit to Corinth (2 Cor. 1: 23 to 2: 1) is excluded by the fact that this latter visit was brief and unplanned. The tone of 2 Corinthians 11: 7–10 betrays the anger of one found out and is explicable only by reference to Paul's boast in an earlier letter that he had been self-supporting during his first visit to Corinth (1 Cor. 9: 15–18). His awareness that this was not in fact the case surfaces in his (unconscious?) use of the compound verb *katachrêsasthai* 'to make full use of' (9: 18b),[94] which implies that he was making partial use of his right to be supported. He had been subsidized from churches in Macedonia, notably Philippi, on the occasion of his founding visit.[95]

Only a single hypothesis integrates all these hints in the epistles. On one and the same voyage Paul founded, in the following order, the churches of Galatia, Philippi, Thessalonica, and Corinth, having passed through Athens, where he accomplished nothing. Confirmation that this voyage must be dated before AD

---

[93] The use by Lüdemann (1984), 104–7, of this phrase to date the journey in question prior to AD 51 is invalidated by the inherent ambiguity of the words.

[94] LSJ 921b; BAGD 420; Zerwick (1953), 376.

[95] So rightly Lüdemann (1984), 103–4.

51 is furnished by a comparison of the superscriptions of 1–2 Thessalonians and Galatians. The absence of any self-identification in the Thessalonian correspondence reveals the confidence of the emissary of a major church, i.e. these letters date from a time when Paul was mandated by Antioch. This relationship, however, came to an end in AD 51–52 (Gal. 2: 11–14). The verbosity of Galatians betrays the nervousness of one who had lost his legitimizing base. It must have been composed on a later journey.

Precisely the same pattern, even to failure in Athens, is manifest in the Acts where, on his first journey into Europe, Paul is presented as moving from Galatia (16: 6) through Philippi (16: 12), Thessalonica (17: 1), and Athens (17: 15) to Corinth (18: 1). It is impossible to avoid the conclusion that Luke had independent information about Paul's first independent missionary journey through Asia Minor and Greece.[96] This being the case, we are entitled to trust the information which complements the data of the epistles, namely that the journey began in Antioch (Acts 15: 30) and that the route to Galatia passed through Syria and Cilicia (Acts 15: 41), Derbe and Lystra (Acts 16: 1), and Phrygia (Acts 16: 6). The accidental character of Paul's visit to Galatia (Gal. 4: 13) indirectly confirms this route, because it would have brought him too far to the west. Were north Galatia his goal, he would have gone north from Lystra through Iconium.

Dating the Stages of the Journey

The next task is to attempt to date the different stages of this journey. The starting-point is Paul's arrival in Corinth in the spring of AD 50, i.e. eighteen months before his departure from Corinth in September AD 51. Given the objection to travelling in winter, he must have spent at least the winter of AD 49–50 in Macedonia. This is the bare minimum, but it is very probable that he stayed longer. The quality of the communities he established in Thessalonia and Philippi[97] betrays protracted careful formation which could hardly be accomplished in a six-month period. The eighteen months Paul spent in Corinth (Acts 18: 11) and the two years and three months in Ephesus (Acts 19: 8–10) are illustrative of the time-span required. Hence, I think it highly probable that he spent at least two winters there, which would mean that he arrived in Philippi sometime in the late summer or early autumn of AD 48.

The journey from Galatia to the coast would have taken Paul most of the summer of AD 48. Where exactly he was in Galatia, and the details of the route he took will be discussed later.[98] Here it is necessary only to note Jewett's

---

[96] Why Luke dated this journey after the Jerusalem Conference is not our concern here, but it may be suggested that he wanted to give the impression that the evangelization of Asia Minor and Greece took place under the aegis of Jerusalem.

[97] See Chs. 5 and 9.

[98] See Ch. 7, 'Galatia and the Galatians'.

estimate of the distance at 771 km. (463 miles), which would not be sensibly different whichever route is chosen, and the time required as six weeks.[99] This is an average of 18 km. (11 miles) per day, which is low in terms of the general daily figure,[100] but may realistically reflect the maximum feasible in the brutal heat of the Anatolian summer.

Paul must have left Galatia in the late spring of AD 48, after the snows had melted, having certainly spent the winter of AD 47–48 holed up with one of his communities.[101] A ministry of six months, however, is probably too short. The impression given by Galatians 4: 13 is that Paul's illness must have been rather serious, and so time must be allowed for convalescence. Moreover, only during summer could he undertake the journeys which may be implied in the fact that he founded a number of communities, 'the churches of Galatia' (Gal. 1: 2). Jewett assigns a year to Paul's stay among the Galatians.[102] But this hypothesis involves obvious difficulties. Paul must have arrived in Galatia at the latest by late September, when it begins to snow on the plateau of Anatolia. His sickness and recovery probably occupied most of the first winter. The following summer he made converts and established churches. By then, however, it would have been too late to undertake a journey to the north.[103] Hence, we must postulate that he spent two winters in Galatia, namely AD 46–48. Not all this time would have been active ministry, owing to his illness and the difficulty, if not the impossibility, of even local travel in winter; the brevity of his ministry by comparison with his stay is confirmed by the fact that the communities in Galatia subsequently proved not as well grounded in the faith as those in Thessalonica and Philippi.

Jewett calculates the journey from Antioch to Galatia to be 1069 km. (641 miles) and estimates the travel time to be forty-three days. This is an average of 25 km. (15 miles) per day, which is feasible, particularly since Paul would have been starting after a winter's rest.[104] In addition Jewett postulates stops of varying duration, namely, ten weeks in Syria–Cilicia–Derbe; eight weeks in

[99] (1979), 60.

[100] The Bordeaux Pilgrim lists twelve 'cities' and 'inns' for the 258 Roman miles between Nicomedia and Ancyra, which is an average of 21 Roman miles (30 km./18 miles) per day.

[101] Jewett (1979), 137 n. 49, appositely quotes Ramsay, 'All travel across the mountains [of Anatolia] was avoided between the latter part of November and the latter part of March; and ordinary travellers, not forced by official duties, but free to choose their own time would avoid the crossing [of the plateau] between October (an extremely wet month on the plateau) and May.' After a battle against the Galatians near the river Halys in autumn the Roman general C. Manlius Vulso had to go all the way back to Ephesus on the coast to find suitable winter quarters for his troops (Livy, *History of Rome* 38. 27).

[102] (1979), 59.

[103] If he did set out, he must have wintered somewhere en route to the Dardanelles, which leaves our relative chronology intact.

[104] A section of the route which Paul would have taken is documented by the Bordeaux Pilgrim. The segment Antioch–Tarsus–Faustinopolis is 202 Roman miles and took nine days, an average of 22 Roman miles (32 km./20 miles) per day.

Lystra–Iconium; and four weeks in Antioch in Pisidia.[105] The arbitrary character of these latter figures is evident, but it is undeniable that Paul would have spent time in each of these places.

If we add Jewett's figures for the segment Antioch–Galatia, which he does not do, a problem immediately becomes apparent. The total is twenty-eight weeks. The earliest that Paul could have left Antioch is the latter part of April, because he would not want to face the Cilician Gates, the narrow pass through the Taurus Mountains behind Tarsus, until well into May. Twenty-eight weeks from mid-April, however, brings us to the middle of November, which is far too late for travellers to be abroad in Anatolia. Paul must have been settled in Galatia by mid-September at the latest. While the travel time cannot be reduced, and might probably be extended, it is perfectly feasible to reduce Jewett's rest/ministry time by two months, because Paul had no responsibility for the churches in the cities mentioned. As we shall see, he participated in their evangelization, but only in a subordinate position.[106] The minimum time, therefore, for the journey from Antioch to Galatia is the summer of AD 46. It is not impossible, however, that it took a year longer.

The results of this analysis and calculations can be tabulated as follows:

Antioch	Winter 45–46
Departure from Antioch	April 46
Journey to Galatia	April–September 46
Ministry in Galatia	September 46–May 48
Journey to Macedonia	Summer 48
Ministry in Macedonia	September 48–April 50
Journey to Corinth	April 50
Ministry in Corinth	April 50–September 51
Journey to Jerusalem	September 51
Conference in Jerusalem	October 51.

These dates, it should be remembered are the rock-bottom minimum.[107] Under no circumstances can less than five to six years be allowed for the journey into Europe which ended with the Jerusalem Conference. It goes without saying, however, that Paul may have travelled more slowly and laboured longer in any one place than my calculations allow. In which case, the nine hidden years (AD 37–46), about which we know only that he spent time in Syria and Cilicia (Gal. 1: 21; Acts 11: 25), could be significantly reduced.

After AD 51

Luke's statements that Paul returned to Antioch after the Jerusalem Conference (Acts 15: 30 = 18: 22) are confirmed by Paul's placing of the episode at

---

[105] (1979), 59.   [106] See Ch. 4, 'A Gap in the Record'.
[107] They are substantially identical with those proposed by Jewett (1979), 100.

Antioch (Gal. 2: 11–14) after his version of the Conference (Gal. 2: 1–10).[108] The 600 km. (360 miles) journey from Jerusalem would have taken between two and four weeks,[109] and Paul would have been comfortably installed in Antioch before winter set in. There would have been no point in going any further until weather conditions improved in the spring. Given Paul's temperament, the sense of urgency he brought to his mission (2 Cor. 5: 14), and the intolerable situation that developed at Antioch due to the interference of James (Gal. 2: 11–14), it is most probable that he left his erstwhile spiritual home and headed north once the the roads had become passable, i.e. April 52 at the earliest.

Ephesus

According to Luke, Paul passed through 'the high country' of central Asia Minor, i.e. Galatia (Acts 18: 23), on his way to Ephesus (Acts 19: 1). Again the letters furnish confirmation. We know that Paul was in Ephesus subsequent to the Jerusalem Conference (1 Cor. 16: 8), because he mentions the collection there agreed upon (Gal. 2: 10), and we know that his route took him through Galatia, because with regard to the collection he tells the Corinthians to do 'as I directed the churches of Galatia' (1 Cor. 16: 1). That Paul had visited Galatia a second time before writing 1 Corinthians and Galatians is perhaps also suggested by the 'formerly' of Galatians 4: 13.[110] Even if we allow a month for his stay among the Galatians, it would have been perfectly feasible for Paul to have reached Ephesus by July, or the latest August AD 52.

Luke gives two figures for the duration of Paul's stay in Ephesus, two years and three months (Acts 19: 8–9), and three years (Acts 20: 31). The latter is a round figure, and therefore suspect. The former is not, and is intrinsically plausible. Ephesus was as big as, if not bigger, than Corinth where Paul had spent eighteen months, and while in the capital of Asia Paul had to concern himself, not only with the teething problems of a new church, but also with the more serious problems of other communities, notably, Galatia, Philippi, Colossae, and Corinth. Thus we can assume with a fair degree of confidence that Paul left Ephesus definitively in October AD 54.

He had planned to leave earlier that year, i.e. at Pentecost (1 Cor. 16: 5–6), which that year fell on 2 June.[111] In fact, however, he spent most of the summer

---

[108] Lüdemann (1984), 75–7, dates the incident at Antioch before the Jerusalem Conference. Why I do not find his reasons convincing will appear in my discussion of this episode in Ch. 6 below.

[109] Jewett (1979), 59.

[110] This is affirmed by Kümmel (1975), 302, but denied by Lüdemann (1984), 91, and Longenecker (1990), 190. The contrast these latter perceive with v. 16 is not the most natural interpretation. The objection of Betz (1979), 10–11, to a second visit is automatically negated by his failure to appreciate that Paul evangelized Galatia before the Jerusalem Conference.

[111] Jewett (1979), 48.

of AD 54 away from Ephesus, because circumstances demanded a quick visit to
Corinth (2 Cor. 12: 14; 13: 1–2); he left Asia by boat in mid-June.[112] His stay in
Corinth was short and painful, and he returned overland via Thessalonica and
Philippi (2 Cor. 1: 23 to 2: 1). 736 km. (460 miles) separate Corinth from
Neapolis (Acts 16: 11), today Kavalla,[113] and it is another 350 km. (210 miles)
from Troas to Ephesus. At an average of 32 km. (20 miles) per day, the journey
would have taken little over five weeks. But time has to be allowed for dealings
with the churches in Macedonia. None the less Paul could well have been back
in Ephesus by early August AD 54. Later that summer he moved from Ephesus
to Troas and thence to Macedonia (2 Cor. 2: 13; 7: 5), presumably just before
the end of the sailing season.

Macedonia

At this point further travel was impossible because winter was setting in, and
we are forced to assume that Paul spent the winter of AD 54–55 in Macedonia.
This inference is confirmed by a reference to the collection. During his sojourn
among them he became aware of the generosity of their contributions to the
collection, and when he could again send a letter to Corinth, in the spring, he
wrote, 'I boast about you to the people of Macedonia, saying that Achaia has
been ready since last year' (2 Cor. 9: 2). The allusion is to the enquiry of the
Corinthians about the collection to which he replied in 1 Corinthians 16: 1–4.
Since 1 Corinthians must be dated in April or May AD 54,[114] we must infer from
the mention of 'one year" that Paul was still in Macedonia in the spring of AD
55.
   Paul's planned third visit to Corinth (2 Cor. 9: 4; 12: 14; 13: 1–2) did in fact
take place, because in Romans he writes, 'At present I am going to Jerusalem
with aid for the saints. For Macedonia and Achaia have been pleased to make
some contribution for the poor among the saints at Jerusalem' (15: 25–6). This
is the fulfilment of his promise made in the spring of AD 55 to go to Corinth in
the near future (2 Cor. 9: 4). Before getting there, however, Paul preached in
Illyricum (Rom. 15: 19) until his ministry there was brutally interrupted by the
crisis at Corinth which triggered 2 Corinthians 10–13. This emergency brought
him to Corinth sometime during the summer of AD 55.
   This minimal hypothesis would place Paul at Corinth during the winter of AD
55–56. From there he planned to return to Jerusalem with the collection (Rom.
15: 25–6). For what happened subsequently we are dependent on Luke, who
informs us that Paul had been a prisoner for two years in Caesarea when the

---

[112] This 'intermediate visit' is denied by some scholars, e.g. Hyldahl (1973), 303–4. The fact and
timing will be justified in the context of our discussion of Paul's relations with Corinth in Ch. 11,
'Contacts with Corinth' and Ch. 12, 'An Unplanned Visit'.
[113] Rossiter (1981), 229, 384, 418, 499, 562.
[114] See Ch. 7, 'An Ephesian Chronology'.

procurator Felix was replaced by Festus (Acts 24: 27 to 25: 1). This shift of authority is tentatively dated to AD 59 or 60. Hence Paul's arrest by the tribune Lysias (Acts 21: 33) should be placed in AD 57 or 58. His transfer to Rome at his own instance (Acts 25: 11) must have occurred before the death of Festus in AD 62. According to Luke, Paul spent two years as a prisoner in Rome (Acts 28: 30), a figure that there are no apparent grounds to question. Subsequently, as we shall see, he made an abortive visit to Spain, which was followed by a circuit of the Aegean Sea to visit his churches. He returned to Rome in order to strengthen the church there in the aftermath of Nero's persecution and was himself arrested and executed, probably in AD 67.[115]

We are now in a position to complete the table on p. 28.

Jerusalem Conference	October 51
Antioch	Winter 51–52
Journey to Ephesus	April–July 52
Ephesus	August 52–October 54
Macedonia	Winter 54–55
Illyricum	Summer 55
Corinth	Winter 55–56
Journey to Jerusalem	Summer 56
Jerusalem–Caesarea	57?–61?
Journey to Rome	September 61–Spring 62
Rome	Spring 62–Spring 64
Spain	Early Summer 64
Around the Aegean	64–66?
Death in Rome	67

[115] For the details, see Ch. 14 'Martyrdom'.

— 2 —

Growing Up in Tarsus

PAUL does not tell us where he was born, but a number of texts contain an important hint. 'I myself am an Israelite, a descendant of Abraham, a member of the tribe of Benjamin' (Rom. 11: 1); 'circumcised on the eighth day, of the people of Israel, of the tribe of Benjamin, a Hebrew born of Hebrews' (Phil. 3: 5). Such concern to affirm his Jewish credentials betrays the expatriate, i.e. a Jew living in the Diaspora.[1] Only the descendants of those who emigrated from Ireland to the United States find it necessary to insist that they are Irish. Those who were born and bred in Ireland take it for granted. If this hypothesis is correct, one would expect a particularly passionate outburst when Paul is challenged by opponents of Jewish Palestinian origin, and this is precisely what we find in 2 Corinthians 11: 2, 'Are they Hebrews? So am I! Are they Israelites? So am I! Are they descendants of Abraham? So am I!'

Where in the Diaspora? The letters offer only one slender clue. After his first visit to Jerusalem as a Christian, Paul goes to 'the districts of Syria and Cilicia' (Gal. 1: 21). His use of 'districts' in 2 Corinthians 11: 10 (cf. Rom. 15: 23) suggests that he is thinking, not in terms of the Roman provinces, but of smaller non-political areas.[2] That he should have been drawn to Syria is understandable. Its capital, Antioch, was in many ways similar to the Damascus which he was forced to leave (2 Cor. 11: 32–3), and it might already have had a Christian presence (Acts 11: 19–21). It certainly offered many opportunities for ministry. Why, then, did he go to Cilicia? The simplest answer is that there was some personal connection.

Such considerations tend to confirm Luke's information that Paul came from Tarsus (Acts 9: 11, 30; 11: 25; 21: 39; 22: 3), which was the capital of Cilicia. Luke, moreover, would have no interest in inventing a Diaspora origin for Paul.[3] From the perspective of his theologico-historical program—'You shall be my witnesses in Jerusalem, and in all Judaea and Samaria and to the end of the earth' (Acts 1: 8)—Luke would have certainly preferred Paul to be a Jerusalemite.[4] He does in fact attempt to make Paul a Jerusalemite by adoption, by suggesting that, although born in Tarsus, Paul was not only educated but

---

[1] Against Hengel (1991), 1.
[2] Betz (1979), 80; Longenecker (1990), 40.
[3] Burchard (1970), 34 n. 41.
[4] So rightly Knox (1950), 34.

nurtured in Jerusalem (Acts 22: 3; cf. 26: 4).[5] The subtlety of the ploy does nothing to enhance its credibility.[6]

THE CITY OF TARSUS

In the fourth century BC Xenophon called Tarsus 'a great and prosperous city',[7] a description which remained true well beyond the time of Paul, as Dio Chrysostom testifies in speaking to the Tarsians,

> Your home is in a great city and you occupy a fertile land, because you find the needs of life supplied for you in greatest abundance and profusion, because you have this river flowing through the heart of your city; moreover, Tarsus is the capital of all the people of Cilicia. (*Discourses* 33. 17; cf. 34. 7; trans. Crosby)

Its merchants had always efficiently exploited both its navigable river and its position 'on one of the great trade routes of the ancient world; the easiest and most frequented land route from Syria and the east to Asia Minor and the Aegean crossed the Amanus by the Syrian Gates, and the Taurus by the Cilician Gates'.[8] The surrounding fertile plain produced cereals and grapes, and above all the flax which provided the raw material for the linen industry, whose product was of such quality that the production of the whole region was named for it.[9]

Tarsus had had a long history before Antiochus IV Epiphanes (175–164 BC) in 171 BC conferred on it the status of a Greek city-state governed by its own elected magistrates and issuing its own coins.[10] Its name was changed to Antiocheia-on-the-Cydnus (which did not last for long), and Greek and Jewish colonists were brought in to increase the productivity of the oriental population.[11] The continuity of Jewish presence into the first century AD is well attested.[12]

Absorbed into the Roman system when Pompey reorganized Asia Minor in 66 BC, Tarsus opposed Cassius, the murderer of its patron Julius Caesar.[13] In 42

---

[5] Van Unnik (1962) has shown convincingly that this is the only interpretation of Acts 22: 3. His only mistake was to believe Luke, an error which has been repeated by Bruce (1977), 43; Richards (1991), 148–9); and Légasse (1991), 34– 5.

[6] See e.g. Haenchen (1971), 624 n. 5.

[7] *Anabasis* 1. 2. 23.

[8] Jones (1971), 191.

[9] Ibid. 206.

[10] 2 Macc. 4: 30–8; Ramsay (1907), 159–61.

[11] Ibid. 165–86. There is no direct evidence for Jewish colonists in Antioch, but there is a precedent in the policy of Antiochus III who settled Jews from Babylonia in Phrygia and Lydia (Josephus, *AJ* 12. 147–53).

[12] The presence of Jews in Cilicia, which is guaranteed by Luke (Acts 6: 9) and Philo (*Legatio ad Gaium* 281), is specified for Tarsus by Philostratus, *Life of Apollonius* 6. 34.

[13] Dio Cassius, *Roman History* 47. 26. 2; 47. 30. 1; 47. 31. 1–4.

BC Mark Antony rewarded its loyalty by granting it freedom and immunity.[14] This rare privilege for a city which was not a colony was renewed after the battle of Actium (31 BC) by Augustus, who conferred upon it 'land, laws, honour, control of the river and of the sea in your quarter of the world, and this is why your city grew rapidly'.[15] Such marks of Roman interest are of considerable relevance for the question of Paul's Roman citizenship. Ramsay points out that they were likely to have been accompanied by grants of Roman citizenship to a certain number of citizens by Julius Caesar, Antony, and Augustus.[16]

In some respects, however, the city was decidedly oriental rather than Western.[17] Whereas the most primitive barbarian, according to Dio Chrysostom, would immediately have discerned the Greek character of Rhodes,[18] he would have had problems at Tarsus, 'Would he call you Greeks, or the most licentious of Phoenicians?'[19] The two features singled out as illustrations of this tendency by Dio Chrysostom are music and women's attire. The first he reprobates, 'Now it is Phoenician airs that suit your fancy and the Phoenician rhythm that you admire most.'[20] The second he heartily approves,

> And yet many of the customs still in force reveal in one way or another the sobriety and severity of deportment of those earlier days. Among these is the convention regarding feminine attire, a convention which prescribes that women should be so arrayed and should so deport themselves when in the street that nobody could see any part of them, neither of the face nor of the rest of the body, and that they themselves might not see anything off the road. (*Discourses* 33. 48; trans. Crosby)

The unlikelihood of the oriental character of Tarsus being a late development—Dio was writing at the beginning of the second century AD—is confirmed by his recognition that the wearing of the all-enveloping black chador there was an ancient custom which must go back to the original indigenous population.

Although firmly rooted in the soil of the east, Tarsus had a Hellenic respect for education, and the means to pay for it. Ramsay rightly considers Philostratus' low estimate of its educational system[21] to be a deduction from the criti-

---

[14] 'He made Laodicea and Tarsus free cities and released them from taxes entirely' (Appian, *History* 5. 1. 7; trans. White). Tarsus on that occasion was also the scene of his first meeting with Cleopatra, which might have influenced the generosity of his mood (see Plutarch, *Anthony* 25–28). Pliny also calls Tarsus 'a free city' (*NH* 5. 92).

[15] Dio Chrysostom, *Discourses* 34. 8. The privileges were never withdrawn (*Discourses* 34. 25).

[16] (1907), 198. According to Hengel (1991), 90 n. 11, between 18 BC and AD 14 the number of Roman citizens increased by almost a million.

[17] Callander (1904), 64–5.

[18] *Discourses* 31. 163

[19] Ibid. 33. 41.

[20] Ibid. 33. 42; cf. 33. 57.

[21] 'Apollonius found the atmosphere of the city harsh and strange and little conducive to the philosophic life, for nowhere are men more addicted than here to luxury: jesters and full of

cisms of Dio Chrysostom and of no evidential value for the early first century.[22] Strabo, on the contrary, was an eyewitness,

> The people at Tarsus have devoted themselves so eagerly, not only to philosophy, but also the whole round of education in general, that they have surpassed Athens, Alexandria, or any other place that can be named where there have been schools and lectures of philosophers. But it is so different from the other cities that there the men who are fond of learning are all natives, and foreigners are not inclined to sojourn there. Neither do these natives stay there, but they complete their education abroad. And when they have completed it, they are pleased to live abroad, and but few go back home.... Further the city of Tarsus has all kinds of schools of rhetoric, and in general it not only has a flourishing population but also is the most powerful, thus keeping up the reputation of the mother-city. (*Geography* 14. 5. 13; trans. Jones)

A close reading of this encomium reveals that what struck Strabo about Tarsus was not the superiority or antiquity of its university, which attracted no students from abroad, but the enthusiastic seriousness with which the Tarsians sought education, even to the extent of leaving their homeland in pursuit of further knowledge. Ramsay is in all probability correct in correlating this situation with the badly needed administrative reforms introduced by Athenodorus[23] around 10 BC and reinforced by his successor Nestor.[24]

The city, therefore, into which Paul was born was well governed and prosperous.[25] Its Greek orientation had to struggle with a strong Eastern spirit. It stood on the frontier of east and west, and its citizens were prepared to function in both.

THE FAMILY OF PAUL

Our information is sparse. Paul tells us that he was 'a Hebrew born of Hebrews' (Phil. 3: 5). Luke adds that he was a Roman citizen by birth (Acts 22: 27–8; cf. 16: 37; 23: 27), in addition to being a citizen of Tarsus (Acts 21: 39), and that he had a sister and a nephew in Jerusalem (Acts 23: 16).[26]

insolence are they all; and they attend more to their fine linen than the Athenians did to wisdom' (*Life of Apollonius* 1. 7; trans. Conybeare).

[22] (1907), 234.
[23] Strabo, *Geography* 14. 5. 14.
[24] (1907), 224–8.
[25] 'In days gone by your city was renowned for orderliness and sobriety, and the men it produced were of like character' (Dio Chrysostom, *Discourses* 33. 48).
[26] Certain novelists with scholarly pretensions, e.g. Ambelain (1972) and Messadié (1991), claim that Paul must have been born of a great noble family because he was a foster-brother of Herod the tetrarch (Acts 13: 1). On this foundation great towers of speculation are raised. The Greek, however, makes it clear that Manaen alone was the foster-brother of Herod.

A Hebrew

The most extensive use of *Hebraios* in first-century Greek is in the works of Philo. In the majority of instances, it means a member of the Jewish people either by birth, e.g. 'the descendants of Hebrews' (*Jos.* 42), or by conversion, e.g. 'an Egyptian by birth but a Hebrew by choice' (*Abr.* 251). An element of contrast is apparent on occasion, thus 'they call Moses an Egyptian, Moses who was not only a Hebrew but of the purest Hebrew blood' (*Mut.* 117; cf. *Mig.* 141). This is intensified in a series of texts in which Hebrew and Greek meanings are contrasted, e.g. 'a place which in the tongue of the Hebrews is called Shinar and in that of the Greeks "shaking out"' (*Conf.* 68).[27]

The hint that 'Hebrew' carried, not merely religious or ethnic overtones, but also a linguistic connotation is confirmed by the response to Ptolemy's request for translators to render the Law into Greek. The high priest 'sought out such Hebrews as he had of the highest reputation, who had received an education in Greek as well as in their native lore, and joyfully sent them' (*Mos.* 2. 32). The implication of this passage, whose emphasis is not knowledge of the Law but on linguistic ability, is that, while few Jews, if any, in the Diaspora knew Hebrew, only some of the Jews in Palestine could write Greek.

Against this background it is difficult to avoid the conclusion that the division of the early church in Jerusalem into 'Hebrews' and 'Hellenists' (Acts 6: 1) was based on the fact that the former spoke Hebrew and the latter Greek. This, however, implies other differences. Since the Twelve, who all come from Galilee, admit their responsibility for the Hebrews (Acts 6: 2), it would seem that 'Hebrew', because of its linguistic connotation, implied a relationship to Palestine in a way which 'Hellenist' did not. Even though these latter may have been Greek-speaking Jews of Jerusalem,[28] use of the ancestral language created a deeper bond with the land.

These considerations create a presumption that when Paul uses 'Hebrew' he intends to imply a positive relationship to Palestine through the use of a Semitic language;[29] it is not a mere synonym for Israelite. This is confirmed by J. B. Lightfoot's perceptive insight that Paul's privileges in Philippians 3: 5 are arranged on an ascending scale.[30] A child circumcised on the eighth day could still be descended from proselytes. But Paul is of the race of Israel. Some Israelites were unable to provide proof of their genealogy.[31] But Paul knew he was of

---

[27] Also *Conf.* 129; *Cong.* 37; *Mut.* 71; *Som.* 1. 58; *Abr.* 17, 27; *Jos.* 28.

[28] They are not necessarily Diaspora Jews or proselytes as Gutbrod maintains (*TDNT* 3. 389).

[29] It is impossible to decide whether he had Hebrew or Aramaic in mind.

[30] (1908), 146. His exegesis, however, which is followed by Gutbrod (*TDNT* 3. 390), is accurate only in part.

[31] 'There were also others besides these, who said that they were of the Israelites, but were not able to show their genealogies' (Josephus, *AJ* 11. 70).

the tribe of Benjamin.[32] The land of Benjamin, however, included Jerusalem[33] where the influence of Hellenism was particularly manifest in the many Jews who spoke Greek. But Paul came of a family which, despite its location in the Diaspora, retained the ancient tongue of the Jews.[34]

The hypothesis of a highly conservative and deeply religious family ever concerned to keep pagan influences at bay is not impossible in itself, but it cannot be harmonized with the type of education that Paul received. One might argue with slightly greater probability that the family needed the language for frequent commercial contacts with Palestine,[35] but it is difficult to conceive of Palestinian Jews in the export–import business failing to learn Greek. The simplest hypothesis is that Paul's ancestors had emigrated from Palestine within living memory.

Certainly it is the one adopted by Jerome to explain Philippians 3: 5 and 2 Corinthians 11: 22,

> We have heard this story. They say that the parents of the Apostle Paul were from Gischala, a region of Judaea and that, when the whole province was devastated by the hand of Rome and the Jews scattered throughout the world, they were moved to Tarsus a town of Cilicia; the adolescent Paul inherited the personal status of his parents.   (*Comm. in Ep. ad Philem.* on vv. 23–4)[36]

In this text it is not clear whether Paul had been born at the time of his parents emigration. The ambiguity no longer exists in Jerome's second reference, written some five years later:[37]

> Paul the apostle, previously called Saul, was not one of the Twelve Apostles; he was of the tribe of Benjamin and of the town of Gischala in Judaea; when the town was captured by the Romans he migrated with his parents to Tarsus in Cilicia.   (*De viris illustribus* 5)[38]

These testimonies have no support in either the letters or the Acts, and contain serious internal contradictions.

In the second text Paul was born in Gischala, in the first probably not. In the first Gischala is a region but in the second a town. In the first the migration of

---

[32] On claims to belong to the tribe of Benjamin, see Jeremias (1969), 277.
[33] The Hinnom valley was the southern boundary of Benjamin (Josh. 18: 16, 28) and the northern boundary of Judah (Josh. 15: 8). Luke apparently was unaware of Paul's Benjaminite connections (Acts 13: 21).
[34] Paul's formulation—'a Hebrew born of Hebrews'—excludes the possibility that Paul learnt the language in Jerusalem, as Lightfoot (1908), 147, seems to imply. On the effect of this Semitic language on Paul's Greek, see van Unnik (1943).
[35] On business relations between Cilicia and Palestine, see Applebaum (1976), 716; Hengel (1991), 99 n. 47.
[36] *PL* 26. 617.
[37] Kelly (1975), 145, 174, dates the commentary on Philemon to AD 387–388, and the *De viris illustribus* to AD 392–393.
[38] TU 14. 9.

Paul's parents appears to have been involuntary, whereas in the second it was voluntary. Manifestly the two accounts cannot be reconciled. A choice, therefore, has to be made. In view of the widespread criticism of *Famous Men*,[39] one's preference must go to the *Commentary on Philemon* as the assertion which merits historical testing.

The note in the *Commentary on Philemon* cannot be dismissed on the grounds that the only known Gischala is located in Galilee not Judaea,[40] because even in the New Testament 'Judaea' is used to mean the whole of Palestine (e.g. Luke 1: 5; 23: 5). This usage was reinforced with the establishment of the Roman province of Judaea after the failure of the First Revolt, and particularly from the 120s when a governor of consular rank controlled two legions, the Sixth *Ferrata* stationed in the north and the Tenth *Fretensis* based in the south.[41] It would have been natural, therefore, for Jerome and his contemporaries to think of Palestine as Judaea.[42]

Apart from his pilgrimage with Paula, which brought them only as far north as Capernaum,[43] Jerome had little personal knowledge of places in Palestine, and certainly did not seek out local traditions.[44] His information, therefore, is secondary. There is evidence that he knew Josephus thoroughly,[45] which might explain how he came across the name of Gischala, whose inhabitants fled when the Romans attacked.[46] But in this case it is almost certain that he would have located Gischala in Galilee, and the hypothesis cannot explain why he associated the city with Paul. Theodore Zahn has argued that Jerome's source was Origen's commentary on Philemon, which is no longer extant.[47] Within the space of a few months in 387 or 388 Jerome wrote commentaries on Philemon, Galatians, Ephesians, and Titus in that order; he explicitly admits his dependence on Origen's commentaries on Galatians and Ephesians, which makes it likely, in Zahn's view, that he also drew on Origen in expounding Philemon and Titus.[48] The argument is rather tenuous, but it has the advantage of explaining why Jerome notes Paul's parentage apropos of Philemon 24–25.

Where Origen got his information is even more mysterious. The likelihood that he or any earlier Christian invented the association of Paul's family with Gischala is remote. The town is not mentioned in the Bible. It had no connection with Benjamin. It had no associations with the Galilean ministry of Jesus. And there is no evidence that it had Christian inhabitants in the Byzantine period. It would seem, therefore, that Origen relied on an oral tradition, whose authority Jerome accepted. His classification of the story as a *fabula* does not necessarily imply that he distanced himself from his source.[49]

---

[39] Kelly (1975), 176–8.   [40] Abel (1938), 2. 338.
[41] Schürer (1973–87), 1. 514.   [42] Adinolfi (1969), 161.
[43] *Letter* 108; Wilkinson (1977), 52.   [44] Wilkinson (1974), 245–57.
[45] Ibid. 254 n. 104; Kelly (1975), 156.   [46] Josephus, *JW* 2. 84–127.
[47] Zahn (1900), 1. 49.   [48] Zahn (1890), 2. 1001–2. Similarly Kelly (1975), 145–9.
[49] On Jerome's varied use of *fabula*, see Adinolfi (1969), 157–8.

Jerome implies that Paul's parents were forced to move to Tarsus by the Romans. The latter took control of Palestine in 63 BC, and subsequently there were a number of occasions (61, 55, 52, 4 BC, AD 6) when Jews from various parts of the country were enslaved and shipped abroad.[50] Terrible as this was, there were advantages, as Philo records, 'The large district of Rome beyond the Tiber was owned and inhabited by Jews. The majority of them were Roman freedmen, who had been brought to Rome as prisoners-of-war and were manumitted by their owners' (*Leg. ad Gaium*, 155). That Paul's father was equally fortunate is the simplest explanation of the Apostle's inherited Roman citizenship.

Roman Citizenship

Luke's assertion that Paul was a Roman citizen cannot be ascribed to his propagandizing intention because he found it in one of his sources, namely, the Travel Document.[51] Moreover, Paul's voyage to Rome, which is presented as a privilege of his citizenship (Acts 25: 11–12; 26: 32; 28: 19),[52] cannot be ascribed to Lucan invention because it is not exploited. Nothing happens in Rome. The one sermon preached there is very mediocre by comparison with others in Acts, and produces only a highly ambiguous result (Acts 28: 24–5). On the other hand, however, nothing in the Pauline letters confirms the Apostle's citizenship. At times they have even been considered to contain a decisive refutation.[53]

Paul notes that three times he was beaten with rods (2 Cor. 11: 25). This punishment is distinguished from the 39 lashes inflicted on him by Jews (2 Cor. 11: 24), and is in fact a specifically Roman punishment whose infliction on Roman citizens was forbidden.[54] At times, however, this law was more honoured in the breach than in the observance, and there are well-documented instances in which individuals whose citizenship is beyond question were beaten and even executed by Roman authorities.[55] The reality of the situation is well formulated by J. C. Lentz, 'A Roman citizen in the provinces was a privileged person. His citizenship could, at times, save him from non-Roman pro-

---

[50] Josephus, *JW* 1. 157–8, 177, 180; 2. 68; Adinolfi (1969), 161–2.

[51] Boismard and Lamouille (1990), 2. 219.

[52] Garnsey (1966), 167–89; Tajra (1989), 144–7.

[53] There are recent discussions of the issue by W. Stegemann (1987), 200–29, and Lüdemann (1987), 249–50. Both conclude in favour of Paul's Roman citizenship, as do Légasse (1991), 25–9; Hengel (1991), 6; and Tajra (1989), 81–9.

[54] 'Also liable under the *lex Julia* on *vis publica* is anyone who, while holding *imperium* or office, puts to death or flogs a Roman citizen contrary to his right of appeal, or orders any of the above mentioned things to be done, or puts (a yoke) on his neck so that he may be tortured' (*Digest* 48. 6. 7). For details see, Schürer (1973–87), 3. 134–5.

[55] The commonly cited exceptions are: Josephus, *JW* 2. 308; Plutarch, *Caesar* 29. 2; Cicero, *Against Verres* 2. 5. 139, 149–51, 170; Dio Cassius, *History* 60. 24. 4. See Sherwin-White (1963), 73–6.

vincial justice. Yet only those citizens who also possessed wealth and prestige as well as the citizenship were in the position to procure any certain legal advantages.'[56]

What Paul says about his social status is also considered an objection. On the basis of certain statistics, which do not derive from Tarsus, it would appear that Roman citizens in the east belonged to the provincial aristocracy.[57] Paul, however, presents himself as an itinerant manual labourer.[58] The postulated incompatibility is severely diminished, if not eliminated, both by Paul's educational attainments, which suggest a background infinitely superior to that of the average artisan,[59] and by his rather upper-class view of manual labour as 'slavish' (1 Cor. 9: 19) and 'demeaning' (2 Cor. 11: 7).[60] This attitude makes it probable that it was the imperative of his missionary strategy which led Paul to master a trade.[61]

Paul's failure to mention his citizenship is also construed as an objection. In this form the argument from silence has no value. Not only was there no reason why Paul should mention his status in letters to communities whom he wanted to convince that 'our citizenship is in heaven' (Phil. 3: 20), but to claim citizenship risked incurring the challenge to prove his right. Documentation had to be produced, and this would not have been easy for someone far from his home base and continuously on the move. The small wooden diptych containing the certificate was too precious to carry around, and if it were contested by the magistrate the original witnesses who signed had to be produced.[62]

'To speculate how and when the family of Paul acquired the citizenship is a fruitless task, though lack of evidence has not deterred the ingenious.' This observation by Sherwin-White remains as true as when he wrote it thirty years ago,[63] but something has to be said in order to counter a more subtle objection to Paul's Roman citizenship. It is seldom explicitly articulated, but is latent in the length of discussions regarding the means of acquiring Roman citizenship and the liberality with which it was accorded. The impression is often given of a complexity so great that no real clarity is possible. In fact in the first century BC when, according to Luke (Acts 22: 27–8), Paul's father or more remote ancestor would have acquired citizenship the matter was not very complicated.[64]

[56] (1993), 127.
[57] W. Stegemann (1987), 225–6.
[58] 1 Thess. 2: 9; 2 Thess. 3: 7–9; 1 Cor. 4: 12.
[59] See Ch. 2, 'Rhetorical Training'.
[60] The arguments of Hock (1978) are in no way affected by the strained objections of W. Stegemann (1987), 227, or Légasse (1991), 41. Compare Cicero, *De Officiis*, 150–1, cited in Ch. 4, 'Learning a Trade'.
[61] See Ch. 4, 'Learning a Trade'.
[62] Sherwin-White (1963), 148–9. See in particular Schulz (1942–3) and Gardiner (1986).
[63] (1963), 151.
[64] Sherwin-White (1972).

The competition for support during the civil wars after 49 BC led to 'liberal offers of individual enfranchisement in the East'.[65] The predominance of the oriental element in Tarsus, of which the Jews would have been a part, has been noted above, and thus it is far from impossible that some leading members of the Jewish community were seduced to Antony's side by the gift of Roman citizenship.[66] The simplest possibility, as already noted, is that Paul's father had been a slave who was set free by a Roman citizen of Tarsus, and who thereby acquired a degree of Roman citizenship which improved with each succeeding generation.[67]

Finally, it has been thought that the obligations of citizenship might conflict with the demands of the Jewish faith of Paul's parents. The reason for thinking them to be strictly religious is Luke's assertion that Paul was 'a son of Pharisees' (Acts 23: 6). We shall see that this is most improbable.[68] Moreover, the Roman tribe in which the new citizen was enrolled had a merely legal existence; its members never met and no liturgies were ever performed.[69] Finally, in Roman law, codified in this respect by Julius Caesar, Jews were exempt from any obligations which conflicted with the demands of their faith,[70] which inevitably gave rise to accusations of having their cake and eating it.[71]

To sum up. Since there is no evidence of Lukan creativity and no objection based on the epistles, Paul's Roman citizenship should be admitted, particularly since the history of his parents constitutes a plausible historical context for its conferral.[72]

A Roman Name

As a Roman citizen Paul had a tri-part Roman name,[73] made up of *praenomen* (= the given name), the *nomen* (= the gens, denoting the ultimate founder of the Roman family), and the *cognomen* (= the family name), e.g. Marcus Tullius Cicero. When a slave or foreigner was granted citizenship, the practice was that he retained his own name as the *cognomen*, and took as his own the *praenomen* and *nomen* of the Roman who obtained the citizenship for him. Thus Cicero's

[65] Sherwin-White (1973), 309.
[66] Ibid. 310. A parallel is furnished by the case of Antipater, who took Julius Caesar's side and was rewarded by citizenship (Josephus, *JW* 1. 193–4; *AJ* 14. 137).
[67] See the section 'Citizenship by Manumission' in Sherwin-White (1963), 151; (1973), 322–36; Duff (1928), 12–35; Dionysius of Halicarnassus, *Antiquitates Romanae* 4. 22. 4–4. 23. 7.
[68] See Ch. 3, 'Where Could Pharisees be Found?'.
[69] Ramsay (1904), 20.
[70] See Schürer (1973–87), 3. 120; Saulnier (1981).
[71] Josephus, *AJ* 12. 126; 16. 27–60.
[72] The question of Paul's citizenship of Tarsus (Acts 21: 39) is a much more complex matter, which fortunately does not concern us here because, in opposition to his Roman citizenship, it played no role in his subsequent career. See most recently, Lentz (1993), 28–43.
[73] The basic study remains that of Harrer (1940).

freedman, Tiro, became Marcus Tullius Tiro, and Demetrius Megas, a Greek of Sicily, became P. Cornelius Megas because P. Cornelius Dolabella had been his sponsor.[74]

Paulos is the Greek form of the Latin *Paul(l)us*, which is attested for the time of Paul both as a *praenomen*, e.g. Paullus Fabius Maximus and Paullus Aemilius Lepidus,[75] and as a *cognomen*, e.g. L. Sergius Paullus (Acts 13: 8). The former is as rare as the latter is common, and in the Roman world the *cognomen* was the name most frequently used because it was the most specific.[76] The force of such observations is to suggest that *Paul(l)us* was Paul's *cognomen*. It is impossible, however, that such a typically Roman name, borne by the great senatorial families of the Aemilii, the Vettenii, and the Sergii, should be the *cognomen* of a Jew, whose family had acquired citizenship only a generation earlier.

Sherwin-White suggests a way out of this dilemma: 'The most likely explanation of the *cognomen* Paulus is that it was chosen as the most similar Latin name to the Hebraic name of Saul.'[77] This hypothesis obviously depends on the reading *Saulos*, which appears in the received text of Acts (7: 58; 8: 1, 3; 9: 1) with the exception of the vocative case where the form is consistently *Saoul* (Acts 9: 4, 17; 22: 7, 13; 26: 14). This latter form, however, is used exclusively in all references in P[45], and the scribe's awareness that he was using an indeclinable non-Greek name is formally indicated by an apostrophe after the last letter.[78] The Semitic name, however, cannot be assumed to be the original form, because *Sa(o)ulos* appears frequently as a proper name in Josephus, despite the connotation of effeminacy attached to the adjective *saulos*.[79] It is not surprising, therefore, that *Silas* should appear, perhaps through Aramaic, as a Greek form of Saul.

The name Saul, however, is known to us only through Luke, whose credibility unfortunately cannot be taken for granted, because his usage smacks of artificiality. In Acts 13: 9 we find the formula 'Saul who is also known as Paul', which is the transition from the exclusive use of 'Saul' previously and the exclusive use of 'Paul' subsequently. The symbolism is evident; a Semitic name while Paul worked among Jews and a Gentile name when he worked among Gentiles. Had Luke known the name 'Paul' and needed to create a Semitic correspondent, 'Saul' would be a rather obvious choice (Acts 13: 21).

---

[74] Harrer (1940), 20.

[75] Ibid. 29.

[76] Pontius Pilatus (Acts 4: 27; 1 Tim. 6: 13), for example, is simply Pilatus in 51 other New Testament passages.

[77] (1963), 153.

[78] Harrer (1940), 24.

[79] According to Chantraine, 'il s'applique en principe à l'allure et à la démarche, dit de la démarche ... de femmes, notamment de bacchantes, de courtisanes ...; volontiers pris en mauvaise part' (1968), 990. Leary (1992), who draws attention to Paul's awareness of punning names (Phil. 4: 3; Philem. 11), evokes this dimension by translating *saulos* as 'slut-arsed'.

Attractive as is this hypothesis, it is not likely that Luke invented the name of Saul. Not only is it found in one of his sources,[80] but it is most probable that a Diaspora-born 'Hebrew of Hebrews' should have a Semitic name; it would have been the obvious way of affirming linguistic identity. Gaelic speakers in Ireland—an endangered species—invariably give their children distinctively Irish names. Paul's parents, of course, had a wide choice, but Saul was the name of the best-known member of the tribe of Benjamin.

Thus there is a lot to be said for Sherwin-White's hypothesis which, however, needs to be refined a little. He does not make it clear that Saul may have been the name of Paul's father or grandfather, or evoke the possibility that Paulus was not a true *cognomen*, but rather a *signum* or *supernomen*, which may have been in the family prior to the Apostle's birth, and was used regularly in their relations with Gentiles. The *signum* began as an informal name used among family and friends, but became so much part of the person's identity that it appeared frequently in public inscriptions where it is introduced by *ho kai* in Greek or *qui et* in Latin.[81]

A beautiful illustration of the preceding discussion is furnished by an undated inscription on a tombstone found in Naples, to which C. J. Hemer has drawn attention,[82]

> To the spirits of the dead. Lucius Antonius Leo, also called Neon, son of Zoilus, by nation a Cilician, a soldier of the praetorian fleet at Misenum, from the century the trireme 'Aesclepius', lived 27 years, served 9 years. Gaius Julius Paulus his heir undertook the work (of his burial).[83]

Like Paul, Leo was both from Cilicia and had an alternative name, whose similarity of sound recalls that between Paul and Saul. Were his heir a kinsman, he would also be a Cilician with a name which might have been Paul's own, since the grants of citizenship in Tarsus were by Pompey, Caesar, and Antony. Consequently, the probable *praenomina* and *nomina* of those freed at that period were Gnaeus Pompeius or Gaius Julius or Marcus Antonius.[84] Lest the importance of this inscription be exaggerated, Hemer is careful to point out that not all who boasted the three names were necessarily Roman citizens. Leo presumably acquired his merely as a matter of naval administration since sailors did not get citizenship on enrolment.[85]

---

[80] Boismard and Lamouille (1990), 2. 66.
[81] Harrer (1940), 21, and see an example below.
[82] (1985), 179–83.
[83] *CIL* 10. 3377.
[84] Ramsay (1907), 198.
[85] In a letter dated at the beginning of the 2nd cent. AD Apion, a naval recruit, announces to his family that, on arrival at Misenum, he was given the service name of Antonius Maximus. In a subsequent letter he uses that name alone; see White (1986), 160.

A Description of Paul

Augustine believed that Paul, 'the least of all the apostles' (1 Cor. 15: 9) chose his name because the Latin adjective *paullus* means 'small, little'. This view has nothing to recommend it, except as an opportunity for rhetorical piety. As we have seen above, *Paul(l)us* had been well known as a proper name for centuries. The phenomenon of proper names which are also adjectives is found in many languages, e.g. Small, Petit, and Klein.

It is far from improbable, however, that the etymology of Paul's name influenced the famous description in the late second century AD *Acts of Paul*.

> He saw Paul coming, a man small of stature, with a bald head and crooked legs, in a good state of body, with eyebrows meeting and a nose somewhat hooked, full of friendliness; for now he appeared like a man, and now he had the face of an angel. (3. 1)[86]

The fact that this description does not conform to our canons of beauty has led many scholars to accept its truth.[87] It was first questioned by R. M. Grant, who argued that the intention was to present Paul as a general.[88] Despite its exaggeration, this study had the value of drawing attention to the idealization inherent in the description. In a world where quickness of response in public debate demanded the ability to sum up the personality of an opponent quickly, there were those who taught their clients how to deduce character traits from physical features. Manuals of physiognomy circulated.[89] The style may be illustrated by Pliny's citation from Paul's contemporary Pompeius Trogus,

> When the forehead is large it indicates that the mind beneath it is sluggish; people with a small forehead have a nimble mind, those with a round forehead an irascible mind. . . . When people's eyebrows are level this signifies that they are gentle, when they are curved at the side of the nose, that they are stern, when bent down at the temples, that they are mockers, when entirely drooping, that they are malevolent and spiteful. If people's eyes are narrow on both sides, this shows them to be malicious in character; eyes that have fleshy corners on the side of the nostrils show a mark of maliciousness; when the white part of the eyes is extensive it conveys an indication of impudence; eyes that have a habit of repeatedly closing indicate unreliability. Large ears are a sign of talkativeness and silliness. (*NH* 11. 275–6; trans. Rackham)

[86] Hennecke and Schneemelcher (1965), 2. 354.
[87] Bruce (1977), 468, approvingly quotes Ramsay's verdict, 'this plain and unflattering account of the Apostle's personal appearance seems to embody a very early tradition'. Preuschen (1901), 191–3, suggested that the traits were designed to identify Paul as Antichrist.
[88] (1982), 1–4.
[89] See in particular the studies of Evans (1935) and (1941).

Applying such criteria to the description of Paul, Malherbe showed that 'Meeting eyebrows were regarded as a sign of beauty, and a person with a hooked nose was thought likely to be royal or magnanimous. Tallness was preferred; nevertheless, since men of normally small height had a smaller area through which the blood flowed, they were thought to be quick.'[90] Bowed legs showed a man to be firmly planted, i.e. highly realistic. Baldness is given no prominence in the manuals of physignomy, but for Pliny it was a distinctively human trait; no animal went bald (*NH* 11. 131). The strong probability of idealization in the *Acts of Paul*'s portrait of the Apostle makes its historical value doubtful.

Paul's Relatives

According to Luke, Paul had a married sister whose son alerted the Roman authorities in Jerusalem to a plot to assassinate his uncle (Acts 23: 16). The historicity of this information is difficult to judge. It is unlikely that Paul was an only child, and the permanent or temporary presence of a married sister in Jerusalem could be explained in a number of plausible ways, e.g. commerce or pilgrimage. It is curious, however, that a grown-up nephew[91] with Roman citizenship should appear at just the moment when it was necessary to have immediate access to, and forceful influence on, the senior Roman officer in Jerusalem. Truth, of course, is often stranger than fiction, but the very neatness of the story, and the facility with which a fictional nephew can be disposed of, leave lingering doubts.[92]

In Romans 16: 13 Paul says, 'Greet Rufus, chosen in the Lord, and his mother and mine'. According to Baslez, the reference is to Paul's natural mother, who after being widowed entered a second marriage which produced Rufus, who was now responsible for her.[93] Attractive as this imaginative portrait may be, it is unlikely to be correct. 'Mother' is well attested in the metaphorical sense as applied to those whose comportment commands respect.[94] If Paul had his real mother in mind one would expect him to mention her at the head of the list, and not to slip her in casually halfway through, as a mere adjunct to her son. Moreover, a significant number of commentators on the epistle to the Romans identify this Rufus as the son of Simon of Cyrene (Mark 15: 21).[95] The consensus among commentators is perfectly brought out by the *NRSV* paraphrase, 'greet his mother—a mother to me also'. She was a woman

---

[90] (1986), 173.

[91] He is described as *neanias* and *neaniskos* (Acts 23: 17– 18), both of which imply that he was over 20; see BAGD 534b.

[92] Similarly Haenchen (1971), 649.

[93] (1991), 34–5.

[94] In the New Testament see Mark 3: 33–4; Matt. 12: 49–50; John 19: 27. For other references, see LSJ 1130a; BAGD 520a.

[95] e.g. Lightfoot, Sanday and Headlam, Lagrange, Cranfield, Dunn, but not Käsemann.

who, like Phoebe (Rom. 16: 2), had befriended Paul; where and when remain a mystery.[96]

Even less likely is the interpretation which Baslez gives to *syngenês* in Romans 16: 7, 11, and 20. She understands it as meaning 'relative' in the strict sense of blood relationship, and on this basis creates an elaborate portrait of an extended family displaying various degrees of assimilation (the different names), but none the less co-operating in the textile business.[97] There is no denying that *syngenês* can have this sense,[98] but the meaning intended by Paul is unambiguously indicated by his first use in Romans 9: 3, 'I could wish that I myself were accursed and cut off from Christ for the sake of my brethren, my kinsmen according to the flesh. They are Israelites.' If Paul can use such intimate language of fellow-Jews with whom he has no blood relationship, then the weaker unqualified 'relative' cannot be interpreted in the strict sense.

EDUCATION

While it is an accurate interpretation of Acts 22: 3, we have seen that van Unnik's view that Paul received all his education in Jerusalem fails to meet the objection that it was in Luke's interest to attach Paul as closely as possible to the city which Luke saw as the culmination of Jesus' ministry and the starting-point of all missions.[99] There is some justification for Luke's sleight of hand, because Paul did receive part of his education in Jerusalem. Where he received the rest is important only from a biographical point of view, because van Unnik's sharp distinction between a 'Jewish' education in Jerusalem and a 'pagan' education in Tarsus is untenable. Jerusalem had been heavily Hellenized for several centuries, and educational facilities similar to those in Tarsus were also available in Jerusalem.[100] Know your enemy and fight him with his own weapons had become a fundamental principle of Jewish apologetic.

In the absence of any evidence regarding Paul's youth, we must presume the normal, namely, that Paul was already grown when he left his home in Tarsus. He ventured out into the world, as young men have ever done, only when he had finished his basic education. Strabo, as we have seen, offers express witness for this custom at Tarsus.[101]

---

[96] Were Rom. 16 a letter to Ephesus, as some have claimed (see below, Ch. 13), the answers would be easy. But the Second Gospel appears to have been written in Rome (most recently Hengel (1985*b*), 28–30), and in this perspective Mark 15: 21 is correctly interpreted as implying that Rufus was a member of the Roman community.

[97] (1991), 30–6.

[98] e.g. Luke 1: 36, 58, 61; 2: 44; 14: 12; 21: 16; John 18: 26; Acts 7: 3, 14; 10: 24.

[99] See above, p. 33 n. 5.

[100] See in particular Hengel (1974), 1. 65–83, and (1989).

[101] See above, p. 35.

Paul himself tells us nothing about his youth. We are forced to make deductions based on the existing educational systems in Tarsus and the traces of Paul's background which can be discerned in the letters.[102]

Early Formation

There were certainly pagan and Jewish elementary schools, which children entered at the age of 6. Both schools trained their pupils in the basic skills of reading, writing, and arithmetic, while at the same time inculcating knowledge of and respect for the institutions of state and religion.[103] As members of a religious minority, however, Jewish children carried a greater burden than their pagan contemporaries; they had to live in two worlds.

On the one hand, they had to learn the observances which were the basis of their identity, and which they were bound to obey from the age of 13,[104]

> All men are eager to preserve their own customs and laws, and the Jewish nation above all others; for looking upon theirs as oracles directly given to them by God himself, and having been instructed in this doctrine from their earliest infancy they bear in their souls the images of the commandments. (Philo, *Legatio ad Gaium* 210; trans. Yonge)

That we have to do with an educational principle is made clear, despite the hyperbole, by Josephus' use of almost identical language,

> Should anyone of our nation be questioned about the laws, he would repeat them all more readily than his own name. The result, then, of our thorough grounding in the laws from the first dawn of intelligence is that we have them, as it were, engraved on our souls.
> (*Against Apion* 2. 178; trans. Whiston and Margoliouth)[105]

It is in this context that Paul would have come to know the Septuagint, the Bible of Greek-speaking Jews. At the beginning it was merely a textbook, from which a little boy learned to read and had to memorize parts (2 Tim. 3: 15), but it became a perennial source of insight which ever informed his teaching;[106] his letters contain almost 90 explicit citations. Given his stress on being a 'Hebrew of Hebrews' (Phil. 3: 5), Paul must also have learnt Hebrew and/or Aramaic. Knowledge of the former was rare in the Diaspora,[107] but commitment, and the availability of personal copies of the Scriptures (1 Macc.1: 56–7), mean that it

---

[102] It should be unnecessary to point out that the schematic and generic character of what follows precludes it from being a description of a concrete situation.

[103] See above all Marrou (1948), 200–22.

[104] *m. Aboth* 5. 21 (cited above, p. 4); *m. Nid.* 5. 6; 6. 11; cf. *m. Yoma* 8. 4.

[105] Cf. also *Against Apion*, 1. 60; 2. 204; *AJ* 2. 211.

[106] Michel (1929); Ellis (1981); Koch (1986); Hays (1989).

[107] Schürer (1973–87), 3. 142–3.

cannot be excluded apriori. Aramaic was current among the Semitic population of Syria and the eastern part of Asia Minor.[108] Whether it was learnt at home or at school remains an open question.

On the other hand, Jewish students in Tarsus had to learn how to function in the Hellenistic world to which they belonged. The Greek they learnt at home had to be refined into the ability to read and write. Their basic curriculum would have been that of pagan children their age. These latter would certainly not have used the LXX as a reader, but Jewish children in addition read Euripides or Homer.[109] If Homer was read in first century Pharisaic circles in Palestine,[110] there can be little doubt that it was on the curriculum of a Diaspora school frequented by the son of a Roman citizen. There Paul learnt to trace letters and eventually to write. There too he learnt to count and presumably mastered the intricate hand signs which enabled his contemporaries to express every number from one to a million.[111]

Secondary studies began as soon as the student could read and write easily, normally about the age of 11.[112] The focus of Hellenistic education was not the development of a critical spirit, but the transmission of a whole culture in the works of such writers as Homer, Euripides, Menander, and Demosthenes. While one might presume that Jewish students in the Diaspora might offer some resistance to the total acceptance of the ideas of the textbooks studied by their pagan contemporaries, the system was so widespread and the method of instruction so consistent that it must have influenced even Jewish teachers.[113]

In a school that had more than one manuscript of a classical author, the first step involved some rudimentary textual criticism if there were differences. Next came the reading of the text, a much more difficult operation than we imagine. It demanded careful preparation because words were not separated from one another and there was no punctuation.[114] Such meticulous analysis created a solid basis for the interpretation of the text, which was expected to contribute to the moral development of the students. As the pupils' knowledge increased they practised literary composition. Though the types of composition were

---

[108] Fitzmyer (1990), §82. 12.

[109] Marrou (1948), 214.

[110] 'The Sadducees say, We cry out against you, O Pharisees, for you say that the Holy Scriptures render the hands unclean but that the books of Homer do not render the hands unclean' (*m. Yad.* 4. 6); on which see Hengel (1974), 1. 75, who plausibly suggests that 'books of Homer' was a stereotyped designation of Greek literature in general, much as Biro, Hoover, and Xerox have become generic descriptives in the twentieth century.

[111] Marrou (1948), 219.

[112] Ibid. 223–42.

[113] The argument from silence—Paul reveals no knowledge of classical Greek prose or poetry—developed by Hengel (1991), 2–3, carries no weight. Paul formally states that he chose not to display his erudition (1 Cor. 2: 2–4).

[114] This explains the development of the liturgical office of Reader. While most could pick their way through a text, it needed experience and well-practised skill to read well in public, particularly at short notice.

varied, each was expected to reflect the four basic qualities of brevity, clarity, probability, and grammatical correctness, at the same time as it integrated agent, action, time, place, mode, and cause.[115]

Rhetorical Training

In first-century Palestine a Jewish boy finished his obligatory studies at the age of 12 or 13, when he technically became a responsible person.[116] The Greek secondary school normally went a year or two longer. Then began the equivalent of today's undergraduate university course. What Philostratus tells us of Apollonius is typical, 'When he reached his fourteenth year, his father brought him to Tarsus, to Euthydemus the teacher from Phoenicia' (*Life of Apollonius* 1. 7). Less typical was the refusal of Apollonius to stay for long. The value of his criticism of the instruction at Tarsus has already been discussed,[117] but the note that his frivolous fellow-students sat like so many water-fowl along the bank of the river confirms Strabo's observation that at Tarsus the Cydnus 'flows past the gymnasium of the young men' (*Geography* 14. 5. 11).

Jewish attitudes towards the gymnasium and the education it offered differed according to the self-confidence of individuals and communities.[118] Philo of Alexandria, a contemporary of Paul, had no doubt about its benefits, and took it for granted that Jews who were of a certain social class would be educated there:

> For who can be more completely the benefactors of their children than parents, who have not only caused them to exist, but have afterwards thought them worthy of food, and after that again of education both in body and soul, and have enabled them not only to live, but also to live well, training their body by gymnastic and athletic rules so as to bring it into a vigorous and healthy state, and giving it an easy way of standing and moving not without elegance and becoming grace, and educating the soul by letters, and numbers, and geometry, and music and every kind of philosophy.
>
> (*Spec. Leg.* 2. 229–30; trans. Yonge)[119]

In theory the gymnasium catered for the whole person, but in practice only the body was well trained; the educational programme was wide but superficial.[120] Tarsus, however, had 'all kinds of schools of rhetoric', according to Strabo, who also noted the consequence, namely 'the facility prevalent among the Tarsians whereby he could instantly speak offhand and unceasingly on any given subject' (*Geography* 14. 5. 13–14).

From this perspective Tarsus is the perfect illustration of Marrou's thesis, 'For the great majority of students higher studies meant attending the lectures

---

[115] Marrou (1948), 240.
[116] Safrai (1976), 953.
[117] See p. 34 above.
[118] Goldstein (1981), 64–87.
[119] For more detail, see Mendelson (1982).
[120] Marrou (1948), 261.

of an orator, and learning with him the art of eloquence.'[121] Oratorical skills
were the key to advancement in an essentially verbal culture. The acquisition of
such skills fell into three parts.[122] The base was the theory of discourse, which
included letter-writing.[123] Techniques, rules, formulae, etc. were discussed ad
infinitum. As divisions and sub-divisions multiplied, the complexity of the
theory made it more and more irrelevant. The second stage was a little more
practical in so far as it involved the study of the speeches of the great masters of
rhetoric. What techniques were used, how did they produce their effects, could
they have been bettered? The final stage was the writing of practice speeches,
for the most part devoted to topics more fantastic than useful.

Was Paul formed in such techniques? His social position argues in the
affirmative, but he himself appears to deny it. He was not sent, he claims, to
preach the gospel 'with eloquent words of wisdom' (1 Cor. 1: 17). He asserts 'my
speech and my proclamation were not in persuasive words of wisdom' (1 Cor. 2:
4), and concedes that 'I am unskilled in speaking' (2 Cor. 11: 6). The truth of
such self-assessment appears to be confirmed by the Corinthians who said, 'his
speech is beneath contempt' (2 Cor. 10: 10).

Neither Paul's protestations, however, nor the criticism of the Corinthians
should be taken at face value.[124] The latter admitted that his letters were
bareitai kai ischyrai (2 Cor. 10: 10). While these adjectives could be rendered
negatively as 'oppressive and severe', the consensus of scholars and translations
is that they should be translated positively, e.g. 'weighty and strong' (*RSV*),
'impressive and moving'.[125] In other words, his vigorous style was reinforced by
the careful presentation expected of a well-trained writer. G. A. Kennedy's
assertion that Paul was 'thoroughly at home in the Greek idiom of his time and
in the conventions of the Greek epistles'[126] is borne out by the evidence of
rhetorical arrangement, not only in the organization of whole letters, but also in
the parts of 1 Corinthians when he is dealing with different subjects.[127] Manifestly he was so well trained that his skill was no longer conscious but
instinctive.

Paul's disclaimer in 2 Corinthians 11: 6 is a rhetorical convention.[128] Note his
assertion that the way he preached was a matter of choice—'I decided' (1 Cor.

[121] Marrou (1948), 269.
[122] Ibid. 272–82. See also Kennedy (1963) and (1972).
[123] For more detail see Stowers (1986), 32–5.
[124] Baslez (1991), 47, and Hengel (1991), 3, have been led astray.
[125] Phillips (1955), 104.
[126] He suggests (1984), 10, however, that Paul could have learnt from one of 'the many hand-
books of rhetoric in common circulation', and adds that it would have been difficult 'to escape an
awareness of rhetoric as practiced in the culture around them, for the rhetorical theory of the
schools found its immediate application in almost every form of oral and written communication'.
One has only to read some of the papyrus letters from Egypt to see that this was not in fact the case.
[127] For details and bibliography, see my (1995), ch. 2.
[128] Betz (1972), 47–69.

2: 2)—and the reason is clearly stated, 'that your faith might not rest in the wisdom of men but in the power of God' (1 Cor. 2: 5). Choice necessarily implies the reality of the alternative. Paul knew that he could have done otherwise; he could have used the persuasive techniques of rhetoric to proclaim the gospel.[129] His conscious control, however, collapsed in the heat of anger, and in the Fool's Speech (2 Cor. 11: 1 to 12: 13) deeply engrained qualities become evident.[130] C. Forbes, after a detailed analysis, rightly concludes, 'What we have seen of Paul's rhetoric suggests a mastery and an assurance unlikely to have been gained without long practice, and possibly long study as well.'[131] It was in the context of the school of rhetoric that Paul was exposed to the various strands of Greek philosophy, which formed part of the intellectual equipment of every educated person. The presence of Stoic teachers in Tarsus is noted by Strabo.[132]

In order to balance this stress on Paul's Hellenistic education, it is important to remember that throughout this whole formative period of his life (age 15–20) he would also have frequented the synagogue of Tarsus. There he was exposed to the tradition of Hellenized Judaism, whose towering figure was his contemporary Philo of Alexandria.[133] How deeply this tradition impregnated his thought is clear from the extensive parallels in his letters to the writings of the Jewish philosopher, despite their very different personalities and concerns.[134]

---

[129] Marshall (1987), 390, astutely observes that the manner and language of Paul's refusal 'indicates that he was more than familiar with the rhetorical traditions he was rejecting. It is feasible to suggest that he may have been trained in rhetoric but had deliberately set it aside' (cf. p. 400).

[130] See Ch. 12, 'Speaking as a Fool'. Cicero once said of Plato, 'It was when making fun of orators that he himself seemed to be the consummate orator' (*De oratore* 1. 11. 47).

[131] (1986), 23. See also Betz (1992), 187. The negative position taken by older authors (e.g. Knox (1950), 75–6) is due to inadequate analysis.

[132] *Geography* 14. 5. 14; see Polenz (1946); Betz (1972).

[133] For a good condensed introduction, see Schürer (1973–87), 3. 809–89.

[134] Chadwick (1965).

— 3 —

A Pharisee in Jerusalem

WE know that it was the custom of young men of Tarsus to leave 'in order to complete their education abroad',[1] but why should Paul have thought it either necessary or appropriate to do so? One can only speculate, but it seems probable that a number of factors were operative. For many the study of rhetoric quickly palled. Rhetorical exercises were incredibly artificial and rules multiplied as subdivisions increased.[2] Only those fiercely ambitious for a place in public life had any incentive to persevere. As a Jew, however, Paul's chances of advancement were limited. The alternative was to plunge himself into the study of his own Jewish tradition, but where? Tarsus had already given him what it could. Babylonia or Alexandria may have flitted across his mind, but neither could compete with Jerusalem, with which all Diaspora Jews were familiar through the synagogue readings, and towards which their minds were directed by precept and their hearts warmed by pilgrimage accounts.[3] It is easy to envisage an enthusiastic young man with a Greek education from a Romanized family desiring to discover for himself the cradle of his religion.

Since the study of rhetoric did not normally last beyond four years, the above line of argument would suggest that Paul left for Jerusalem about the age of 20,[4] i.e. around AD 15.[5] Since his conversion is to be dated in AD 33,[6] this means that he lived in Jerusalem for over fifteen years before becoming a Christian. In any person's life the years between 20 and 35 are a crucial period when reality puts its grip on ambition, and when the speculations and dreams of youth solidify and settle into the mature perspective of adulthood.

A STUDENT IN JERUSALEM?

Paul, as we shall see, recognized the importance of this period of his life, but he does not tell us where it was lived. For that information we are indebted to Luke

---

[1] Strabo, *Geography* 14. 5. 13; trans. Jones.   [2] Marrou (1948), 278–81.
[3] For the details, see Safrai (1974); Jeremias (1969), 58–84.
[4] Rabbi Eliezer ben Hyrcanus was 21 when he went to study in Jerusalem (*Aboth of Rabbi Nathan*; Version A, 6; Version B, 12–13).
[5] For the calculation of this date, see above, p. 4.   [6] See above p. 7.

who refers to pre-Christian activity of Paul in Jerusalem (Acts 8: 1, 3; 9: 1–2) and makes him confess that he was 'brought up in this city at the feet of Gamaliel' (Acts 22: 3; cf. 26: 4). Luke's concern to bind Paul as closely as possible to Jerusalem has already been mentioned, and here it forces us to ask whether he is inventing a background for the Apostle or merely emphasizing a fact.

Paul opposed to Luke?

We have already noted John Knox's methodological principle that in cases of conflict priority must be given to the letters against the Acts of the Apostles.[7] Thus, because Damascus is the earliest point in Paul's career mentioned by the letters, Knox insinuates that Paul, perhaps for business reasons, moved directly from Tarsus to Damascus.[8] To maintain this position he has to demolish Luke's thesis that Paul had previously been a student in Jerusalem. His weapons are three points in the letters.[9]

First, Knox points out, even though Paul had to defend the authenticity and orthodoxy of his Judaism on more than one occasion, he never claims Gamaliel as his teacher. The reason for this silence, of course, is Paul's awareness that the well-taught may be traitors and sinners. The most impeccable pedigree, academic or otherwise, does not protect against base behaviour. The only convincing proof of Paul's Jewishness was the one he provided, namely, the zeal with which he observed the commandments, 'as to righteousness under the Law blameless' (Phil. 3: 6; cf. Gal. 1: 14).

The second argument is no more convincing. Had Paul lived in Jerusalem, Knox claims, he would have spoken of 'returning' there in Galatians 1: 18, as he does in the case of Damascus in the previous verse. The answer to this is simple. Even though Luke makes Jerusalem Paul's home, he only once speaks of him 'returning' there[10] (Acts 22: 17); normally he employs 'going up',[11] which is precisely the phraseology Paul himself uses in both Galatians 1: 18 and 2: 1. In both cases we have to do with stereotypical language consecrated by Jewish usage,[12] which tells us more about Paul's formation than his domicile. Moreover, once Paul had become convinced of his vocation as apostle to the Gentiles, a sojourn in any Jewish city could only be a visit.

---

[7] See p. vi above.

[8] (1950), 76. Similarly Saldarini (1988), 292. 'That man [Paul] will set out from the land of Cilicia to Damascus in Syria to tear asunder the church which you must create' (*Epistula Apostolorum*, 33).

[9] (1950), 34–6.

[10] Acts 22: 17; cf. 13: 13, where the connotation is 'turning back' (15: 37–8). Elsewhere Luke uses 'return' of a journey to Jerusalem only when the starting-point is higher than the Holy City, e.g. Mount of Olives (1: 12); Samaria (8: 25); Gaza road (8: 28).

[11] Acts 18: 22; 21: 12, 15; 24: 11.

[12] To the references in BAGD s.v. *anabainō*, one should add Pss. 120–34, the 'Songs of Ascents'.

For his third argument Knox invokes Galatians 1: 22, 'I remained personally unknown to the churches of Christ in Judaea'[13] to demonstrate that Paul's persecuting activity did not take place in Judaea.[14] Rom. 15: 31 precludes an answer based on a distinction between Judaea and Jerusalem, as if Paul had operated only in the Holy City, and not in the countryside. The true response emerges from Paul's insistence that on his first visit to Jerusalem after his conversion he met only Cephas and James. If he 'saw none of the other apostles' (Gal. 1: 19), it is highly unlikely that he made the acquaintance of any other members of the church. Galatians 1: 22 implies nothing more; it merely serves as the introduction to what the churches of Judaea heard about their converted persecutor.[15] Obviously Paul had no time for the dramatic gesture of a public apology. He always had difficulty in admitting a mistake (e.g. 2 Cor. 11: 7–11), and it is very much in keeping with his character that he should be totally focused on the future, the mission in Syria and Cilicia (Gal. 1: 21).

Thus Knox's attempt to find elements in the letters which contradict Luke cannot be considered successful. The next step is to question whether Knox is correct in asserting that as far as the letters are concerned Paul's career begins in Damascus. In the process we shall also look for data which might confirm Acts, a type of control which Knox did not employ.

A Pharisee

Paul asserts that prior to his conversion he had been 'with respect to the Law a Pharisee' (Phil. 3: 5). In other words, his obedience to the Law was that which characterized the Pharisees. Who were they? Where were they based?

In order to answer this question we can draw on only two sources, Josephus, the first-century Jewish historian who claims that he himself was a Pharisee (*Life* 12), and the first-century traditions incorporated into later rabbinic compilations. The former shows the Pharisees from without and the latter from within. To extract historically valid information from either is a complex and difficult task.[16]

Josephus mentions the Pharisees in three works, *The Jewish War* (published AD 75–79), *The Antiquities of the Jews* (published AD 93–94), and the *Life* (published c. AD 95). Not only are these accounts difficult to reconcile with one another, but some details are implausible, and the most extensive treatment is highly tendentious. From a careful analysis, however, the Pharisees emerge as a political interest group deeply embroiled in the conflicts of the Hasmonean

---

[13] For the translation, see Betz (1979), 80; Longenecker (1990), 41.

[14] Similarly Bultmann (1964), 132; Bornkamm (1971), 15; Becker (1989), 18.

[15] See in particular Hultgren (1976), 105–7. Hengel (1991), 76–7, and Dunn (1993), 81, suggest that Paul's persecuting activity had been directed against the 'Hellenists' (Acts 6: 1), who subsequently left the city (Acts 11: 19), and that he was unknown to the 'Hebrews'.

[16] An excellent brief and reliable introduction is by Michel and LeMoyne (1965).

period, and whose involvement in public life continued into the first century AD but to a markedly lesser degree.[17]

From the mass of rabbinic traditions, J. Neusner confesses himself unable to extract very much regarding the history of the Pharisees prior to the destruction of the temple in AD 70. They were 'primarily a society for table-fellowship, the high point of their life as a group'.[18] This is a deduction from the fact that 'approximately 67% of all legal pericopae deal with dietary laws: ritual purity for meals and agricultural rules governing the fitness of food for Pharisaic consumption. Observance of Sabbaths and festivals is a distant third.'[19] Only after the fall of Jerusalem were the Pharisees forced to deal with a much wider range of problems. Neusner would see their relative withdrawal from the political arena as due to the influence of a contemporary of Herod the Great (37–34 BC), namely, Hillel the Elder, whose dominant role in the reformed party is thereby explained.[20] It would be highly unusual, however, for a group which believed in the divinely sanctioned importance of their way of life not to wish to transform the society of which they were a part.[21] The tactical abandonment of any aspirations to imposition of their view via political control is unlikely to have affected their long-term strategy.

A perfect illustration of their revised approach is the career of Gamaliel I (or the Elder). He was the successor of Hillel (*m. Aboth* 1. 18); the claim of later texts that he was also his son or grandson is rightly viewed with scepticism.[22] His dates cannot be established with any certitude, but his years of activity are thought to be roughly AD 20–50,[23] or more narrowly, but still approximately, AD 30–40.[24] Apart from the fact that he had a son and a daughter we know nothing about him personally. Unlike the case of Hillel, there is no tendency to elaborate data about Gamaliel I. In consequence, Neusner is inclined to attach a high degree of authenticity to the legal decisions attributed to him. Neusner moreover points out that they focus on issues other than those central to Pharisaic concerns.[25] A specific location is a characteristic of Gamaliel I materials,[26] and very often this is the Temple.[27] The image that comes across is of a teacher who played an active role in the deliberations of the administration of the Temple,[28] which is precisely the portrait painted by Luke, 'a Pharisee in the council

---

[17] Neusner (1972). The general thrust of this seminal study remains valid; see Goodblatt (1989).
[18] (1971), 3. 318.
[19] Ibid. p. 3. 304. Criticism of the relevance of such statistics is met by Dunn (1992), 257–60.
[20] Neusner (1971), 3. 305.
[21] So rightly Saldarini (1988), 284–7.
[22] Neusner (1971), 1. 294; 3. 306.
[23] Fitzmyer (1990), 18.
[24] Jeremias (1961), 55.
[25] (1971), 1. 375–6.
[26] Ibid. 1. 344.
[27] The list is given in ibid. 1. 373.
[28] Ibid. 3. 314.

named Gamaliel, a teacher of the Law, held in honour by all the people' (Acts 5: 34).[29] According to the *Mishnah*, 'When Rabban Gamaliel the Elder died, the glory of the Torah ceased, and purity and abstinence died' (*m. Sotah* 9. 15). Neusner doubts that this was actually said of him in his lifetime,[30] but the mere fact of Gamaliel's appearance in an early third century AD list of rabbinic greats underlines how high his reputation must have been.

The details of Gamaliel's teaching are not relevant here, apart from his answer to the question: How many Torahs were given to Israel? 'Rabban Gamaliel said, "Two, one in writing and one orally."'[31] Not only was the written Law binding on the Pharisees but also its traditional understanding and expansion. This is the one point on which the rabbinic traditions and Josephus agree.[32] The latter says, 'The Pharisees have imposed on the people many laws from the tradition of the fathers not written in the Law of Moses.'[33]

It is at this point that we find confirmation (all the more valuable because it is indirect) of Paul's assertion that he was a Pharisee. 'Being extremely zealous for the traditions of my fathers' (Gal. 1: 14) is precisely the sort of language a Pharisee would use.

Where could Pharisees be found?

Where would it have been possible for Paul to come into contact with Pharisees? Had we only Josephus and the Fourth Gospel the answer would be simple, because both locate the Pharisees exclusively in Jerusalem. According to Saldarini, however, 'Mark sees them [the Pharisees] as active only in Galilee',[34] and he accepts this portrait as historical because 'it is very unlikely that the early followers of Jesus would have placed Pharisees in Galilee if their presence there would be manifestly contrary to the first century situation'.[35] The cogency of this argument depends on the solidity of the factual base. Regretably it is non-existent. With the exception of Mark 12: 13 (cf. 11: 15) and 10: 2, not a single Markan story in which the Pharisees figure contains any element which would permit us to identify the location.[36] The reader of the gospel is given the impression of a location in Galilee by juxtaposition in some but not all cases; the Galilean location of 2: 16 depends on 2: 13; that of 3: 6 on 3: 7; that of 7: 1 on 6: 53; and that of 8: 11 on 8: 13, which is a redactional introduction to the follow-

---

[29] Jeremias (1971), 1. 347, 376.   [30] Ibid. 1. 351–52.
[31] *Sifré Deut.* 351, quoted by Neusner (1971), 1. 343. For similar answers by Hillel and Shammai, see ibid. 1. 322, 329.
[32] Ibid. 3. 304. See Urbach (1979), ch. 12 ('The Written Law and the Oral Law').
[33] *AJ* 13. 297. 'They prided themselves on the exact interpretation of the law of the fathers' (*AJ* 17. 41). 'The Pharisees and all the Jews do not eat without washing their hands observing the tradition of the elders' (Mark 7: 3; Matt. 15: 2).
[34] (1988), 147.   [35] Ibid. 291.
[36] Mark 2: 16, 18, 24; 3: 6; 7: 1, 3, 5; 8: 11; 10: 2; 12: 13.

ing story. The evangelist's mode of composition, which associated stories on grounds other than their locale, did not intend transposition from the original context (in all probability Jerusalem), and it is most unlikely that the audience of the gospel would have understood it in this way. For Diaspora readers it would have been irrelevant, and knowledgeable Palestinians would have subconsciously made the necessary corrections. In their parallels to the instances discussed neither Matthew nor Luke can be considered to have had access to independent historical information. In addition both exhibit a tendency to multiply references to Pharisees as stereotypical opponents of Jesus.[37]

Luke 13: 31 is an exception to this rule, because the Pharisees give Jesus a friendly warning, 'Some Pharisees came and said to him, "Get away from here because Herod wants to kill you."' The reference to Galilee is unambiguous and integral to the story because only in his own territory did Herod Antipas have the power of life and death. This single witness, however, does not demonstrate a permanent Pharisaic presence in Galilee.[38] On the contrary, the Pharisees' knowledge of the king's intention would rather suggest that they were on a mission to the court from Jerusalem.[39]

A rabbinic text has also been interpreted as implying the presence of Pharisees in Galilee and elsewhere, which is important because of Luke's assertion that Paul's parents were Pharisees (Acts 23: 6):

> The story is told concerning Rabban Gamaliel and [the] Elders, who were sitting on steps on the Temple Mountain, and Yohanan, that scribe, was before them. He said to him, 'Write:
>
> To our brethren, men of Upper Galilee and men of Lower Galilee. May your peace increase. We inform you that the time of the burning has come, to bring out the tithes from the vats of olives.
>
> And to our brethren, men of the Upper South and men of the Lower South. May your peace increase. We inform you that the time of burning has come, to remove the tithes from the sheaves of wheat.
>
> And to our brothers, men of the Exile of Babylonia and men of the Exile of Medea *and the men of the Exile of Greece* and the rest of all the Exiles of Israel. May your peace increase. We inform you that the pigeons are tender, and the lambs are young, and the time of spring has not come, and it is good in my view and in the view of my colleagues, and we have added to this year thirty days.'[40]

The texts of these letters also appear in the Jerusalem Talmud (*Maaser Sheni* 5. 4 and *Sanh.* 1. 2) and in the Babylonian Talmud (*Sanh.* 11b).[41] With one

---

[37] Bultmann (1963), 52–4. [38] Against Freyne (1980), 322.

[39] Similarly Saldarini (1988), 296, but in the context of a far too imaginative reconstruction of Pharisaic activity in Galilee.

[40] *Tos. Sanh.* 2. 6, cited from Neusner (1971), 1. 356–7. According to the calendar it was not spring, but according to nature it was, so the calendar had to be corrected by the intercalation of a month.

[41] Translations and synopsis in Neusner (1971), 1. 360, 361, 368, 372–3.

exception—the italicized phrase found in both versions in the Jerusalem Talmud is lacking in the Babylonian Talmud—the differences are minor. Neusner accepts the authenticity of these letters because they are evidently those referred to in the phrase 'our fathers used to write to your fathers', which concludes the letter of Simeon ben Gamaliel (d. 70) and Yohanan ben Zakkai (d. c. 80) to Upper and Lower Galilee and Upper and Lower South.[42] Neusner finds it plausible that 'brethren' in the first two letters should mean Pharisees, but rejects the same meaning for the term in the third letter, because there is no confirmatory evidence of Pharisees in the areas mentioned, and the Pharisees had no authority to regulate the calendar.[43]

Even with Galilee and the South, however, there are problems. Upper Galilee and Lower South are precisely the sort of non-urbanized areas which Pharisees avoided. Moreover, Pharisees were by definition scrupulous about tithing, so why should they have to be reminded? Finally, 'brethren' when used in the Simeon and Yohanan letter can only mean Jews in general, because they had not paid their tithes as promptly as Pharisees would.[44] Hence, it is best to think of the above letters as having been written at the behest of the Temple authorities, because 'The Pharisees' stress on tithing and priestly piety for the laity could have been attractive to the Jerusalem authorities who desired to collect tithes from all Jews in Palestine and who could have met resistance from Jews in Galilee, outside their political control'.[45] In this perspective the three letters would have been addressed to all Jews. Only this interpretation brings out the homogeneity of the third letter with the other two.

Individual Pharisees may have gone to Galilee sporadically to check crops which they intended to buy, but there is no evidence of anything remotely resembling a Pharisaic movement in Galilee.[46] This makes it even less likely that there was permanent and significant Pharisaic presence in the Diaspora.[47] Their ambition to live the Law as perfectly as possible would have been frustrated by a Gentile environment. Neusner's scepticism has already been noted, and Saldarini's argument that Paul's identification of himself to the Philippians as a Pharisee implies that these must have been familiar with Pharisees from personal contact carries no conviction.[48] Thus there is no reliable evidence for Pharisees permanently based outside Jerusalem.[49] Luke's claim that Paul was 'a son of Pharisees' (Acts 23: 6) must be dismissed as a rhetorical flourish without historical value.[50]

---

[42] Text in Neusner (1971), I. 378.   [43] Ibid. I. 357–8.

[44] Freyne (1980), 282.   [45] Saldarini (1988), 296.

[46] Freyne (1980), 319. Saldarini (1988), 292, speaks of 'a minor presence', which is still a little too strong.

[47] *Pace* Becker (1989), 43.   [48] (1988), 292.

[49] So rightly Hengel (1991), 29–31, but he then goes on to accept Acts 23: 6 at face value (p. 122 n. 173)!

[50] Lentz (1993), 52–5.

If Paul joined a Pharisaic group, and there is no reason to doubt his word (Phil. 3: 5), it must have been a personal decision made after his arrival in Jerusalem. Furthermore, if Paul arrived in Jerusalem around AD 15,[51] his sojourn in the city would have coincided with that of Gamaliel I, and it is extremely improbable that Paul or any other Pharisee would have escaped his influence. Such confirmation does not necessarily imply that precise historical information stands behind Acts 22: 3. It could be based on a series of deductions parallel to mine. They therefore reinforce one another as independent estimates of historical probability.

PHARISAIC STUDIES

In rabbinic tradition devotion to study appears as one of the fundamental characteristics of a Pharisee. The mild Hillel is reported as saying, 'he who does not learn [the Law] is worthy of death', whereas the strict Shammai merely counselled, 'make your [study of the] Law a fixed habit', and Gamaliel I proffered the practical advice, 'provide yourself with a teacher and remove yourself from doubt' (*m. Aboth* I. 13–16). The authenticity of such statements, of course, cannot be taken for granted; they fit too well with the attitude of rabbinic circles after the destruction of the Temple. This does not mean, however, that the ideal they embody should be dismissed as anachronistic in terms of the first century. What has been said above concerning Jewish education, in particular the citations from Philo and Josephus,[52] highlights the continuity into the first century of a venerable tradition of exhortation to study.[53] Moreover, if the Pharisees prided themselves on meticulous observance, detailed knowledge of the commandments as articulated in both the written and oral Torah was obviously indispensable.[54] Even though Hillel may not have said 'an ignorant man cannot be saintly' (*m. Aboth* 2: 6), the sentiment reflected Pharisaic values at all times.

Understandably, therefore, the Pharisees gathered in groups.[55] For those concerned with the ritual purity of food, it simplified life when those of like

---

[51] See p. 52 above.   [52] p. 47.

[53] Although perhaps a little extreme, Essene legislation (dating from the 1st cent. BC) is a good illustration, 'In the place where the ten are, let there not lack a man who studies the Law night and day, continually, concerning the duties of each towards the other. And let the Many watch in common for a third of all the nights of the year, to read the Book and study the Law, and bless in common' (1QS 6. 6–8). For an excellent summary of the whole tradition, see Viviano (1978), 111–57.

[54] The *Mishnah* with wry humour underlines the problem, 'The rules about the Sabbath, Festal-offerings and Sacrilege are as mountains hanging by a hair, for [the teaching of] Scripture [thereon] is scanty and the rules many' (*m. Hagigah*, I. 8).

[55] Defects in the analysis of a series of texts should not impede common sense assumptions, *pace* Saldarini (1988), 216–20. Such gatherings, of course, do not imply complete common life in anything resembling the monastic sense.

mind ate together; each was trusted to respect the standards of the other. The consensus of interpretation regarding what is demanded, which this implied, necessitated common study and discussion. The hothouse atmosphere characteristic of such élite groups was intensified by the presence of young men. The dangers must have been as obvious to first-century teachers as they are today. It is not at all surprising that Hillel (or someone equally experienced and perceptive) should have warned,

> [1] Do not separate from the community. [2] Do not trust yourself until the day of your death. [3] Do not judge your fellow until you stand in his place. [4] Do not say of a thing which cannot be understood that it will be understood in the end. [5] And do not say, 'When I have leisure I will study!' Perhaps you will never have leisure.   (*m. Aboth* 2. 5; trans. Danby adapted)[56]

At one end of the scale are those who permit themselves to be carried along by the group, who feel sure that a solution to thorny problems will be found—by someone else (4)—because they lack a deep personal interest in study (5). At the other end are those so intensely committed that they become arrogant in judgement (2) and contemptuous of those slower or less insightful (3). These run the risk of developing a superiority complex expressed in leaving the community in search of greater perfection (1). The feverish environment is perfectly evoked by a saying attributed to Simeon the son of Gamaliel,

> All my days I have grown up among the Sages and I have found naught better for a person than silence; and not the expounding [of the Law] is the chief thing but the doing [of it]; and he that multiplies words occasions sin.
>   (*m. Aboth* 1: 17; trans. Danby)

Viviano's commentary is perfectly apposite, 'It breathes a certain weariness with the interminable and excessive wrangling with the overproduction of secondary refinements, which characterized one side of the Sages' learned activity.'[57]

It is only against this background that Paul's description of his youth becomes intelligible, 'I was advancing in Judaism beyond many Jews of my own age, so extremely zealous was I for the tradition of my fathers' (Gal. 1: 14). The Pharisaic ring of 'the tradition of my fathers' has already been noted.[58] The comparison is with Paul's contemporaries generally and does not define his position with respect to other Pharisees. The combative tone and competitive spirit are equally characteristic of élite groups. Paul was proud to belong to such a minority.

[56] See the commentary in Viviano (1978), 36–9. I am responsible for the numbers in the text.
[57] (1978), 29.
[58] See above, p. 56.

His stress on 'contemporaries' is suggestive; he uses the term only here. He claims not an absolute but a relative victory. The simplest explanation of such diffidence is that he was aware of having started late. As a newcomer from the Diaspora he had a lot to make up. What Paul retained from this period is difficult to determine. Not only did it blend with his generalized Jewishness—Palestinian and Hellenistic Judaism had a great deal in common—but it was subsequently absorbed into a radically different pattern of religion when he became a Christian.[59] None the less certain details in the letters reflect the training he received at the feet of Gamaliel.[60] This is not really surprising because his involvement must have been so single-minded that he became oblivious to everything extraneous.

It is from this perspective that we must deal with an objection. Were Paul in Jerusalem from AD 15 to 33, he would have been in the city when Jesus came on pilgrimage, and when he was crucified there in AD 30.[61] Under such circumstances, we are told, he could not avoid encountering Jesus and/or having first-hand knowledge of the crucifixion. The fact that he never even hints at any personal connection,[62] we are told, proves either that he was never in Jerusalem before his conversion or arrived there after AD 30.

The weakness of this objection is that it presumes to know what Paul should or should not have written. In addition it overestimates the importance of Jesus during his lifetime. The gospels give the impression that he commanded widespread support at least in Galilee (e.g. John 6: 14–15), but this is contradicted by Herod Antipas' shift from suspicion deep enough to plot assassination (Luke 13: 31) to total lack of interest (Luke 23: 8–12). He was less paranoid than his father, Herod the Great, but he would never have released Jesus unless he was convinced that the latter was absolutely harmless. The fact that Antipas had John the Baptist imprisoned and executed underlines the difference between the two.[63] If Jesus' impact in Galilee was so minimal, it must have been even less among the general population in Jerusalem. His execution was likely to have been carried out without fanfare by the authorities amidst the indifference of a city preoccupied with preparations for Passover. Even if it were *the* event of the spring of AD 30, which I very much doubt, there is no guarantee that it would have impinged on the attention of a Paul passionately committed to the study of the Law. Despite the ubiquity of radio, television, and gossip, Hassidic students in Jerusalem today manage to maintain their principle of avoiding

---

[59] So correctly E. P. Sanders (1977), 542.

[60] The fundamental study remains that of Davies (1962); see also Bonsirven (1939) and Daube (1956).

[61] The objection is half-heartedly formulated by Fitzmyer (1990), 18, and avoided by Baslez (1991), 75–6, who assumes that Paul arrived in Jerusalem only after the death of Jesus.

[62] 2 Cor. 5: 16 must be translated 'we knew Christ in a fleshly way' and cannot be rendered 'we knew Christ in the flesh'. See in particular Fraser (1970).

[63] Mark 6: 17–29; Josephus, *AJ* 18. 116–19.

knowledge of secular events; it is reported that they were completely unaware of Anwar Sadat's peace visit to Israel, which brought the whole of Jerusalem into the streets.

A MARRIED MAN

When Paul wrote 1 Corinthians he was single. His formulation, however, draws attention to a latent ambiguity, 'To the unmarried and widowed I say that it is good for them remain as I am' (1 Cor. 7: 8; cf. 9: 5). Was he a widower or had he never been married? Older commentators assumed the latter option,[64] while modern scholars tend towards the former,[65] or merely note the problem without committing themselves.[66] The situation moreover has been unnecessarily complicated by misuse of data and the introduction of extraneous problems such as rabbinic ordination.[67]

That Paul had never married cannot be excluded apriori. Jews placed a high value on marriage (Gen. 1: 28; 38: 8–10; Deut. 25: 5–10),[68] and texts specify the marriagable age for a man to be between 18 and 20.[69] Thus, it is natural to conclude that most Jewish men married young. There were, however, exceptions in both fact and date.[70] Some married much later, and some not at all.

The prophet Jeremiah refused to marry as a symbolic gesture (Jer. 16: 1). According to a rabbinic tradition, Simeon ben Azzai (c.AD 110) justified his celibate lifestyle by the words, 'My soul is in love with the Torah. The world can be carried on by others.'[71] Josephus married at the age of 30, and then only at the behest of Vespasian.[72] This would have been ten years too young to meet the ideal of Philo who wrote, 'The 40th year is the right time for the marriage of

---

[64] e.g. Robertson and Plummer (1914), 138; Allo (1956a), 162.

[65] e.g. Fee (1987), 288 n. 7; Légasse (1991), 45.

[66] e.g. Fitzmyer (1990), 19.

[67] e.g. the controversy between J. Jeremias (1926; 1929) and Fascher (1929).

[68] For the understanding of these texts as implying a binding precept, see Billerbeck (1922–8), 2. 372–3. Josephus (JW 18. 21), Philo (Apol. 14), and Pliny (NH 2. 276) are not correct in imputing celibacy to the Essenes. According to H. Stegemann (1994), 267–74, all Essenes married very young women, who were expected to give birth to a child every year. Most of these women did not live beyond their early twenties. The vast majority of the Essenes, therefore, were single men who by rule had married later than other Jews, and who as widowers could not remarry. The unusual proportion of single men gave the impression that they had chosen celibacy as a matter of principle, whereas in fact they obeyed a rigorous interpretation of Gen. 1: 28.

[69] See the texts cited above. Ch. 1, 'The Jewish Tradition'.

[70] Thornton (1972).

[71] b. Yebam. 63b. The historicity of the episode is irrelevant. The important point is that commitment to study was recognized as exceptionally exempting from obedience to Gen 1: 28. According to Philostratus, Apollonius of Tyana renounced marriage (Life 1. 13).

[72] The marriage is dated to the early summer of AD 67 (Life 414), and he was born in the winter of AD 37–38 (Life 5; AJ 20. 267). He subsequently married twice (Life 415, 427).

the wise man.'[73] According to the *Testament of the XII Patriarchs*, Levi married at 28 and Issachar at 30 or 35.[74]

At the time of his conversion in AD 33 Paul would have been about 40, certainly in his late thirties. While it is possible that he had either postponed marriage or renounced it completely, this cannot be assumed to be the natural meaning of 1 Corinthians 7: 8. Such decisions were made only by a tiny minority, who certainly were considered abnormal by their contemporaries. Only those bolstered by a high degree of personal security rooted in past achievement, and the sympathetic understanding of friends (e.g. Simeon ben Azzai), could take the risk of flouting public opinion so radically. As a young immigrant from the Diaspora, Paul met none of these conditions. The complacently competitive tone of Galatians 1: 14, on the contrary, betrays the stranger who has successfully integrated into Jewish Jerusalem through perfect conformity to the norm. It is much more probable, therefore, that Paul cheerfully bowed to the expectation that young men should marry in their early twenties.

This is perhaps the best point at which to draw attention to a bizarre tradition concerning Paul found in the Judaeo-Christian *Ascension of James*,

> Paul was a man of Tarsus and indeed a Greek, the son of a Greek mother and a Greek father. Having gone up to Jerusalem and having remained there a long time, he desired to marry a daughter of the (high?) priest and on that account submitted himself as a proselyte for circumcision. When nevertheless he did not obtain the girl, he became furious and began to write against circumcision, the sabbath and the Law.[75]

Windisch believes that this story has a historical kernel in so far as it provides a plausible motive for Paul's celibacy.[76] In fact its manifest anti-Pauline bias robs it of all historical value.[77] Paul is made a gentile, who converts to Judaism for the worst of all reasons, namely, sex, and when frustrated becomes violently anti-Semitic. The crudity of the attack is of a piece with the accumulation of improbabilities. No explanation is offered as to why such a person should have taken up residence in Jerusalem, or how he might have become acquainted with the high priest's daughter. In consequence, since well-known facts about Paul are deliberately falsified, there is no reason to suppose that the author had any independent knowledge of Paul's marital status. On the contrary, the tenor of the rest of the account would suggest that his celibacy was also fabricated as a criticism.

---

[73] *Quaest. in Genesin* 4. 154.
[74] *T. Levi* 11. 1; 12. 5; *T. Iss.* 3. 5.
[75] Preserved only in Epiphanius, *Panarion* 30. 16. 9; cf. 30. 25; modified translation from Hennecke and Schneemelcher (1965), 2. 71.
[76] (1934), 133.
[77] So rightly Lüdemann (1989), 180 n. 43.

It is most probable, therefore, that Paul had a wife. Is she ever mentioned? According to Eusebius, Clement of Alexandria claimed that she is alluded to in Philippians 4: 3,

> Peter and Philip had families, and Philip gave his daughters in marriage, while Paul himself does not hesitate in one of his epistles to address his yoke-fellow, whom he did not take round with him for fear of hindering his ministry.[78]

This long dominant interpretation has now rightly fallen out of favour, for a simple grammatical reason. Paul wrote *gnêsie*, the masculine form of the adjective meaning 'true, genuine'. Had he a woman in mind he would have written *gnêsia*.[79] It is a question, therefore, of a man named Syzygus, and the play on his name is well brought out by the paraphrase, 'I ask you, Syzygus, really to be a "partner" and help them' (*NJB*).

What happened to Paul's wife? It is possible that he divorced her, and thereafter maintained a single lifestyle. It seems more likely, however, that what happened was much more traumatic. This would not only justify his silence, but would go a considerable way towards explaining his persecution of the church. Most, if not all, biographies of Paul take this latter activity for granted. We are invited to assume that, if Pharisees were so opposed to Jesus during his ministry, then it would be natural for them to persecute Christians. This gospel portrait of the Pharisees is now recognized as being without historical foundation; for the evangelists the name 'Pharisee' had become a code-word for an irreconcilable opponent. When we look at other sources a very different picture emerges, one which only serves to make more urgent the question of why Paul persecuted Christians.

All the conflicts recorded in pre-AD 70 Pharisaic material are internal.[80] There is no hint of any agression directed against those who disagreed with them. The hostility documented by the New Testament always comes from the Sadducees. When these conspired to execute James, the leader of the church in Jerusalem, and others,[81] it was the Pharisees—'the most fair-minded and strict in observance of the Law'—who protested.[82] Why did Paul, who had made such strenuous efforts to conform (Gal. 1: 14), suddenly break away from the pattern of tolerance established by the Pharisaic movement?

A well-known psychological mechanism switches anger from unacceptable to acceptable channels of expression. As a Pharisee, Paul believed God had a hand in all that happened in history.[83] Jerusalem is sited in an earthquake zone, and it cannot have been immune to the domestic tragedies of fire and building

---

[78] *History of the Church* 3. 30. 1, quoting *Stromata* 3. 6. 52; trans. Williamson.
[79] Lightfoot (1908), 159.
[80] Neusner (1971), 3. 304.
[81] Josephus, *AJ* 20: 201.
[82] Hengel (1985a), 73-4.
[83] Josephus, *AJ* 18. 13; *JW* 2. 163.

collapse, which were so frequent at Rome.[84] Had Paul's wife and children died in such an accident, or in a plague epidemic, one part of his theology would lead him logically to ascribe blame to God, but this was forbidden by another part of his religious perspective, which prescribed complete submission to God's will. If his pain and anger could not be directed against God, it had to find another target. An outlet for his pent-up desire for vengance had to be rationalized.

Christians could be seen as a danger, not only to the reforming hopes of the Pharisees, but to the fabric of the Jewish people, whose nationalistic aspirations were inevitably going to bring it into conflict with Rome, which needed absolute control of its eastern frontier. Unity was imperative if the Jews were to survive such a conflict. From this perspective Christians were both a religious and a social threat; their existence flouted God's will. I am not saying that only a bereaved husband could have thought in this way. Undoubtedly many perceived the danger, perhaps even some of the Pharisees. All of these latter would have subscribed to the theological reasons which are sometimes suggested as Paul's motive,[85] but they did not react as he did. Some extraneous historical factor must be invoked to explain Paul's uniqueness.[86] Redirected anger is but a possible answer, whose plausibility none the less is enhanced by its ability to explain Paul's silence regarding his wife.

PERSECUTOR OF THE CHURCH

Luke is prolific with detail in describing Paul's persecution of the church. He is first noticed as a youth (Acts 7: 58) looking on with satisfaction at the grisly execution of Stephen (Acts 8: 1; cf. 22: 20). His next appearance is as the archpersecutor, bursting into Christian houses and throwing their occupants into prison (Acts 8: 3). His authority is confirmed when, at his request, the high priest issues letters enabling him to bring prisoners from Damascus (Acts 9: 1–2; cf. 22: 4–5). The nature of this authority becomes apparent only when Paul is made to say, 'I not only shut up many of the saints in prison, by authority from the chief priests, but when they were put to death I cast my vote against them' (Acts 26: 10). Since only the Sanhedrin was competent in capital cases,[87] and

---

[84] Juvenal, *Satires* 3. 190–203; Carcopino (1981), 43–5. It would be naive to imagine that the only disasters to occur in Jerusalem were those listed by Jeremias (1969), 140–4.

[85] e.g. Hengel (1991), 69–71.

[86] According to Taylor (forthcoming), when Paul arrived in Jerusalem he became a religious nationalist, i.e. a Zealot, whose theological opinions were identical with those of the Pharisees (Josephus, *AJ* 18. 23), and so persecuted Christians as a pacifist threat to the national struggle. The weakness of this attractive hypothesis is that it may exaggerate the importance of 'zealot' in Gal. 1: 14 and 'zeal' in Phil. 3: 6.

[87] The historical facts are disputed, see Schürer (1973– 87), 2. 221–2; there is no doubt about Luke's intention.

since only full members could vote, Paul is undoubtedly here presented as a member of the Sanhedrin.

The historicity of this picture is compromised by a number of factors. The transition from 7: 58 to 8: 3 is never explained; its abruptness hints at artificiality.[88] Paul's authority to make arrests in Damascus is variously described as derived from the high priest alone (Acts 9: 2) or in conjunction with the council (Acts 22: 5), and from the chief priests (Acts 9: 14; 26: 12). The differences are disturbing, but pale into insignificance beside the fact that neither the high priest nor the Sanhedrin had judicial authority outside the eleven toparchies of Judaea proper.[89] Their moral authority might be persuasive, but they could not empower Paul to make arrests, particularly on the territory of a Roman province. Recognition of the strength of this position is apparent in the desperate lengths to which conservative scholars have to go to defend Luke's veracity. I. H. Marshall, for example, accepts Hanson's view that Paul's authorization was to kidnap Christians if he could get away with it![90]

As regards Paul's membership of the Sanhedrin; in the first century Pharisees were certainly members,[91] and by AD 25 or thereabouts Paul would have reached the age at which a Jewish male was permitted to function as a judge.[92] Thus it is within the bounds of possibility that he was a member. But so many other points in the Lukan presentation of the persecution have been seen to be improbable, if not impossible, that it is difficult to accept Luke's unsupported statement on this point at face value. The silence of the letters is also significant. Had Paul in fact been a member of the Sanhedrin, the competitive spirit manifested in Galatians 1: 14 makes it certain that he would have mentioned it there; it would also have been an effective component in the argument of Philippians 3: 5 and 2 Corinthians 11: 22.

Once it is noticed that the strongest statements concerning Paul's pre-Christian activity always occur as introductions to narratives of his conversion, it becomes obvious that it was in Luke's artistic interest to exaggerate certain negative traits of Paul the persecutor in order to set in greater relief the miracle of his conversion and the success of his apostolate. It enhanced the dramatic impact of his book to have the perfect persecutor transformed into the ideal apostle.

The only element of Luke's presentation which is confirmed by Paul himself

[88] Haenchen (1971), 294, 298.

[89] Ibid. 320–1; Schürer (1973–87), 2. 218. Recognition of this fact might explain why Hengel (1991), 85, gratuitously speculates that it was from certain synagogues in Jerusalem that Paul accepted a mission to Damascus.

[90] (1980), 168. Bruce (1977), 72, relies on 1 Macc. 15: 21, which, however, does not empower the high priest to seek Jewish wrongdoers abroad, but invites the ruler of Egypt to send such persons to the high priest.

[91] Schürer (1973–87), 2. 210.

[92] See the texts cited above, Ch. 1, 'The Jewish Tradition'.

is the fact of persecution; the general picture is very different. The brevity of 'As to zeal a persecutor of the church (Phil. 3: 6) and 'I persecuted the church of God' (1 Cor. 15: 9) is amplified slightly in Galatians, 'You have heard of my former life in Judaism how intensely (*kath' hyperbolēn*) I was persecuting the church and trying to destroy it' (1: 13; cf. 1: 23).

The influence of the Acts of the Apostles is evident in the translation of *kath' hyperbolēn* by 'violently' (*RSV, NRSV*), or 'savagely' (*NEB*) and somewhat less obviously in an alternative rendering 'I went to extremes' (*NAB*). All such renderings suggest physical attacks culminating in injury if not death. Paul's use of the adverb elsewhere,[93] however, and the context here, combine to indicate that it articulates the quality of the Apostle's commitment, not the means he employed. He was completely dedicated to, and totally involved in, what was for him an habitual activity.[94] It was not a one-off event, and it was not without success. The addition of 'to devastate, destroy, annihilate' to 'to persecute' in Galatians 1: 13, 23 removes the remote possibility that the latter connoted merely an attempt. Paul did real damage over a period of time impossible to estimate.

Even though by the time of Paul the meaning of *ho diōkōn* had progressed from the generic 'hunter' to the specific 'prosecutor',[95] neither context supports the claim of Acts that Paul was acting in official juridical capacity. Implicitly in Galatians, but explicitly in Philippians, Paul's mention of persecution is to demonstrate how seriously he took his Jewishness. A duty would not have had the evidential value which Paul manifestly ascribes to this activity. Thus it was something undertaken on his own initiative. We should think in terms of an officious little zealot rather than of a calm, objective prosecutor.

It is within this framework that we have to try to work out what Paul meant by 'persecution'. Hultgren astutely perceived that the best clue would be provided by Paul's use of 'to persecute' and 'persecutor' elsewhere.[96] The noun and verb are used of the persecution of believers (Rom. 12: 14; Gal. 4: 29; 6: 12) and of Paul himself (1 Cor. 4: 12; 2 Cor. 4: 9; 12: 10; Gal. 5: 11). No details are given in the former, but in his own case, if we except the wrestling metaphor of 2 Corinthians 4: 8–10 whose details cannot be pressed,[97] Paul mentions being reviled, slandered, and insulted in addition to unspecified hardships and calamities. These latter can be illustrated from 2 Corinthians 11: 25b–27. We cannot, however, extrapolate from this passage that Paul was able to order Christians to be scourged (2 Cor. 11: 24), because such punishment could be

---

[93] Rom. 7: 13; 1 Cor. 12: 31; 2 Cor. 1: 8; 4: 17.

[94] Both verbs in Gal. 1: 13 are in the imperfect, as is *eporthei* in Gal. 1: 23. The aorist in 1 Cor. 15: 9 merely looks at the span of time from a different perspective.

[95] E. Burton (1921), 45.

[96] (1976), 108–9.

[97] Despite the objection of Pfitzner (1976), 76, this remains the only unified hypothesis capable of explaining the series of pairs; it was first proposed by Spicq (1937).

administered only by qualified authorities,[98] and not by private individuals such as Paul was.[99]

Private individuals, however, could denounce individuals to the authorities, and it is entirely possible that Paul employed this tactic. Intellectual zealots are often tattletales. Even though Luke alone mentions them (Acts 6: 9; 24: 12), there is no doubt that there were a number of different synagogues in Jerusalem, where Jews met to study the Law, as there were in any city with a sizeable Jewish population, e.g. Alexandria and Rome.[100] Josephus highlights the importance of Jewish legislation that 'every week the people should set aside their other occupations and gather together to listen to the Law and learn it accurately',[101] and Philo confirms that 'Even now this practice is retained, and the Jews every seventh day occupy themselves with the philosophy of their fathers'.[102]

In this situation, where one could challenge and be challenged, as the Gospels report concerning Jesus,[103] it is easy to visualize Paul popping up to denounce abrasively those whose whose divisiveness, in his view, threatened the survival of the Jewish people. Social customs which deviated from the norm (e.g. Acts 2: 46) would initially have drawn attention to the fact that Christians were different, but Paul could later challenge on the basis of specific information. The simplest technique to flush out Christians would have been that mentioned by Luke, 'he tried to make them blaspheme' (Acts 26: 11), e.g. by demanding assent to formulations which implied denial of Jesus, such as an oath that the Messiah had not yet come.[104] A similar elementary ploy was used by Pliny the Younger in Asia Minor at the beginning of the second century to ferret out Christians (*Letters* 10. 96). What action the synagogue authorities might have been induced to take if a Christian refused such an oath is a matter for speculation. It was within their power to excommunicate,[105] and they could certainly ensure that the life of the person so accused would be less than pleasant. On a less official level, Paul could have made individuals' lives a misery by frequent challenges, harrassment, and threats.[106]

---

[98] Deut. 25: 1–3; the detailed specifications of how the sentence is to be carried out are given in *m. Makkot* 3. 10–14. According to Josephus, the lash was to be wielded by the public executioner (*AJ* 4. 238).

[99] This factor is ignored by Hengel (1991), 71–2, who mistakenly insists on the literal sense of 'to destroy' in order to force Paul into line with Luke. Similarly Dunn (1993), 58.

[100] Philo, *Legation to Gaius* 132 and 156.   [101] *Against Apion* 2. 175

[102] *Vita Mosis* 2. 216; cf. Acts 15: 21.

[103] Mark 1: 21–8; 3: 1–6; Luke 13: 10–17.

[104] It has been suggested that Paul invited believers to repeat the formula 'Cursed be Jesus' (1 Cor. 12: 3). While not impossible, this hypothesis is improbable; see my (1990), 243–4.

[105] All the first-century evidence is from the New Testament (Luke 6: 22; John 9: 22; 12: 42; 16: 2); see the discussion in Schürer (1973–87), 2. 431–3.

[106] A much more elaborate scenario is developed by Hengel (1991), 85, who without adequate evidence claims that Paul 'took the initiative and brought about a "pogrom" within the limited sphere of the "Hellenistic" synagogues of Jerusalem'.

Once Paul's persecution of the church is seen in this light, i.e. as an immature religious bigot working out his personal problems, his journey to Damascus becomes highly problematical. As long as there are victims close at hand this sort of person does not go far afield. That it (unwittingly) accounts for this point is perhaps the most that can be said in favour of the hypothesis that the Damascus to which Paul went was not the great city on the Orontes, but a place within a day's walk of Jerusalem.

One of the texts found at Qumran is called the *Damascus Document* (abbreviated as *CD*) because it mentions 'the land of Damascus' as a place of exile where the Essenes made a new covenant.[107] The majority of commentators take this Damascus as a symbolic name for Qumran.[108] In consequence, some scholars claim that Paul merely went to Qumran, thereby rehabilitating Luke's assertion that he was operating under the aegis of authorities in Jerusalem whose writ certainly ran in all of Judaea.[109]

This facile solution cannot stand for three reasons. First, 'Damascus' in *CD* is not a symbol for Qumran, but for Babylon.[110] Secondly, the descriptions of Paul's undignified escape from Damascus imply a city surrounded by a high wall pierced by a number of gates (2 Cor. 11: 33; Acts 9: 24–5). From the excavations it is certain that nothing of the sort ever existed at Qumran. Thirdly, Paul had to flee Damascus because a Nabataean governor sought him (2 Cor. 11: 32). At this time the Nabataeans had no authority over the Jericho area in which Qumran is located; it was under Roman control.

Recognition of this impass makes Knox's hypothesis less arbitrary than it first sounds. Paul, he maintains, did not go to Damascus: he lived there. The Syrian city, not Jerusalem, was the scene of his persecution of the church. We have already seen, however, that the arguments he invokes to support this hypothesis do not stand up under close examination.[111] The natural implications of Paul's own witness confirm the location specified by Luke.

Unless we are prepared to admit that we do not know why Paul went to Damascus, we have to postulate a private journey with an unknown objective. Thus Baslez, for example, proposes that Paul went there as a tourist when en route to a vacation with his parents in Tarsus.[112] The trouble with such speculation is that the temptation to go further is almost irresistible. The fact that the ice becomes progressively thinner is ignored. Baslez provides the perfect illustration of such danger. She suggests that when the Sanhedrin became aware of Paul's plans they entrusted him with letters to synagogues in the areas through

---

[107] *CD* 6. 5, 19; 8. 21; 19. 34; 20. 12.

[108] e.g. Jaubert (1958); Cross (1961), 83; de Vaux (1973), 113–14.

[109] Sabugal (1976), 221; Eisenman (1983), 68–70; Lapide (1993), 104–26.

[110] The point is argued in my (1974a), 221, and my conclusion has been accepted by P. R. Davies (1983), 122.

[111] See above, Ch. 3 'Paul Opposed to Luke?'.

[112] (1991), 68.

— 4 —

Conversion and its Consequences

B Y contrast with Luke's three circumstantial accounts of Paul's conversion,[1] the Apostle himself is dismayingly discreet. There are only glancing allusions to the most shattering experience of his life.[2] The earliest appear in I Corinthians. The first is an indignant rhetorical question, 'Have I not seen Jesus our Lord?' (I Cor. 9: I), whereas the second presents Paul as a witness of the resurrection, 'Last of all, as to one untimely born, he appeared to me' (I Cor. 15: 8). Galatians uses a different language in speaking of the event as 'a revelation of Jesus Christ' (Gal. I: 12) and 'God was pleased . . . to reveal his Son to me' (Gal. I: 16). The connotation of communicated knowledge is reinforced by the fact that, strictly speaking, what is revealed in v. 12 is the 'good news', whereas the purpose of the revelation in v. 16 is 'to preach good news'. In Paul's case conversion and call to ministry are inseparable.

This minimal list of allusions, on which all agree, is greatly expanded by some authors. Seyoon Kim, for example, approvingly reports that various scholars have found formal references to the Apostle's conversion (in Rom. 10: 2–4; I Cor. 9: 16–17; 2 Cor. 3: 16; 4: 6; 5: 16; Phil. 3: 4–12; Eph. 3: 1–13; Col. I: 23–9; in the opening verses of Rom., 1–2 Cor., Gal., Eph., and Col.; and in all instances of the formula 'the grace given to me' Rom. 12: 3; 15: 15; I Cor. I: 4; 3: 10; Gal. 2: 9; Eph. 3: 2, 7; Col. I: 25).[3] Some of these are manifestly irrelevant; the value of others will be tested in the contexts in which they are deemed to contribute.

This is very little. The effect has been that elements from Acts are either consciously invoked to re-create Paul's conversion, or unconsciously permitted to influence our understanding of this episode. In an attempt to avoid this danger here no attempt will be made to account for the material provided by Luke.

---

[1] Acts 9: 1–19; 22: 4–16; 26: 9–18. From a vast literature, see Stanley (1953), 315–38; Hedrick (1981), 415–32. The most detailed tradition history of the three accounts is that given by Boismard and Lamouille (1990), 2. 120 ff., 182 ff., 341 ff., 372 ff. In arguing that Luke knew Gal. I: 13–25 (p. 185), they reinforce the position of C. Masson (1962), 161–6.

[2] Since Paul in his own mind did not cease to be a Jew, and since at this point Judaeo-Christianity was a party within Judaism, some scholars have pedantically denied that Paul was converted, e.g. Stendahl (1977), 7–23; Betz (1979), 64; Georgi (1991), 19. On the contrary, given the radical shift in his perception of God and of the divine plan of salvation implicit in his acceptance of Jesus as the Messiah and the dramatic change in his life-style which ensued, the term is perfectly justified; see Segal (1990). [3] (1984), 3–31.

CONVERSION

The two allusions in 1 Corinthians betray that Paul saw his conversion in a very specific context. 1 Corinthians 9: 2 has very close parallels in Mary Magdalene's experience, 'She saw Jesus' (John 20: 14), and announced to the disciples, 'I have seen the Lord' (John 20: 18). They in turn proclaimed, 'We have seen the Lord' (John 20: 25). The use of the verb 'to see' in immediately post-paschal contexts is well attested.[4] The hint that Paul understood his conversion as a post-paschal apparition is confirmed by 1 Corinthians 15: 8 in which he lists himself as the last of those privileged to have seen the Risen Lord.

Recognition Appearances

The post-paschal apparition stories exhibit an extraordinary variety. It is imperative, therefore, to reduce them to some sort of order. Various classifications have been proposed, but the most helpful is that of M. Albertz, who distinguishes between 'private' and 'apostolic' Christophanies.[5] The terminology is unfortunate because it suggests that the basis is the type of the recipient. In fact, Albertz' concern is to separate appearances which contain a commission from those which do not. Hence, it is perhaps preferable to speak of 'recognition appearances' (Matt. 28: 9– 10; Luke 24: 13–42; John 20: 11–16, 24–9) and 'mission appearances' (Matt. 28: 16–20; John 20: 19–23). The former, in which disciples recognize that the Jesus who died is now alive, are at least logically prior to the latter, as Paul's conversion was to his missionary call.

'Recognition appearances' vary significantly in detail but all the stories reflect the same basic pattern. This may be artificial to the extent that it is schematic, but its spread and consistency indicate that it certainly reflects the conversion experience of many in the early church. The pattern consists of four elements in the following order:

(1) *The disciples acknowledge that the death of Jesus is the end.*
Mary weeps at the tomb (John 20: 11). Cleopas confesses deep disappointment (Luke 24: 21). Disciples hide in fear (John 20: 19). Thomas mocks (John 20: 25).
(2) *Jesus intervenes.*
He calls Mary (John 20: 16). He joins Cleopas (Luke 24: 15). He appears in the midst of the disciples (John 20: 19). Jesus came (John 20: 26).
(3) *Jesus offers a sign of his identity.*
He shows his hands and side (John 20: 20). He shows his hands and feet (Luke

---

[4] Matt. 28: 10, 17; Luke 24: 37, 39; John 20: 20, 27, 29; Acts 1: 9.
[5] (1922), 259–69.

24: 40). He breaks bread (Luke 24: 30). He offers his hands and side (John 20: 27).

(4) *Jesus is recognized.*
They worshipped him (Matt. 28: 9). Mary says 'Rabboni' (John 20: 16). Their eyes were opened and they recognized him (Luke 24: 31). They saw the Lord (John 20: 20). 'My Lord and my God!' (John 20: 28).

These narratives flatly contradict the common assumption that the disciples were so enthusiastically convinced of Jesus' continuing life that they invented the resurrection to confess their belief. On the contrary, the death of Jesus dashed all their hopes. They had nothing to look forward to; they expected nothing. It took an initiative of Jesus to lift them out of their pessimistic lethargy. A sign of identity was required by the difference between the earthly and resurrection body (cf. 1 Cor. 15: 51–3; John 20: 19). Yet it was not a proof, because acknowledgement of the risen Jesus as the one who died was not unanimous (Luke 24: 41; Matt. 28: 17).

This grid provides the framework in which, according to Paul, his conversion should be analysed.

A Pharisee's Knowledge of Jesus

Before we do so, however, it is important to try to determine what Paul knew about Jesus of Nazareth, the one whom he was about to encounter. Inevitably this controlled and channelled his perception. That he did know something is certain, for he later confessed, 'we have known Christ in a fleshly way' (2 Cor. 5: 16).[6] At one time, manifestly prior to his conversion, he thought about Jesus in a way of which he was later ashamed.

What sort of knowledge of Jesus might a first-century Jew have had, and particuarly a Pharisee? There are two approaches to an answer, what was actually said and what might be deduced.

Josephus, the first-century Palestine-based Jewish historian who claimed to be a Pharisee, mentions Jesus twice.[7] The authenticity of one reference is accepted by all, whereas the genuineness of the other is bitterly disputed. The former reads,

> This younger Ananus took the highpriesthood. He was a bold man in his temper, and very insolent; he was also of the sect of the Sadducees, who are very rigid in judging offenders above all the rest of the Jews, as we have already observed. When, therefore, Ananus was of this disposition, he thought he had

[6] Both Pauline usage and the context indicate that *kata sarka* must be understood as an adverb. Were it an adjective it would have occupied a different place in the sentence (cf. Rom. 4: 1; 9: 3; 1 Cor. 1: 26; 10: 18).
[7] Recent discussions, with full bibliographies, which argue in favour of the positions adopted here are Schürer (1973–87), 1. 428–41, and Meier (1990).

now a proper opportunity to exercise his authority. Festus was now dead, and Albinus was but upon the road; so he assembled the sanhedrin of judges, and brought before them the brother of Jesus, who was called Christ, whose name was James, and some others. And when he had formed an accusation against them as breakers of the law, he delivered them to be stoned.

(*AJ* 20. 199–200; trans. Whiston and Margoliouth)

The high priest filled the power vacuum between two Roman procurators by asserting his authority in a particularly brutal way. Josephus' interest in the episode was that it led to the deposition of Ananus (*AJ* 20. 203). James was important merely as the only identified victim. The name 'James' however, was so common that Josephus was forced to specify him by reference to his better-known brother Jesus, whose name was also so widespread that it too demanded a qualifier. Josephus' choice of 'Christ', rather than 'of Nazareth' indicates that statements about Jesus' role as Messiah enjoyed sufficiently wide circulation to be understandable even among those who rejected it.

The second reference in Josephus is known as the *Testimonium Flavianum*, and opinions of its authenticity range from complete acceptance to flat rejection. Opposed to such extremes is an intermediate position, which I am convinced is correct; it maintains that a Christian editor added to (and perhaps deleted from) a note on Jesus composed by Josephus.

> Now, there was about this time Jesus, a wise man, *if it be lawful to call him a man*, for he was a doer of wonderful works, a teacher of such men as receive the truth with pleasure. He drew over to him both many of the Jews and many of the Gentiles. *He was the Christ.* And when Pilate, at the suggestion of the principal men among us, had condemned him to the cross, those that love him at first did not forsake him; *for he appeared to them alive again at the third day, as the divine prophets had foretold these and ten thousand other wonderful things concerning him.* And the tribe of Christians so named from him are not extinct at this day. (*AJ* 18. 63–4; trans. Whiston and Margoliouth)

The italicized portions represent the Christian additions to Josephus' text, and the arguments are obvious from their content. Without them we have a bland report written in language which may be deliberately ambiguous. Even when interpreted in the most positive way, the Christology is so low that it certainly cannot be attributed to a follower of Jesus of the patristic or medieval period.

The meaning of the attribute 'wise man' is furnished by two phrases expressing what he did in word and deed. His works are qualified as *paradoxa*; the basic meaning of this adjective is 'contrary to expectation' whence 'incredible', which Josephus may have intended positively ('exceptional')[8] or negatively ('unbelievable'). A slight clue to his intention emerges in what he says of Jesus' teaching.

---

[8] This appears to be the force of the adjective in *AJ* 9. 182; 12. 63.

Instead of summarizing its content he merely notes the character of the audience.

There are two possible versions. According to the received text (quoted above) Jesus' hearers were seekers after truth. This could be to condemn with faint praise, because Josephus offers no guarantee that they were not misled. But there is something that does not ring quite true here. Josephus goes on to say that Jesus won a considerable number of adherents. If all were seekers after truth, this could leave the impression that what Jesus preached was in fact the truth. Is this likely to stem from Josephus? It harmonizes better with the concerns of the Christian interpolator. What, then, might Josephus have written?

A plausible suggestion involves the change of only one letter, and offers a more fitting context for *hêdonê*, which normally connotes physical pleasure rather than intellectual gratification. The corrected text reads *didaskalos anthrôpôn tôn hêdonê ta aêthê* (in place of *talêthê*)*dexomenôn* 'a teacher of those with an appetite for novelties'.[9] This rendering has a certain intrinsic probability. Even in a writer as sloppy as Josephus, one would expect at least a hint of why some leading Jews delated Jesus to Pontius Pilate.

Jesus merited a place in the history of Josephus merely because, against all expectations, he acquired a following which survived him. When the assertion that 'He won over many Jews and Gentiles' is compared with the Gospels, it is manifest that Josephus is reading back into the lifetime of Jesus what was true only much later in the first century. Jesus in fact converted few Jews and fewer Gentiles. Only in the post-paschal period did they grow into 'a tribe of Christians'. This name should be translated 'Messianists' in order to bring out its resonance for anyone with a Jewish upbringing. Josephus, it will be recalled, was aware that Jesus' followers thought of him as the Messiah (*AJ* 20. 200).

Equally, it was the existence of Christians (though they were not yet known as such; Acts 11: 26) which directed Paul's attention to Jesus. It is inconceivable that he should have persecuted Christians without learning something about the founder of the movement. Paul the Pharisee certainly was in a position to discover as much as Josephus did. Thus we can safely assume that Paul knew (1) that Jesus had been a teacher to whom wonders were ascribed; (2) that he had been crucified under Pontius Pilate as the result of Jewish charges; and (3) that his followers thought of him as the Messiah. It is unlikely, however, that he would have been content with such bare bones. Pharisaic interests would have driven him to flesh them out.

Given their concern to transform the Jewish people through more exact instruction in the written and oral Law, the Pharisees would have been extremely sensitive to the fact that Jesus had disciples whom he taught (John 7:

---

[9] Or 'those who accept the abnormal with delight'. This emendation has no textual support, but then neither does any MS of the *Antiquities* lack the Christian interpolations.

15). Any success by other teachers threatened their hoped-for monopoly. The natural response would have been to challenge what Jesus was saying, particularly in areas where they sensed vulnerability.

The touchstone of Jewish observance has always been the sabbath, and there can be little doubt that the complexity of Pharisaic legislation which culminated in the 39 types of work forbidden on the sabbath had already begun in the time of Jesus (*m. Sab.* 7.2). Thus it is not surprising that the Gospels record a number of controversies in which Pharisees challenge Jesus on what is permissible on the sabbath (Mk 2: 23–8; 3: 1–6; Luke 6: 6–11; 14: 1–6; John 9: 1–40). The basis of their objection to his healings was that illnesses which he treated were not life-threatening; they could have been deferred for a day. Jesus, on the contrary, saw his cures as a matter of life and death. By his action and response precisely on God's day, the sabbath, he was criticizing current Jewish halakha in order to emphasize that God's love expressed in healing power was available at each and every moment, and not merely when permitted by the Law. Why should a sick person have to wait for relief when it was available now? Jesus' attitude was less a repudiation of the sabbath than an affirmation of the imminence of the kingdom of God.[10]

Presumably the Pharisaic version of the results of such encounters differed from that of Jesus' followers. His attitude of unperturbed authority, however, would have hinted at an attitude towards the Law embodying a personal claim so extravagant as to make even closer attention to his teaching imperative.

Through infiltration or, less dramatically, through questioning of verbosely enthusiastic supporters, Pharisees could easily have come to learn that Jesus' sabbath actions were confirmed and reinforced by his relativization of the Law. Even the simplest of his followers must have realized the implications of assertions such as 'It was said to those of old [in the Law] ... but I say to you ...' (Matt. 5: 21, 27, 33, 38, 43),[11] particularly when accompanied by a claim that Jesus was the touchstone of salvation (Matt. 10: 32–33).[12] Jesus thought of himself as the Messiah empowered to articulate God's will; the Law was no longer the sole or final authority.

Finally, there was one aspect of the gossip about Jesus which would have been of particular interest to Pharisees. In opposition to the Sadducees who denied any afterlife,[13] the Pharisees believed in resurrection of the body.[14]

---

[10] B. Schaller, 'Jesus and the Sabbath'; a lecture delivered at the École Biblique on 20 December 1991.

[11] See in particular Käsemann (1964), 37–8. His acceptance of only three antitheses (Matt. 5: 21, 27, 33) as authentic rests on the arguments of Bultmann (1963), 134–6, which have been refuted by Jeremias (1971), 251–3.

[12] The authenticity of this saying is confirmed by its dilution in Luke 12: 8–9 by the introduction of two intermediaries (the Son of Man and angels). The criterion of dissimilarity could hardly be more perfectly verified; against Perrin (1976), 185–91.

[13] Acts 23: 8; Josephus, *JW* 2. 165.

[14] Acts 23: 6–8; Josephus, *JW* 2. 163; 3. 374; see Cavallin (1974), 171–92.

Fundamental to the preaching of the first Christians was the assertion that God had raised Jesus from the dead; it appears in the earliest formulation of the faith of the church (1 Cor. 15: 3–5). The resurrection was the great sign which validated the mission of Jesus and guaranteed his teaching. No Christian could avoid speaking of it and, once heard, it would rankle in the memory of a Pharisee.

While there may be some hesitancy in determining what Paul knew of Jesus while still a Pharisee, there can be no doubt as to what he thought of Christian claims. To his way of thinking it was ridiculous to maintain that God had intervened to raise from the dead a false teacher whose blasphemous claim to be the Messiah went hand in hand with deliberate subversion of the authority of the Law. It now becomes clearer why Paul tried to turn Christians from their beliefs. They had been disastrously misled.

Recognizing the Risen Lord

Given this attitude, it is certain that Paul was in no way disposed to expect anything to happen en route to Damascus. His reaction paralleled the initial response of Jesus' followers for whom his crucifixion was the end of hope. Jesus, Paul was convinced, had died a fitting death, and all that remained was the return of his supporters to the fold of authentic Judaism.

Paul explicitly reports that Jesus took the initiative in the encounter; there had been no preparation on his part. The most important passage is in his addition to the earliest creed,[15] which needs to be looked at closely. The ambiguity of *eschaton de pantôn hôsperei tô ektrômati ôphthê kamoi* (1 Cor. 15: 8) is brought out by the variety of translations, e.g. 'Last of all, as to one untimely born, he appeared also to me' (*NRSV*); 'Last of all he was seen by me, as one born out of the normal course' (*NAB*). There is a significant difference between 'he appeared to' and 'he was seen by'. The latter takes *ôphthê* as passive voice, whereas the former treats it as middle voice, 'he showed himself' (cf. John 21: 1).

Which is correct? The active meaning is demanded by Acts 26: 16, and strongly recommended by LXX usage, e.g. *ôphthê kyrios tô Abram* (Gen 12: 7 = Acts 7: 2), which must be translated 'he showed himself to Abram' because it renders *wayyera' Yahweh 'el Abram*, where the particle of motion or direction *'el* unambiguously indicates the active agent. Philo's comment is very apposite,

> God, by reason of his love for humanity, did not reject the soul which came to him, but went forward to meet it, and showed to it his own nature as far as it was possible that he was looking at it could see it. For which reason it is said not that 'the wise man saw God' but that 'God appeared to the wise man'. (*De Abrahamo* 79–80; trans. Yonge).[16]

[15] See my (1981*b*), 582–9.   [16] See Pelletier (1970).

In the case of Paul the active meaning is made certain by other references in which the stress on the initiative of God/Christ is unequivocal. 'He was pleased . . . to reveal his son to me' (Gal. 1: 16). 'I was laid hold of by Christ Jesus' (Phil. 3: 12). The weight of these texts is not countered by the exceptional 'Did I not see the Lord?' (1 Cor. 9: 1), which simply reflects the natural shift towards the graphic which is also found in the gospels, where in a secondary phase of the tradition Jesus is 'seen' by those whom he 'met' or 'stood among' or 'journeyed with'.[17]

The most difficult element of the recognition appearance grid to account for is the sign of identity because, in opposition to Jesus' disciples, Paul had not met Jesus during his earthly ministry. By definition, therefore, Paul could not have recognized Jesus on the same basis as those who had come with him from Galilee to Jerusalem. We can be sure, however, that Paul had a mental image of Jesus. Many create a portrait in their minds of authors whose books they happen to be reading. If simple interest can produce such images, then the intense anger which Paul directed against the one who had led Jews astray was capable of the same effect.[18] The stress under which Paul was operating would have interfered with his rationality and would have heightened his susceptibility to anyone or anything associated with the focus of his emotion.[19] What actually happened must remain a mystery unless we are prepared to invoke the vivid details of Luke's accounts, in each of which, incidentally, Jesus has to identify himself (Acts 9: 5; 22: 8; 26: 15). In any event, the reality and the mental image fused and Paul's world was turned upside down.

Paul now knew with the inescapable conviction of direct experience that the Jesus who had been crucified under Pontius Pilate was alive.[20] The resurrection which he had contemptuously dismissed was a fact, as undeniable as his own reality. He knew that Jesus now existed on another plane. This recognition is all that was necessary to his conversion, because it completely transformed his value system.

If one of the resonances that the name of Jesus set up in his Pharisaic mind was true (i.e. resurrection), then the others automatically had to be viewed in a completely different perspective. No longer were they the blasphemous pretentions of a madman and his dupes, but utter truth.

---

[17] Active use of the verb 'to see': Matt. 28: 10, 17; Luke 24: 37, 39; John 20: 14, 18, 20, 25, 27, 29; Acts 1: 9. Physical encounters: Matt. 28: 9; Luke 24: 15, 36; John 20: 19, 26; Acts 1: 3.

[18] Knox (1950), 126, who is one of the few to raise the identification question, answers it thus, 'Christ had begun to make himself known to Paul—perhaps against the latter's will—as the Spirit of the persecuted *koinonia* before he made himself known in the visual experience in which Paul's conversion culminated.'

[19] This point is developed by Gager (1981), 699–700.

[20] Kim (1984), 7, 108, 223–7, deduces from 2 Cor. 4: 6 (which in fact is neither explicitly nor exclusively concerned with Paul's conversion) that the risen Christ appeared in divine glory, i.e. 'he saw him exalted by God and enthroned at his right hand' as the physical embodiment of divinity. See the understated critique by Dunn (1987), 256–62.

Jesus, therefore, must be precisely what he implicitly, and his disciples explicitly, claimed he was, namely, the Messiah. Equally the attitude of Jesus' towards the Law must be correct; the Law was not the definitive expression of God's will. What the Law laid down as the prerequisites of salvation had no further validity. As grace had been made available to Paul, despite his efforts to thwart the divine plan of salvation as revealed in Jesus, so it could be made accessible to those whom the Law had excluded.

Only when it is conceded that Paul's conversion consisted essentially in the revaluation of ideas which he already possessed does it become possible to understand how he can write, 'For I would have you know, brethren, that the gospel preached by me is not according to man, for I did not receive it from man nor was I taught it but [it came] through a revelation of Jesus Christ (Gal. 1: 11–12; cf. 1: 1). Sandnes is typical of many commentators in taking this statement at face value; he concludes, 'His gospel was not dependent on information given him by others. He has received it direct from Jesus in a revelation.'[21] In reality the text embodies a slight deviation from the absolute truth which is excused by the polemic context. No one convinced of the truth of Jesus had taught Paul about Christ or Christianity. He had never studied them in the way that he had applied himself to understanding the Law.

None the less, as 2 Corinthians 5: 16 shows, he had assembled information about the Jesus movement. His point, therefore, can only be that as gospel such concepts were not as he had acquired them. He had not heard of Jesus of Nazareth as Lord and Messiah. He had not been taught that the Law was merely a source from which one could choose to draw or not.

His encounter with Christ revealed the truth of what he had once taken as falsehood by forcing a new assessment of what became the Christological and soteriological poles of his gospel.[22] Christ was the new Adam, the embodiment of authentic humanity. The Law was no longer an obstacle to the salvation of Gentiles; they could be saved without becoming Jews.

Apostle to the Gentiles

According to Paul, his conversion was for the Gentiles, 'But when he who had set me apart from my mother's womb, and had called me through his grace, was pleased to reveal his son to me, in order that I might preach him among the nations' (Gal. 1: 15–16).[23] The opening words are immediately evocative of two

---

[21] (1991), 53.

[22] Against Betz (1979), 64, who maintains that a verbal revelation is implied. According to the *Pseudo-Clementine Homilies*, 'But if you [Simon, a surrogate for Paul] were visited by him [Christ] for the space of an hour and were instructed by him and thereby have become an apostle' (17. 19. 4). The point is to contrast the brevity of Paul's experience with the year-long instruction given by the Risen Lord to his authentic disciples; see Lüdemann (1989), 187.

[23] In addition to the commentaries, see Denis (1957).

celebrated Old Testament vocations, the Isaian Servant of Yahweh and the prophet Jeremiah. The version cited is that of the LXX which Paul knew, and the underlined words are those he used in narrating his own vocation. 'From *my mother's womb he called* my name. . . . He said to me . . . I will give you as a light to the *nations*' (Isa. 49: 1, 6). 'Before I formed you in the *womb*, I knew you, and before you came forth from [your] *mother* I hallowed you. I appointed you a prophet for [the] *nations*' (Jer. 1: 5).

The repetition of the three key terms cannot be coincidental. As in the case of his two great predecessors, Paul saw his conversion as the working out of a plan devised much earlier by God. The goal of that plan was the extension of God's grace to the Gentiles. Thus he was called precisely in order to bring the good news to those who did not belong to the Jewish people. Both Galatians 1: 11–12 and 1: 15–16 unambiguously indicate that Paul's mission to the Gentiles was not a late development, nor a mere extension of a presumed outreach of Hellenists in Jerusalem. It should be unnecessary to stress this obvious point, but it has in fact been challenged.

J. P. Bercovitz has argued that when Paul employs *kalein* absolutely it always means an efficacious call to faith. In consequence, he maintains, the aorist participle *kalesas* in Galatians 1: 15 should be understood as a parenthetical clause referring to Paul's conversion to Christianity, which is prior to the apostolic commissioning mentioned in the main clause.[24] Were Bercovitz concerned to highlight a mere logical priority, one might agree, but he stresses a temporal gap between conversion and commission. Paul, he insists, was already a believer when Jesus appeared to him. This would imply, however, that the Apostle had been instructed in Christianity, which is precisely what he formally denies in Galatians 1: 11–12.[25] Moreover, Paul uses *kalein* with explicit reference to his commission. More significant than 1 Corinthians 1: 1 is Romans 1: 1, where he combines the two verbs of Galatians 1: 15, 'Paul a slave of Christ Jesus, *called* to be an apostle, *set apart* for the gospel of God'.

F. Watson contends that Paul initially belonged to a Judaeo-Christian group which he perceived as a reform movement within Judaism. His mission was to make mediocre Jews better. That ambition met with no response. It was then, Watson maintains, that Paul, who could not live with failure, decided to turn to the Gentiles. In order to make the gospel more attractive to them he repudiated portions of the Law, and thereby transformed Christianity into a sect bitterly opposed to the synagogue.[26]

Not only does this hypothesis lack any textual foundation but it is flatly contradicted by Paul himself who tells us that his first act after his conversion was to 'go away into Arabia' (Gal. 1: 17).

---

[24] (1985), 28–37.
[25] So rightly Knox (1987).
[26] (1986), 28–38.

ARABIA

Where was Arabia? Strabo gives the geographers' answer, 'Arabia Felix is bounded by the whole extent of the Arabian Gulf [= Red Sea] and the Persian Gulf.'[27] The extent of this huge land mass underlines the need for a more tightly focused question. What would the term 'Arabia' have suggested to a Jew who lived in first-century Judaea?

Location and Mission

Josephus provides a very clear answer. Arabia could be seen to the east from the tower Psephinus in Jerusalem.[28] Thus, it lay on the desert side of the three easternmost cities of the Decapolis, Damascus, Raphana, and Philadelphia.[29] More specifically, it was contiguous to Herodian territory running along the southern border of the Roman province of Syria,[30] and south and east of the great fortress Machaerus.[31] Petra was the royal seat of Arabia.[32] Whence the name 'Arabia Petrea',[33] or 'Arabia belonging to Petra'.[34] This mountain-encircled city, however, was the capital and chief city of the Nabataeans.[35] Whence another name 'Arabia of the Nabataeans'.[36]

Paul, therefore, went into Nabataean territory,[37] which at that period ranged from the Hauran down through Moab and Edom and expanded on both sides of the Gulf of Aqaba.[38] What was his purpose? Some have thought that he sought a quiet place for reflection and study.[39] The Law had ceased to be the centripetal force which held the different facets of his life together. That power was now exercised by the Risen Lord and, it is suggested, he needed time and tranquility in order to assimilate a change of such magnitude.

Plausible as this suggestion is, it does not adequately account for what happened subsequently. Paul must have been doing something to draw attention to himself and arouse the ire of the Nabataeans because he had to return to Damascus, and even three years later the Nabataean authorities still

---

[27] *Geography* 2. 5. 32.
[28] *JW* 5. 159–60.
[29] Pliny, *NH* 5. 16. 74. Cf. Strabo, *Geography* 16. 2. 20.
[30] *AJ* 16. 347.
[31] *JW* 7. 172.
[32] *JW* 1. 125; cf. 1. 159, 267; 4. 454.
[33] *AJ* 18:109.
[34] Dio Cassius, *History* 68. 14. 5.
[35] Strabo, *Geography* 16. 4. 21.
[36] Ibid. 17. 1. 21. Similarly Plutarch, *Anthony* 36. 2.
[37] 2 Cor. 11: 32–3, in addition to the texts of Pliny and Strabo cited in n. 29, excludes the hypothesis of Bietenhard (1977), 255, that Paul preached in the Decapolis, possibly at Pella.
[38] The most detailed ancient treatment of the life-style of the Nabataeans is that of Strabo, *Geography* 16. 4. 26. The classic study remains that of Starcky (1966). Supplementary information on the period 30 BC–AD 70 is provided by Negev (1977). See also Graf (1992).
[39] Most recently Baslez (1991), 101; Longenecker (1990), 34; N. Taylor (1993), 73, but without suggesting, as Lightfoot (1910), 87–90, did on the basis of Gal. 4: 25, that Arabia was Sinai (Gal. 4: 25).

wanted to arrest him (Gal. 1: 17; 2 Cor. 11: 32–3). The only explanation is that Paul was trying to make converts.[40] This first act subsequent to his conversion confirms his understanding of his conversion as a commission to preach the gospel among pagans.[41]

Nabataeans and Jews

In order to understand the violence of the Nabataean reaction, the salient points of their stormy relations with the Jews must be recalled. Things started well, when Antipater of Idumea sealed an alliance between Hyrcanus II and Aretas III by marrying Kypros, who came from an eminent Nabataean family and later became the mother of Herod the Great.[42] Against his will, in 32–31 BC, the latter was forced into a war with the Nabataeans, which he won after suffering heavy losses.[43] He again defeated them c.9 BC.[44] It is not at all surprising, therefore, that the Nabataeans enthusiastically provided auxiliaries to aid P. Quinctilius Varus, the governor of Syria, in his brutal suppression of the revolt which followed the death of Herod around 4 BC.[45]

In order to calm the tensions between the two peoples Herod Antipas married the daughter of Aretas IV,[46] possibly at the suggestion of the emperor, Augustus, if Suetonius' report of his policy is correct (*Augustus* 48). In time, however, Antipas tired of her and divorced her in order to marry Herodias, the wife of his half-brother Philip.[47] This marriage is probably to be dated in AD 23.[48] Its criticism by John the Baptist is reported both by the Gospels and Josephus.[49] Their divergent emphases (moral for the former; political for the latter) are in fact complementary, and adequately explain John's arrest and imprisonment in Machaerus, probably around AD 28.[50] Herod Antipas had moved there from Galilee in order to be prepared for an attack by Aretas in revenge for the insult to his daughter. The latter in fact made a disputed area on his northern border a pretext for war.[51] In the battle, which probably should be dated c.AD 29,[52] the troops of Antipas were routed. Whereupon, according to

[40] So recently Bruce (1977), 81–2; Betz (1979), 74; Légasse (1991), 72.
[41] Only by an extremely tendentious treatment of a series of texts can N. Taylor (1993), 92–3, conclude that Paul's missionary consciousness began only when he was mandated by Antioch (Acts 13: 1–3).   [42] *JW* 1. 181.
[43] *JW* 1. 364–85.   [44] *AJ* 16: 282–5.
[45] *JW* 2. 68.   [46] *AJ* 18. 109.
[47] *AJ* 18: 110.   [48] Saulnier (1984), 365–71.
[49] Mark 6: 17–18 and par.; Josephus, *AJ* 18. 118.
[50] *AJ* 18. 119. Although presented as the date of the beginning of John the Baptist's ministry, 'the 15th year of the reign of Tiberius Caesar' (Luke 3: 1) is more likely to be the date of his arrest.
[51] The text of *AJ* 18. 113 is defective and Gabalis is the most probable restoration; see Hoehner (1972), 254–5; Schürer (1973–87), I. 350. It is not impossible that the conflict inspired Jesus' parable of two kings going out to war (Luke 14: 31–2). Were the allusion certain it would date the battle before AD 30, the year in which Jesus died.
[52] Saulnier (1984), 375.

Josephus, he indignantly complained to Rome.[53] The historicity of this complaint cannot be guaranteed, but a war could not be kept secret and the news would certainly come to the ears of the emperor.

The Situation when Paul Arrived

Aretas IV had every reason to feel anxious, because he had both indirect knowledge, and direct experience, of the anger of a Roman emperor, when peace was disturbed on the eastern frontier of the empire. The following are the essentials of a rather complicated story which took place some forty years earlier.[54]

With the authorization of C. Sentius Saturninus, the governor of Syria, Herod the Great went into Arabia to arrest criminals from his territory in Trachonitis. A skirmish ensued when the Nabataeans intervened to protect them, and some soldiers on both sides died. Syllaeus, who represented the Nabataeans in Rome, presented the affair to Augustus as an unwarranted breach of the peace. The emperor's extreme displeasure explains why Herod had been careful to secure prior Roman approval for his military action. It did him no good, however, and he lost the imperial favour completely. As Augustus put it in a severe reprimand, Herod had been relegated from the status of a friend to that of a subject. The emperor refused a first embassy from Judaea, and only reluctantly heard a second embassy led by Nicolaus of Damascus, who with the support of ambassadors from Aretas IV, proved Syllaeus' version of the episode to be false. Herod was restored to favour, but Aretas found himself in serious trouble, because on the death of Obodas he had assumed the throne of Arabia without the permission of Rome. Augustus had planned to entrust Arabia to Herod, and it was only the latter's refusal that enabled Aretas to succeed after being reproved for his rashness.

Aretas, therefore, knew from personal experience that Rome had little patience with warlike actions between the client kings who guarded the eastern frontier of the empire. It would be most surprising if he had not feared some reaction on the part of Tiberius, as he had once dreaded the response of Augustus. The retirement of Tiberius to Capri in AD 26 has been interpreted as a loss of interest in the affairs of state. While there may be some truth in this as regards internal affairs, it is not so as regards the provinces.[55] Philo's judgement that in his twenty-three years of rule Tiberius 'did not let the smallest spark of war smoulder in Greece or the world outside Greece',[56] while not completely accurate,[57] is borne out (for the part of the world with which we are concerned) by his vigourous and effective responses in matters large, e.g. the Parthian

---

[53] *AJ* 18. 115.   [54] *AJ* 16. 271–355.
[55] See Charlesworth (1950).   [56] *Leg.* 141.
[57] There were uprisings in Africa (AD 17), in Thrace (AD 19, 21, 25), and in Gaul (AD 21). See Scullard (1982), 278–80.

occupation of Armenia in AD 34,[58] and small, e.g. when Pilate to annoy the Jews placed shields with the emperor's name in Herod's palace in Jerusalem.[59] His vigilance regarding the security of the eastern frontier is perfectly illustrated by the way he responded to the death of Herod Philip in AD 33/34. He immediately attached the territory to the province of Syria, while leaving the revenues to accumulate for a successor.[60] Despite his age and weariness, Tiberius was perfectly capable of reacting quickly and decisively. In the case of Aretas it only needed an order to the governor of Syria, who had four legions at his disposition.[61]

As Aretas waited tensely for something to happen, his attitude towards Jews was certainly anything but benign. They (in the person of their king) were responsible for the desperate anxiety which weighed upon him. A Roman reprisal would be but the latest in the series of disasters which they had brought upon his people. His subjects presumably shared his apprehension and his anger, both of which intensified as the years passed. By the time Paul arrived *c.* AD 33 the tension would have been building for some three years. It was certainly not a propitious moment for a Jew to begin preaching what to an outsider was but a new variety of Judaism. To those Nabataeans who were the objects of his ministry it could only appear as an attempt to infiltrate, divide, and weaken them. What they saw as an invitation to betrayal would have prompted an immediate and violent reaction. Paul, however, escaped. Otherwise there would have been no point in drawing the authorities into the affair and painting him in such colours that he was remembered as dangerous three years later (cf. 2 Cor. 11: 32–3).

If the above assessment of the situation is correct, it is unlikely that Paul penetrated very deeply into Arabia. He may not even have reached Bosra; there were three Nabataean towns further north, Phillopolis, Kanatha, and Suweida.[62] If Aretas contemplated armed resistance to Rome, he would certainly have had troops in that area, and Paul would have been a figure of suspicion once he opened his mouth. This makes it improbable that Paul stayed long.[63] His silence as to the duration suggests that it was very short, since he lists his two weeks in Jerusalem and his three years in Damascus (Gal. 1: 18).

---

[58] Josephus, *AJ* 18. 96–105; Dio Cassius, *History* 58. 26. 1–4; 59. 27. 2–4; Tacitus, *Annals* 6. 31–7.

[59] Philo, *Leg.* 299–308; Josephus, *JW* 2. 169–74.

[60] *AJ* 18. 108.

[61] Who precisely was in charge in Syria at this stage is problematic; see Schürer (1973–87), I. 260–2, 362. It is improbable that L. Vitellius, who became governor of Syria in AD 35, was ever ordered by Tiberius to attack Aretas as Josephus reports (*AJ* 18. 115, 120–6). The latter gives a completely different explanation for the presence of Vitellius in Jerusalem in *AJ* 18. 90–5. See Saulnier (1984), 373–4.

[62] See the map in Negev (1977), 550.

[63] Against Meeks (1983), 10, who believes that Paul preached for three years in such cities as Petra, Gerasa, Philadelphia, and Bosra.

The imprudent gesture is important only in so far as it indicates that from the beginning he was convinced that his mission was to Gentiles.

DAMASCUS

Paul gives us no information on how he passed the next three years in Damascus (Gal. 1: 18). According to Luke, Paul's ministry there was devoted to the conversion of Jews (Acts 9: 20). Not only is this part of the build up to Luke's explanation of why Paul had to leave the city (Acts 9: 23), which is contradicted by the Apostle (2 Cor. 11: 32–3), but it is incompatible with Paul's conviction that his mission was to the Gentiles.

Learning a Trade

Before touching on this issue, a more fundamental question needs to be raised: how did Paul support himself? For some reason this question is never asked. Perhaps it is taken for granted that he was independently wealthy, even though subsequently he had to work for a living, or that the church there granted him free room and board for life, even though he had no claim upon it, having contributed nothing to its foundation. Moreover, no church could assume the financial burden of guaranteeing subsistence to converts. Not only was it the road to financial difficulties,[64] but it was most unwise to give the appearance of buying converts. It seems probable, therefore, that Paul supported himself. The key questions then become: what skill did he have, and where did he acquire it?

Paul himself tells us only that he worked with his hands (1 Thess. 2: 9; 2 Thess. 3: 7–9; 1 Cor. 4: 12). His attitude towards such labour is all the more significant in that it emerges only indirectly. He lists it among the unfair hardships of his life (1 Cor. 4: 12; 2 Cor. 6: 5; 11: 23, 27) and qualifies it as 'slavish' (1 Cor. 9: 19) and 'demeaning' (2 Cor. 11: 7). No one bred to a craft would speak of it in this way.[65] Paul's stance is that of those whose inherited status preserved them from physical work.

This conclusion conflicts with the widespread view that Paul owed his trade to his Jewish background. Texts such as 'He who does not teach his son a craft teaches him brigandage' (*b. Kidd.* 33a) are adduced as statements of principle designed to explain the fact that rabbis were self-supporting. This practice,

---

[64] When the money of wealthy members of the church of Jerusalem (Acts 2: 45; 4: 34–5: 11) ran out, help from abroad became necessary (Gal. 2: 10).

[65] The arguments of Hock (1978) are in no way affected by the strained objections of Stegemann (1987), 227, or Légasse (1991), 41.

however, is neither as clear nor as well attested as is often thought.[66] Ben Sira begins his developed contrast between the tradesman and the scholar (38: 24 to 39: 11) with the words 'The wisdom of the scribe depends on the opportunity of leisure; only the one who has little business can become wise,' and goes on to exclude from access to wisdom 'every artisan and master craftsman who labours by night as well as by day' (38: 27; cf. 2 Cor. 6: 5; 11: 27). Any occupation was a distraction from the study of the Law. All the evidence of rabbis practising trades dates from the post-AD 70 period when conditions in Jerusalem had changed radically for the worst. Then the rabbis had to work in order to survive, and necessity was transformed into a virtue, 'All study of the Law without [worldly] labour comes to naught at the last and brings sin in its train' (*m. Aboth* 2. 2) Presumably it was at this stage that trades were attributed to Hillel and Shammai.[67]

Thus there is no evidence to suggest that when Paul was a Pharisaic student in the Holy City he was under any pressure from his masters to learn a trade. Moreover, there was neither need nor incentive. His intense commitment to his studies (Gal. 1: 14) precluded the distraction of other interests, and even if he received nothing from his family (which appears unlikely) he would not have starved, because there were many who sought merit by alms-giving.[68]

On his conversion to Christianity, however, Paul would no longer be an acceptable recipient of institutionalized Jewish charity, and he may have lost contact with his family. When, speaking of his family background, he says, 'Whatever gain I had, I counted as loss for the sake of Christ' (Phil. 3: 8). It is natural to understand that he had been disinherited. In any case, it was during his stay in Damascus, and perhaps because of his travels in Arabia, that Paul is most likely to have become conscious of the need to be self-sufficient. His mission demanded a mobility which would enable him to reach out to the whole Gentile world. Only financial independence could give him such freedom, and it was impossible without a marketable skill.[69] According to Luke, he decided to become a 'tent-maker' (Acts 18: 3).[70]

Although the letters furnish no direct confirmation, the statement has a very definite intrinsic plausibility. When contemplating which trade to choose, Paul must have established a number of criteria for himself. The skill to acquire had to be in demand throughout the Roman empire, in the cities as well as on the road; it had to bring him into contact with all levels of the population; its tools

[66] So rightly Hock (1978), 557. The speculative character of the considerations advanced by Hengel (1991), 15–16, dramatically underlines the lack of hard evidence to support the view that pre-AD 70 rabbis had trades.

[67] The texts are assembled by Billerbeck (1922–8), 2. 745–6.

[68] See 'The Subsidized Sections of the Population' in Jeremias (1969), 111–19.

[69] The other options were to acquire a patron or to beg as he travelled. The former would have impaired his mobility, and the latter would have compromised his credibility.

[70] See further below, Ch. 11, 'Working with Prisca and Aquila'.

had to be easily portable; and it had to be quiet and sedentary so that he could preach and work at the same time. When judged from the standpoint of twentieth century life, tent-making would appear to fail on all counts. What need, for example, had urban dwellers of tents?

In first-century Rome a number of inscriptions attest the existence of an organization known as 'The Tent-makers Association'.[71] Pliny the Elder describes what work they did in a way which answers the above question.

> Linen cloths were used in the theatres as awnings, a plan first invented by Quintus Catulus when dedicating the Capitol.[72] ... Next even when there was no di splay of games Marcellus, the son of Augustus's sister Octavia, during his period of office as aedile in the eleventh consulship of his uncle, from the first of August onwards afixed awnings of sailcloth over the forum, so that those engaged in lawsuits might resort there under healthier conditions. What a change this was from the stern manners of Cato the ex–censor, who had expressed the view that the forum ought to be paved with sharp pointed stones.[73] Recently awnings actually of sky blue and spangled with stars have been stretched with ropes even in the emperor Nero's amphitheatres. Red awnings are used in the inner courts of houses and keep the sun off the moss growing there.[74] (*NH* 19: 23–4; trans. Rackham)

Tent-makers, therefore, could expect both public and private commissions in furnishing protection from the glaring summer sun, which was much more intense in the Middle East than in Italy. It should also be kept in mind that linen was one of the prime products of Tarsus.[75]

Awnings, however, were not the tent-maker's only product. On certain occasions the inns of Rome were not capable of handling all those who flocked to the city. Thus in 45 BC when Julius Caesar celebrated the defeat of all his enemies by magnificent displays of all sorts, 'Such a throng flocked to all these shows from every quarter that many strangers had to lodge in tents pitched in the streets or along the roads' (Suetonius, *Caesar* 39. 4; trans. Rolfe).

Whether such tents were also of linen is an open question. Certainly a distinction must be made between the light linen of summer beach pavilions designed to provide shade without impeding the breeze,[76] and the much stouter linen used as sailcloth,[77] and for hucksters' booths.[78] The difference in weight

---

[71] *CIL* 6. 5183b, 9053, 9053a.

[72] The technique is illustrated in Macaulay (1974), 104–5.

[73] To discourage loiterers.

[74] The moss grew in a rectangular basin in the middle of the courtyard which collected rainwater. The covering prevented evaporation in the heat of summer.

[75] A guild of linen-workers is mentioned by Dio Chrysostom, *Discourses* 34. 21 and 23.

[76] Cicero, *Against Verres* 2. 5. 30 and 80.

[77] According to Pliny, 'Cleopatra had a purple linen sail when she came with Mark Antony to Actium, and with the same sail she fled' (*NH* 19. 22) after the victory of Octavian.

[78] 'In wintertime, when the arcades are crammed with canvas market-stalls' (Juvenal, *Satires* 6. 153–4). The reference is to the feast of the Saturnalia celebrated 17–19 December.

and flexibility between such canvas and leather is negligible, and the water-proofing of the latter is superior. It is doubtful, therefore, that leather tents were used exclusively by the military.[79]

Since it was to a tentmaker's economic advantage to be able to work all year and in all climates, it must be assumed that Paul was equally at home in sewing together strips of leather or different weights of canvas. There is little difference in technique in joining two thicknesses of leather or heavy canvas. It takes an awl to make the hole in a rolled-over canvas seam as it does in leather, and in both cases the curved needle must be slipped through before the hole closes.

With this silent skill Paul needed only a moon-shaped knife, an awl, needles, and waxed thread, and could be sure of finding jobs on every road he travelled and on every sea he sailed. He could reinforce a sail and remake the tents that passengers and crew used for shelter on deck and for accommodation on shore at night.[80] He could repair the canvas roof of a wagon or the harness of the draught animals. He could put a stitch or two in any of the multifarious articles of leather used by travellers, sandals, gaiters, belts, cloaks, and gourds.

Every town of any size had its festival for which booths and tents were necessary.[81] If a traveller timed his visit right, the local workshops would be glad of a skilled hand. At Corinth, for example, Paul found work with Prisca and Aquila (Acts 18: 3),[82] who catered to the perennial need for awnings, but who also profited from the fact that the biennial Isthmian Games meant continuous business in the repair and creation of tents.[83] There was no town at Isthmia, and the tents set up around the sanctuary of Poseidon catered for vast numbers of visitors from far and near, as well as for the hucksters of Corinth who went out to fleece them.[84]

Thus in terms of his missionary strategy Paul chose wisely. He acquired a skill whose products many needed. It enabled him to travel widely, although it would never make him rich, even though he worked 'night and day' (1 Thess. 2: 9; 2 Thess. 3: 9).[85] It enabled him to survive, but only barely, because he never stayed in one place long enough to build up a stable clientele. As his ministry ate into his time, subsidies became necessary (Phil. 4: 15–16; 2 Cor. 11: 9). But that day was still a long way in the future.

The disadvantage of Paul's choice was that it stigmatized him as belonging to a group, which was despised by a social class from which he had to recruit assistants. In a letter designed to contribute to the education of his son Marcus, then a student in Athens, Cicero speaks for the world in which Paul lived:

---

[79] *Pace* Lampe (1987), 256–61. Hock (1980), 20–1 is also too categorical in claiming that all tents were made of leather, and that Paul in consequence would be better described as a leather-worker.

[80] Casson (1979), 154.   [81] Ibid. 91.

[82] See Ch. 11, 'Working with Prisca and Aquila'.   [83] See my (1992e), 14–17.

[84] For a description of the crowds, see Dio Chrysostom, *Discourses* 8. 9.

[85] On the poverty of the artisan, see Hock (1980), 34–5.

In regard to trades and other means of livlihood, which ones are to be considered becoming to a gentleman and which ones are vulgar, we have been
taught, in general, as follows: First, those means of livelihood are rejected as
undesirable which incur people's ill-will, as those of tax-gatherers and usurers.
Unbecoming to a gentleman, too, and *vulgar are the means of livelihood of all
hired workmen whom we pay for mere manual labour, not for artistic skill; for in
their case the very wages they receive is a pledge of their slavery.* Vulgar we must
consider those also who buy from wholesale merchants to retail immediately;
for they would get no profits without a great deal of downright lying; and verily
there is no action that is meaner than misrepresentation. *And all artisans are
engaged in vulgar trades; for no workshop can have anything liberal about it.*
Least respectable of all are those trades which cater to sensual pleasures: 'Fish-
mongers, butchers, cooks, and poulterers and fishermen,' as Terence says. Add
to these, if you please, the perfumers, dancers, and the whole vaudeville crowd.
But the professions in which either a higher degree of intelligence is required or
from which no small benefit to society is derived—medicine and architecture,
for example, and teaching—these are proper for those to whose social position
they are appropriate.

(*De Officiis* 150–1; trans. Miller adapted; emphasis added)

The élite found various reasons to justify their prejudice. The bent-over position in which craftspeople worked indicated servility. Manual labour coarsened
not only the body but the spirit. Dirt under the nails stained the soul. Lack of
time excluded the acquisition of virtue or learning. Poverty engendered
venality.[86]

Naturally those who belonged to the working class did not think of themselves in this way. To work for reward was as integral to their self-understanding as their parentage. Their pride in their craft, however lowly, is underlined
by the fact that they often had it inscribed on their tombstones.[87] As one of
them, Paul had easy access to the vast majority of his fellow-citizens, but in
each city he also needed one or two from among the élite, if only to provide a
space large enough for the believers to meet. That he succeeding in impressing
such people—and at Corinth his first converts certainly belonged to the
élite[88]—says much about the quality of Paul's personality.

Ministry to Gentiles

Even if Josephus exaggerates the number of Jews slain at the outbreak of the
First Revolt in AD 66,[89] there can be little doubt that Damascus had a sizeable

---

[86] MacMullen (1976), 114–18; Hock (1980), 35–7.
[87] MacMullen (1976), 120. On retirement, scribes dedicated their equipment to Hermes (*Greek
Anthology* 6. 63 and 65).
[88] See Ch. 11, 'The First Converts'.
[89] The hesitation is due to the fact that one cannot harmonize the two figures he offers in different parts of the same work, 10,000 (*JW* 2. 561) and 18,000 (*JW* 7. 368).

Jewish population. None the less its ethos was essentially pagan.[90] As a founding member of the Decapolis, it was an independent Greek city whose culture was strongly Hellenized; its coins exclusively represent Greek deities. It owed its prominence and wealth[91] to its position at one of the great crossroads of the ancient world. The trade routes from Anatolia and Mesopotamia joined there before splitting again to go down the plateau into Arabia or out to the coast and south to Egypt. If in the second century BC its merchants went as far as Delos, we can be sure that those of other nationalities had a base in Damascus. In other words, Paul would have had little difficulty in fulfilling his missionary vocation in Damascus.

Presumably he preached during his trade apprenticeship as the opportunity arose. A three-year stay would suggest that he found plenty of work to do, and he must have been extremely disconcerted when the Nabataeans assumed control of the city in the latter part of AD 37[92] and moved to arrest him. Why they bothered remains a mystery. The lapse of time had made it clear that the Romans had no intention of bringing Aretas to book for the war against Herod Antipas. There is no mention in Josephus of any Nabataean reprisals against the Jewish community. More importantly, the emperor Tiberius was dead and his successor Gaius was their friend. All the anxieties which had led the Nabataeans to see the Apostle as a threat had dissipated. Perhaps Paul exaggerated the danger!

JERUSALEM

In any case Paul fled Damascus, never to return. The journey to Jerusalem would have taken about a week. This time there is no mystery about his purpose; he came *historêsai Kêphan* (Gal. 1: 18). Two translations are possible: (a) 'to visit Cephas';[93] 'to get acquainted with Cephas';[94] and (b) 'to get information from Cephas'.[95] The majority opt for the former rendering on the grounds that Paul's defence in Galatians demands that he show complete independence of Jerusalem. While this was certainly Paul's objective in the letter, I suspect that Paul deliberately chose an ambiguous term,[96] because one cannot imagine

[90] See Schürer (1973–87), 2. 36–7, 127–30.
[91] Strabo calls it 'remarkable, noteworthy', and mentions merchants from Arabia Felix (*Geography* 16. 2. 20). Its best-known citizen at this period was Nicolaus of Damascus, friend of Herod the Great, and Josephus' best source; see Wacholder (1962) and Schürer (1973–87), 1. 28–32.
[92] See Ch. 1, 'Date of Departure from Damascus'.
[93] Betz (1979), 76.
[94] Longenecker (1990), 37.
[95] Kilpatrick (1959), 144–9.
[96] So rightly Dunn (1992), 73, 'Paul was evidently concerned neither to claim too much nor to deny too little.'

that Paul's conversation with Peter focused exclusively on the weather,[97] the health of the latter's mother-in-law, or his nostalgia for fishing on the Sea of Galilee!

Knowledge of the Historical Jesus

It takes neither imagination nor intelligence to recognize how Paul must have reacted in the presence of one who had lived with Jesus from the time that both were disciples of John the Baptist. The centrality of Christ in Paul's conversion experience and his theology, and the natural curiosity engendered by the hints he picked up during his three years in the Christian community at Damascus, make it extremely improbable that he did not avail himself to the utmost of Peter's knowledge of the historical Jesus. At this point Peter had been preaching for seven years, and through repetition his story would inevitably have acquired the fixed form of a gospel, with a beginning, middle, and end. Having lived for two weeks with the prime eyewitness of the earthly ministry, Paul certainly learnt much about the historical Jesus.[98]

A number of features in his letters tend to confirm this conclusion. The historical Jesus is fundamental to Paul's theology. The disciple who wrote Ephesians caught the Apostle's approach perfectly when he presents Jesus as the truth of Christ (Eph. 4: 21).[99] When his converts attempted to separate the Christ of faith from the Jesus of history, Paul resisted by insisting that the Lord of Glory was the crucified Jesus (1 Cor. 2: 6), and by stressing that Christ had been received 'as Jesus the Lord' (Col. 2: 6).[100] The implication that Paul preached the historical Jesus is formally confirmed by his condemnation of anyone 'who preaches a Jesus other than the one we preached' (2 Cor. 11: 4).[101]

There are two references to sayings of Jesus in First Corinthians, the prohibition of divorce (7: 10–11) and the directive concerning the livelihood of pastors (9: 14). It is emphasized by some that these are not direct quotations but rather allusions or reminiscences. This is done in order to bring them into line with the rest of Paul's correspondence, where the situation has been rather precisely described by F. Neirynck, 'Possible allusions to gospel sayings can be noted on the basis of similarity of form and context but a direct use of a gospel saying in the form in which it has been preserved in the synoptic gospels is hardly possible.'[102] The negative thrust of such a judgement should not be exaggerated.

---

[97] According to Lüdemann (1984), 70, 'W. D. Davies once picked up on a statement by C. H. Dodd and remarked humorously, "Certainly Paul and Peter did not spend their time talking about the weather".'
[98] Similarly Dunn (1985), 138–9.
[99] See de la Potterie (1963).
[100] Abbott (1897), 244; Lightfoot (1904), 174.
[101] See my (1990).
[102] (1986), 320.

Formally attributed direct quotations were the exception rather than the rule in the age and world in which Paul lived. Use acknowledged value; one borrowed only from the rich. One should expect, therefore, that if Paul knew the teaching of Jesus it would have informed the Apostle's thought to the point where any distinction of source and personal elaboration would be, not only impossible, but meaningless.

Recent studies, moreover, suggest that Paul knew not just the dominical saying but the context in which it appears in the synoptic tradition.[103] One example must suffice. The theme of the support of pastors appears in Luke 10 and it has been shown that this chapter is linked to 1 Corinthians 9 by a whole series of shared terms: an 'apostle' who is ('to sow' and) 'to reap' has the 'right' to a 'reward' for his 'preaching the good news' because a 'workman' has a right 'to eat' and 'to drink'. The contacts are too numerous to make coincidence a credible explanation, particularly since the same type of contacts are to be found in other blocks of material.[104] The influence of the historical Jesus on the Pauline parenetic tradition has also been demonstrated in Romans. 'The echoes of the Jesus tradition are not all of the same strength, but together they build into an impressive case for saying that Paul must have known a substantial amount of the Jesus tradition which was later committed to the present Gospel form by the Evangelists.'[105]

It has also been pointed out that, although Pharisaism was essentially an urban movement and Paul a city man, the Apostle uses an unusually high proportion of metaphors which reflect a rural environment and an agrarian culture.[106] H. Riesenfeld has persuasively argued that these show that Paul was familiar with the language of Jesus' parables, because the contacts are too specific to be explained by common dependence on the Old Testament.[107]

Yet when we come to tabulate the references to the historical Jesus in the Pauline letters all we learn is that he was a Jew (Rom. 9: 4–5) of the line of David (Rom. 1: 3), who had a mother (Gal. 4: 4), who was betrayed (1 Cor. 11: 23) and crucified (1 Cor. 2: 2 and *passim*), as a result of which he died and was buried (1 Cor. 15: 3–4).

The meagreness of this result and the obscurity of the allusions have led many to deny that Paul had any detailed knowledge of the gospel tradition. They argue that had Paul known any more about the historical Jesus, he would have used it. This argument from silence only looks strong. It is meaningless without the unprovable and unwarranted assumption that Paul would have reacted in the same way as we would, if we had access to first-hand information

---

[103] Dungan (1971).

[104] Fjärstedt (1974). See also Allison (1982).

[105] Dunn (1989), 205.

[106] Rom. 1: 13; 6: 21; 7: 4–5; 11: 17–24; 15: 28; 1 Cor. 3: 6–9; 9 (*passim*); 15: 36–44; 2 Cor. 9: 6–10; Gal. 5: 22; 6: 7–9; Phil. 1: 22; 4: 17.

[107] (1960), 47–59.

about the historical Jesus. The real question is, why did not Paul display all the knowledge he had? The search for an answer leads to important insights into his Christology.[108]

A Missionary Agreement

We have seen that Paul's conversion experience, in addition to a new vision of Jesus of Nazareth, also embodied the conviction that hitherto his life was to be dedicated to preaching the gospel to the Gentiles. One would expect this issue to have surfaced also in his discussions with Peter. The present thrust of Galatians is no objection, because it is evident that Paul only subsequently transformed his original *de facto* independence of Jerusalem into a matter of principle.

Confirmation of this obvious assumption has been sought by G. Lüdemann who argues that the phrase, 'just as Peter had been entrusted with the gospel to the circumcised, for he who worked through Peter for the mission to the circumcised worked through me also for the Gentiles' (Gal. 2: 7b–8), refers to an agreement made between Peter and Paul on the occasion of the latter's first visit to Jerusalem (Gal. 1: 18).[109]

These verses have always been a problem to commentators because they stand in tension with their context. Verses 7b–8 use 'Peter', whereas in the context he is called 'Cephas' (v. 9, 11, 14; cf. 1: 18). In verses 7b–8 'Peter' is the sole authority figure to negotiate with Paul, who is alone, but in verse 9 the latter is accompanied by Barnabas, and 'Cephas' is but number two in a triumvirate led by James. Exegetes aware of these tensions, who have also observed that 'knowing the grace given me' (v. 9a) is a *reprise* of 'seeing that I have been entrusted with' (v. 7a), postulate that Paul is here quoting part of the minutes of the Jerusalem Conference.[110] Lüdemann with eminent common sense points out that such an official report would not use the first-person singular and would certainly mention Peter before Paul.[111] Equally untenable is the solution proposed by G. Klein, namely, that verse 7 reflects the situation in Jerusalem at the time of the Conference, while verse 9 was added at the time of the composition of Galatians in order to take into account the change in leadership in the Jerusalem church which had taken place in the interval.[112] How would Paul have known of the change? And, if he did, why would he have bothered to note it, particularly in a situation where he was desperately trying to prove his independence of Jerusalem?

---

[108] See my (1982*b*), 33–57.
[109] (1984), 64–71. Similarly A. Schmidt (1992), 149–52.
[110] e.g. Cullmann (1953), 20; Dinkler (1953), 182–3.
[111] (1984), 68.
[112] (1960), 286–7.

The hypothesis that verses 7b–8 refer, not to the Conference whose conclusion is described in verse 9, but to a different meeting, is the only explanation that does full justice to the evidence.[113] When did this meeting take place? It cannot have taken place after the Conference. That would have been pointless, since Peter had been a party to the decision of the Conference. Moreover, a comparison of Galatians 1: 18–19 with Galatians 2: 9 shows that James had superseded Peter within the Jerusalem church—a shift in the authority structure which is confirmed by Galatians 2: 12 and Acts—so that after the Conference Peter was not in a position to act alone. Hence the meeting with Peter in verse 7b must have taken place before the Conference.

In this span of time, however, we have at most three possibilities. Galatians 2: 2 may evoke two meetings,[114] but both can be excluded immediately, because in each case Paul was confronted by a group, not a single individual as in verse 7b.[115] Thus we are forced to locate it during Paul's first visit to Jerusalem. There is no other known possibility within the framework of the letters.[116]

Why did Paul transpose an agreement made during his first visit to Jerusalem (Gal. 2: 7–8) into an account of his second visit (Gal. 2: 1–10)? He gained a number of significant advantages.[117] First, by separating the fact of the first meeting (Gal. 1: 18) from its content (Gal. 2: 7–8) he avoided giving the impression that the missionary work done subsequent to his conversion was carried out under the aegis of a Jerusalem commission, and he was able to fix in the mind of his readers the value of that first meeting as a purely exploratory encounter; the ambiguity of the *historeô* has already been noted. Secondly, by juxtaposing the contents of the two meetings he managed to insinuate that the equality that emerged from the Conference (Gal. 2: 6, 9) was also true of the first meeting, where Paul certainly lacked the authority in the church which Peter enjoyed.[118]

Practice was the basis of Paul's agreement with Peter at this initial meeting in the autumn of AD 37. Peter had in fact been preaching to Jews, just as Paul had been preaching to Gentiles in Arabia and Damascus. What success they had, they attributed to God; note the formulation of Galatians 2: 8, 'because he who was *at work in Peter* for the apostolate to the circumcised also *worked in me*

---

[113] Compare the vacillation of Betz (1979), 97–8; Longenecker (1990), 55–6.

[114] Betz (1979), 86; Longenecker (1990), 48.

[115] Against Légasse (1991), 79 n. 14.

[116] According to the present text of Acts, Paul made six visits to Jerusalem (9: 26–8; 11: 29–30; 12: 25; 15: 1–2, 11; 18: 22; 21: 1, 5–17). The first and the last can be equated with Gal. 1: 18 and Rom. 15: 25, respectively. Source criticism suggests that the other four are all to be equated with Gal. 2: 1; see Benoit (1959); Lüdemann (1984), 149–57.

[117] Quintilian denies that the 'statement of facts' (*narratio*) should always depict the real order of events; the decision should be based on what is most advantageous for the defendant in the circumstances and nature of the case (*Institutio Oratoria*, 4. 2. 83–4).

[118] The difference between the two is in fact underlined by the attribution of 'apostleship' to Peter but not to Paul (Gal. 2: 8).

for the Gentiles'. The initial 'because' justifies the divine passive of the preceding verse, 'I was entrusted with the gospel,' and explains why Paul says that his responsibility was visible, 'they saw' (Gal. 2: 7). Each ministry was authenticated in the same way, namely, by the effectiveness of grace, the power of God made visible. Only later did such simplicity yield to the complications of institutionalization.

THE MISSING YEARS

A Gap in the Record

From Jerusalem Paul went to 'the regions of Syria and Cilicia' (Gal. 1: 21). One can deduce that he visited Antioch on the Orontes, the capital of the Roman province of Syria, and his home town of Tarsus in Cilicia. From this point there is an eight-year gap in the record, i.e. until early in AD 46 when the letters again furnish us with information about Paul's career.[119]

In Acts this gap is to some extent filled by a mission of Paul and Barnabas in Cyprus and southern Asia Minor (Acts 13–14). While some scholars dismiss this mission as a fictional creation of Luke, the majority tend towards the view that his source contained at least a list of the places visited by Paul and Barnabas.[120] The basis for this opinion is the allusion to 'what befell me at Antioch, at Iconium, and at Lystra' in 2 Timothy 3: 11, which at best is confirmation by Paul himself[121] of a mission in southern Asia Minor (Acts 13: 14; 14: 1, 8), or at worst a Pauline tradition completely independent of Acts.[122]

The letters unambiguously confirm that at one time Paul worked with Barnabas. They appear together at the Jerusalem Conference (Gal. 2: 1) as representatives of the mission to the Gentiles (Gal. 2: 9). The fact that Paul singles out Barnabas in 1 Corinthians 9: 6 ('Is it only Barnabas and I who have no right to refrain from working for a living?'), even though he did not accompany Paul to Corinth (2 Cor. 1: 19; Acts 15: 36–41), underlines the position of Barnabas as a senior well-known missionary whose status was in some way comparable to that of other apostles, the brothers of the Lord and Cephas (1 Cor. 9: 5). The nature of the allusion suggests that Paul knew Barnabas' attitude towards financial support because they had acted in the same way in the same circumstances.

Paul's treatment of the incident at Antioch (Gal. 2: 11–21) suggests that both he and Barnabas were members of that community.[123] They are simply

---

[119] See Ch. 1, 'Prior to AD 57'.
[120] See Haenchen (1971), 438–9.
[121] See the discussion of the authenticity of 2 Tim. in Ch. 14.
[122] See Lüdemann (1984), 180 n. 2.
[123] Gal. 2: 11–21 is dealt with more fully in Ch. 6, 'The Incident at Antioch'.

there, whereas the 'coming' of Peter and the 'arrival' of the emissaries of James are explicitly mentioned (Gal. 2: 11–12). There is also a slight hint in the letters that Paul's early missionary journeys were undertaken as an emissary of Antioch, as Luke suggests (cf. Acts 13: 1–3). It has already been pointed out that the lack of any self-justification in 1–2 Thessalonians dates them to a period when Paul enjoyed the security of a missionary mandate, which can only be that of Antioch.

The initiative displayed in his abortive expedition into Arabia (Gal. 1: 17) makes it impossible to assume that Paul did not obey the imperative of his commissioning conversion in Syria, Cilicia, and elsewhere. His letters show that he considered himself the solely responsible for the communities he founded during the great journey which began at the latest in AD 46. The necessary implication of this banality is that Paul did not feel the same sort of responsibility for other converts made previously to whom he did not write. Why not? The obvious answer is that he was sure that someone else had the responsibility and was faithfully exercising it. This means that during the years AD 37–46 the Apostle's position resembled that of Timothy when he was Paul's assistant. Who, then, was in charge? On the basis of the letters the only name which can be suggested is that of Barnabas.

These epistolary clues of somewhat unequal value can be combined to create the following picture of some, if not all, of Paul's career in the years AD 37–46. He joined the community at Antioch, whence he was sent as assistant to Barnabas to evangelize the southern part of Asia Minor, notably the cities of Antioch in Pisidia, Iconium and Lystra.[124] The mission would have been completely in accord with his own understanding of his responsibility to the Gentiles. Unless Barnabas established a rhythm notably slower than that of Paul when he became independent, this mission would have occupied at most two years. It is not impossible, however, that the mission field may have been more extensive than Luke suggests. But where precisely the rest of the time was spent we shall never know.

Dangers on the Road

Of two things, however, we can be certain. Paul was not idle and he kept moving. Some idea of the distances he covered has been given in Chapter 1, but of his experiences on the road Paul gives but a very summary account, 'During my frequent journeys I have been exposed to dangers from rivers, dangers from brigands, dangers from my own people, dangers from Gentiles, dangers in the town and in the country, dangers at sea, dangers at the hands of false brothers' (2 Corinthians 11: 26).[125] The conditions of travel in the world of his day are

---

[124] Thus there is a definite historical basis to Luke's account in Acts 13–14.
[125] Translation from Martin (1986), 378.

well documented,[126] and permit us to amplify his hints. The point is not merely of historical interest. The integration of his experience into a world-view had a significant impact on his theology.

The surprising feature of 2 Cor. 11: 26 is the emphasis, not on difficulty, but on 'danger'. The element of risk, not struggle, is evidently uppermost in his mind. There would have been little peril from rivers on the great Roman arterial roads furnished with bridges.[127] Secondary roads were another matter. In the east they were built for the long dry season, when stream beds had little or no water. In the spring the runoff of the winter rains turned crossings into dangerous fords whose violence can still be experienced on the banks of the Dead Sea and in the Arava valley.

Such danger, however real, was sporadic. Robbers were a much more consistent threat. Casson reflects a rather common but mistaken view in writing, 'the through routes were policed well enough for him [any traveller] to ride them with relatively little fear of bandits. . . . Wherever he went, he was under the protective umbrella of a well-organized, efficient legal system.'[128] In Italy, on the contrary, as a consequence of the anarchy of the civil wars, brigandage was endemic, and travellers faced the additional risk of being shanghaied as slaves by owners of land bordering the roads. Augustus reacted by stationing troops on the roads,[129] and by delegating Tiberius to inspect the slave-prisons to ensure that no freemen were being held.[130] When the latter became emperor he had to concentrate garrisons even closer together in order to make the roads safe.[131] Yet at the end of the first century, Pliny the Younger could write of the disappearance without trace of a Roman knight and of a centurion and their parties on main roads in Umbria as if it were a not unheard of occurrence.[132]

If such was the case at the centre of the empire, one can infer with a high degree of probability that conditions were much worse in distant provinces.[133] The legions stationed in Syria, Asia, and Macedonia did not double as police forces in anything like the modern sense. Detachments might be sent to deal with a particularly troublesome robber band, and even then not as a matter of policy, but in response to pressure from an influential person.[134] In senatorial

---

[126] The basic study remains Casson (1974). It should be supplemented, however, by Millar (1981), to which I am heavily indebted. See also André and Baslez (1993). Graphic illustration of how little things have changed in two thousand years is provided by Lithgow (1974). Much of the some 36,000 miles (57,600 km.) which the unprivileged Lithgow (1582–1654) covered alone and on foot was in Pauline territory, and his experience reflects the lot of those who travelled without the protection of rank and/or wealth.

[127] Chevallier (1972), 103–15.

[128] (1974), 122.

[129] Suetonius, *Augustus* 32. 1.

[130] Suetonius, *Tiberius* 8.

[131] Ibid. 37. 1.

[132] *Letters*, 6. 25

[133] For for a catalogue of horrors in Mysia, see S. Mitchell (1993), 1. 166.

[134] Apuleius, *Metamorphoses* 7: 7.

provinces the proconsul normally had at his disposition some auxiliary units,[135] which accompanied him as he travelled throughout his territory. In principle a proconsul was expected to hold court in various cities within his charge during his year of office, but the size of provinces and the length of proceedings ensured that even an energetic and competent administrator (a rare species) could intervene only sporadically and in the more important centres, where, we can safely assume, access was monopolized by prominent figures.[136] The poor had no recourse,[137] and the vast majority of small towns and villages never saw a Roman official. Apuleius catches the reality of the situation in the warning a friend gives Lucius in Hypata (modern Ipati) in Thessaly,

> Don't stay too long at the party. Come back as soon as you can, for in the early hours Hypata is terrorized by a gang of young thugs who think it amusing to murder whoever happens to be passing by, and to leave the streets strewn with corpses. They are members of the first families in town, and the Roman barracks are so far away, so nothing can be done to end the nuisance.
>
> (*Metamorphoses* 2. 18; trans. Graves adapted)

Security in the countryside was much worse. Poverty forced many into brigandage.[138] Any robber anywhere could be sure that all travellers had money on them; they had to pay their way and no credit cards or cheque books were available.[139] Those obliged to travel alone did so with fear and trembling, certainly in lonely wooded stretches of the road,[140] where in Greece wild animals were as much a danger as bandits. Apuleius mentions bears (*Metamorphoses* 4. 13; 7. 24), wild boar (8. 4), but reserves his goriest language for wolves,

> The authorities requested us not to continue our journey that night or even the following morning, because the district was overrun by packs of enormous wolves, grown so bold that they even turned highwaymen and pulled down travellers on the roads or stormed farm-buildings, showing as little respect for the armed occupants as for their defenceless flocks. We were warned that the road we wished to take was strewn with half-eaten corpses and clean-picked skeletons and that we ought to proceed with all possible caution, travelling only in broad daylight—the higher the sun the milder the wolves—and in a compact body, not straggling along anyhow. (*Metamorphoses* 8. 15; trans. Graves)

Travellers voyaged in groups whenever possible and, given the double danger, it seems reasonable to assume that many were armed, at least with staves.[141]

---

[135] Ritterling (1927).

[136] Burton (1975).

[137] This is wittily brought out by the ass Lucius, who can only pronounce the 'O' of the introduction to his appeal 'O Caesar' (Apuleius, *Metamorphoses* 3. 29).

[138] Ibid. 4. 23. On brigandage in general, see MacMullen (1966), 255–68.

[139] Horace, *Epistles* 1. 17. 52–3.

[140] Apuleius, *Metamorphoses* 1. 7.

[141] Ibid. 8. 16.

This meant that they could be seen as a threat by any village they approached. The consequences were predictable.

> When we reached a small village, the inhabitants very naturally mistook us for a brigade of bandits. They were in such alarm that they unchained a pack of large mastiffs which they kept as watch-dogs, very savage beasts, worse than any wolf or bear, and set them at us with shouts, halloos and discordant cries. (*Metamorphoses* 8: 17; cf. 9: 36; trans. Graves)

The villagers had to rely on themselves for protection. If self-help was not sufficient, only neighbours could be relied on for aid.[142] Slaves could loot the house of their dead owners, and escape retribution by moving to another town.[143] Inevitably all strangers came under suspicion, whose degree was the inverse of the size of the place.[144] This was not xenophobia but the fruit of hard experience.

A moment's absence from a cottage carried the risk of pilferage, and those who tried to defend their poor possessions endangered their lives.[145] Though protected by walls, stout gates, and numerous slaves, who could be armed in an emergency, large houses were not immune. Thieves used all sorts of tricks to infiltrate the premises, both to spy out where valuables were stored and to open doors for accomplices.[146] The alternative was a silent forced entry.[147] When they thought they could get away with it, brigands in a frontal attack simply broke down the main gate, and held off the inhabitants with swords while looting the house.[148]

Inns were even less secure. Across the Roman province of Asia, they were spaced a day's journey apart—25 Roman or 22 English miles (35 km.)—with a small establishment (*mutatio*) where dispatch riders could change horses roughly halfway between two inns.[149] The rooms were grouped around three or four sides of a courtyard with public rooms on the ground floor and sleeping accommodation above.[150] Those with money to spend could buy privacy, but those with slender purses had to share a room with strangers; how many depended on the number of beds the landlord could cram in, or on his or her attitude to guests sleeping on the floor.[151] Unless they wanted to cart their baggage with them, guests had to leave it unguarded while they visited the baths and a restaurant.[152]

The ease of theft needs no emphasis. Roman legislation made innkeepers responsible for the acts of their employees,[153] but not all guests were honest,

---

[142] Ibid. 4. 3, 10
[143] Ibid. 8. 15–23.
[144] Ibid. 7. 23.
[145] Ibid. 4. 12.
[146] Ibid. 4. 14–18; cf. 7:1.
[147] Ibid. 4. 9
[148] Ibid. 3. 28.
[149] *Itinerarium Burdigalense* 571–81.
[150] Casson (1974), 201–3.
[151] *Acts of John* 61–2 in Hennecke and Schneemelcher (1965), 2. 243–4.
[152] Apuleius, *Metamorphoses* 1. 24.
[153] Casson (1974), 205.

and in a crowded room at night one had only to stretch out a hand to appropriate something from another's baggage. If an inn was isolated and the bandits numerous, they did not hesitate to attack it.[154]

Each town, of course, had its magistrates who were responsible for public order, and who carried out their duty through servants of the court.[155] Only the wealthy, however, could be elected to municipal offices. In smaller towns, therefore, office circulated among the dominant families, who effectively ran such towns in their own interest, which could of course be directed by public demonstrations. As F. Millar notes, 'The cities ran themselves. Or rather—and this is one of the most vivid impressions left by the novel [The *Metamorphoses* of Apuleius]—they were run by a network of local aristocratic families, whose doings, public and private, were the subject of intense observer participation—approbation, curiosity, indignation, incipient violence—on the part of the lower classes of the towns.'[156] The latter appear as victims of the violence of the ruling élite, who in this respect had no interest in restraining its own members. The wealthy could rob,[157] or murder with impunity.[158] If theft from their constituents demanded a victim, any outsider would do.[159]

This brief and generalized description is valid for that part of the Graeco-Roman world in which Paul was active, namely, the provinces of Syria, Asia, Macedonia, and Achaia. It takes little imagination to visualize the tension set up within him by such an environment. His conversion had made him a follower of Jesus who had given his life for the salvation of humanity. That totally other-directed mode of existence became Paul's ideal. His goal was to make it transparent in and through his own comportment, 'always carrying in the body the dying of Jesus so that the life of Jesus may be manifested in our bodies' (2 Cor. 4: 10).[160] Yet every road he travelled forced him to worry about his personal safety. Every inn he visited obliged him to consider others as potential thieves, at least in so far as he had to take measures to protect the precious tools on which his livelihood depended. Circumstances conspired to push the self to the centre of his consciousness, whereas he wanted to be totally focused on the other. His life became a perpetual struggle against the insidious miasma of egocentricity.

It is in this tension that we find the roots of Paul's concept of Sin. When he says 'all, both Jews and Greeks, are under (the power of) Sin' (Rom. 3: 9) he is obviously speaking of something other than personal sinful acts, hence my use of the capital letter in translating *hamartia*. This inference is confirmed by a

[154] Apuleius, *Metamorphoses* 7. 7.
[155] Apuleius mentions 'a captain of the night watch' (*Metamorphoses* 3. 3).
[156] (1981), 69.
[157] Apuleius, *Metamorphoses* 9. 35.
[158] Ibid. 2. 18.
[159] Ibid. 3. 28; 7. 11–2.
[160] For further details, see my (1990).

series of other texts in which Sin is personified.[161] Manifestly Sin in these texts is a symbol or a myth expressive of a world in which individuals were forced to be other than they desired to be; the authentic self was alienated (Rom. 7: 20). From his own experience as a travelling missionary, Paul learned that people were not selfish because they chose to be. They were forced to be egocentric in order to survive. Their pattern of behaviour was dictated by irresistible societal pressures. They were controlled by a force greater than any individual, namely the value system which had developed within their society. The power of system became clear to Paul in the difficulty he experienced in being true to himself as the model of Jesus Christ (1 Cor. 11: 1). Hence his anguished cry, 'Who is not weak, and I am not weak? Who is made to fall, and I do not burn with anger?' (2 Cor. 11: 29).

---

[161] See Ch. 13, 'Sin, Law, and Death'.

— 5 —

Learning with the Thessalonians

In our discussion of the chronology of Paul's life, his initial ministry in Macedonia was dated between September 48 and April 50, the two years being equally divided between Philippi and Thessalonica.[1] There I emphasized that one year in each place was the absolute minimum; it is entirely possible that Paul stayed longer. The two Macedonian churches were perhaps the communities that gave Paul the greatest happiness. The divisions which marred other foundations were virtually non-existent. More importantly, the quality of their communal life made them stand out as beacons of life and hope (1 Thess. 1: 6–8; Phil. 2: 14–16). They were apostolic in precisely the way he wished, and generous to a fault (2 Cor. 8: 1–4). If they had assimilated his teaching so well, he must have spent a considerable time among them.

This fact is the decisive refutation of Luke, for whom Paul spent only three weeks in Thessalonica (Acts 17: 2) before being hounded out by Jewish agitators (Acts 17: 5–10). In addition to the considerations just mentioned, such a short visit is explicitly contradicted by the character of his initial letter (1 Thess. 2: 13–4: 2). The profound affection and confidence therein displayed argues unambiguously for a prolonged acquaintance.[2] This is confirmed by Philippians 4: 16 in which Paul thanks the Philippians for the financial aid they sent him more than once after he had left them to minister in Thessalonica.[3] His gratitude for such subsidies, despite the fact that he had found work (1 Thess. 2: 9; 2 Cor. 3: 8), indicates an extended and successful ministry. The hours he had to devote to new converts eroded his income, and to live he needed financial support from elsewhere.

Philippi and Thessalonica were linked by the Via Egnatia, which ran from the Adriatic Sea to the Bosphorus (see Fig. 7).[4] Leaving Philippi Paul followed it to the west, as did the Bordeaux Pilgrim on his way back from the Holy Land in AD 333. The latter gives the distances in Roman miles, as does the *Antonini*

---

[1] See Ch. 1, 'Prior to AD 51'.

[2] So rightly Kümmel (1975), 216.

[3] Morris (1956) has argued, but unconvincingly, that *kai hapax kai dis* does not imply more than one subsidy to Paul in Thessalonica.

[4] Strabo, *Geography* 7. 7. 4; Oberhummer (1905); Radke (1973), 166–7; O'Sullivan (1972).

Itinerarium, but in the opposite direction.[5] The following list summarizes their data:

Antonini Itinerarium	Bordeaux Pilgrim
Philippis	civitas Philippi
	12
	mutatio Ad Duodecimum
30	7
	mutatio Domeros
	13
Amphipoli	civitas Amphipholim
	10
	mutatio Pennana
17	10
	mutatio Peripidis
	11
Apollonia	mansio Appollonia
20	11
	mutatio Heracleustibus
Mellissurgin	14
27	mutatio Duo dea
	13
Thessalonica	civitas Thessalonica

Manifestly the list of the Bordeaux Pilgrim is the more complete, because it told the dispatch riders where they could change horses (a *mutatio*) and ordinary travellers where they could find lodging for the night (in a *civitas* or a *mansio*). His distances are also more realistic.[6] Normally one would expect to get from one *civitas/mansio* to another in one day's march. As a general rule only one *mutatio* intervenes. In this section there are two, giving stages of 32, 31, and 38 Roman miles. These are unusually long, as are three others going west out of Thessalonica, and may suggest that travellers were expected to make good time on this part of the road. The Philippians, therefore, were only three days' good walking away from Paul in Thessalonica. To stay in touch and and to respond to his needs was not difficult. A round trip could be made in a week.

With Paul's ministry in Thessalonica I shall deal in a moment, because it is known only as reflected in letters written later. Here the question of why Paul left Thessalonica demands attention. Luke blames it on the hostility of the

---

[5] Kuntz (1929).
[6] He gives the distance between Philippi and Thessalonica as 101 Roman miles (against 94 for *Ant. Itin.*); on the modern road it is 166 km. (103 miles). In opposition to the *Ant. Itin.* which gives 17 miles for one section and 47 for another, the Bordeaux Pilgrim correctly shows that Amphipolis and Appollonia broke the route into three roughly equal parts.

Jewish community there, but this cannot be taken at face value. It could be true, but Luke elsewhere invokes it as an explanation when it is demonstrably false, e.g. Paul himself tells us that the Nabataeans caused his hasty departure from Damascus (2 Cor. 11: 32–3), but for Luke the Jews were responsible (Acts 9: 23–5).[7] Moreover, the only hostility documented by Paul in Thessalonica came from Gentiles. The majority of the community was of pagan origin ('you turned to God from idols', 1 Thess. 1: 9), and the persecution they suffered originated with 'your own countrymen' (1 Thess. 2: 14).[8] The implication of 1 Thess. 3: 1–3 is that this persecution had begun before Paul left, and it is the simplest explanation of his departure.

It is not impossible, as some have argued, that when Paul moved west along the Via Egnatia from Philippi to Thessalonica his plan was to follow the great highway to its terminus on the Adriatic coast where it would have been easy to find a boat for Italy and Rome (Rom. 15: 23).[9] Had he to flee because of trouble with the municipal authorities in Thessalonica, however, it would have been most imprudent to stay on the main road. He needed to get to an area where the municipal writ did not run, particularly if the authorities had invoked Roman assistance.

Strabo's description of Macedonia in book 7 of his *Geography* is preserved only in fragments, in one of these (frag. 10) the Via Egnatia is identified as the southern border of the Roman province. This is certainly wrong. A whole array of indicators combine to fix that border on average some 70 km. south of the Via Egnatia.[10] The most efficient escape route for Paul would have been a ship to bring him south into a different Roman province, Achaia. Thus there may be a historical reminiscence in Luke's note that Paul headed for the coast after further trouble in Beroea (Acts 17: 14).[11] Be that as it may, Paul safely made it to Athens (1 Thess. 3: 1; Acts 17: 15), where he was able to take breath and to reflect on the situation of his converts at Thessalonica.

CONTACTS WITH THESSALONICA

The New Testament contains two letters to the Thessalonians. Some scholars, however, claim they were originally one missive, whereas others detect four

[7] See the discussion in Ch. 1, 'Date of Departure from Damascus'.
[8] For reasons behind the persecution, see p. 118 below.
[9] Bruce (1982a), p. xxvi.
[10] See the discussion and map in Papazoglou (1979), 334–7; Hammond (1974).
[11] It is not difficult to develop a speculative justification for the Lucan itinerary after Thessalonica. Paul could have set out on the Via Egnatia, but decided that it was too dangerous, and after Pella took the road south to Beroea (modern Veroia), whence there was a road which reached the coast at modern Katerini.

letters.[12] I am obliged, therefore, to try to justify my own view of the letters because it underpins my historical reconstruction.

A characteristic feature of all Pauline letters is the thanksgiving; it immediately follows the address and subtly and indirectly introduces the main themes of the letter.[13] 1 Thessalonians stands out among the letters of Paul in that it contains two thanksgivings.[14] Even a quick reading reveals the close similarities between 1: 2–10 and 2: 13–14. Schubert disputes the fact that there are two thanksgivings, maintaining that formally the one thanksgiving of 1 Thessalonians runs from 1: 2 to 3: 10, and in fact constitutes the body of the letter.[15] Not only is his argument from form confused and implausible, but it has been contradicted on both formal[16] and rhetorical grounds.[17] Moreover, the vast majority of commentators, who divide the letter on thematic principles, limit the first thanksgiving to 1: 2–10.[18] Finally, if 1: 2 to 3: 10 was in fact intended to be an unusually long thanksgiving, one would expect Paul to heap up different reasons for gratitude. This is not the case, which doubles the abnormality of the reiterated thanksgiving form.

Schmithals also discerned two conclusions in 1 Thessalonians, namely 3: 11 to 4: 2 and 5: 23–8.[19] In fact the former contains a number of elements normally found in the concluding portion of a Pauline letter. The desire to see the recipients (3: 11) is found in 1 Cor. 16: 5; 2 Cor. 13: 10; and Philem. 22. The prayer in 3: 12–13 has a very close parallel in 5: 23. The phrase 'finally believers' (4: 1) is evocative of 2 Cor. 13: 11; Phil. 4: 9; 2 Thess. 3: 1; cf. Gal 6: 17.

Such data permits only one conclusion. An originally independent letter, 2: 13 to 4: 2 with its own thanksgiving and conclusion, has been inserted into another letter constituted by 1: 1 to 2: 12 and 4: 3 to 5: 28. In other words, 1 Thessalonians as it now stands is a compilation of two letters. This phenomenon is not unique among the Pauline letters, it is also to be observed in Philippians (three letters) and 2 Corinthians (two letters).[20] Inevitably when

[12] The opinions are surveyed by Jewett (1986), 19–46. Malherbe (1990) finds no evidence that Paul sent a letter to Thessalonica with Timothy, but considers it probable that the latter brought one back to Athens.

[13] The classic study remains Schubert (1939). On 1 Thess. in particular see Lambrecht (1990).

[14] This point has been argued most effectively by Schmithals (1964), 298. Pearson (1971) has attempted to prove that 1 Thess. 2: 13–16 was not originally part of the letter. His arguments, however, have force only for vv. 15–16, and they have not convinced everybody; see W. D. Davies (1977), 6–9; Donfried (1984), 242–53; Gillard (1989).

[15] (1939), 26. He is followed by Jewett (1986), 32, whose only objection to Schmithals is the unmotivated assertion that 1 Thess. 2: 13 is 'an integral portion of the lengthy but formally normal thanksgiving that extends from 1 Thess. 1: 2 to 3: 13'.

[16] J. T. Sanders (1962), 355.

[17] Kennedy (1984), 142; Hughes (1990), 109; Wuellner (1990), 128–9

[18] Jewett (1986), 216–18.

[19] Critics of the compilation theory often forget this crucial point. Their exclusive focus on the thanksgiving is misplaced, because two thanksgivings alone do not indicate two letters, which is why Schmithals's proposal for 2 Thess. is unacceptable; it has only one conclusion.

[20] See below, pp. 215 and 254.

two letters are combined one or both are truncated, because the beginning and/ or ending become superfluous, if the fiction of a single letter is to be preserved. Thus it is not surprising that 2: 13 to 4: 2 lacks both the address and the elements which constitute the normal ending of a Pauline letter, namely, the peace wish, the greetings, and the grace-benediction.[21]

Before appealing to the contents of the two letters for confirmation of this conclusion, a preliminary question must be dealt with. Why was 2: 13 to 4: 2 inserted into the middle of 1: 1 to 2: 12 and 4: 3 to 5: 28 rather than being attached to the end of the letter, as 2 Cor. 10–13 was appended to 2 Cor. 1–9? The similarity between 2: 11–12 and 4: 1–2 guaranteed that 4: 1–2 would integrate perfectly with the list of directives (4: 3–12) which originally followed 2: 11–12. The match was even better between 2: 12 and 2: 13 because the 'call' of God in the former is taken up in the latter by 'the word of God which you heard from us'. At both ends, therefore, the harmonization was too neat not to be capitalized upon. Hence, the impression of a single letter.

If 1 Thessalonians is in fact two letters, which came first? Here the judgement is necessarily much more subjective, but I think that a serious case can be made for the priority of 2: 13 to 4: 2, which in consequence I call *Letter A*.

Letter A

This letter is permeated by a profound sense of relief. Paul had been deeply worried (2: 17) that persecution (2: 14) would force the Thessalonians to abandon Christianity (3: 2–3, 5, 6, 7). He had wanted to come to their aid personally, but had been prevented (2: 17). Instead he sent Timothy (3: 2), and the good news the latter brought of the steadfastness of the Thessalonians (3: 6) was the cause of the joy Paul now experienced (3: 9), and the occasion of this letter.

The whole tone of *Letter A* suggests that it was Paul's first communication with the Thessalonians since his flight. It originated in his need and was not a response to any communication from the Thessalonians. The persecution, which had begun prior to Paul's departure, remained his major preoccupation. His uncertainty with regard to the outcome is the sole explanation of the emotionally charged language of virtually every verse. *Letter A* is precisely the sort of letter one would have expected Paul to write in reaction to Timothy's good news, whereas 1 Thessalonians in its present form definitely is not. Some commentators see an allusion to persecution in 1: 6. This is not at all certain.[22] If it were, however, it is psychologically impossible that such a detached reference

---

[21] Gamble (1977), 83.

[22] Malherbe (1987), 48, has argued correctly that the distress is related explicitly, not to persecution, but to the reception of the gospel, and thus is intended to evoke the new converts' break with their past and dear, familiar things.

to persecution should belong to the same letter as the desperate involvement implied by the allusions in *Letter A*. Paul, it should be emphasized, dealt immediately with what was uppermost in his mind, e.g. the defection of the Galatians (Gal. 1: 6) and the lack of unity among the Corinthians (1 Cor. 1: 10). That he should here restrain his emotional response to the fidelity of the Thessalonians for two whole chapters would be totally out of character. Moreover, his vision of them is so rosy that their steadfastness became sufficient evidence of their moral probity. The euphoric optimism of *Letter A*'s 'just as you are doing' (4: 1) stands in vivid contrast, not only to the tone of *Letter B* but to its list of ethical directives in 4: 3–8, and especially to its hints that all is not well in the community (5: 13–14). Other differences will come to the fore when we discuss *Letter B* in detail.

What time-interval should we postulate between Paul's leaving Thessalonica and writing *Letter A*? The time he spent worrying before sending Timothy into danger cannot be computed with any precision, but Paul's ardent temperament repudiated procrastination, and I doubt that more than two weeks should be allowed. The distance between Thessalonica and Athens is roughly 500 km. (320 miles).[23] This figure must be tripled in order to include Paul's flight south, and Timothy's round trip to Thessalonica and back. Sixty days would be necessary to cover 1,500 km. (960 miles) at an average of 25 km. (16 miles) per day. Hence a minimum of some ten weeks must have elapsed before Paul got Timothy's good news. His anxiety certainly had plenty of time to intensify.

In the interval Paul preached at Athens. His ministry was not a success. His silence regarding any converts confirms the basic thrust of Luke's account in Acts 17. It is not impossible that Paul's distracted state—his preoccupation with the fate of Timothy and the Thessalonians—contributed to his failure at Athens. His anxiety inhibited the wholeheartedness that effective preaching demands.

1 Thessalonians 3: 1 is usually understood to imply that Paul had already left Athens when the letter was written,[24] but this is in no way demanded by either formulation or context. The interpretation is inspired by a conscious or unconscious desire to harmonize *Letter A* with Acts 17: 15 and 18: 5. Common sense, on the contrary, suggests that Paul's profound anxiety obliged him to remain where Timothy would be able to find him without any difficulty. Were Timothy's return route known (he could have come by land or sea), Paul's reaction would have been to move north to meet him as soon as possible, as he did on another occasion, when he was equally anxious for news brought from Corinth by Titus (2 Cor. 2: 12– 13). It is out of the question that Paul should have delayed his reunion with Timothy, by going in the opposite direction,

---

[23] Rossiter (1981), 499.
[24] e.g. Kümmel (1975), 257; Bruce (1982*a*), 60.

towards Corinth, a strange city in which Timothy would have had enormous difficulty in finding him.[25]

The Move to Corinth

There is no doubt that Paul eventually moved from Athens to Corinth, but only after he had been joined by Timothy. The move is noted by Luke (Acts 18: 1), and confirmed by 2 Corinthians 1: 19 which puts the three co-authors of *Letter B* (Paul, Timothy, and Silvanus) in Corinth during Paul's founding visit there. It is difficult to say with certitude why he made the move. Two linked reasons, however, can be suggested. Corinth offered advantages which Athens lacked, and these facilitated Paul's missionary planning.

Up to this point Paul gives the impression of merely wandering west, with Rome perhaps as the vague long-term objective. Once he had established a community he felt free to move on, leaving its development to the guidance of the Holy Spirit. The situation at Thessalonica, however, forced him to recognize the need (both his and theirs) to stay in contact with his foundations. Thus for the first time he had to think in terms of a base, which, at the minimum, had to fulfil two conditions; it must be relatively easy to establish a church there, and communications with the surrounding area must be excellent.

In the mid-first century these conditions were met much more satisfactorily at Corinth than at Athens. The latter was an old sick city whose past was infinitely more glorious than its present.[26] The implications of the delicately nuanced description of the city by Strabo[27] are spelled out succinctly by Pausanias, 'Athens was badly hurt by this war with Rome, but flowered again in the reign of Hadrian.'[28] The contrast drawn by Horace between the quiet of Athens and the tumult of Rome highlights its lack of vitality.[29] Athens was no longer either productive or creative. Essentially a mediocre university town dedicated to the conservation of its intellectual heritage, it viewed new ideas with reserve. Tradition, enshrined in a rigid hierarchy, was its one safeguard against the threat of novelty. As a centre of learning it had been surpassed even by Tarsus.[30] The poverty of its economy is shown by the dearth of new buildings.

Corinth, on the contrary, was a wide-open boomtown. San Francisco in the days of the California gold rush is perhaps the most illuminating parallel. The decision of Julius Caesar in 44 BC to re-establish the city, which his predecessors had sacked a century earlier, was motivated by the legendary character of the

---

[25] See the quotation from Terence, Ch. 14, 'Rome and Spain'.
[26] Geagan (1979), 378–89.
[27] *Geography* 9. 1. 15–16.
[28] *Description of Greece* 1. 20. 4. The allusion is to the war against Mithridates in 88–86 BC.
[29] *Epistles* 2. 2. 77–86.
[30] Strabo, *Geography* 14. 5. 13, quoted in Ch. 2, 'The City of Tarsus'.

city as 'wealthy'.[31] The new colony quickly justified his expectations, and within two generations it had become the most important trading centre in the eastern Mediterranean. All its wealth was new money. Even those who in Paul's time had inherited riches were close enough to their origins to know where it came from. In opposition to the complacency of Athens, Corinth questioned. It was still a city of the self-made, and lived for the future. New ideas were guaranteed a hearing, not necessarily because of intellectual curiosity, but because profit could be found in the most unexpected places.

This atmosphere was to Paul's advantage. He must also have been aware that the establishment of a church at Corinth would carry weight elsewhere as an argument for the value of Christianity. The proverb 'Not for everyone is the voyage to Corinth' was widely known.[32] The bustling emporium was no place for the gullible or timid; only the tough survived. What better advertisement for the power of the gospel could there be than to make converts of the pre-occupied and sceptical inhabitants of such a materialistic environment (cf. 2 Cor. 3: 2)?

Over and above such advantages, Corinth offered Paul both outreach and superb communications with all points of the compass. It was one of the great crossroads of the ancient world and, even if we discount the flowery language, the praise lavished on Corinth by Aelius Aristides contains a great deal of truth, '[Corinth] receives all cities and sends them off again and is a common refuge for all, like a kind of route or passage for all humanity, no matter where one would travel, and it is a common city for all Greeks, indeed, as it were a kind of metropolis and mother in this respect' (*Orations* 46. 24).

Traffic in, out, and through the city was intense. It stood on the land bridge linking Greece to the Peloponnese. Boats shuttled between Asia and Europe. Paul had the possibility of influencing people from a great variety of different areas, and converts could carry the gospel back to their own people. Travellers going in all directions offered some security for Paul's messengers.

Paul tells us nothing about his founding visit to Corinth, with the exception of the fact that he was accompanied by Silvanus and Timothy (2 Cor. 1: 19; cf. Acts 18: 5). All that can be deduced from his letters regarding his relations with the church there belongs to a later period and will be dealt with in the context of the Corinthian correspondence, which is also the appropriate place to deal with the information provided by Luke in Acts 18: 1–22. It must suffice here to note that he lodged with Prisca and Aquila, from whose home he wrote his next letter to the Thessalonians, namely, *Letter B*.

---

[31] This is the adjective associated with Corinth ever since Homer wrote the famous Catalogue of Ships in the *Iliad* 2. 570; see Salmon (1984).

[32] Strabo, *Geography* 8. 6. 20; Horace, *Epistles* 1. 17. 36; Aulus Gellius, *Attic Nights* 1. 8. 3–4.

Letter B

This letter (1: 1 to 2: 12 and 4: 3 to 5: 28) makes no allusion to persecution. A calm didactic tone replaces the effervescent warmth of *Letter A*. Compliments are doled out carefully instead of being scattered broadside. More importantly, Paul's attention is no longer entirely concentrated on the Thessalonians. His image among members of the community has become a major preoccupation, if we are to judge by its position in the letter immediately after the thanksgiving (2: 1–12). When his focus again shifts back to the Thessalonians, it is to spell out the demands of Christian living (4: 3–12; 5: 12–22), and to deal with issues concerning the Day of the Lord (4: 13 to 5: 11).

Particularly noteworthy in *Letter B* is the complete evaporation of Paul's desire to see the Thessalonians. This is easily explained if his affectivity has acquired a new object. Evidently he has become progressively more absorbed in the nascent Corinthian community. He recognizes his continuing responsibility for the church at Thessalonica, but it is no longer his primary concern. The emotional distance between *Letter A* and *Letter B* implies that some considerable time separates them.[33]

This inference is confirmed by indications in *Letter B* that there have been contacts between the churches of Thessalonica and of Corinth independently of Paul. The existential witness mentioned in 1: 7–8 ('you became an example to all the believers in Macedonia and in Achaia ... your faith in God has gone forth everywhere, so that we need not say anything') demands direct contact. Thessalonians who came south could only speak of the exemplary quality of their community life (cf. 4: 9); it became visible only to Corinthian believers who went north.[34] It is these latter who reported to Paul how favourably he was still remembered at Thessalonica (1: 9; cf. 3: 6). Presumably it was they who also brought the information with which he deals in the body of the letter.

2 Thessalonians

Whereas the authenticity of 1 Thessalonians is accepted without question, that of 2 Thessalonians is still a matter of debate. For a significant number of scholars it was written, not by Paul, but by one of his followers towards the end of the first century.[35] They invoke differences of style and vocabulary, but in a highly selective way which prejudges the conclusion. When used objectively,

---

[33] Unwitting confirmation of this hypothesis is provided by Best (1979), 8–11. All the observations he uses to date 1 Thess. not long after Paul's departure from Thessalonica come from *Letter A*, whereas those which suggest a longer interval are found in *Letter B*.

[34] The first person plural ('we') in 1 Thess. 1. 8 includes Timothy, one of the co-authors, and thereby excludes him as a source of information on post-persecution conditions in Thessalonica.

[35] In this sense, see in particular Trilling (1972); Bailey (1978); Giblin (1990), 2–9.

however, such evidence proves that 2 Thessalonians is more at home in the Pauline corpus than 1 Thessalonians or 1 Corinthians.[36] The cold impersonal tone of 2 Thessalonians is often contrasted with the warmth of 1 Thessalonians. In reality, however, there is a much greater difference in tone between *Letter A* and *Letter B* (see above) than there is between the latter and 2 Thessalonians.

Ever since the synoptic presentation of 1 and 2 Thessalonians by W. Wrede,[37] it is generally agreed that the strongest argument against the authenticity of 2 Thessalonians is its identity of structure and often of language with 1 Thessalonians. Such an extensive overlap, we are told, is unlikely in a single author, but a forger would tend to copy the framework and vocabulary of 1 Thessalonians in order to enhance the credibility of a different eschatological vision of which 2 Thessalonians is the vehicle. The validity of this conclusion and the observations on which it is based have not gone uncontested.

The argument from the sequence of material is drastically weakened by a glance at the actual arrangement of the two letters, even when *Letter A*, which has no parallel in 2 Thessalonians, is removed:

	Letter B	*2 Thessalonians*
1: 1	Address	1: 1–2
1: 2–10	Thanksgiving	1: 3–12
2: 1–12	Apostolic Defence	
4: 3–12	Ethical Exhortation	
4: 13–18	Reunion with Beloved Departed	
5: 1–11	Times and Seasons	2: 1–12
	Encouragement and Prayer	2: 13–17
	Mutual Prayer	3: 1–5
5: 12–22	Ethical Exhortation	3: 6–15
5: 23–8	Conclusion	3: 16–18

If we exclude the address, thanksgiving, and conclusion, which are the most stereotypical features of all Pauline letters, the argument from sequence boils down to the fact that 'Times and Seasons' is followed at some point by 'Ethical Exhortation'.[38]

The arguments against the authenticity of 2 Thessalonians are so weak that it is preferable to accept the traditional ascription of the letter to Paul. In this case, the most natural explanation of the similarity of order is that circumstances forced him to return to the same subjects, and we know that it was his custom to deal with doctrinal points before turning to exhortation. It should be unnecessary to emphasize that in treating identical issues it is inevitable that the same language should reappear.

---

[36] Kenny (1986), 98; Neumann (1990), 213.
[37] (1903), 3–27.
[38] Best (1979), 53, vigourously criticizes the so-called identity of structure.

At this point we must confront the question of the order of *Letter B* and 2 Thessalonians. It has been argued that 2 Thessalonians antedates 1 Thessalonians,[39] but once again the reasoning is anything but apodictic; dubious observations are given forced interpretations.[40] 2 Thessalonians, on the other hand, contains one objective argument showing that it is posterior to *Letter B*. In 2 Thessalonians 2: 15 Paul writes, 'So then, believers, stand firm and hold to the traditions which you were taught either by word of mouth or by letter.' The meaning is unambiguous; the preaching of Paul has been supplemented by a previous letter.[41] Since *Letter A* contains nothing that can be described as 'traditions', there can be little doubt that the allusion is to *Letter B*. Confirmation of this conclusion is provided by 2 Thessalonians 3: 17, 'I, Paul, write this greeting with my own hand. This is the mark in every letter of mine.' The implication is that the Thessalonians had at least one other letter from Paul. Even though the Apostle does not identify himself in 1 Thessalonians 5: 27, the sudden appearance of the first-person singular suggests the handwriting of the author guaranteeing the work of his secretary.[42] By drawing attention to this feature of his letters Paul gave the Thessalonians the criterion to distinguish his authentic letters from forgeries.

Whether Paul consciously envisaged the possibility of forgery is not absolutely certain. Some have deduced it from 2 Thessalonians 2: 2,[43] but Bruce's observation on this verse deserves to be quoted,

> The particle *hôs* does not definitely deny the writers' authorship of the epistle in question: the misunderstanding may or may not have arisen from an epistle, and if it has so arisen, the epistle may or may not be authentic. If the reference is to an authentic epistle (and the genuineness of 2 Thessalonians itself be accepted), we should have to think of a misunderstanding of 1 Thessalonians.[44]

This insight has been exploited most effectively by Jewett.[45] He discerns a series of five passages in 1 Thessalonians, which he considers were susceptible of misinterpretation by the Thessalonians. Three of them (1 Thess. 2: 16, 18; 3: 11– 13) appear to me to be rather implausible, and moreover belong to what I have called *Letter A*. He has perceptively recognized, however, that two texts from *Letter B* easily lend themselves to misunderstanding.

The first is 'You yourselves know that the day of the Lord comes as a thief in the night' (1 Thess. 5: 2). Manifestly Paul intended his words to be understood

---

[39] The argument was given its classical form by Manson (1952).

[40] Kümmel (1975), 263–4; Best (1979), 42–5; Jewett (1986), 24–6.

[41] The improbability of the aorist *edidachthête* being merely epistolary is noted by Bruce (1982a), 193. Best (1979), 318, notes that, were the allusion to 2 Thess., one would expect the definite article, as in 2 Thess. 3: 14.

[42] Richards (1991), 179.

[43] References in Frame (1912), 247.

[44] (1982a), 164.

[45] (1986), 186–90.

in a future sense, but the present tense 'comes' could be read by those in a fever of intense eschatological expectation as meaning that the day of the Lord had already arrived but secretly. This interpretation would be facilitated by a diminution or cessation of persecution which, in opposition to *Letter A*, is not mentioned at all in *Letter B*. Those who wanted to believe could have seen a momentary halt in persecution as a sign of Christ's all-powerful presence among them. This line of reasoning, or something very similar, is necessary to explain why the Thessalonians believed that the day of the Lord had come (2 Thess. 2: 2).

Paul also laid himself open to misunderstanding by writing 'God has not destined us to wrath but to obtain salvation through our Lord Jesus Christ who died for us so that whether we wake or sleep we might live with him' (1 Thess. 5: 9–10). The main clause is 'God has destined us for salvation'. What God decides, however, will necessarily take place. The Thessalonians could legitimately conclude, therefore, that their salvation was guaranteed. The corollary was equally obvious. How they lived was irrelevant, because nothing that they did could modify the divine decision. They could even find explicit justification for this further deduction in the phrase 'whether we wake or sleep'. Paul intended this to mean 'whether we are alive or dead' (cf. 1 Thess. 4: 13–15). But just previously in this context he had used the same words in a different sense, 'Let us not sleep as others do, but let us keep awake and be sober' (1 Thess. 5: 6), which encouraged the interpretation 'whether we are vigilant or careless', namely, with respect to moral observance.

Those whose words have been misunderstood will have little difficulty in identifying with the mystified irritation with which Paul would have reacted to such reports of what he was supposed to have said in *Letter B*. Whatever the explanation—honest misunderstanding, deliberate distortion, forgery—it was not something that he could afford to let pass. He had to react, and did so with somewhat icy clarity in 2 Thessalonians. As Jewett has noted,

> The addition of new material in 2 Thessalonians, designed to clarify the nature of the eschatological signs that must precede the parousia, does not indicate a changed eschatological perspective on the author's part but rather the urgent need to demolish the belief that the parousia could be present while this evil age is still so clearly in evidence.[46]

It would have been natural for whoever brought the news of the situation at Thessalonica to Paul (2 Thess. 3: 11) to stress the increased potential for disorder in the community. Such unruliness had already concerned Paul in *Letter B* (1 Thess. 5: 14); a second and more vigourous admonition (2 Thess. 3: 6, 11) would have been entirely appropriate.

[46] (1986), 191–2. This factor is not adequately dealt with by Holland (1990).

This scenario is admittedly speculative, but it has the merit of dealing adequately with all the relevant aspects of *Letter B* and 2 Thessalonians. In weighing its value, its intrinsic plausibility must be contrasted with the utter improbability of the scenarios sketched to justify 2 Thessalonians as a pseudepigraphic construction.[47] It is easy to ascribe motives to a post-Pauline Christian author, but impossible to explain how and why the newly created letter was accepted as Pauline.

We are now in a position to sum up. Paul wrote three letters to the Thessalonians. The first, *Letter A*, was written from Athens, some ten weeks or so after Paul had fled from Thessalonica, hence, in the spring of AD 50. *Letter B* was written next, but from Corinth. The interval between these two letters is difficult to calculate, because an unspecifiable length of time has to be allowed for Corinthian converts to visit Thessalonica and return. This could have happened during the summer of AD 50.[48] How and when news of the misinterpretation of *Letter B*, which created the need for a third communication, 2 Thessalonians, reached Paul is impossible to determine, but nothing demands that a great interval separated them. 2 Thessalonians, therefore, could have been written in the late summer or early autumn of AD 50 or, at the latest, the following spring.

THESSALONICA AND ITS CHRISTIANS

Thessalonica owes its name to Thessalonike, a half-sister of Alexander the Great.[49] Her husband, Cassander, founded the city in 316 BC by amalgamating a number of villages on the best natural harbour in northern Greece at the head of the Thermaic Gulf.[50] It is frequently mentioned in the war which concluded with the conquest of Macedonia by Rome in 167 BC; its shipyards were important.[51] When Macedonia was made a Roman province in 146 BC, Thessalonica became the capital. Its economic prosperity was greatly enhanced in 130 BC when the Via Egnatia was constructed, linking it to Neapolis in the east and the Adriatic Sea on the west (see Fig. 7).[52] Strabo called Thessalonica

---

[47] A representative selection has been assembled by Jewett (1986), 5–10.

[48] The argument of Donfried (1990), 8, in favour of dating 1 Thess. to AD 43, involves acceptance of Lüdemann's hypothesis of a visit by Paul to Corinth shortly after AD 41, against which I have argued in my (1982a), 86–91.

[49] Historical surveys are to be found in Elliger (1978), 78–114; Hendrix (1992b).

[50] Strabo, *Geography* 7, frags. 21 and 24.

[51] Livy, *History* 44. 10, 12, 32.

[52] The best map is that in Papazoglou (1979), 304. Cicero notes how busy the road was (*Att.* 3. 14. 2).

'the mother city of what is now Macedonia' because it had the biggest population.[53] A native son, Antipater, addressed it as 'the mother of all Macedonia'.[54]

The Ethos of the City

Thessalonica was not permitted to stand aloof from the confusion of the Roman civil wars. Cicero spent six miserable months in exile there (May–November 58 BC) but none of the eighteen letters he wrote in this period tell us anything about the city. From the spring of 49 BC to August 48 it became a second Rome.

> For not only the consuls, before they had set sail, but Pompey also, under the authority he had as proconsul, had ordered them all [the senators] to accompany him to Thessalonica, on the ground that the capital was held by enemies and they themselves were the senate and would maintain the form of government wherever they should be. For this reason most of the senators and the knights joined them, some at once, and others later, and likewise all the cities that were not coerced by Caesar's armed forces.
>
> (Dio Cassius, *History* 41. 18. 4–5; cf. 41. 43. 1–4.; trans. Cary)

Julius Caesar put paid to such ambitions as effectively as his own were demolished by Brutus and Cassius. These latter in turn fell to Octavian and Antony. Thessalonica so admired Antony that its rulers created a new era in his honour. The degree of embarassment this caused after his defeat by Octavian at Actium in 31 BC is evident in the erasure of the dates from inscriptions.[55] Octavian, however, held no grudge, and the honours accorded him by the city were reciprocated by an acknowledgement of Thessalonica's status as a free city.[56]

The vitality of the population eventually would have repaired the damage of the civil wars, but Augustus speeded up the process by establishing Roman colonies 'at Dyrrachium, Philippi and elsewhere'.[57] This development could only have benefited Thessalonica, given its port at Neapolis and its position on the Via Egnatia. A brief report of Tacitus that in AD 15 'since Achaia and Macedonia protested against the heavy taxation it was decided [by Tiberius] to relieve them of their proconsular government for the time being and transfer them to the emperor'[58] is significant on two counts. The fat pickings expected by the governors underlines the increasing prosperity of the province, but this is perhaps of less importance than the inference that the financial élite, which

---

[53] *Geography* 7, frag. 21. Livy also called it 'extremely populous' (*History* 45. 30).
[54] *Greek Anthology* 9. 428.
[55] Hendrix (1992*b*), 524.
[56] Pliny, *NH* 4. 36.
[57] Dio Cassius, *History* 51. 4. 6.
[58] Tacitus, *Annals*, 1. 76. 4; 1: 80. 1.

must have been concentrated in the capital, Thessalonica, had enough clout in Rome, not merely to have one or two rapacious governors disciplined, but to have the system changed. This meant intimate and continuous contact with Rome, in addition to access at the highest level. In AD 44 Claudius made Macedonia independent once again by detaching it from Moesia and restoring it to senatorial control.[59]

The special relationship between Thessalonica and Rome is graphically illustrated by the city's acceptance of the divinity of Julius Caesar, which subsequently found expression in the establishment in Thessalonica of a temple of Caesar directed by a priest and agonothete of the emperor Augustus.[60] Roman benefactors were already being honoured by a priesthood early in the first century BC. Sometime later the goddess Roma was associated with them; the benefactors were the channels through which her bounty reached the city.[61] Inscriptions reveal a consistent hierarchy of cults; the priest of Augustus, the priest of 'the gods' (the tutelary deities of the city), and the priest of Roma.[62] The pre-eminence of Roman-oriented cults is a clear indication that civic life was dominated by those whom the native inhabitants of the city considered outsiders. This is confirmed by the fact that official decrees of the city were sometimes enacted in conjunction with the association of Roman traders.[63]

Even though Romans exercised political and ideological control, they were only one component of the city's élite. There was also a municipal cult of the Egyptian gods, whose functionaries created a dining club of some social pretensions under the patronage of Anubis.[64] An inscription records a divine imperative to propagate the cult of these gods in the interior of the country.[65] If Thessalonica had such contacts with Egypt, one can be sure that it entertained equally intimate relations with the other trading cities of the eastern Mediterranean, whose pantheons are well represented.[66] Profit was the unifying factor in a merchant class whose membership was drawn from everywhere but the city in which they made their living. The local indigenous population found itself blocked by foreigners from access to the decision-making process, and cut off from the sources of real wealth, in which they participated as salaried employees or even more remotely as slaves.[67]

[59] Dio Cassius, *History* 40. 24. 1; Suetonius, *Claudius* 25.
[60] Hendrix (1992*b*), 524.
[61] Donfried (1985), 345.
[62] Ibid. 346; Hendrix (1992b), 525.
[63] Papazoglou (1979), 356–7.
[64] Edson (1948), 181–8.
[65] Hendrix (1992*b*), 525; Meeks (1983), 19.
[66] Rigaux (1956), 15, lists Zeus, Neptune, Pallas-Minerva, Apollo, Diana, Venus, Mercury, Bacchus, Sun, Pan, Nemesis, Roma, Isis, Anubis, Victoria, Janus, Cabirus, Heracles, the Dioscuri, Perseus, Nike, Dionysios, and Serapis. Donfried (1985), 337–8, adds Asclepios, Aphrodite, and Demeter.
[67] Jewett (1986), 121–2.

Converting the Proletariat

The letters indicate that the Thessalonian Christians were drawn from this latter group. The admonition 'to work with your [own] hands' (1 Thess. 4: 11) assumes that the community was recruited essentially, if not exclusively, from the working class.[68] The hint that manual labour was the normal occupation is confirmed by 2 Thessalonians 3: 12, where those who have ceased to practice their trades are advised to return to working and earning a living. The same conclusion emerges from another line of argument based on the poverty of the Thessalonian church. The 'extreme poverty' of Macedonian believers (2 Cor. 8: 2) explains why, despite unusually intensive labour (1 Thess. 2: 9; 2 Thess. 3: 7–9),[69] Paul needed to be subsidized more than once by the Philippians (Phil. 4: 16). In the letters there is not the slightest trace of the wealthy patrons—the householder Jason and certain prominent women—who dominate the scene in Acts 17: 1–10. It has also been pointed out that the injunction 'If anyone does not want to work, let him/her not eat' (2 Thess. 3: 10) is most at home in a situation where everyone in the community was expected to make a contribution to the common meal.[70] In the absence of an individual capable of hosting the community (Rom. 16: 2, 23), the members had to entertain themselves. Thessalonian Christians met in tenements, not in villas.

Correspondingly, the workshop was the scene of Paul's ministry in Thessalonica.[71] It is not known who employed him,[72] or whether the workshop was part of his patron's residence, or how big the establishment was. It is clear, however, that there must have been a considerable demand for tents and other leather articles in a city which had so many travelling merchants. We should envisage a setting which provided Paul with both a stable base and a web of ready-made contacts focused on his patron, the clients, and his fellow-workers.[73] All three groups had family and friends, and unless the patron was a very wealthy person, which does not seem probable for this type of business, there must have been continuous interchange on a variety of different levels, all of which Paul could put to use ('working we proclaimed', 1 Thess. 2: 9). Such a workshop would be in a busy street or market, another world which made demands upon Paul's energies.

The average artisan had to work twelve hours a day seven days a week in

---

[68] Rigaux (1956), 521.
[69] Hock (1980), 31, points out that the formulation of 1 Thess. 2: 9 means that Paul exceeded the normal working day which was from sunrise to sunset.
[70] Jewett (1993).
[71] So rightly Hock (1980), ch. 3.
[72] In order to reconcile Luke and Paul inevitably it has been suggested that Jason provided Paul with work, as did Prisca and Aquila at Corinth (Acts 18: 2–3).
[73] Meeks (1983), 29–30; Malherbe (1987), 17–18.

order to barely make ends meet.[74] How did Paul manage to make converts in this bustling preoccupied milieu? The only serious attempt to answer to this question is Jewett's hypothesis that Paul's preaching of Jesus filled a spiritual vacuum.[75] Earlier scholars had noted the importance of the mystery cult of Cabirus at Thessalonica, its distinctive features, and its progressive development into an official religion, but Jewett is the first to exploit the consequences for the labouring class from which the first Christians were recruited.

The Cabirus legend tells of a young man, murdered by his two brothers, who was expected to return to aid the powerless and the city of Thessalonica.[76] His symbol was the hammer, and his blessings were invoked for the successful accomplishment of manual labour. He was the god to whom the Greek working class looked for security, freedom, and fulfillment. For some unknown reason in the Augustan age Cabirus was taken up by the ruling élite and incorporated into the official cult. This left the artisans and workers of Thessalonica without a benefactor. They naturally assumed that he, like other gods, was more responsive to the appeals and gifts of the wealthy. The sense of alienation was intensified by the fact that the members of the ruling élite were perceived as outsiders. Not only did they deny to the indigenous population the democratic equality which Greeks felt to be their birthright, but they had monopolized the sources of profit, and now they had taken away the one traditional divine friend of the poor.

Given these circumstances, it is easy to see how attractive Paul's preaching would be to the dispossessed; it reproduced the broad lines of a theology which they had thought lost.[77] He proclaimed a murdered young man, who had in fact risen from the dead, and who, in consequence, had the power to confer all benefactions in the present. Moreover, he would assume all his followers into a very different world.

It is not difficult to surmise, as Jewett has emphasized,[78] that the hint of a new 'god', who would radically transform the situation of the underprivileged, would have been perceived by the municipal authorities as subversive. Were the movement to take root and grow, it would threaten the fabric of society. However ridiculous the crucified Jesus of Nazareth might appear as a 'god' to sophisticated Romans or Greeks, the ruling class was politically astute enough to recognize the danger of an uncontrollable 'god' outside the structures of civic religion. He could serve as a rallying-point for a proletariat which by definition was unsatisfied. His message could be the magnifying glass to give inflammatory focus to frustrated ambitions. Under his inspiration, stirrings of unease could become revolutionary action.

---

[74] See in particular Hock (1980), 35.
[75] (1986), 127–32.
[76] The fundamental study remains that of Hemberg (1950).
[77] Jewett (1986), 131.   [78] (1986), 132.

Paul was used to being misunderstood and to being abused for it, and this was the type of persecution about which he warned his converts (1 Thess. 3: 4). He must have been as astounded as his converts when the authorities moved against them for very different reasons. He knew his message to be no threat to the security of the city. His converts, on the other hand, had assumed that in their new state they would be exempt from the violence that was endemic to their previous existence. Their reaction to persecution was not fear or coward- ice but mental perturbation (1 Thess. 3: 3).[79] Recognition of the potentially disastrous consequences of such disorientation, namely disappointment so pro- found as to lead to the abandonment of the faith, explains why Paul was so anxious about the steadfastness of the Thessalonians. If they felt that they had been deceived, all was lost.

MAINTENANCE DIFFERS FROM MISSION

We have seen above the exuberant relief expressed by Paul in *Letter A* after Timothy informed him that the Thessalonians, though bewildered, had remained faithful. We do not know whether Timothy's report went beyond the single issue which had been the *raison d'être* of his mission to alert Paul to other features of church life at Thessalonica. If it did, the latter's euphoria at the survival of the community effectively blocked assimilation of other details of their life. Only in *Letter B* do we find out what was really going on in the com- munity, as Jewett has so brilliantly demonstrated. What does the way Paul handled the situation tell us about him at this stage of his career?

Remembering his own Experience as a New Convert

A. J. Malherbe has very astutely used texts from the philosophical tradition to bring to light the emotional state of the members of the nascent church at Thessalonica.[80] Founders of philosophical sects were very explicit about what their recruits were going through. The commitment to a new vision of life brought in its train 'social, as well as religious and intellectual dislocation, which in turn created confusion, bewilderment, dejection, and even despair in the converts. . . . This distress was increased by the break with the ancestral reli- gion and mores, with family, friends, and associates, and by public criticism.'[81] Malherbe also draws attention to the concern of a sect to define its specific identity and to the efforts made by the group to assimilate new members. His detection of similar features in 1 Thessalonians leads him to the conclusion that

---

[79] Chadwick (1950), 156–8, and all subsequent commentators.
[80] (1987), 34–60.　　　　　　　　　　　　　　[81] (1986), 45.

'Paul consciously used the conventions of his day in attempting to shape a community with its own identity, and he did so with considerable originality'.[82]

Given the quality of Paul's secular education, it is not at all impossible that he should have been acquainted with the philosophical tradition, but I find it difficult to concede that he deliberately adopted its techniques. Even if he knew of them, he must have considered them inappropriate, because the community he desired to create was different from all other groupings in that an indispensable feature was mutual love (1 Thess. 4: 9).[83] Is it not much more probable that Paul drew on his own experience? He needed no one to tell him what the Thessalonians were going through. He had been converted twice, the first time to Pharisaism, and the second time to Christianity.

Thus when Paul speaks of the Thessalonians as 'having received the word in much affliction and joy inspired by the Holy Spirit' (1 Thess. 1: 6), we catch a glimpse of his own initial ambivalence on two occasions. The profound satisfaction of doing what he believed to be God's will was mixed with a gnawing sense of loss. The past still bound him emotionally, while the future had not yet established its claim. What got him through this stressful period was the support of the community he had joined, Pharisaic in Jerusalem,[84] and Christian in Damascus. This is the obvious source of the kinship language which Malherbe has highlighted as one of the features of 1 Thessalonians.[85] As both 'father' (2: 11) and 'nurse' (2: 7), Paul relates to his converts as 'children', whose bonding is evoked by the unusually frequent use of 'brethren' (eighteen times). He tries to create for them the calm rooted in a sense of security, which he had himself experienced.

Role models had been an important factor in promoting Paul's own stability. The conviction that led him to a new life had nothing to do with rational evidence. It was a leap of faith rooted in an unknowable impulse. Yet he would not have been human had he not felt the need for some justification. This he found in those whose lives exhibited, not only the fruits of the effort he was making (and perhaps considered inadequate), but the pattern of behaviour appropriate to the new mode of existence he had chosen. It was because he knew the importance of the satisfaction of his own need to see the gospel vindicated by comportment, in other words to see grace at work here and now, that he recognized his responsibility to be a model to the Thessalonians (1 Thess. 1: 6; 2: 9; 2 Thess. 3: 7, 9).[86]

---

[82] Ibid. 109.

[83] The force of the phrase 'you are taught by God' is to underline that mutual love is of the very nature of the church. Malherbe (1986), 104, suggests that Paul coined the expression precisely to distinguish Christians from Epicureans.

[84] Provided that the Pharisees obeyed the injunction reported by Philo (*Virt.* 102–4) that converts should be loved by those who had been born Jews.

[85] (1986), 48.

[86] The point is made with admirable clarity by Seneca, 'Of course, however, the living voice and

Malherbe is certainly correct that this is the fundamental perspective in which Paul's presentation of his ministry in 1 Thessalonians 2: 1–12 must be read, but he is unrealistic in divorcing it entirely from the situation at Thessalonica.[87] It would be totally out of character for Paul to waste time depicting himself to believers as the ideal philosopher (cf. 1 Cor. 1: 19–20). Were he convinced that his example was being followed, why should he justify in detail his conduct when among the Thessalonians?

On what ground was Paul attacked? One cannot simply assume that each statement he makes is the refutation of a specific charge. That would imply a condemnation of the Apostle so thorough and radical that a completely different and much more vigourous response along the lines of 2 Corinthians 10–13 would be expected. A narrowly focused hypothesis is provided by Jewett who, following the lead of Lütgert and Schmithals, argues that Paul was reacting to an accusation that he had failed to display the ecstatic behaviour which would mark him out as a true 'spiritual'.[88] This view, however has no foundation in the text.[89] Others have suggested that Paul was criticized for having decamped when the persecution began, leaving the Thessalonians to face the music alone. Again this is unfounded. Not only does Paul make no effort to justify his departure, which would be the only adequate response to such an accusation, but any such resentment on the part of the Thessalonians is excluded both by *Letter A* (1 Thess. 3: 6) and by *Letter B* (1 Thess. 1: 9) which note the respect in which Paul is held at Thessalonica.

The tension between the sweeping defence and the passionless tone suggests that Paul had become aware that he was the object of criticism, but knew none of the details. Only this explains his concern to refute all possible accusations without giving weight to any one in particular. His insistence in response to what must have been a hint rather than a report betrays a sensitivity to criticism, which is rather curious in a man of 50 who has been an active missionary for over ten years. This may have been a personality trait, but it was also the other side of his identification with his message. If he exemplified the gospel (2 Cor. 4: 10–11), then any attack on him was an affront to the word of God, which justified—necessitated?—a response.

---

the intimacy of a common life will help you more than the written word. You must go to the scene of the action, first, because men put more faith in their eyes that in their ears, and second, because the way is long if one follows precepts, but short and helpful, if one follows patterns' (*Epistulae morales* 6. 5; trans. Gummere).

[87] (1986), 74.
[88] (1986), 102.
[89] For detailed criticism, see Best (1979), 19–21.

Learning on the Job

The most probable source of criticism of Paul on the part of the Thessalonians is associated with his teaching on the eschaton. Their reading of *Letter B* (see above) is incomprehensible, unless they were convinced that they had been taught a realized eschatology. A predisposition to millenarianism is not an adequate explanation for their systematic transposition of all Paul's future statements into present ones. They were hard-headed working people with little time for idealism, and a very limited capacity for self-deception. No doubt they heard what they wanted to hear, but for them to have continued listening to Paul, there must have been some relation between his teaching and their desires. For their tinder to have caught flame, he must have proposed fire, not water. There is, of course, the possibility that they misunderstood the Apostle. Even in that case it is most probable, as we shall see in other letters, that the seeds of such misunderstanding were sown by Paul himself. He tended to assume that his audience would know what he meant, no matter what he actually said, and his impetuous temperament often led him to overstatement and the use of ambiguous language.

The possibility of such misinterpretation can be illustrated by the traditional fragment which Paul cites in 1 Thessalonians 1: 9b–10, and which is commonly accepted as reflecting the basic tenor of his preaching. The creed is composed of two strophes, each containing three lines:

> [We] turned to God from idols
>   to serve the living and true God
>   and to wait for his Son from heaven
> Whom he raised from the dead
> Jesus who delivers us
>   from the wrath to come.[90]

At first reading the meaning appears unambiguous, but a little reflection reveals that a simple shift of emphasis can change the interpretation radically. To put the stress on 'to *wait* for his Son from heaven' yields a future eschatology, but to highlight 'Jesus who *delivers* us' leads to a realized eschatology. One could even argue, on the grounds that the second strophe is designed to bring out the meaning of the first, that the latter interpretation is the dominant one. The waiting must be over, because Jesus is here and now delivering us; the coming wrath has been side-tracked. Thus even if Paul contented himself with reciting traditional doctrine—it is much more probable that he enthusiastically embroidered it—the Thessalonians could have heard him proclaiming a realized eschatology, intensified by his own fervour.

[90] See in particular Best (1979), 86–87; Rigaux (1956), 388–97. On the wider issue of other traditional material in the letters to Thessalonica, see Gundry (1987); Tuckett (1990).

Whether Paul was in fact preaching a realized eschatology, as C. L. Mearns maintains,[91] or whether he was merely thought to be so doing, developments at Thessalonica forced him to recognize the dangers of such a world-view when applied literally to daily life. The death of some members to whom a glorious assumption had been promised created intolerable problems for those left behind (1 Thess. 4: 13); it was something that should not have happened.[92] Others, there is no indication of how many, exhibited a typical millenarian disregard for the demands of normal living, perhaps indulgence in sexual excesses, certainly the cessation of productive labour (1 Thess. 4: 11; 2 Thess. 3: 6–12).[93]

Charity demanded that Paul provide an answer for the bereaved, and his concern for the witness value of the community (1 Thess. 1: 6–8; 4: 12) made it imperative for him to exclude practices which would bring the church at Thessalonica into disrepute. Even if he had previously understood the creed (1 Thess. 1: 9b–10) as implying a realized eschatology, the need to move the gaze of the Thessalonians from the present to the future forced him to recognize that the creed does not necessarily proclaim a realized eschatology. The function of the second strophe could be merely to identify the Son whose advent is expected. He is none other than the Jesus who is now at work in the community through his Spirit (1 Thess. 1: 5–6; 4: 8; 5: 19). Be that as it may, *Letter B* contains a series of allusions to a future Parousia (1 Thess. 1: 10; 5: 2, 23), at which, Paul assures his readers, the beloved dead will not be left behind abandoned but, once raised from the dead, will be assumed with the living (1 Thess. 4: 13–18).

It was perhaps inevitable that the shift from a realized to a futurist eschatology should be accompanied by the conviction that the Parousia would take place within the life-span of the present generation (1 Thess. 4: 15, 17). This could be read as the implication of the creed's choice of 'to wait'. Those who said this creed would be alive when the waiting ended. On the subjective level, this belief facilitated Paul's internalization of the new (or refined) perspective imposed on him by circumstances. As far as the Thessalonians were concerned, he hoped that it would minimalize the disconcerting dislocation caused by the substitution of a futurist for a realized eschatology. Paul, however, had failed to recognize the extent to which the Thessalonians had become imbued with the realized eschatology, which they believed to be his message. Their reaction was to move the proximate future, on which the Apostle now wanted to fix their gaze, back into the present (2 Thess. 2: 2). Thus another letter—2 Thess.— became necessary, in which Paul is forced to spell out the signs which will precede the in-breaking of the eschaton (2 Thess. 2: 3–12). Contrary to what he

---

[91] (1981), 137–57.
[92] Plevnik (1984).
[93] Jewett (1986), 172–6.

said in 1 Thessalonians 5: 2–3, the Day of the Lord will not come suddenly or quietly; it will be prefaced by major social upheavals.[94]

Exemplary Behaviour

The doubt as to whether Paul actually preached a realized eschatology at Thessalonica, or was mistakenly assumed to have done so, is not resolved by the fact that he instructed converts in ethical behaviour during his initial visit (1 Thess. 1: 11–12; 4: 1, 6, 11; 2 Thess. 3: 10). Moral teaching was not an afterthought dictated by the delay of the Parousia. Even at the stage when his eschatological expectation was most intense, Paul's perspective was radically apostolic. No matter how limited the time remaining, his mission was to convert the Gentile world. From his Jewish background he learnt that the word of God differs from all others in that it is intrinsically effective; it is laden with a power that transforms proclamation into performance.[95] If the gospel really was the word of God, then it could not be ineffective. This insight was reinforced by his own experience, which taught him that the one essential apologetic argument was the demonstration of the power of the Spirit in the lives of the ministers of the gospel. They did not convince by carefully crafted persuasive arguments (1 Cor. 2: 4–5), but by revealing the effectiveness in their personalities of the word they proclaimed (1 Thess. 1: 5; cf. 1 Cor. 9: 2; 2 Cor. 3: 2).

An important factor in the imitation which Paul expected of his converts (1 Thess. 1: 6) was to be the prolongation of his apostolic mission. The Thessalonians fulfilled their duty as Christians to extend the range of the gospel (1 Thess. 4: 12) by being 'an example' (1 Thess. 1: 7); the existential proclamation of their lives manifested the power of the word at work within them (1 Thess. 2: 13; 2 Thess. 1: 4, 11).

While such teaching is beautiful and impressive, it is too vague to be practical. Without some specification it would be ignored. While at Thessalonica Paul had to indicate at least the broad lines of the type of behaviour he considered conducive to the diffusion of the gospel. In *Letter A* Paul presumes that the Thessalonians recall these directives (1 Thess. 4: 2). He is less sanguine in his next letter, and the close parallels between 1 Thessalonians 1: 11–12 and 4: 2 strongly suggest that 1 Thessalonians 4: 3–7 exemplifies the type of oral instruction he considered appropriate.

---

[94] Thus 2 Thess. 2: 5 cannot mean that on his initial visit Paul gave the Thessalonians repeated apocalyptic instruction, as Best (1979), 290, maintains; similarly Rigaux (1956), 662. Had Paul in fact rigorously inculcated an apocalyptic timetable, the realized eschatology of the Thessalonians becomes inexplicable. Mearns (1981), 154, suggests that Paul had spoken of an Anti-Christ, but not in the context of the Parousia; Mearns identification of a specific reference to Caligula, however, is highly problematic.

[95] For a collection and analysis of biblical and extra-biblical references to the power of the divine word, see my (1964), 146–96.

The limits of the section are clearly defined by an inclusion: 'this is the will of God your sanctification' (4: 3) is echoed by 'God has called us in sanctification' (4: 7). Best and Bruce reflect the opinion of most commentators by entitling 1 Thess. 4: 3–7(8) 'Sex' and 'On Sexual Purity', respectively. Neither title, however, accurately reflects Paul intention which is to draw attention to the difference between the life-style of believers and that of non-believers. The contrast is made explicit in the first and last verses.

The life-style of believers is qualified as 'sanctification' (vv. 3, 7), which in the first place does not denote personal sanctity but rather having been 'set apart' by God, and thereby 'dedicated' to God. Christians are 'saints in virtue of a divine call' (Rom. 1: 7; 1 Cor. 1: 2; cf. 2 Cor. 1: 1; Phil. 1: 1; Col. 1: 20); the complete absence of 'saint' in Galatians underlines that in Paul's lexicon it is anything but a banal formula.

The alternative to 'sanctification' is described as *porneia* (v. 3) and *akatharsia* (v. 7). The latter means 'uncleanness' and is the antithesis of 'sanctification' used in the cultic sense just defined. Thus for a Jew it functioned as the definition of a pagan life-style.[96] Rigaux's claim that in Paul it always has a sexual connotation[97] is excluded both by 1 Thessalonians 2: 3[98] and by the reference to the children of Corinthian parents at Corinth; their unbaptized state should make them 'unclean' but in fact they are 'holy' (1 Cor. 7: 14).[99] *Akatharsia*, of course, can be used of sexual immorality. It is associated with *porneia* 'unchastity' and *aselyeia* 'licentiousness' as works of the flesh in Galatians 5: 19 (cf. 2 Cor. 12: 21), and with *epithymia* 'desire' in Romans 1: 24. In both of these contexts, however, Paul is describing unredeemed humanity, and in a way which merely reflects the standard Jewish association of pagan cults and sexual debauchery (Hos. 6: 10; Jer. 3: 2, 9; 2 Kgs. 9: 22), which is made explicit here by 'not in the passion of desire like the pagans who know not God' (v. 5). It is in this same context that the contrast of *porneia* with 'sanctification' is best understood; its connotation here is not specifically sexual (i.e. 'fornication') but symbolic, and in this sense is best rendered by 'immorality' (*RSV*).

That Paul is not thinking in terms of particular sexual problems at Thessalonica is confirmed by the admonition 'that each of you learn to acquire his own *skeuos*' (v. 4). *Skeuos* is literally a 'vessel' (2 Cor. 4: 7), and the two standard interpretations here understand it as meaning 'body' (*NRSV*) and 'wife' (*RSV*). The basic argument which has prompted the adoption of 'wife' is well stated by Best, 'No one can be said to "gain his body".'[100] This, however, is to

---

[96] 'Their works are unclean, and all their ways are a pollution and an abomination and uncleanness' (*Jubilees* 22. 16). Cf. Isa. 52: 1; Amos 7: 17; Acts 10: 28.

[97] (1956), 513.

[98] So rightly Best (1979), 93–4.

[99] For details, see my (1977*b*).

[100] Best (1979), 161.

forget that, for Paul, unbelievers are not their own masters; they are manipulated by social and economic forces to the point where they are 'under the power of Sin' (Rom. 3: 9) or 'enslaved to Sin' (Rom. 6: 17).[101] Thus liberation from Sin meant that one had to learn self-mastery in a much more profound and wide-reaching sense than physical self-control.[102] The sexual language ('not in the passion of desire', v. 5) in which the antithesis is expressed should not be permitted to blur the real contrast between the chosen commitment of 'dedication to God' and the servitude of the unbeliever, who is the victim of socially conditioned desires, 'Do not let Sin reign in your bodies to obey its desires' (Rom. 6: 12).

The mention of 'brother' in verse 6 recalls the specific identity of the group to which the believers belong. The attitude that one should have is described negatively by two verbs '(not) to go beyond' and '(not) to take too much'. In essence this is a warning against 'covetousness, greed', which caused the Fall (Rom. 7: 7), and which remains the dominant characteristic of fallen humanity (Num. 11: 34; 1 Cor. 10: 6). For most commentators the nature of the injury is limited to sexual matters (i.e. adultery) by 'in the matter', because the demonstrative article must refer to what has been mentioned previously. As we have seen, however, the sexual dimension of the preceding verses is merely a symbolic representation of the disordered and disorderly life-style of pagan non-believers.

The function of 1 Thessalonians 4: 8 ('whoever disregards this, disregards not man but God who gives his Holy Spirit to you') is to underline the fundamental importance of the distinction between the two modes of being. The gift of the Spirit as the source of sanctification is the effective implementation of the 'call' which articulates the 'will of God'. It enables discernment and empowers the choice of good. The directives in 1 Thessalonians 4: 3–7 are so generic that they set a direction without imposing specific obligations. They do no more than alert the Thessalonians to the fact that they must discover a life-style appropriate to their new being in Christ. The function of the directives is educative. Designed to orient those who have moved from an egocentric form of existence to an other-directed mode of being, they are the counsels of a wise father to his children (1 Thess. 2: 11–12).

The assumption underlying *Letter A*, namely, that the Thessalonians had grasped what Paul wanted to convey, could no longer be maintained when he wrote *Letter B*. He found himself forced to offer more explicit guidance. The way he handled this aspect of his problems with the Thessalonians reveals him to be much more consistent and clear-minded in the domain of Christian living than in that of eschatological speculation. The directives he gives are a mixture of advice and precepts. The latter, however, are entirely generic (1 Thess. 5:

[101] See Ch. 4, 'Dangers on the Road', and my (1982b), 89–105.
[102] Against Bruce (1982a), 83.

13b–22)—they concern values rather than structures—whereas the former are very detailed (1 Thess. 4: 10b–12; 5: 12–13a). Thus in *Letter B* Paul does not impose or prohibit any specific act. More significantly his list of commands embodies the crucial 'test everything' (5: 21a), which throws back to the Thessalonians the responsibility for their moral decisions. In the last analysis it is their judgement that counts. The implication that the Thessalonians themselves are responsible for the running of their own community is also significant. Paul did not consider it his role to tell them what to do.

This high-minded approach to morality came under severe pressure, when it became clear that certain members of the community continued to lead disorderly lives. *Ataktos* 'disorderly' is found once in *Letter B* (1 Thess. 5: 14); its cognates appear three times in 2 Thessalonians (3: 6, 7, 11). Correspondingly, the single instance of *parangellō* in *Letter B* (1 Thess. 4: 11) jumps to four in 2 Thessalonians (3: 4, 6, 10, 12). The meaning of this verb ranges from 'to give advice, to notify, to inform' at one end of the scale to 'to order, to command' at the other end.[103] Which did Paul intend?

Despite a number of commentators, 'to command' is not required in 2 Thessalonians 3: 4 and 6; 'to instruct' is perfectly adequate.[104] Paul is formally indicating the line of action he wants the community to adopt, namely to ostracize the unruly, thereby making it quite clear to outsiders that true Christians do not act in ways condemned. Such restraint, however, breaks down at the very end of the letter. The correlation of the imperatival infinitive 'do not mix with' and 'if anyone does not obey' in 2 Thessalonians 3: 14 unambiguously indicates that Paul expects them to do precisely what he says. To mandate a moral decision concerning the effective exclusion of a community member is a definite deviation from Paul's practice as revealed in the two previous letters.[105] His justification for the exception can only have been the hope that the punishment will effect the reformation of the erring brother (2 Thess. 3: 15). A more mature and sophisticated Paul will achieve the same result without compromising his principles in 1 Corinthians 5: 1–5, but by then he will have worked out most if not all of the implications of the incident at Antioch (Gal. 2: 11–14).[106]

It is highly indicative of Paul's understanding of the nature of his authority, and of how it should be exercised, that he does not install a representative at Thessalonica to report back to him, and to ensure that his wishes are carried out. This signals his recognition of the autonomy of the local church. It is responsible for itself. In consequence, it must evolve its own leadership. The

---

[103] Spicq (1978–82), 2. 647–9.

[104] Best (1979), 333, against Rigaux (1956), 703, and Bruce (1982a), 204.

[105] The imperatives concerning work in 2 Thess. 3. 10–12 are administrative rather than moral, but even there note how Paul waters down the force of 'we command' by the addition of 'we exhort' (v. 12).

[106] See Ch. 6, 'Pastoral Instruction'.

most that Paul can do is to hint at what qualities he considers necessary in such leaders. The Thessalonians should 'acknowledge those who labour among you, taking the lead in caring for you in the Lord and admonishing you' (1 Thess. 5: 12).[107] Such total dedication to the good of others can only be the fruit of love. Hence the only appropriate response is love (1 Thess. 5: 13). The leaders whom Paul hopes will emerge are not identified by social position or special skills, and the relationship of others to them is not one of obedience or deference. In this we catch a further hint of Paul's awareness that the Christian church is radically different in nature from any secular grouping. His perception of its true identity will grow. At this stage in his career all that comes across is that it is a community of love which radiates love (1 Thess. 3: 12; 4: 9–10).

The Paul of the First Letters

The eschatological issue at Thessalonica brought to light traits of Paul's character which will emerge with some consistency in other situations. He was not very good at working out what was going on in other peoples' minds. Certainly he never developed much insight into the mentality of the Thessalonians, even though the unusual problem of cessation of work had already manifested itself during his visit (1 Thess. 4: 11; 2 Thess. 3: 10). Presumably his delight at their response to his preaching—for which he would certainly have given the credit to divine power—made it impossible for him to grasp how exactly he was coming across. Inevitably he was mystified at the practical outcome of his words, and deeply hurt at criticism of his changeability, when he attempted to correct what he perceived as egregious misinterpretations of his teaching. His first attempt to rectify the realized eschatology of the Thessalonians was not successful. The alternative futurist version was not presented with sufficient vigour and clarity.

The hesitancy may be due to the belated recognition of the extent of his own responsibility. He found himself in the unhappy position of attempting to controvert well-received ideas, which the Thessalonians believed he accepted when he lived among them. In order to avoid the impression that he was making a complete about-face, he merely insinuated the new perspective by parenthetical allusions to the Parousia (1 Thess. 1: 10; 3: 13; 5: 2, 23), while at the same time giving the Thessalonians latitude to persevere in their error by phrases such as 'But as to the times and seasons, believers, you have no need to have anything written to you' (1 Thess. 5: 1).

Paul's naïveté in interpersonal relations is highlighted by his shock at the

---

[107] The translation of *proïstamenoi* by 'taking the lead in caring for' attempts to bring together the two attested meanings on which opinions are divided, namely, 'to preside, lead' (*RSV, NRSV, JB*) and 'to protect, care for' (Best (1979), 224–5; Bruce (1982a), 118–19). A similar rendering is appropriate in Rom. 12: 8, on which see Dunn (1988), 731.

failure of the Thessalonians to respond to what he saw as his gentle but firm and unambiguous invitation in *Letter B*. He had learnt a lesson, however, and his voice in 2 Thessalonians is clearer and much more forceful. But in order to get out of a corner he had to adopt an apocalyptic scenario (2 Thess. 2: 1–12) as an *ad hominem* argument. We do not know with what degree of conviction he accepted the scenario, whose meaning may have been just as obscure to the Thessalonians as it is to contemporary exegetes,[108] but he never used it again.

The assurance of his moral teaching, on the contrary, is noteworthy and betrays a clear vision of the nature of the Christian community. Evidently he was much more concerned with what the community did than what it thought, and had worked out a strategy in advance. From the beginning he realized that if the Thessalonian church was to have the sort of witness value that would reinforce and prolong his mission, its members would have to exhibit an attractive, freely chosen life-style. This intuitive insight would soon be strengthened by the conviction that to impose binding precepts would be to re-create the Mosaic law for believers. To acquire that conviction, however, he had to live through a crucial meeting in Jerusalem, and an agonizing conflict at Antioch.

---

[108] After noting Augustine's confession 'I admit that the meaning of this [2 Thess 2: 1–12] completely escapes me', Bruce (1982*a*), 175, comments that 'guesses at its meaning are all that the exegete can manage even today'.

6

Meetings and Meals:
Jerusalem and Antioch

PAUL himself does not tell us what happened in Thessalonica after the writing of 2 Thessalonians. It is difficult to imagine that this letter solved all problems of the self-absorbed Thessalonians, but we next hear of them some four years later when he lays out a plan to pass through Macedonia en route from Ephesus to Corinth in the summer of AD 54 (1 Cor. 16: 5). A lot was to happen in the interval.

Luke's estimate that Paul's stay in Corinth lasted eighteen months (Acts 18: 11) enjoys solid probability; it is the figure that one would have to postulate to explain the nature of the Apostle's relationship to the Corinthians. From Corinth, we are told, he sailed for Syria and, having landed at Caesarea, went up to Jerusalem,[1] and eventually returned to Antioch (Acts 18: 18–22). This sea voyage should be dated to the late summer of AD 51, because mid-summer of that year is the only date for Paul's encounter with Gallio (Acts 18: 12),[2] and Luke gives the impression that the voyage took place before the close of the sailing season in September.

Paul himself tells us only that fourteen years after his first visit to the Holy City as a Christian he returned to Jerusalem (Gal. 2: 1). But this visit, as we have seen,[3] must also be dated to AD 51. The simplest, and in fact the only adequate hypothesis, is to recognize that the accounts of Paul and Luke are references to the same visit; no valid objection can be raised against it.[4] The letters furnish slight and indirect confirmation in so far as they invite us to assume that, on his return from Corinth, Paul dropped off Prisca and Aquila at Ephesus, precisely as Luke says (Acts 18: 19, 24, 26).

---

[1] Some have disputed that 'to go up' (Acts 18: 22) is intended to evoke Jerusalem, but the verb is entirely inappropriate to describe a visit anywhere in the city of Caesarea situated as it is on the flat plain around the port. Equally the use of 'to go down' is appropriate only as an account of a journey from Jerusalem. So rightly Haenchen (1971), 547–8, who none the less thinks that Luke drew a false inference from an unplanned visit of Paul to Caesarea which was occasioned by the destination of the cargo ship or unfavourable winds.

[2] See Ch. 1, 'Paul's Encounter with Gallio'.

[3] See Ch. 1, 'Date of Departure from Damascus'.

[4] *Pace* Haenchen (1971), 544 n. 6. See Lüdemann (1984), 149–57.

Prisca and Aquila were with Paul in Corinth. The warmth of their greeting to the church there (1 Cor. 16: 19) permits of no other explanation. Subsequently they are living in Ephesus (1 Cor. 16: 19), and arrive in Rome prior to Paul (Rom. 16: 3–4). In other words, Prisca and Aquila appear in the same cities as Paul and in the same order. If, as seems likely, their role in Rome was to prepare for Paul's arrival, is it not likely that he placed them in Ephesus for the same reason?

THE JERUSALEM CONFERENCE

From Antioch to Jerusalem

Did Paul go directly to Jerusalem from Ephesus? The text we have just seen (Acts 18: 22) gives a clear affirmative answer. But this is what one would expect of Luke. By placing the journey into Europe (Acts 16–18) after the Jerusalem Conference (Acts 15), Luke intended to co-opt Paul, that is, to detach him from Antioch and make him an extension of the missionary effort of Jerusalem. Thus it was imperative that Paul should return to Jerusalem after having established Christianity in Greece.

The journey into Europe, however, antedates Paul's second visit to Jerusalem,[5] and in Galatians he makes it perfectly clear that at no time was he ever an emissary of Jerusalem. He did go there after his conversion, but simply to make a brief visit to Cephas (Gal. 1: 18), and the agreement they made regarding their respective spheres of activity did not in any way imply that he was subordinate to Peter, at least as far as Paul was concerned (Gal. 2: 7b–8).[6] The implication of the difference between the address of 1 and 2 Thessalonians, and those of all subsequent letters, was also noted, namely, when Paul wrote to the Thessalonians his missionary work was under the aegis of the church of Antioch.[7] This relationship continued until the incident narrated in Galatians 2: 11–14. One must assume, therefore, that Paul in fact returned from Greece to Antioch his home base.[8] There would have been no reason for a detour to Jerusalem.

This inference is supported, and the possibility of an accidental visit due to a boat sailing to Antioch being driven off course excluded, by the fact that Barnabas was with Paul in Jerusalem (Gal. 2: 1). Had he accompanied Paul on the long journey across Asia Minor into Greece, Paul's failure to mention him in 2 Corinthians 1: 19 and in 1 and 2 Thessalonians is inexplicable. The only feasible inference, namely that he was not with Paul at the foundation of Philippi, Thessalonica, and Corinth, is confirmed by Luke, according to whom,

---

[5] See Ch. 1, 'Prior to AD 51'.
[6] See Ch. 4, 'A Missionary Agreement'.
[7] See Ch. 4, 'A Gap in the Record'.
[8] Similarly Haenchen (1971), 464.

Paul and Barnabas had planned a joint missionary journey, but quarrelled over John Mark, and thereafter went their separate ways (Acts 15: 36–41). Whatever the value of his explanation,[9] the important thing is that Luke was aware that Barnabas was not Paul's companion on the 'second journey'.

Where, then, could Paul have encountered Barnabas? The speculative possibilities are almost limitless—e.g. Paul's ship from Ephesus put in at Cyprus, the homeland of Barnabas (Acts 4: 36–7) to which he had returned (Acts 15: 39)—but the most plausible place for the meeting is Antioch, which was the home base of both missionaries. Formally stated by Luke (Acts 13: 1–3), this is implied by Galatians 2: 11–14; whereas Peter and the people of James 'come' to Antioch, Paul and Barnabas are simply there.

The Occasion of the Conference

It will gradually become clear that Paul and Barnabas went to Jerusalem as delegates of the church of Antioch. Paul, however, insists that he went up to Jerusalem on account of a revelation (Gal. 2: 2). His reason for putting this interpretation on his voyage is to head off the accusation made by his opponents in Galatia that by going to Jerusalem, he acknowledged the superiority of the Jerusalem apostles, and thereby at least implicitly put himself under their orders.[10] There must have been a serious practical reason which forced two committed missionaries to divert from their real task to participate in a meeting in an area which was not their concern.

Lüdemann finds this reason in the dispute between Peter and Paul at Antioch in Galatians 2: 11–14, which he dates prior to the conference in Jerusalem.[11] While he correctly argues that a strict chronological order need not be followed in the *narratio*,[12] his hypothesis is excluded by one simple observation. If Lüdemann is right, at the Jerusalem Conference Paul and Barnabas should have been opposed to one another, because the latter ceased to eat with Gentile Christians (Gal. 2: 13), thereby implying that they should conform to Jewish law. In fact, however, at the conference Paul and Barnabas were on the same side because the pillars 'gave to me and Barnabas the right hand of fellowship that we should go to the Gentiles' (Gal. 2: 9). Moreover, the issue at Jerusalem concerned the circumcision of Gentile converts, not the problem of dietary laws, which subsequently became the issue at Antioch.[13]

---

[9] See Haenchen (1971), 475–7.

[10] See Ch. 8 'Discrediting Paul'.

[11] (1984), 75. He rightly dismisses the suggestion of Suhl (1975), 123 n. 102; 127 note 116, that the reason for the visit was to hand over a collection. This would imply the identification of Gal. 2: 1–10 with the famine visit of Acts 11: 30, a view espoused among others by Bauckham (1970). Such harmonization with Acts ignores the way Luke combines his sources; see in particular Benoit (1959).   [12] Quintilian, *Institutio Oratoria*, 4. 2. 83–4.

[13] So rightly E. Burton (1921), 82.

What, then, forced Paul to go to Jerusalem? His own answer is hidden in the confused language of Galatians 2: 4–5. Certain facts are clear but not their precise relationships. He describes those who insisted on the circumcision of Gentile converts in most derogatory terms; they are 'false brethren' who were 'secretely smuggled in' in order 'to spy'. Unfortunately he does not tell us where this 'infiltration' took place. Such language, however, implies that his opponents 'are alien to the body into which they have come'.[14] Thus a *terminus a quo* and a *terminus ad quem* have to be determined. Within the framework of a Jerusalem–Antioch axis there are two possibilities; either conservatives from Jerusalem infiltrated Antioch or conservatives from Antioch infiltrated Jerusalem.[15] Given that Jerusalem was much more conservative than Antioch (cf. Gal. 2: 12), the latter hypothesis is most unlikely. Hence, it is probable that Paul had in mind pro-circumcision believers from Jerusalem, who created trouble in a community to which they did not belong, namely, Antioch.[16] It goes without saying, of course, that, once the debate had been transferred to Jerusalem, these people would also make their case there.

Thus if one had to imagine a scenario to explain the conference at Jerusalem on the basis of the letters alone, it would be difficult to better Luke's,

> But some men came down from Judaea and were teaching the brethren, 'Unless you are circumcised according to the custom of Moses, you cannot be saved.' And when Paul and Barnabas had no small dissension and debate with them, Paul and Barnabas and some of the others were appointed to go up to the apostles and elders in Jerusalem about the question. (Acts 15: 1–2)

For Luke, therefore, Paul and Barnabas went to Jerusalem because they were selected to go as members of an official delegation. This is the antithesis of what we find in Galatians, where by his silence, his stress on a revelation as the motive of his visit, and his use of the first-person singular, Paul insinuates his independence of the church of Antioch. In this instance, however, Luke's version is preferable. The needs of the letter to the Galatians forced Paul to distance himself from Antioch as much as from Jerusalem. As we shall see when dealing with Galatians 2: 11–14, the shift in the position of Antioch with respect to Gentile converts brought it into line with the practice of Jerusalem. Under such circumstances for Paul to acknowledge his dependence on Antioch, while at the same time disagreeing with its policies, would have been to give arms to his opponents in Galatia. The agreement with Peter made fourteen

---

[14] Ibid. 78.

[15] Burton (1921), 83–4, evokes a further possibility that the opponents were Jews, who had feigned to become Christians in Jerusalem precisely in order to subvert the Jesus movement. This hypothesis is exluded by Gal. 2: 4b. They were anti-Paul, not anti-Christian. Though 'false' they were none the less 'brethren'.

[16] So rightly Burton (1921), 79; Lüdemann (1989), 35; Walker (1992), 503–10, against Betz (1979), 89, and Longenecker (1990), 50.

years earlier (Gal. 2: 7 b–8)[17] could not have been invoked because it concerned only the fact of a Gentile mission, and not the conditions under which Gentiles could be received into the church.[18]

Paul manages to give the impression that the appearance of aggressive missionaries of the Law-observant Jerusalem church in Antioch, and later in Galatia, Macedonia, and Corinth, was inspired by unworthy motives. But he never specifies what they were. He must have had a severe shock when he found his missionary practice called into question when he returned to Antioch. Manifestly he saw only the danger to his life-work. Under such conditions tolerant understanding of the concerns of those with a different theology was not a psychological option. Subconsciously, his resistence may also have had roots in an awareness that those who differed from him did so on grounds that were difficult to dispute.

At this stage in the history of the church it was taken for granted by all, including Paul, that salvation was related to the chosen people, who worshipped the one God, and to whom he had sent his Messiah. The salvation question as far as Gentiles were concerned was: how can they be integrated into God's messianic people?[19] Paul's adversaries could point to situations in which Jesus not only obeyed the Law (e.g. when he went on pilgrimage to Jerusalem) but proclaimed its eternal value (Matt. 5: 18–19) and recommended obedience to it (Mark 1: 40–5). Not unnaturally, therefore, they took it for granted that converts to Christianity should accept the same obligations as converts to Judaism. This point of view is documented in the Jewish Christian pseud-epigraph *The Epistle of Peter to James*:

> Some from among the Gentiles have rejected my [Peter's] lawful preaching and have preferred a lawless and absurd doctrine of the man who is my enemy [Paul]. And indeed some have attempted, while I am still alive, to distort my words by interpretations of many sorts, as if I taught the dissolution of the law and, although I was of this opinion, did not express it openly. But that may God forbid! For to do such a thing means to act contrary to the law of God which was made known by Moses and was confirmed by our Lord in its everlasting continuance. For he said, 'The heavens and the earth will pass away, but one jot or one tittle shall not pass away from the law'. (2. 3–5)[20]

Given the intense eschatological expectation of the beginnings of the Jesus movement, it is most unlikely that anyone among the first generation of Christians thought that Jerusalem would ever lose its centrality in determining the orientation of Christianity. The imminence of the Parousia, it was felt, guaranteed that the authority of Jerusalem would not be overwhelmed. Tradi-

---

[17] See Ch. 4, 'A Missionary Agreement'.
[18] So rightly Lüdemann (1989), 43–4.
[19] Stendahl (1963).
[20] Translation from Hennecke and Schneemelcher (1965), 2. 112.

tionally a massive influx of Gentiles would take place only in the eschaton.[21] In the present there simply would not be enough time for great numbers of pagans to be converted.

This projection, based on the painful slowness of the mission to Jews, failed to take into account the appeal the gospel would have for pagans. The tremendous success of the missionary effort of the church at Antioch, which demanded only faith in Jesus Christ for conversion, brought home to some Law-observant Jewish Christians in Jerusalem that their vision of the church as the flowering of Judaism was in serious danger. If things were permitted to continue as they were, they foresaw themselves becoming an ever smaller minority in an institution whose only ties to Judaism were (1) the racial identity of its founder and of the first generation of his disciples, and (2) recognition of the Old Testament as the record of God's preparatory work for the advent of Jesus Christ. This, they decided, must not be permitted to happen.[22]

Such Law-observant Jewish Christians had only two options in order to fight back. On the one hand, they could contest the validity of Paul's approach, while on the other they could attempt to convert Gentiles who would accept circumcision. The alternatives were not mutually exclusive. And a two-pronged attack developed. The first, as we have seen above, is documented by Galatians 2 and Acts 15. The second, as J. L. Martyn has pointed out,[23] is attested by the Clementine *Recognitions*:

> It was necessary that the Gentiles should be called into the place of those [Jews] who did not believe [in Jesus as the Messiah], so that the number might be filled up which had been shown [by God] to Abraham. Thus the preaching of the blessed Kingdom of God is sent into all the world. (1. 42. 1)

The mention of Abraham is manifestly an allusion to the promise 'in you shall all the tribes of the earth be blessed' (LXX Gen. 12: 3), which is explained by Ben Sira, 'The Lord therefore promised him on oath to bless the nations through his descendants' (44: 21; cf. Jer. 4: 2). Law-observant Jewish Christians would have had no difficulty in considering this promise adequate legitimization for a mission to Gentiles. The advent of the Messiah in the person of Jesus of Nazareth signalled the providential moment when the privileges of election and covenant should be extended to pagans, but obviously on the same conditions which governed their enjoyment by Jews, namely, observance of the

---

[21] Isa. 2: 2–4; 49: 6; 56: 6–7; 60: 4–7; Zech. 2: 11; 8: 20–3; Tobit 14: 6–7.

[22] The imputation of selfishness as a motive for action is the consequence of historical scepticism, but the possibility cannot be excluded that some Law-observant Christians were genuinely worried about the salvation of the Gentiles; if they did only what Paul advised they might not really be saved.

[23] (1985), 310–11.

Law.[24] Thus there began a movement among Jewish Christians to invite Gentiles to a Jewish life-style rooted in belief in Jesus.[25]

The Meeting in Jerusalem

What must have been a long, complex, and stormy meeting in Jerusalem is compressed by Paul into two verses which succinctly articulate the problem and its solution.

First the problem: 'I laid before them the gospel which I preach among the Gentiles—but privately before the men of eminence—lest somehow I should run or had run in vain' (Gal. 2: 2b). 'The gospel' is a very broad concept, but here it can only mean that faith in Jesus Christ is the one indispensable condition for salvation; everything else is secondary and fundamentally irrelevant. This, as far as Paul was concerned, was the one item on the agenda. Paul, it must be stressed, at this stage is not saying that obedience to the Law is wrong, but only that it is unnecessary.

It is impossible to decide whether there were one or two meetings, i.e. with the church as a whole and subsequently with the leadership group identified as James, Cephas, and John (Gal. 2: 9),[26] or only with the latter.[27] In any case, it was the troika who made the critical decision in the name of the community whose 'pillars' they were.[28]

The official tone of 'to submit something for consideration to somebody' (Gal. 2: 2)[29] is implicit recognition of the authority of the Jerusalem church,[30] which Paul attempts to attenuate by calling its leaders 'men of eminence', which could be taken in a derogatory or ironic sense. Paul was aware that he could not force, but only await, a decision. The extreme level of his anxiety is betrayed by the confession that, if the decision went against him, all that he had done so far would be in vain. The language may reflect what Paul's opponents were currently telling the Galatians, namely, without obedience to the Law their conversion was ineffective in terms of salvation.[31] Schlier, thus, opts for what at first sight seems to be the most natural interpretation, and the one sup-

---

[24] On the relation of election and law as covenant nomism, see E. P. Sanders (1977), 422–3.

[25] The view that Jewish Christians simply followed the Jewish practice of aggressive proselytization is without foundation, because there was no systematic Jewish outreach to pagans; see my (1992a).

[26] So Betz (1979), 86.

[27] So Longenecker (1990), 48.

[28] The metaphorical sense of 'pillar' is well attested in both Greek and Jewish sources (Longenecker (1990), 57). Aus (1979a) has drawn attention to a particularly relevant parallel, namely, the identification of Abraham, Isaac, and Jacob as the three 'pillars' supporting Israel and the world.

[29] Betz (1979), 86 n. 268.

[30] Note also the implicit recognition in Gal. 1: 16 that the apostles in Jerusalem were the appropriate people to judge his revelation.

[31] Betz (1979), 88.

ported by Tertullian and Jerome,[32] i.e. that Paul had doubts about the validity of his gospel which could be assuaged only by confirmation by the mother church.[33]

Others, however, have seen that this interpretation is excluded by Galatians 1: 11–12, where Paul formally articulates the certitude he experienced in the revelation of his encounter with Christ. He needed no confirmation from any external authority; he was utterly convinced that he was right. The troika could never persuade him that he was wrong, but they could destroy what he had achieved, and they could systematically oppose any future ministry.[34] Envoys could be sent to the communities he had founded to inform them that Paul was an isolated, unrepresentative maverick, since all authentic followers of Jesus observed the Law. The threat was very real. Paul had few illusions about his converts' loyalty to his theological principles. Many misunderstood his teaching, and those who did understand could easily be persuaded that they were in error. He knew the difficulties of living in freedom, and he had experienced the spurious but seductive security that rules and regulations offered.[35]

The solution: 'But not even Titus, who was with me and was a Greek, was compelled to be circumcized' (Gal. 2: 3). Paul dramatizes the response of Jerusalem by personalizing it. Titus found himself the test case which decided a question of principle. The fact that Titus was not required to undergo the operation made it clear that circumcision was not necessary for salvation.[36]

Some have detected a certain ambiguity: was Titus not compelled or not circumcized? The former interpretation could imply that he had freely accepted circumcision once the principle had been established. It arose because some witnesses to Galatians 2: 5 lack the negative particle *oude* and so have Paul saying 'to them [those demanding that Titus be circumcized] we yielded for a moment'. In other words, having won his point that circumcision was not necessary, Paul made a graceful concession to his opponents by permitting the circumcision of Titus. And thereby showed his consistency, because according to Acts 16: 3 he had Timothy circumcized.[37] Lagrange, however, has shown that it

---

[32] Cited in Lagrange (1925), 26.

[33] (1962), 66–9.

[34] Lagrange (1925), 27. Others stress a concern for the unity of the church and the consistency of its mission which is unlikely to have been a major preoccupation of Paul at this stage; so rightly Betz (1979), 86.

[35] In dealing with the Thessalonian correspondence we noted Paul's unwillingness to impose precepts. Whether at the time of the Jerusalem meeting he already identified the Law with 'enslavement' (Gal. 2: 4) is open to doubt. It is more likely to have been the consequence of his post-Conference experience at Antioch, on which see the section 'Pastoral Instruction' below.

[36] Nothing is known about Titus' antecedents or how and where he met Paul; see Barrett (1969); Gillman (1992). Subsequently he became one of Paul's assistants and was chosen to bring the Severe Letter (2 Cor. 2: 4) to Corinth precisely because, in the face of Judaizing opposition, he could report authoritatively on what happened at the Jerusalem Conference; see my (1991b), 42.

[37] The view of Betz (1979), 89, that Timothy was Jewish because his mother was Jewish is not supported by any 1st-cent. evidence; see Cohen (1986).

was precisely the desire to harmonize Acts and Galatians that led to the omission of the negative particle.[38] Paul in fact made no concession with respect to Titus, because none had been demanded of him; 'the men of eminence added nothing to me' (Gal. 2: 6). They imposed no conditions on his ministry. He could continue as he had begun.

WHY DID JAMES AGREE WITH PAUL ON CIRCUMCISION?

We have seen the reasons why certain Law-observant Jewish Christians wanted to impose circumcision on Paul's converts. Why did others object to this proposal? In other words, how did Paul persuade James, Cephas, and John? The possibility that they were predisposed in his favour would seem to be excluded by the subsequent activity of James in Antioch, where he is strongly in favour of the maintenance of Jewish practices. The arguments used by those who insisted on the circumcision of all converts must have appealed to James, and we have seen that their force is considerable. Circumcision was the traditional sign of belonging to the covenant people which was still seen by all Christians as the divine channel of salvation.

It is unreasonable to assume that the troika accepted Paul's position against the opposition of the entire Jerusalem church whose 'pillars' they were. The fact that Paul singles out 'false brethren' for provoking the crisis insinuates that others in the Jerusalem church favoured his liberal view. Hence, James could be assured of backing from certain members of by far the biggest block in the Jerusalem church, Jewish Christians. This, however, simply pushes the question back a stage; it does not answer it. How and why had these come to the conviction that Gentiles should not be circumcised?

It is not sufficient to appeal to a division within Judaism on whether proselytes should be circumcised.[39] Not only is it unreasonable to assume that a similar division should automatically follow within the Christian church, but there is no real evidence that there was a signficant body of opinion within Judaism opposed to circumcision. No one had any doubt as to the mandatory character of circumcision,[40] even if there were those who attempted to hide it,[41] or who spiritualized it for converts.[42] Moreover, when conversion without cir-

---

[38] (1925), 28–31. See Metzger (1971) 591–2; Barrett (1985), 112.

[39] As does Betz (1979), 89. See also McEleney (1973).

[40] Gen. 17: 10–14; 1 Macc. 2: 46; *Jub.* 15. 25–34; Philo, *Migr. Abr.* 89–92; Josephus, *AJ* 13. 257–8, 318; Nolland (1981); Collins (1985).

[41] 1 Macc. 1: 15; Josephus, *AJ* 12. 241; *Jub.* 15. 33–4; 1 Cor. 4: 18; Martial, *Epigrams* 7. 35, 82; Hall (1988).

[42] Philo, *Quaest. Exod.* 2. 2; *Som.* 2. 25; *Spec. Leg.* 1. 305; *Sibylline Oracles* 4. 163–70; Arrianus, *Diss.* 2. 9. 20.

cumcision was contested, circumcision was imposed.[43] As far as Jewish Christians were concerned, circumcision was the traditional sign of belonging to the covenant people, which was seen as the divine channel of salvation.

Nor is it sufficient to claim that James and the others started from the premiss that 'the law was given solely to Israel', and thus were led to the conclusion that it could not be applied to Gentile converts.[44] Not only does this fail to respect the intrinsic link, admitted by all in the early church, between salvation through faith in Jesus and belonging to the messianic people, but it makes the position of Paul's opponents inexplicable.

If the religious situation of Judaism throws no light on the issue, perhaps its political situation might be more illuminating. In the Roman empire the Jews had certain rights which were clearly and precisely defined in law.[45] Such privileges, however, were enjoyed at the good pleasure of the emperor, and never stood in the way of imperial action against Jews. Thus in AD 19 Tiberius expelled the Jews from Rome.[46] When the Nabataeans attacked and routed the troops of Herod Antipas, probably in AD 29, Rome exacted no vengence.[47] The situation deteriorated seriously when Gaius (Caligula) came to power in AD 37. His weakness permitted a violent outburst of anti-Semitism in Alexandria in the middle of AD 38.[48] Synagogues were burnt or desecrated, and the mob persuaded A. Avillius Flaccus, the prefect of Egypt, to downgrade the status of Jews in Alexandria to that of aliens without right of domicile. Many Jews were massacred, and those who survived were forced into an overcrowded ghetto. Violence ceased with the arrest of Flaccus and the arrival of a new prefect, C. Vitrasius Pollio, in October, but traditional Jewish rights were not immediately restored.

Jews in Palestine can hardly have been unaware of what was happening to their co-religionists in Egypt, and feared for themselves. They had good reason. In the spring of AD 40, in reprisal for Jewish destruction of an altar of the imperial cult set up in Jamnia, the emperor Gaius ordered the legate of Syria, Publius Petronius (AD 39–41), to transform the Temple in Jerusalem into an imperial shrine by erecting a giant statue of the emperor as Jupiter in the Holy of Holies. He was authorized to use two of his four legions to enforce the decision.[49] Petronius managed to delay implementation of his orders until Agrippa I in late summer persuaded Gaius to change his mind. For the Jews of Palestine it

---

[43] Josephus, *AJ* 20. 38–48.

[44] So Haenchen (1971), 468.

[45] Saulnier (1981).

[46] Josephus, *AJ* 18. 65–84; Tacitus, *Annals* 2. 85. 5; Suetonius, *Tiberius* 36. 1; Dio Cassius, *History* 57. 18. 5. These texts are discussed in detail by Smallwood (1981), 202–10.

[47] Josephus, *AJ* 18. 109–15; see Saulnier (1984), 365–71.

[48] The detailed documentation furnished by Philo in *Legatio ad Gaium* and *In Flaccum* is discussed by Smallwood (1981), 235–42.

[49] Philo's report—*Legatio ad Gaium* 188, 198–348—is preferable to the different versions of Josephus, *AJ* 18. 261–309; *JW* 2. 184–7, 192–203. See Smallwood (1981), 174–80.

must have been a nerve-wracking six months as they prepared to sacrifice themselves rather than submit. They could never be fully at ease while Gaius lived. In fact he was planning to go back on his word when he was assassinated on 24 January 41.

On his accession Claudius (AD 41–54) moved quickly to undo the damage caused by the madness of Gaius.[50] The emperor made it very clear, however, that he considered the Jews a disruptive ferment throughout the empire, and that their enjoyment of their privileges was conditional on good behaviour.[51] Thus, though the right of religious assembly was guaranteed, when a disturbance broke out in a Roman synagogue in AD 41 Claudius closed the synagogue and expelled the agitators from the city.[52] The Jews were served unambiguous notice that they were on probation.

The next Roman move came in response to the appearance of a false Messiah.[53] In the spring of AD 45 the procurator of Judaea, Cuspius Fadus (AD 44–?46) ordered that the vestments of the High Priest (without which he could not function), which had been released to the Jews by Vitellius in AD 36,[54] should be restored to Roman custody and housed in the Antonia fortress.[55] The Jews persuaded Claudius to rescind the order, but once again they were made to feel fortunate. Whatever their rights, the decision could very easily have gone against them. Their awareness of the fragility of their position was intensified by two episodes which took place when Ventidius Cumanus (AD 48–52) was procurator of Judaea.[56] Both involved senseless deliberate provocation by individual soldiers. The first was permitted to escape unscathed, even though a great many Jews died. The second was executed, but a scroll of the Law had been ripped apart and burnt.

The inevitable consequence of such repeated incidents—many others may not have been recorded—was a profound sense of insecurity among Jews. If the Romans could not be trusted, then there was nothing for it but for Jews to take matters into their own hands. This is precisely what happened on one of the pilgrimage feasts in AD 51.[57] When Cumanus did not arrest the Samaritans who had slaughtered Galileans en route to Jerusalem, their friends and other Jews took their own vengence on the Samaritans.[58] Things had reached such a pass

---

[50] On the complex issue of what constitutes reliable data in this matter, see in particular Schwartz (1990), 90–106.

[51] The *Letter to Alexandria* is conveniently available as n. 48 in Barrett (1987), 47–50. It is discussed by Smallwood (1981), 245– 50, 360–1.

[52] Suetonius, *Claudius* 25; Dio Cassius, *History* 60. 6. 6; Orosius, *History* 7. 6. 15–16. See the discussion above, Ch. 1, 'The Edict of Claudius'.

[53] Smallwood (1981), 259–60.

[54] Josephus, *AJ* 18. 90; cf. 15. 405.

[55] Ibid. 20. 6–14

[56] Ibid. 20. 105–17; *JW* 2. 223–31.

[57] Smallwood (1981), 265 n. 29.

[58] Josephus, *AJ* 20. 118–24.

that any perceptive observer could have predicted growing tension between the Jews and Rome with an ever increasing potential for violence. Clearly it was imperative for Jews to stand together. Only if they were totally united could they survive. Any diminution of commitment could be fatal.

The dilemma in which this placed politically conscious Jewish Christians is obvious. They were first and foremost Jews. All that separated them from their brethren was their acceptance of Jesus of Nazareth as the Messiah. Even without pressure from their co-religionists, their own instincts would have told them that the beginning of the 50s was a time to affirm, not to dilute, Jewish identity. Which end would the circumcision of Gentile converts achieve? Manifestly the latter. To circumcise Gentile converts was to accept them publicly as Jews, even though they had no attachment to Judaism; they were followers of Christ not of Moses. What loyalty to the Jewish people could be expected of such individuals when hostile pressures began to take their toll? In a crisis could any nationalistic Jew really trust them? Would such nominal Jews be prepared to sacrifice their lives for the Temple and the Law?[59]

Questions such as these must have occurred to the more far-sighted members of the Jerusalem church. What seemed to be right in the present could be seen to be a dangerous threat in the not too distant future. James, I suggest, was one of these. As the leader of the Jerusalem church he was swayed, not by theological reasons, but by practical considerations. Those who demanded the circumcision of Gentile converts might be correct in theory, but it was not the moment to insist on principle. Whatever his personal inclinations, historical circumstances conspired to make James want to find justification for not circumcising Gentile believers. This need made him receptive to Paul's personality and arguments. No more than he had fourteen years earlier (Gal. 1: 19), could he doubt the sincerity with which Paul explained the implications of his conversion. Nor could he deny the grace manifested in the number of Gentiles who accepted the Pauline gospel (Gal. 2: 9a). Similar success, presumably, was duplicated by Barnabas elsewhere. Such evidence of the presence of the Holy Spirit manifested the divine will that Gentiles should be admitted to the church as Gentiles.[60]

If this line of argument is correct, it should have as a corollary a concern on the part of James to strengthen the identity of Christians who were of Jewish origin by insisting on more exacting observance of Jewish practices. As we shall see, this is precisely what happened at Antioch (Gal. 2: 11–14).

---

[59] These questions highlight the implausibility of the hypothesis of Jewett (1970), 205, that Jewish Christian desire to circumcise Gentile converts was motivated by the desire to avoid reprisals from Zealots, who insisted on complete separation from non-Jews. The zenophobic Zealots would have been the last to be bluffed by such a transparent tactic.

[60] This is also the argument used by Luke in Acts 15: 8, 12.

THE AGREEMENT

Paul expresses the agreement reached in Jerusalem thus: 'James and Cephas and John, who were reputed to be pillars, gave to me and Barnabas the right hand of fellowship, that we should go to the Gentiles and they to the circumcision' (Gal. 2: 9). The inclusion of Barnabas underlines, not only that Paul was but one of a number of missionaries to the Gentiles, but also that the exception from circumcision accorded his converts was valid for all others.

The frustration and anger which Paul experienced when Law-observant Jewish Christians appeared in his communities in Galatia and Corinth suggests that the terms of the agreement might not be as unambiguous as one would wish. The possibility that both sides could have read it in different ways is confirmed by the variety of interpretations current among scholars.

E. Burton argues for a geographical meaning,

> The use of *eis ta ethnê* rather than *tois ethnesin*, therefore, favours the conclusion that the division, through on a basis of preponderant nationality, was nevertheless territorial rather than racial. This conclusion is, moreover, confirmed by the fact that twice in this epistle (1: 16; 2: 2) Paul has spoken unambiguously of the Gentiles as those among (*en*) whom he preached the gospel, and that he has nowhere in this epistle or elsewhere used the preposition *eis* after *euangelizomai* or *kêryssô* to express the thought 'to preach to'. . . . The whole evidence, therefore, clearly indicates the meaning of the agreement was that Paul and Barnabas were to preach the gospel in Gentile lands, the other apostles in Jewish lands.[61]

Abstracting from the spurious clarity of the philological argument, one has only to ask the precise meaning of 'Jewish lands' to see the weakness of this position. In Judaea alone, possibly in Galilee and Perea, was there a preponderance of Jews. Yet all together they numbered less than a million,[62] whereas estimates of the Jewish population of the Roman empire range from four to eight million.[63] It is highly unlikely that James and the others intended to cede all these potential converts to Paul.[64]

Those who appreciate the force of this objection go to the other extreme and understand the agreement in exclusively ethnic terms. Paul and Barnabas could approach Gentiles anywhere but not Jews, whereas missionaries from Jerusalem could preach to Jews anywhere but not to Gentiles.[65] This inter-

---

[61] (1921), 98. Similarly Lagrange (1925), 38; Holmberg (1978), 30.
[62] Hamel (1990), 137–40; Broshi (1979).
[63] Tcherikover (1959), 504–5 n. 86. A survey of the Jewish Diaspora is provided by Philo, *Legatio ad Gaium*, 281–2.
[64] So rightly Haenchen (1971), 467.
[65] So Betz (1979), 100; Lüdemann (1984), 72; (1989), 37.

pretation gives rise to two serious problems. The cogency of the arguments justifying a Judaizing approach to Gentiles has been emphasized above. It is difficult to think that Paul's opponents would abandon such a well-founded position. Moreover, if they were prepared to give way on Paul's gospel, they had a right to expect that both he and the 'pillars' would accept their version of the Gentile mission. It is not as if it would become a squabble over a small number of possible converts. The world was vast and the number of Gentiles uncountable. Secondly, the ethnic understanding of the agreement would deny Paul access to Diaspora synagogues, and there is no hint that he felt so restricted. On the contrary, many arguments in his letters are unintelligible without a reasonable knowledge of the Jewish Scriptures, which implies that there were at least some Jews, and certainly God-fearers in his communities. There is also another side to this objection. Is it reasonable to think that Judaizers would have felt themselves bound to ignore 'God-fearers', who had shown themselves so sympathetic to Judaism that they participated in its prayers and study, just because they were Gentiles?[66]

The fact that neither the geographical nor the ethnic interpretation of the Jerusalem agreement can explain what actually happened in the missionary expansion of the early church forces us to look at the agreement from another perspective. The issue at the meeting was not the legitimacy of a mission to Gentiles, but the conditions under which Gentiles could be accepted as members of the church. Jerusalem accepted that Paul need require nothing more of them than faith in Jesus Christ. What the agreement meant as far as Paul and Barnabas were concerned was that they need not circumcise their converts. One would assume, in the light of the points made above, that other missionaries were free to circumcise their recruits. The agreement, in other words, concerned neither territory nor race, but missionary practice.[67]

Thus Paul and Barnabas were free to accept converts from both Judaism and paganism, as were their opponents. Both parties to the agreement, therefore, recognized and accepted mixed communities. Whether the implications were as clear to Paul and Barnabas as they were to their rivals remains to be seen. The Judaizers could look forward to churches that were mixed only in theory, since Gentile converts who accepted circumcision would naturally also accept other Jewish observances. For all practical purposes they were Jewish communities (cf. Acts 2: 42-7). Since Paul and Barnabas had resisted attempts to force Gentiles to live like Jews, it must be assumed that they recognized that they had no mandate to force Jewish converts to live like Gentiles. All that they demanded of each convert was belief in Christ. Jewish converts, therefore, were at liberty to continue to obey the Law and, if they wished, to circumcise their

---

[66] The Aphrodesias inscription lists two God-fearers as members of a Jewish 'decany of the students of the law, also known as those who fervently praise God'; see my (1992c), 421.
[67] So rightly Haenchen (1971), 467.

children. Such communities were truly mixed and inherently unstable, because their components followed different rules. Often what was important to one part of the community was irrelevant to the other. If they blended to the point of creating a genuine unity, it can only have been because of conscious concessions by both sides.[68] Such arrangements were a permanent source of tension because they were continuously renegotiable, as the case of the church of Antioch illustrates.

THE COLLECTION

Paul concludes his account of the meeting in Jerusalem with the words, 'all they asked was that we should remember the poor, which very thing I was eager to do' (Gal. 2: 10). A number of commentators capitalize 'The Poor', and understand 'remember' in the sense of an acknowledgement of personal merits. Thus, we are told, the agreement 'stipulated that the Gentile Jesus believers were to give recognition to the exemplary performance on the part of their fellow believers in Jerusalem'.[69] It is perhaps not impossible that this was what the Pillars of the Jerusalem church had in mind when they invited Paul to accept the condition, but it is most improbable that it was Paul's interpretation.[70]

With the exception of the Christological statement in 2 Corinthians 8: 9, Paul always uses 'poor' (2 Cor. 6: 10; Gal. 4: 9) and 'poverty' (2 Cor. 8: 3) in their natural material sense. The socio-economic meaning is confirmed by his reference to 'the poor among the saints in Jerusalem' (Rom. 15: 26), where it is most improbable that the genitive is anything but partitive.[71] The natural reading is that some believers were in need. How many they cannot be determined, but the formulation does not exclude a high proportion of the community.

Those who espouse a more spiritual interpretation do in fact recognize an economic dimension, but they formulate the problem in such a way as to make need a by-product of unrealistic detachment.[72] This is to wilfully ignore what J. Jeremias has established regarding social conditions in Jerusalem in the first century:

> Jerusalem in the time of Jesus was already a centre for mendicancy; it was encouraged because alms-giving was regarded as particularly meritorious

[68] Recognition of this necessity subsequently inspired the promulgation of the so-called Apostolic Decree (Acts 15: 23–9); see Haenchen (1971), 468–72.

[69] So e.g. Georgi (1992), 38. For Holmberg (1978), 55–6, the request was an exercise of power and authority designed to demonstrate the inferiority of Gentile Christians. The collection, however, would have produced the opposite effect, because it would have made Jerusalem the client and the Gentiles the patrons; see Ch. 12 'Financial Assistance'.

[70] So rightly Lüdemann (1984), 79.

[71] The view that the genitive is epexegetical—'the poor who are the saints in Jerusalem'—is rightly rejected by Dunn (1988), 875.

[72] Georgi (1992), 35.

when done in the Holy City. . . . Jerusalem had already in Jesus' time become a
city of idlers, and the considerable proletariat living on the religious importance of the city was one of its most outstanding peculiarities.[73]

That a number of Christians belonged to this class is shown by the note in Acts
to the effect that wealthy members of the community sold land and houses in
order to subsidize needy members of the community (Acts 2: 45; 4: 34–5).
Unless rich new members were regularly recruited, this system could have only
one result; the community would run out of money. Since Christians were
persecuted by at least some Jews (Gal. 1: 22–3), the possibility of aid from traditional Jewish sources steadily diminished. That left only the burgeoning Gentile
church, whose members, though not rich, were almost certainly better off than
the majority of the Jerusalem community.

The shift from the plural to the singular in Galatians 2: 10 is not without significance.[74] The basic agreement was between churches. Jerusalem had made a
fundamental concession to Antioch and intended to profit from it. As a mere
agent of Antioch, Paul had no personal responsibility for the collection of
funds, and once he had broken with Antioch (Gal. 2: 11–14), he was in no way
officially involved. His conscience, however, thought otherwise. Paul had lived
in Jerusalem long enough to be fully aware of the social conditions of the city.
Coming from a well-established Diaspora community, he may even have
wished, while still a Pharisee, that Jews abroad would do more for their co-
religionists in Jerusalem. Now that he had the opportunity to invite others
indebted to him to alleviate some of the misery of poor Christians in the Holy
City did he have any choice? Common sense and the subsequent witness of 1
Corinthians 8 combine to exclude the possibility that Paul might have used his
differences with the authorities in Jerusalem as an excuse to avoid a simple
imperative of charity.[75] The opportunity to pour burning coals on the head of
his enemies (Rom. 12: 20) might have been an added attraction!

Once the business of the Antioch delegation had been completed in
Jerusalem, there was no need for Paul to hang around there. Fortified by the
affirmation he had received, he would have been eager to strike out into new
mission fields. Already autumn, it was too late in the year for boats to be still at
sea. In any case sailing to Europe was always an extremely slow and laborious
business. The prevailing wind was from the west, and boats of the period were
not rigged to sail into the wind efficiently. They had to anchor when the wind
was contrary, and then scramble to take advantage of any favourable breeze.
The decision to go north overland was the obvious one, even if Paul were not

---

[73] (1969), 116, 118.
[74] See Wedderburn (1988), 37–41.
[75] The refusal of N. Taylor (1992), 198–9, to take Gal. 2: 10 seriously pales into insignificance
beside his assertion that Paul undertook the collection in order to win for his churches the same
status that Antioch enjoyed.

obligated to return to Antioch with Barnabas to report the outcome of the conference in Jerusalem. By November the first snows had already fallen on the
Anatolian plateau, and passage beyond the Cilician Gates, if not impossible,
was highly dangerous. Hence, we must assume that Paul and Barnabas spent
the winter of AD 51–52 in Antioch.

ANTIOCH AND ITS JEWS

At the time with which we are concerned Antioch-on-the-Orontes—so-called to
distinguish it from the fifteen other cities endowed with the same name by a
single founder, Seleucus I Nicator (311–281 BC)—was the third largest city in
the Roman empire, surpassed only by Rome and Alexandria.[76] The prime position at the intersection of north–south and east–west trade routes, which
Corinth enjoyed as a birthright, was created for Antioch. The north–south
route already existed, but the building of a harbour, Seleucia Pieria, 20 miles
(32 km.) to the west at the mouth of the Orontes river, encouraged exploitation
of the river valley as a trade route through the mountains to the Fertile
Crescent.

The original population was artificially assembled from Macedonians,
Athenians, and Jews, plus some native Syrians who were very much second-
class citizens.[77] Under the Seleucids the grid-plan original city grew by the progressive and systematic addition of three further areas. The most conservative
estimate puts the population of Antioch in the first century AD at 100,000.[78]
When the legions tramped into the east in 64 BC, Pompey made Antioch the
capital of the new Roman province of Syria. Emperors and kings vied for the
honour of augmenting its splendour and beauty. Herod the Great, for example,
gave it the most majestic *cardo maximus* in the known world by paving the 9.5
metre (31 feet) wide street with marble, and building 9.8 metre (32 feet) wide
covered sidewalks along both sides of its entire 3.2 km. (2 miles) length.[79]

By Herod's time the Jewish community was well established. According to
Josephus, the Jews in Syria were particularly numerous and were concentrated
in Antioch (*JW* 7. 43–5). The validity of his claim that Jews as such were full
citizens of Antioch is suspect because of the lack of any independent confirmation.[80] Some may well have acquired citizenship, but as a group the Jews would
have been resident aliens organized as a separate *politeuma* within the body
politic.[81] This gave them an official position in the city, but without having a

---

[76] For the historical background, see Festugière (1959); Downey (1961); Liebschutz (1972);
Lassus (1977); Meeks and Wilken (1978). A good map is to be found in Finley (1977), 222.
[77] Josephus, *AJ* 12. 119.   [78] Norris (1992), I. 265.
[79] Josephus, *AJ* 16. 148; *JW* 1. 425.
[80] See Schürer (1973–87), 3. 126–7.
[81] Kraeling (1932), 138–9; Smallwood (1981), 359–60.

voice in its affairs. None the less they were masters of their own affairs with clearly defined rights, which at Antioch were displayed publicly on bronze tablets.[82]

The Foundation of the Church

Paul tells us nothing about the evangelization of Antioch. Luke's rather detailed account (Acts 11: 19–26) is a complex mix of tradition and redaction.[83] His source attributes the foundation of the church to 'certain people from Cyprus and Cyrene who, having come to Antioch, spoke to the Greeks' (Acts 11: 20).[84] Luke himself created Acts 11: 19, but J. Taylor convincingly argues that he drew on an ill-defined tradition of a mission from Jerusalem to Jews in Antioch.[85] How these two missions were related to one another remains obscure, but a mixed church certainly existed there by the end of the 30s.[86] It was in this community that Paul and Barnabas ministered for at least a year. According to the continuation of Luke's source, Barnabas arrived in Antioch from Jerusalem, and subsequently recruited Saul from Tarsus (cf. Gal. 1: 21).[87]

The implication of Luke's source that the evangelization of Antioch was an unimpeded success is accentuated by his redactional additions. Only one brief note hints at savage currents roiling beneath the placid surface of the narrative. He tells us that 'in Antioch the disciples were for the first time called Christians' (Acts 11: 26). J. Taylor has drawn attention to the fact that in non-Christian first-century sources the names 'Christ' and 'Christians' are invariably associated with public disorders and crimes,[88] and linked this fact to three reports of events in Antioch all of which are dated to the same year, namely, AD 39–40.[89] The first is the note by the Byzantine chronicler John Malalas that many Jews were killed in a pogrom, whose improbable cause is said to have been a dispute between two circus factions.[90] The second is the synthesis of a series of hints in Josephus that the affair of Gaius' statue[91] provoked a pogrom similar to that which occurred at the same time in Alexandria under Flaccus.[92]

---

[82] Josephus, *JW* 7. 110.
[83] For details, see Boismard and Lamouille (1990), 3. 165–8.
[84] The source is reconstructed by Boismard and Lamouille, (1990), 2. 66 .
[85] (1994*b*), 65.
[86] Ibid. 70.
[87] Boismard and Lamouille (1990), 2. 66.
[88] '[Claudius] expelled from Rome the Jews who were constantly causing disturbance at the instigation of Chrestus' (Suetonius, *Claudius* 25. 4; cf. *Nero* 16); 'Nero punished the culprits, whom, hated for their shameful acts, the populace called "Christians". The author of this name, Christus, had been put to death by the procurator Pontius Pilate during the reign of Tiberius' (Tacitus, *Annals* 15. 44. 2).
[89] (1994*a*), 75–94.
[90] *PG* 97. 373–5.
[91] See above, pp. 139–40.
[92] Kraeling (1932), 148–9; Smallwood (1981), 360–1.

The third is the notice in the *Chronicle* of Eusebius that the founder(s) of the church in Antioch left the city for Rome.[93]

The natural impression that the four events are somehow related is given expression by Taylor in a simple but eminently plausible hypothesis: 'the disciples of Jesus were first called Christians at Antioch in connexion with a disturbance among the Jewish population of the city in the third year of Gaius (AD 39–40), and that this disturbance had to do with the preaching of the Gospel and the beginning of the church at Antioch'.[94] To be more specific is to be more speculative, but the chain of events has a certain inevitability. The Christian missionaries preached Jesus as the Messiah, which could be understood in political terms as a call to liberation. In the extremely tense atmosphere created by the announcement of Gaius' proposed desecration of the temple, Jewish extremists seized this as a pretext to whip up opposition to Rome. In the process they alienated the Antiochenes who rose against them and killed many. When Petronius succeeded in stopping the violence, he looked for the instigators. Realizing that they were likely to be blamed, the founders of the church departed, leaving behind converts who found themselves landed with a name which identified them as troublemakers capable of sedition.

If this estimate of the situation in Antioch in the spring of AD 40 is correct, it provides a very simple explanation of why the Jerusalem church sent Barnabas to Antioch. In view of the intense communication between Jerusalem and Antioch engendered by the affair of the statue of Gaius, there is nothing implausible in news of the fracas for which Christians were blamed reaching Jerusalem very quickly. The fate of the young community, which had lost its leadership, naturally would be a matter of great concern to believers in Jerusalem. The fraternal response would be to fill that gap by sending an experienced Jewish Christian from the Diaspora (Acts 4: 36). Barnabas did not go to Antioch to inspect or correct, but to stabilize a demoralized community. His name, it will be recalled, means 'son of encouragement' (Acts 4: 36); this quality might explain why he was selected for the mission, or the name could be due to what he achieved at Antioch. In this scenario Barnabas' recruitment of Saul was a most astute tactical move. For a persecuted community, the symbolic value of a converted persecutor of the church could not be over-emphasized. The presence of Saul among the distressed believers of Antioch verified the power of grace promised in the gospel. God was all-powerful. There was hope for the future.

---

[93] Schoene (1875), 150, 152–3. See the discussion in Taylor (1994a), 89 n. 42.
[94] Ibid. 91.

THE PROBLEMS OF A MIXED COMMUNITY

There is no doubt that, once it overcame its initial difficulties, the church at Antioch flourished. The missionary outreach attested by Acts reveals a level of energetic commitment to the Good News that betrays a confident and vital community. Given the mixture of Jews and Gentiles (Gal. 2: 12–13), this was no mean achievement, and it is necessary to assess the factors which contributed to it.

Unlike the Jews whose synagogues were legally recognized public meeting-places, the first Christians had to make do with the hospitality offered by the more affluent members of the community. There is no evidence that any of these belonged to the patrician class which owned vast mansions.[95] In consequence, space became a problem as the size of the church increased. The number of Christians in Antioch cannot be determined, but at the very least it cannot be less than the minimum of 50 postulated for Corinth.[96] It would have been difficult to fit all these into the public space of the average house of a moderately wealthy person. Presumably this is why Paul speaks so rarely of a meeting of 'the *whole* church' (Rom. 16: 23; 1 Cor. 14: 23). If believers met only as a single group the adjective 'whole' is unnecessary. Its use necessarily implies the existence of sub-groups, 'the church in the home of X' (Rom. 16: 5; 1 Cor. 16: 19; Col. 4: 15; Philem. 2).[97] Hence, we must assume, that at Antioch for purely practical reasons the Christian community was made up of a number of house-churches.

Such an arrangement had the advantage of offering converts a choice. While in theory they were joining a single community, in practice they had to opt for one particular house-church among a number. Many and highly diverse factors no doubt influenced selection, but it would be unrealistic to assume that individual house-churches had both Jewish and Gentile members. The trend must have been towards the creation of Gentile and Jewish house-churches, which were grouped together under the umbrella of one *ekklêsia*. Unless the umbrella was to be a complete fiction, however, there had to be strong and regular links between the different house-churches.

The most important of such links was table-fellowship. In the ancient Near East a formal meal was the prime social event. To share food was to initiate or reinforce a social bonding which implied permanent commitment and deep ethical obligation.[98] In the eyes of their contemporaries there would have been

---

[95] According to Meeks (1983), 73, 'The extreme top and bottom of the Greco-Roman social scale are missing from the picture. It is hardly surprising that we meet no landed aristocrats, no senators, equites, nor (unless Erastus might qualify) decurions.'

[96] See my (1992*e*), 164–6.   [97] Banks (1980), 38.

[98] See Smith (1992), 6. 302–4.

no genuine community among Christians unless, in addition to the ritual of the Eucharist, they gathered around a common table.

Nowhere was the significance of the meal more accentuated than in Judaism. As we have seen above, 67 per cent of Pharisaic legislation which can be dated with some plausibility to the pre-AD 70 period is concerned with dietary laws, 229 specific rulings out of 341.[99] Not all Jews would have been as scrupulous as the Pharisees. It is equally certain, however, that the vast majority would have observed the fundamental distinction between clean and unclean food, and would have insisted on the former being entirely drained of blood (cf. Acts 10: 14). It was a matter of principle for which their ancestors had died (1 Macc. 1: 62–3), and it was one of the most obvious identity markers of the Jewish religion. 'Separate yourselves from the nations, and eat not with them' (Jub. 22. 16). What this meant in practice for relations between Jews and Gentiles is well spelt out by E. P. Sanders, 'All the Jewish evidence thus far considered presents the *legal* situation perfectly clearly: There was no barrier to social intercourse with Gentiles, as long as one did not eat their meat or drink their wine.'[100]

How then did the Jewish and Gentile house-churches of Antioch maintain any semblance of unity? Dunn rightly dismisses the two extreme possibilities, namely, that the Jews created no difficulties for Gentiles by ignoring their own laws, or that the Gentiles created no problems for Jews by adopting a Pharisaic level of dietary observance.[101] In this latter case Peter could not have been said to have 'lived like a Gentile' (Gal. 2: 14) simply because he ate with believers of pagan origin. The most probable scenario lies somewhere in the middle.

When Gentile believers dined with Jews they accepted the food offered them, even though kosher meat might not have been to their taste.[102] When Jews dined in a Gentile house, they trusted their fellow-believers to offer them Jewish food and drink. From a Jewish perspective such trust was a significant concession. Most if not all the meat available outside Jerusalem would have been part of a pagan sacrifice, and the common assumption was that Gentiles would pollute Jewish food and drink if they got the slightest chance (*m. Abodah Zarah* 5. 5). Hence Jews regularly brought their own food when dining with Gentiles.[103]

The plausibility of this compromise is enhanced by the number of God-fearers at Antioch (*JW* 2. 463; 7. 45). If, as seems probable, the majority of Gentile converts to Christianity at Antioch were drawn from such people, whose attraction to Judaism found expression in the adoption of Jewish

---

[99] See Ch. 3, 'A Pharisee'.

[100] (1990), 178.   [101] (1983), 31.

[102] The king of Egypt said to the Jewish translators of the Bible, 'Everything shall be prepared in keeping with your usages, and for me also along with you' (*Letter of Aristeas* 181).

[103] When going out to Holofernes, Judith 'gave her maid a skin of wine and a flask of oil, and filled a bag with roasted grain, dried fig cakes, and fine bread; then she wrapped up all her dishes and gave them to her to carry' (Judith 10: 5; cf. 12: 2).

practices,[104] it would have been very easy for them to make the relatively minor concession which made table-fellowship with their Jewish fellow-believers possible. In practice all they needed to do was to buy at a Jewish shop when they had Jewish guests and to accept the added expense. 'Not only would tithes have had to have been paid on [such produce] to conform with the Law, but it is an universal economic reality that any produce required to meet specifications over and above what is normative in the market will accordingly be more expensive.'[105]

Outside Interference

This delicate balance was disturbed by a delegation from Jerusalem; 'certain people from James came' (Gal. 2: 12). Prior to their arrival Peter had had no difficulty eating regularly in Gentile house-churches.[106] He continued for a while, but he gradually drew back and ended up by stopping completely,[107] 'and the rest of the Jewish believers joined him in playing the hypocrite—so that even Barnabas was led astray by their hypocrisy' (Gal. 2: 13; trans. Longenecker). A barrier rose between the Jewish and Gentile house-churches.

What had the people sent by James insisted on to precipitate this crisis? No prohibition of mutual hospitality was necessary. All they had to do was to assert that Jewish believers should no longer assume that Gentile Christians would offer them Jewish food. Such blanket and unwarranted criticism of their standards of honour and decency must have proved extremely offensive to Gentile church members. Those who were prepared to accept the slur, and who believed that communion with Jews was essential to preserve the ideal of unity would have had to hand over control of their kitchens to Jews.[108]

One's judgement as to whether even this would have satisfied James depends on why he intervened at Antioch. The nationalistic reasons which led him to refuse the circumcision of Gentiles also obliged him to insist on the observance of dietary laws for Jewish converts.[109] In both cases it was a question of conserving Jewish identity, in one by refusing dilution, and in the other by positive reinforcement. In such circumstances no matter what concessions Gentile converts might be prepared to make, others would be demanded. Separation was the real objective, and Judaization only the means.

---

[104] Esther 8: 17 LXX; Josephus, *JW* 2. 454 show that this is the sense of *ioudaizein*; see Dunn (1983), 26.   [105] N. Taylor (1993), 126.

[106] E. Burton (1921), 104, insists that the use of the imperfect tense 'implies that he did this, not on a single occasion, but repeatedly or habitually'.

[107] Ibid. 107.

[108] So rightly Dunn (1983), 31. Much less probable is the hypothesis of E. P. Sanders (1990), 186, that James was concerned exclusively with the damage to Peter's reputation that would result from frequent association with Gentiles.

[109] Dunn (1983), 32.

Whatever his personal feelings, such consistency would have been imposed on James by those in Jerusalem, who had had to accept his position on circumcision. Peter found himself on the horns of a dilemma. His actions had declared the table-fellowship of the church at Antioch unobjectionable, but he had sided with James at the meeting in Jerusalem, and he was responsible for the mission to Jews (Gal. 2: 8). He was now in a situation where he could not have it both ways. He had to make a public decision, and he opted for his Jewish roots. For Paul his motive could only be unworthy, and he postulates 'fear' (Gal. 2: 12).[110] It is entirely possible, however, that Peter read the situation clearly, and in great agony of mind decided for those who needed him most. The strength of the Gentile church was apparent at Antioch, and it had dynamic leaders in Paul and Barnabas. The Jewish church was struggling, and would be shattered by the defection of one of its most revered figures.

Peter's decision reinforced the authority of the Jerusalem delegation, and naturally the Jewish Christians followed his lead. What is surprising is that Barnabas also did so. This pained Paul grievously. The pathos of 'even Barnabas' (Gal. 2: 13) reveals the depth of his disappointment. They had soldiered together in the mission field (1 Cor. 9: 6), and in defence of the freedom of the Gentiles at Jerusalem (Gal. 2: 1–10). Why did Barnabas now deny everything that he had stood for? The verb 'to carry off with' bears the connotation of irrationality and suggests that Barnabas was swayed by emotion.[111] The nationalistic appeal of James touched his Jewish heart and blinded him to the consequences for the church at Antioch.

For Paul this development was manifestly unchristian. James, on the contrary, did not see it as at all incongruous. In fact he must have been surprised and offended by Paul's reaction. It is a tragic paradox that James's inherited conviction that separation was the only way to preserve Jewish identity was reinforced by the very argument on which Paul had insisted so passionately during the circumcision debate, namely, that belief in Jesus as the Messiah was the one essential condition for membership in the church. James could hardly be blamed for drawing the conclusion that social contacts between Jewish and Gentile converts were irrelevant.

THE LAW A RIVAL TO CHRIST

For Paul the shock of being hoist with his own petard proved to be the providential incentive to rethink his vision of a mixed Jewish and Christian local

---

[110] Certain exegetes take him seriously and debate whether it was fear of the political consequence of losing his position of power (so Betz (1979), 109) or fear of the consequences for the Jerusalem church if he, one of its 'pillars', were known to fraternize with Gentiles (so Longenecker (1990), 75).   [111] Betz (1979), 110; Longenecker (1990), 76.

church in at least two respects. First, what he had always acted on in practice (cf. 1 Thess. 4: 9), he now was forced to articulate as a principle. Faith in Jesus was basic, but it alone did not make a person a Christian. A believer had to live the truth in love ('Without love I am nothing'; 1 Cor. 13: 2), and a believing community had to 'put on love which is the bond of perfection' (Col. 3: 14). Second, whereas previously Paul had been content to permit Jewish members of the church to continue to observe the Law, he now recognized that if the Law was given the tiniest toe-hold in a local church it would ultimately take over, as it had in fact done at Antioch. If Paul did not immediately become antinomian, he was well on the way to perceiving the fundamental incompatibility of the Law and Christ.

The vivid urgency of his criticism of Peter in Galatians 2: 14–21 strongly suggests that Paul is reliving the crucial argument. What he writes, therefore, is probably an adequate reflection of the thrust of what he actually said.[112] Though his logic is not ours, the central thrust of his argument is unambiguous.

The action of the delegation from Jerusalem said in effect that, though Gentile believers were 'in Christ', they none the less remained 'sinners', because to be a Gentile and to be a sinner were one and the same thing (Gal. 2: 15b). Paul understood them to assert that the death of Christ was meaningless (Gal. 2: 21b). Not only did it change nothing for the better, in fact it made the situation worse. If Jewish believers are also 'in Christ', Paul prolongs the logic, they are in an intimate union with Gentiles, and so they too must be 'sinners'. Hence Christ is nothing but an 'agent of sin' (Gal. 2: 17).

To any believer in Christ the two conclusions are absurd. In consequence, the premises from which they flow must be false. Neither Jewish nor Gentile believers are 'sinners'. What is true for one, however, is true for the other. If Gentiles are justified, Paul asserts, it cannot be in virtue of 'works of the Law', because they neither know nor execute its demands. It must be solely in virtue of their faith in Jesus Christ that they are saved (Gal. 2: 16). If faith alone is adequate for Gentiles, then it is also sufficient for Jews (Gal. 2: 21).

Were Paul, or any other Jewish believer, to accord the Law the absolute authority it enjoyed when he was a Pharisee, he would in effect be denying Christ, the true source of authentic life (Gal. 2: 20). It is no longer the Law which speaks for God, but Christ alone. Henceforth obedience is defined by reference to Christ (Gal. 6: 2). Paradoxically, therefore, to obey the Law is to make oneself a transgressor (Gal. 2: 18).

Whereas he had once seen the Law simply as another factor in the human situation, Antioch taught Paul that it was a dangerous rival to Christ. He saw for the first time that, if the Law was given a foothold in any community, it

---

[112] It goes without saying that Paul has the Galatians in view in this section, and is not interested in a historicizing reconstruction of his actual words at Antioch.

would assume a dominant role.[113] Once one imperative was obeyed, the increasing insistence of other demands would deflect attention away from Christ. It was only in his letter to the Romans, written some five years later, that Paul spelled out in detail his criticisms of the Law. But one can see in his comportment,[114] and in the pastoral instructions he gave his communities in the interval that he became convinced, not merely that nothing in the Law was binding on believers,[115] but that 'law as such is no longer valid for the Christian'.[116]

PASTORAL INSTRUCTION

When dealing with the ethical directives which Paul gave the Thessalonians, attention was drawn to his recognition that the witness value of believers depends on freely chosen behaviour.[117] He instinctively refrained from imposing or prohibiting any particular act. Only when he felt he had no other choice did he issue a command ordering the community not to associate with the undisciplined (2 Thess. 3: 14). No doubt he regretted the necessity, but he could still justify the precept in terms of his concern to ensure the positive impact of the church on its environment (1 Thess. 4: 12).

After the incident at Antioch, this was no longer possible. Could Paul have intended his precepts to have a coercive force which he denied to the commandments of God in the Law? Could he have insisted on being obeyed, while arguing that to submit to the Law was to become a transgressor (Gal. 2: 18)? In this sphere Paul proved to be totally consistent, both as regards his own practice, which had exemplar value for his converts (Gal. 4: 12; 1 Cor. 11: 1), and in what he said to his churches.

Paul twice quotes commands of Jesus.[118] The first is the prohibition of divorce in 1 Cor. 7: 10, which Paul accepted in a particular case (1 Cor. 7: 11), not because he felt bound by it, but because he disagreed with the reasons for the divorce.[119] In another instance, however, he found the reasons compelling and permitted a divorce (1 Cor. 7: 15),[120] thereby revealing that, despite the

---

[113] Barrett (1994), 82 (see 94, 104), is one of the few to have seen that 'it is very difficult (perhaps not in the end impossible) to have a law without legalism'.

[114] 1 Cor. 9: 19–23 formally articulates Paul's sense of freedom. The comment of Fee (1987), 427, is very much to the point, 'when he was among Jews he was kosher; when he was among Gentiles he was non-kosher—precisely because, as with circumcision, neither mattered to God (cf. 7: 19; 8: 8). But such conduct tends to matter a great deal to the religious—on either side—so that inconsistency in such matters ranks among the greatest of evils.'

[115] As Westerholm (1988), 205–9, has reaffirmed.

[116] Knox (1962), 99.

[117] See Ch. 5, 'Exemplary Behaviour'.

[118] Dungan (1971).

[119] See my (1981a), 901–6,

[120] Everything necessary for a correct interpretation of this verse is contained in the Jewish legislation on divorce, 'The essential formula in the bill of divorce is "Lo, thou art free to marry any

imperatival form, he refused to give the prohibition of Jesus the force of a constraining precept. His attitude towards the second command is even clearer. The form in which he quotes it—'I command those who proclaim the gospel to live from the gospel' (1 Cor. 9: 14)—makes it an obligation for the minister to receive, not for the community to give.[121] Yet Paul immediately goes on to insist that he has not obeyed and will not obey; he will continue to earn his own living (1 Cor. 9: 15–18). The citation of the two dominical commands underlines their value, and the respect in which they should be held, but Paul's practice indicates that he did not see them as imposing an obligation.

Since habits of speech are not automatically altered by ideological conversion, it was perhaps inevitable that Paul should occasionally command that something be done. In some cases he catches himself and introduces a correction, but in others he does not. In this he cannot be accused of inconsistency. A distinction can be drawn between the two sets of situations. He speaks in the imperative mood regarding conjugal relations (1 Cor. 7: 5), and generosity in giving to the poor of Jerusalem (2 Cor. 8: 7), but in both instances he immediately adds, 'I say this not as a command' (1 Cor. 7: 6; 2 Cor. 8: 8).[122] The issues on which he does not correct himself concern change of social status subsequent to conversion (1 Cor. 7: 17), issues raised by the Corinthians (1 Cor. 11: 34), and the mechanics of the transmission of the collection to Jerusalem (1 Cor. 16: 1). Paul, in other words, is careful to avoid imposing strictly moral judgements, but has no hesitation in making administrative decisions. The latter concern purely practical matters, whereas the former involve interpersonal relations which are of the essence of Christian life. On basic moral issues, Paul will only offer advice, 'I say this for your advantage, not to lay any restraint upon you' (1 Cor. 7: 35).

The assumptions behind this attitude should be clear from what has already been said about the Law, but Paul none the less makes them explicit in two passages. He refuses to oblige anyone to contribute to the collection for Jerusalem because 'Each one must give as he has decided in his heart, not reluctantly or under compulsion, for God loves a cheerful giver' (2 Cor. 9: 7). The freedom of the decision is stressed both positively and negatively. It must come from the 'heart', which in biblical terms is the core of the personality. The

---

man." R. Judah says: "Let this be from me thy writ of divorce and letter of dismissal and deed of liberation that thou mayest marry whatsoever man thou wilt." The essential formula in a writ of emanciption is, "Lo thou art a freedwoman; lo, thou belongest to thyself"' (*m. Git.* 9. 3; trans. Danby). The understanding of a writ of divorce as a 'deed of emancipation' justifies the juxtaposition of the two cases (wife and bondwoman) and explains why Paul here wrote 'is not enslaved' rather than 'is not bound' (cf. 1 Cor. 7: 27, 39). No mention is made of remarriage because that right is the very essence of a divorce, *pace* Fee (1987), 303.

[121] So rightly Klauck (1984), 66, against Robertson and Plummer (1914), 187; Barrett (1968), 208.

[122] See esp. Fee (1987), 283.

choice must well up from within. It cannot be forced in any way. What one is compelled to give will always be given with regret, and cannot be pleasing to God (Prov. 22: 8).

The second passage is even more explicit. Paul writes to Philemon, 'Even though I have full authority in Christ to order you to do what is fitting, yet for love's sake I rather beseech you' (vv. 8–9a). Paul knew that he had the personal authority to command Philemon to do the right thing concerning Onesimus, his runaway slave, namely, to receive him back without any punishment. In a subtle *captatio benevolentiae* Paul expects Philemon to recognize that only love is always 'fitting' for Christians. But there was also another reason, 'I preferred to do nothing without your consent in order that your good act might not stem from compulsion but from your own free will' (v. 14). The opposition between 'compulsion' and 'free will' is absolute; the same act cannot be both voluntary and forced. To be bound by a precept is to be incapable of acting freely. The constraint of a command makes a free choice impossible. If Philemon is to love Onesimus, the decision must be entirely his.

Only when the pattern of such Pauline passages is perceived does it become clear just how radical Paul's antinomian stance was.[123] He would not give obedience to any law, and he would not exact submission from his converts. He would indicate what he expected of them. He would attempt to persuade them to modify their behaviour. He would propose his own example (e.g. 1 Cor. 8: 13). It would have been much easier for him to have forcibly imposed the comportment he desired. But his experience at Antioch had taught him that to operate through binding precepts would necessarily bring him and his converts back into the orbit of the Law.

Henceforth for Paul there was only 'the law of Christ' (Gal. 6: 2b; cf. 1 Cor. 9: 21). The complex debate about the meaning of this phrase is fuelled, not by its intrinsic difficulty, but by an obstinate desire on the part of certain eminent exegetes to find in Paul the basis of a binding code of Christian conduct.[124] The vanity of this quest should be clear from what has been said above. The genitive, in consequence, is appositive (BDF §167), and the underlying idea is clearly articulated by Philo, 'The lives of those who have earnestly followed virtue may be called unwritten laws' (*De Virt.* 194; cf. *Vita Moysis* 1. 162; trans. Yonge). *Nomos Christou* should be translated as 'the law which is Christ'. The Jewish Law no longer enshrines the will of God for humanity (Rom. 2: 18). Now God's will is embodied in the comportment of Christ, who both exemplifies the demand made on, and models the response of, humanity. As the immediate con-

---

[123] This conclusion could be confirmed by an analysis of many other passages, as Furnish (1968), 187, has recognized, 'This survey of passages shows, surprisingly, that the apostle nowhere speaks directly about obedience to "the law" or its "commandments," or to God's "will".' See also my (1974*b*), 99–144.

[124] E. Burton (1921), 329–30; Dodd (1953), 96–110; Davies (1962), 73; Longenecker (1990), 275–6). See Furnish (1968), 51–98.

text indicates—'Bear one another's burdens' (Gal. 6: 2a)—love is the sole binding imperative of the new law. It was the salient feature of Christ's humanity,[125] and is the content of the one true precept which remains (Gal. 5: 14; Rom. 13: 8–10; 1 Cor. 7: 19), because it is of the very essence of Christian life (1 Thess. 4: 9; 1 Cor. 13: 2).

[125] See my (1982*b*), 45–8.

7

The Years in Ephesus

PAUL does not inform us who prevailed in his dispute with Peter at Antioch. His silence, however, tells its own story. Had he won, he could hardly have failed to mention it in Galatians.[1] To have been able to assert that Peter had eventually sided with him rather than with James on Jewish practices would have been an important argument against the Judaizing tendency of the churches of Galatia.

The fact that Barnabas had aligned himself with the delegation from Jerusalem left Paul completely isolated. He no longer felt at home in Antioch. The new pattern of its community life reflected an understanding of the gospel with which he could not identify. Its faith was no longer his because, as he saw it, Christ had been moved from his position of absolute centrality. Moreover, Antioch had in effect become a Jewish church. It now mirrored the radical separation between Gentile and Jewish churches, which was the ambition of the nationalistic Jewish Christians, but which was anathema to Paul. He wanted freedom for the Gentile church, but not at the expense of its historical roots in Judaism. He also feared for the Jewish church. His experience as a Pharisee enabled him to foresee what legalism would do to a religious community.

Since the church at Antioch no longer embodied the power of grace, he could not in conscience continue to be its representative in the mission fields to the west (Acts 13: 1– 3). We must assume that this troubled Paul on the human level, but it did not paralyse him. From the beginning he had understood his conversion to be a call to preach among the Gentiles. Even if he was no longer the emissary of a church, the divine commission, which had inspired his abortive mission among the Nabataeans, would validate his subsequent career. He was 'an apostle, not from men or through a man, but through Jesus Christ and God the Father' (Gal. 1: 1).

Sometime in the spring of AD 52, therefore, when the gorge through the Taurus mountains known as the Cilician Gates was passable, and most of the snow had melted on the plateau, Paul left Antioch. He was never to return.

---

[1] So the majority of commentators.

TWO JOURNEYS THROUGH ASIA MINOR
TO THE WEST

Paul had crossed Asia Minor at least once before, on the journey that brought him to the coast of the Aegean Sea, which he crossed to land at Neapolis the port of Philippi (Acts 16: 11). When dealing with that journey earlier in Chapter 1, my concern was with the time it would have taken him.[2] Since the distances were substantially the same, it was not necessary to choose among the routes he might have followed. Only at this stage does the precise route become important because it determines the identity of the converts who most disappointed him, the churches of Galatia (see Fig. 1).

Galatia and the Galatians

Where was Galatia? Paul tells us only that there was such a place (Gal. 1: 2; 1 Cor. 16: 1), that the inhabitants not surprisingly were called Galatians (Gal. 3: 1), and that his first visit there was the result of an accident (Gal. 4: 13).

We know that Augustus created a Roman province of Galatia. Dio Cassius notes in his report for the year 25 BC, 'On the death of Amyntas he [Augustus] did not entrust his kingdom to his sons but made it part of the subject territory. Thus Galatia together with Lycaonia obtained a Roman governor, and the portions of Pamphylia formerly assigned to Amyntas were restored to their own district.'[3] Amyntas was the last in a series of Celtic rulers stretching back to the third century BC when tribes from the Pyrenees pushed their way into Anatolia.[4] The Roman province, however, was greater than the tribal territories. Its southern border englobed Pisidia, Isauria, Lycaonia, and part of Pamphylia.[5] In effect, the province was a strip averaging some 200 km. wide running almost the full way across the centre of Asia Minor from north-east to south-west. This means that four towns evangelized by Paul and Barnabas on what Luke presents as the first missionary journey, namely, Antioch in Pisidia, Iconium, Lystra and Derbe (Acts 13: 13 to 14: 28) belonged to the province of Galatia. 2 Timothy 3: 11 confirms that Paul did in fact minister in these towns.

Opinion is divided as to whether by 'Galatians' Paul intended converts from these towns (the South Galatia, or province, hypothesis) or from the tribal areas (the North Galatia, or territorial, hypothesis). At one stage it was thought

---

[2] See Ch. 1, 'Dating the Stages of the Journey'.

[3] *History* 53. 26. 3; trans. Cary. What is said of Pamphylia is only partially true; see Sherk (1980), 959 n. 17.

[4] Strabo, *Geography* 4. 1. 13.

[5] 'Galatia also touches on Cabalia in Pamphylia and the Milyae about Baris; also on the Cyllanicum and the district of Oroanda in Pisidia, and Obizene which is part of Lycaonia' (Pliny, *NH* 5. 42. 147). See Sherk (1980), 959 and the map facing p. 960; S. Mitchell (1993), map facing 40.

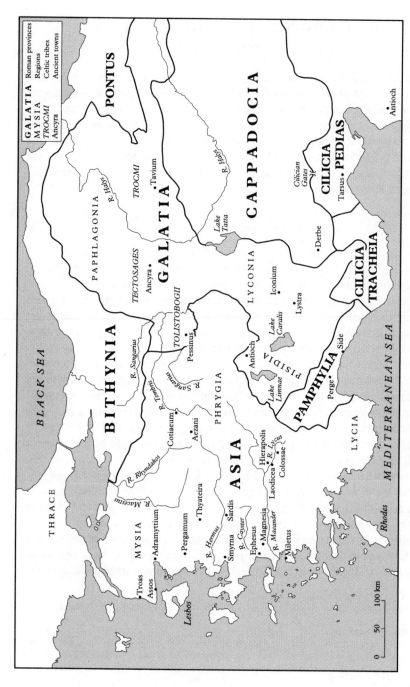

Fig. 1 Asia Minor at the Time of Paul (Source: *Tübinger Atlas des Vorderen Orients*, BV7 (1983))

GALATIA | Roman provinces
MYSIA | Regions
TROCMI | Celtic tribes
Ancyra | Ancient towns

BLACK SEA

THRACE

PONTUS

PAPHLAGONIA

TROCMI

R. *Halys*

Tavium

BITHYNIA

TECTOSAGES

GALATIA

Ancyra

TOLISTOBOGII

R. *Sangarius*

R. *Tembris*

Cotiaeum

Aezani

R. *Sangarius*

Pessinus

PHRYGIA

R. *Rhyndakos*

R. *Macestus*

MYSIA

Adramyttium

Pergamum

Thyateira

ASIA

Sardis

R. *Hermus*

R. *Coyster*

Smyrna

Ephesus

Magnesia

R. *Maeander*

Miletus

Hierapolis

Laodicea

R. *Lycus*

Colossae

Troas

Assos

Lesbos

LYCIA

Rhodes

MEDITERRANEAN SEA

Lake Tatta

CAPPADOCIA

R. *Halys*

Cilician Gates

CILICIA PEDIAS

Tarsus

CILICIA TRACHEIA

Derbe

LYCONIA

Iconium

Lystra

PISIDIA

Lake Caralis

Antioch

Lake Limnae

PAMPHYLIA

Perge

Side

Antioch

0   50   100 km

that Paul could not have been so crude as to use the essentially ethnic names of Galatia and Galatians of those who were not Celts.[6] Respectable citizens of Pisidia and Lycaonia, it was implied, would not appreciate being identified as wild barbarians! More recent studies, however, have removed this apparently plausible argument from contention. 'There are even some hints, contrary to views often repeated, that the term "Galatian" was a correct and honourable title especially acceptable to the more Hellenized or Romanized people in the province.'[7] As far as linguistic usage is concerned, therefore, Paul's letter to the Galatians could have been written to a group of communities anywhere in the Roman province.

This point is developed very adroitly by Burton, 'if the churches addressed [in Gal.] were those of Derbe, Lystra, Iconium and Antioch, which he [Paul] founded on his first missionary journey, he could not address their members by any single term except Galatians.'[8] We are invited to admire how cleverly Paul glosses over the differences between these churches by finding a common denominator in their belonging to the same Roman province. But is it really likely that Paul would have written a single letter to so many and so diverse churches? The improbability is accentuated by their dispersion. Derbe is 286 km. (172 miles) from Antioch in Pisidia.[9] In Macedonia, even though Thessalonica and Philippi are only half that distance apart (166 km., 100 miles), and even though the same type of Judaization was a threat (Phil. 3: 2), Paul deals with each church individually.

Moreover, according to Galatians 4: 13, Paul's first visit to the Galatians was not planned. It was the result of an accident; he fell sick and they nursed him back to health. This fact has not escaped the partisans of the North Galatia hypothesis. But they develop it into an argument only by contrasting Paul's account of his condition with Luke's presentation of missionaries so dynamic that they are taken for gods (Acts 14: 12). While it is easy to imagine reasons why Luke would have passed over an illness in discreet silence, it is much more difficult to explain why Paul himself would omit such a trial in his list of the difficulties he experienced in South Galatia (2 Tim. 3: 11). The way in which he speaks of the problems occasioned by an intensely active ministry there excludes a period of illness so serious that he was a grave burden to the Galatians (Gal. 4: 14).

If the Galatians to whom Paul writes are unlikely to be the believers in the southern part of the Roman province, where did they live? Luke provides a possible answer. With regard to Paul's journey to Ephesus, he tells us that 'he went through the region of Galatia and Phrygia strengthening all the disciples'

[6] e.g. Lightfoot (1910), 19–20; Kümmel (1975), 298; Haenchen (1971), 483 n. 2.
[7] Hemer (1989), 304–5.
[8] E. Burton (1921), p. xxix.
[9] Jewett (1979), 59.

(Acts 18: 23).[10] The impression given is that he retraced the route he and Timothy (at least) had taken previously from Lystra, 'they went through the region of Phrygia and Galatia having been forbidden by the Holy Spirit to speak the word in Asia' (Acts 16: 6). The inversion of the words Phrygia and Galatia and the fact that both can be used as adjectives strongly suggests that Luke has in mind territory which was both Phrygian (by language and culture) and Galatian (by Roman administrative fiat).[11]

The description applies perfectly to the territory in which Antioch in Pisidia and Iconium were located, but it is unlikely that this is what Luke had in mind. Paul must have reached Antioch in Pisidia, the last major town in Galatia, when he realized that he was not going to be able to cross over into Asia. The alternative he chose was to go through Phrygio-Galatian territory, namely, the border area between Asia and Galatia north of Antioch in Pisidia, which would bring them to a point east of Mysia and south of Bithynia (Acts 16: 7).[12]

What is known of the routes, however, indicates that it would be more natural to travel on the Phrygian side of the border.[13] Only south-east of Pessinus (modern Balahisar) would it have been easy to make a diversion to the east, which would have brought him within Galatian tribal territory. But why would Paul make a turn diametrically opposed to his planned journey to the west? It is impossible to find motivation for a change of plan. Something must have happened to force Paul to abandon temporarily his project to work his way around Asia. The illness he mentions (Gal. 4: 13) is such an explanation, but to speculate on what it was and how it changed his plans is fruitless. Hence it seems most probable that the Galatians to whom Paul wrote were inhabitants of the north-eastern corner of Galatia, the almost square territory bordered on three sides by the immense bend of the river Sangarios (modern Sakarya).

From Galatia to Troas

We have seen that the letters suggest that after Galatia Paul evangelized Macedonia.[14] The Apostle gives no hint of the route he took. Luke, however, tells us that 'passing through Mysia they came down to Troas' (Acts 16: 8).[15] Even though Mysia was part of Asia, its relation to the province paralleled that

[10] With regard to the reliability of Luke's account, Taylor (1994b), 239, manifests extreme scepticism. It is clear from the letters, however, that Paul made two east–west journeys through Asia Minor, and informed speculation can determine with some probability the route he followed.
[11] See Ibid. 236.
[12] So rightly Haenchen (1971), 484 n. 3; S. Mitchell (1992), 870–2, here 871.
[13] In 189 BC Gnaeus Manlius marched from Dyniae in Phrygia (some 20 miles west of Antioch in Pisidia) to Abassion (modern Jüsgad Ören) which was at the frontier of the Tolostobogii, the most western of the three Galatian tribes (Livy, *History of Rome*, 38. 15).
[14] See above, Ch. 1, 'Galatia and a Journey into Europe'.
[15] This translation of the *JB* and *NAB* (against the *NRSV*) is justified by Haenchen (1971), 484 n. 4; *JB*.

of Judaea to the province of Syria; it was treated as a separate district and ruled by a procurator.[16] Thus Paul and his companion(s) could go through Mysia without encountering any official obstacles, if it were such which prevented them from entering Asia. The route they took can only be a matter of speculation, even if one were sure that they were heading for Troas.[17] From Luke we know only that this is where they ended up. There were three possible western routes from Galatia.

A northern route from Cotiaeum followed the valley of the Rhyndacos to reach the Sea of Marmora east of Cyzicus whence there was a road to the west linking the little harbours along the Hellespont.[18] If Paul's intention was to cross over to Macedonia, his reason for bypassing these ports can only have been lack of shipping and/or contrary winds. The least probable is a postulated central route also starting in Cotiaeum and running almost due west through what later became Hadriania, Hadrianutherae, and Scepsis whence it followed the valley of the Scamander to Illium and the coast road.[19] The terrain is very difficult and the eastern part was notorious for its brigands.[20] The best route apparently was the southern one. From Cotiaeum it ran south-west to Aezani and then across to the headwaters of the Macestus, which the traveller followed to Hadrianutherae, whence there was a good road to the port city of Adramyttium.[21] Both it and Assos (Acts 20: 13) would have been convenient only for a sea voyage south along the coast of Asia. Anyone wishing to sail north or west would have made for Troas.

Some years later when Paul was concerned about news from Corinth which he expected to come through Macedonia, he went to Troas; when the messenger was delayed, he crossed from there (2 Cor. 2: 12–13). Evidently both Paul and Luke knew that Troas was the normal departure point for Europe, and the port of entry into Asia for those sailing from Neapolis the eastern terminal of the Via Egnatia, which crossed northern Greece. The absence of any letters to churches in Mysia suggests, either that Paul made no attempt to evangelize that area, or that he failed disastrously. Only his lack of success at Athens confers any plausibility on the latter hypothesis. That failure, however, was completely atypical, and was due to a combination of circumstances, the arrogance of the Athenian closed mind and Paul's anxiety concerning the Thessalonians.[22] We should assume, therefore, that Paul marched through Mysia without stopping to preach. If this is correct, it necessarily implies that when he left Galatia he had decided to go to Troas precisely in order to take ship for Europe.[23]

[16] Carroll (1992), 941.
[17] See below, p. 300.   [18] So Ramsay (1897), 197.
[19] So Munro and Anthony (1897), 256–8.
[20] Strabo, *Geography* 12. 8. 8; Lucian, *Alexander* 2.
[21] So Broughton (1937), 137.   [22] See Ch. 5, 'The Move to Corinth'
[23] As Bowers (1979) has argued on rather inadequate grounds.

Maintenance not Mission

If at first sight shocking and inexplicable, on reflection Paul's decision to leave Asia Minor reveals a deliberate missionary strategy and provides confirmation of the accuracy of the basic thrust of Luke's narrative (Acts 16: 6–12). The communities at Antioch in Pisidia, Iconium, Lystra and Derbe were well established. Paul had high hopes for his new foundations around Pessinus. The faith, in his estimation, was well planted in Galatia, the central province of Asia, whence it could radiate out in all directions. And he was convinced it would. Imbued as he was with his own profound sense of mission, he could not but take it for granted that his converts would be aggressively apostolic. The best course for him, therefore, was to move west beyond the furthest missionary reach of the Galatian churches.

Paul's strategy in Greece, it will be recalled, was essentially the same. There, however, instead of capturing the middle, he established bases at the extremities, in Macedonia (Thessalonica and Philippi) and in the Peleponnese (Corinth). Even though he walked the length of the country several times, he made no attempt to evangelize Thessaly. In his view, central Greece was the missionary responsibility of the churches which bracketed it on the north and south. At a later stage, when he had done all he could for Corinth, Paul planned to leap-frog over Rome to preach in Spain (Rom. 15: 24), the western edge of the known world.

The optimism of this vision was not justified by events. Paul's experience of the growing pains of the church of Thessalonica had made him conscious that founding churches was not enough. They had to be nurtured. Children in the faith needed time to grow, and in the process 'proclamation' had to be complemented by 'teaching'. What had happened at Antioch-on-the Orontes brought it home to him even more clearly that the development of a church could not be taken for granted. Thus when he crossed Asia Minor for the second time maintenance had become more important than outreach, at least for the time being.

Luke's economy with the truth as regards Paul's reason for leaving Antioch gives way to perfect accuracy when he depicts him as meticulously visiting each place in Galato-Phrygia where a community had been established (Acts 18: 23).[24] We know from the letters that Paul passed through Galatia at least twice (Gal. 4: 13) prior to writing Galatians. He founded churches there prior to the conference in Jerusalem (Gal. 2: 5) and subsequently organized the collection for the poor of Jerusalem which was agreed on at the conference (1 Cor. 16: 1).

It is, of course, possible that he invited the Galatians to participate in the col-

---

[24] Common sense demands this sense for *kathexês* (BAGD 388; *NRSV*). The alternative is to take the adverb as meaning that Paul first went through Galatia and then through Phrygia (so Haenchen, (1971) 545), a fact so obvious that to state it is pointless.

lection by letter, but it is much more probable that he did so in person.[25] The wind pattern during the sailing season made it preferable to cross Asia Minor from east to west by foot rather than sail round it by ship. The long dusty roads might be wearying, but one could make steady progress. Ships dispensed the traveller from personal effort, but the winds were predominantly adverse for those coming from the east, and headwinds forced boats to anchor for days, if not weeks, on end as they inched their way along the south coast of Asia Minor.[26] It was the opposite for those going from west to east, which is why Paul always returned home by ship (Acts 18: 18–22; 20: 6 to 21: 3).[27]

It is virtually certain, therefore, that since Paul had been in Galatia (1 Cor. 16: 1) prior to his arrival in Ephesus (1 Cor. 16: 8) his route to the west took him through the *anôterika merê*. Literally 'the upper parts',[28] this unparelleled expression may simply be a way of speaking about the Anatolian plateau, which is also the interior of Asia Minor, but Hemer follows Ramsay in giving it a more specific sense as evoking 'the traverse of the hill-road reaching Ephesus by the Cayster valley north of Mt. Messognis, and not by the Lycus and Maeander valleys, with which Paul may have been unacquainted'.[29] This route becomes plausible only if it is assumed that Paul was coming, not from Antioch in Pisidia, as Hemer's South Galatia hypothesis requires, but from somewhere much further north, such as the region around Pessinus.

What has been said above regarding Paul's missionary strategy prohibits adopting Hemer's confirmatory argument, namely, that Paul could not have passed through the Lycus valley because he had not preached there (Col. 2: 1). At this point in his career, Paul was concerned with stabilizing established existing communities, not with founding new ones. Thus even if he had taken the great common highway to the west via the Lycus and Meander valleys,[30] he would not have stopped along the way to evangelize. His goal was to reach Ephesus.

---

[25] So rightly Fee (1987), 812.

[26] Casson (1979), 151–2.

[27] The emperor Gaius advised Agrippa I not to return from Rome to Palestine overland but 'to wait for the Etesian winds and take the short route through Alexandria. He told him that the ships are crack sailing craft and their skippers the most experienced there are; they drive their vessels like race horses on an unswerving course that goes straight as a die' (Philo, *In Flaccum* 26; trans. Casson).

[28] BAGD 77; 'the upper country' (*RSV*); 'the interior regions' (*NRSV*); 'the interior of the country' (*NAB*); 'overland' (*NJB*); 'le haut-pays' (*BdeJ*).

[29] (1989), 120, cf. 187.

[30] 'There is a kind of common road constantly used by all who travel from Ephesus towards the east, Artemidorus traverses this too: from Ephesus to Carura, a boundary of Caria towards Phrygia, through Magnesia, Tralleis, Nysa, and Antiocheia, is a journey of seven hundred stadia' (Strabo, *Geography* 14. 2. 29; trans. Jones).

EPHESUS

Paul's choice of Ephesus for his second long-term base was as well thought out as his earlier selection of Corinth. The centrality of this city on the western coast of Asia Minor with respect to churches he had previously founded is well illustrated by some simple statistics. As the crow flies, Ephesus is equidistant from Galatia and Thessalonica (480 km., 288 miles). Corinth (400 km., 240 miles), Philippi (445 km., 267 miles), and Antioch in Pisidia (330 km., 198 miles) fit easily within the same circle. Paul himself does not tell us how long he stayed in Ephesus. Luke gives two figures, first two years and three months (Acts 19: 8–10), and later three years (Acts 20: 31). The latter is a round figure, but the specificity of the former inspires confidence. If Luke's information were not available, a similar figure would have to be postulated in order to allow time for the activities of Paul which can be deduced from the letters.

The oldest remains of the city date from the mid-second millenium BC. Its antiquity is confirmed by the mythical origin of its name. According to Pausanias, it was called after Ephesos, who was believed to be the son of the river Cayster.[31] Around 286 BC the city was given its present location (between the hills now named Bülbül Dagh and Panayir Dagh) by Lysimachus (360–281 BC), a companion and successor of Alexander the Great. The majesty of the wall he built (7 m. high, 3 m. wide, and 9 km. long) was accentuated by its position on the crests of the hills; sections still exist. His purpose in moving the city to the west was to compensate for the silting up of the river valley.[32] He did not solve the problem which, according to Strabo,[33] was exacerbated by one of his successors, who, by ordering a badly placed breakwater, intensified the silting up of the harbour called Coressos.[34] There was another harbour named Panormus further to the west.[35] A functioning port was essential if Ephesus was to realize its full economic potential, and inscriptions[36] confirm that the proconsul Barea Soranus in AD 61 was not the only one to have seen the necessity of periodically dredging the harbour.[37]

In 133 BC by the testament of Attalus III (170–133 BC) Rome acquired the

---

[31] *Description of Greece* 7. 2. 4.

[32] 'From these [the tributaries of the Cayster] comes a quantity of mud which advances the coastline and has now joined the island of Syrie on to the mainland by the flats interposed' (Pliny, *NH* 5. 31. 115; trans. Rackham). See the map in PW 5. 2780.

[33] *Geography* 14. 1. 24.

[34] Herodotus, *Histories* 5. 100.

[35] 'Then comes the harbour called Panormus, with a temple of the Ephesian Artemis, and then the city of Ephesus' (*Geography* 14. 1. 20; trans. Jones). See the map in PW 5. 2780, which is simplified in *DBSup* 2. 1087.

[36] Oster (1992), 543, draws attention to *IvEph* 23, 274, 2061, 3066, 3071.

[37] Tacitus, *Annals*, 16. 23.

kingdom of Pergamum, which became the province of Asia.[38] It had a stormy history until Octavian succeeded in establishing control. Even though Pergamum remained the titular capital, Ephesus was in fact the most important city.[39] This undoubtedly influenced the decision of Augustus to make it the seat of the proconsul. He also saw the opportunity to enhance his own glory by ensuring that Ephesus blossomed with magnificent buildings.[40] The Attalids had made Pergamum one of the most beautiful of Greek cities, whereas the possibilities of the grid plan of Lysimachus at Ephesus had never been fully exploited.[41]

A number of notable buildings are dated to the reign of Augustus,[42] and were a dominant feature of the city at the time of Paul (see Fig. 2). One complex in the southern part of the city contained the town-hall, the double temple dedicated to Rome and Julius Caesar, and a 200 metre-long open basilica which filled the northern side of the State Agora. Ceremonial gates gave dignity to the south entrance of the Square Agora (112 × 112 m.) and to the west end of the great street running from the theatre to the harbour. Three new aqueducts, the *Aqua Julia* and the *Aqua Troessitica,* to which the emperor contributed, and the aqueduct of C. Sextilius Pollio both improved the quality of life in the city and provided for an expansion of the population.[43] They also facilitated the construction of baths cum gymnasia (six are known), in which the social life of the Roman city was concentrated.

The stage area of the great 25,000–seat theatre (Acts 19: 31) was expanded not long before Paul's arrival.[44] The only other monument to rival it in size was of course the temple of Artemis (Acts 19: 35). Rebuilt many times since its foundation in the seventh century BC,[45] it quickly found its place in the earliest lists of the seven wonders of the world.[46] In the early second century BC Antipater of Sidon wrote,

> I have set eyes on the wall of lofty Babylon, on which is a road for chariots, and the statue of Zeus by the Alpheus [at Olympia], and the hanging gardens [of

[38] In addition to the standard dictionary articles, see especially Knibbe and Alzinger (1980).

[39] In his report on 29 BC Dio Cassius noted, 'Caesar [Augustus], besides attending to the general business, gave permission for the dedication of sacred precincts in Ephesus and in Nicaea to Rome and to Caesar, his father, whom he named the hero Julius. These cities had at that time attained chief place in Asia and in Bithynia respectively' (*History* 51. 20. 6; trans. Cary).

[40] Knibbe and Alzinger (1980), 759.

[41] Ibid. 811–814.

[42] Ibid. 815–18. The city plan given opposite p. 760 appears in larger scale in PWSup 12. 1584 with 1600.

[43] Alzinger (1970), 1604–5.

[44] Erdemgil (1989), 96.

[45] A schematic historical overview is given by Strabo (*Geography* 14. 1. 22–3) who, though familiar with the canon of 'the seven wonders' (14. 2. 5; 14. 2. 16; 16. 1. 5; 17. 1. 33), does not classify the temple of Artemis among them.

[46] Lanowski (1965), 1020–30.

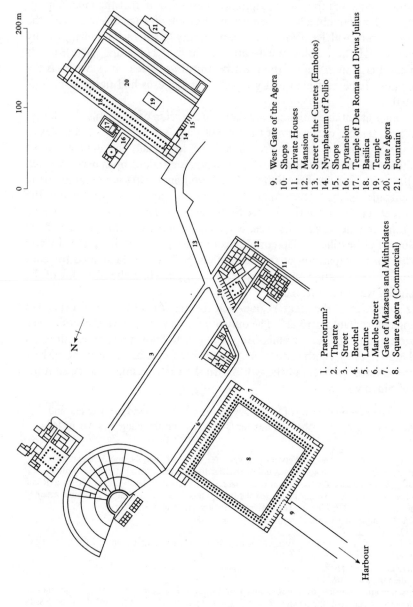

1. Praetorium?
2. Theatre
3. Street
4. Brothel
5. Latrine
6. Marble Street
7. Gate of Mazaeus and Mithridates
8. Square Agora (Commercial)

9. West Gate of the Agora
10. Shops
11. Private Houses
12. Mansion
13. Street of the Curetes (Embolos)
14. Nymphaeum of Pollio
15. Shops
16. Prytaneion
17. Temple of Dea Roma and Divus Julius
18. Basilica
19. Temple
20. State Agora
21. Fountain

Harbour

N

0   100   200 m

Fig. 2 Central Ephesus c. AD 50 (*Sources*: W. Alzinger, PW Sup. XII; W. Alzinger, *ANRW* II, 7/2 (1980))

Babylon], and the colossos of the Sun [at Rhodes], and the huge labour of the
high pyramids [in Egypt], and the vast tomb of Mausolus [at Halicarnassus],
but when I saw the house of Artemis that mounted to the clouds, those other
marvels lost their brilliancy, and I said, 'Lo, apart from Olympus, the Sun
never looked on aught so grand'. (*Greek Anthology* 9. 58; trans. Paton)

This temple measured 115 × 55 m., and the 98 colums in the double row surrounding the building were 17.65 m. high.[47] The pilgrims it attracted to Ephesus
were an important factor in the economy. It survived until the third century AD.

If such majestic structures contributed to the ethos of the city, they were not
where the inhabitants lived. Private dwellings tell us much more about the conditions under which Paul operated. Two complexes have been excavated on the
slope of Bülbül Dagh, south of the street linking the Square Agora and the State
Agora.[48] First built in the first century BC, the quality of construction was such
that they were still in use 600 years later, though often repaired. The ground
floor of the eastern complex spread over 3,000 sq. m. and was the house of a
single very wealthy family.[49] Its spacious and numerous public rooms would
have been a boon to the nascent Christian church, but such magnates were
rarely if ever to be found among its members.

Those who could afford to host the liturgical assemblies of the community
were much more likely to have lived in a house similar to one of the seven two-
storey dwellings in the western complex (see Fig. 3). The ground floor area of
House A is 380 sq. m.[50] From the right of the vestibule a door leads to the
Roman bath, which also heated the house. Directly ahead is the atrium (7.5 × 5
m.) with its impluvium (3 × 3.75 m.). A small room on one side gives access to
the dining room (3 × 5.5 m.); the kitchen is nearby. A much larger room (6.5 ×
4.25 m.) is entered directly from another side of the atrium through a large
arched doorway.

These were the public rooms; the rest of the house was off limits to casual
visitors. The size of the rooms meant that once the Christian community
reached a certain size complications were inevitable. When it became impossible to get everyone into the same room, there was a danger of creating first-
and second-class members, which in fact happened in Corinth (1 Cor. 11: 17–
34).[51] Inevitably there was a tendency to meet in smaller groups, such as the
house-church which assembled in the home of Prisca and Aquila (1 Cor. 16: 19).

The heterogeneous character of the population of Ephesus needs no
emphasis. It was the door to the west for the Anatolian hinterland, and the
opening to Asia for Greece and Rome. Many went no further. It was as much a

---

[47] Erdemgil (1989), 30.
[48] See the map in Alzinger (1970), 1600.
[49] Knibbe and Alzinger (1980), 824.
[50] The measurements which follow are taken from the plan in Erdemgil (1988), 14.
[51] See my (1992e), 161–9.

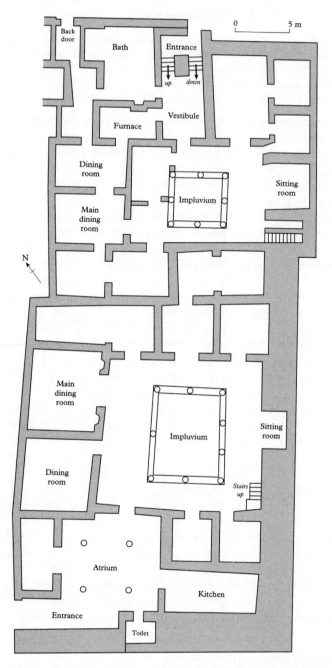

FIG. 3 Ephesus: Private Houses (*Source*: S. Erdemgil *et al.*, *La maisons du Flane à Ephese* (Istanbul, 1988))

melting-pot as Rome itself.[52] The population is estimated at a quarter of a million. Citizens in the strict sense (i.e. with voting rights) were perhaps a quarter of that figure. In the light of our present knowledge, it seems that there were nine 'Tribes' each with six 'Thousands'.[53]

THE FOUNDING OF THE CHURCH

When Paul headed for Ephesus in the summer of AD 52,[54] he was not venturing into the unknown. According to Luke, he had made a brief stop there en route to Palestine, after founding the church at Corinth (Acts 18: 19). More importantly he had prepared his welcome by leaving there Prisca and Aquila (Acts 18: 24–8), the couple who had provided him with a base when he first went to Corinth (Acts 18: 2–3). This means that they had been there for a year prior to Paul's arrival after the Jerusalem Conference. It is extremely improbable that they had devoted all their energy exclusively to re-establishing their tent-making business. They had been Paul's partners in the evangelization of Corinth, and preaching the gospel would have become second nature to them by now. Even while at work, they would have availed of every opportunity to proclaim the good news. Prisca, and Aquila, therefore, were the real founders of the church at Ephesus.

The reliability of Luke's information regarding the precedence of the couple in Ephesus is confirmed by the fact that it embarassed him. This is clear from the curious note, 'They [Paul, Prisca and Aquila] came to Ephesus, and he left them [Prisca and Aquila] there, but he himself went into the synagogue and lectured to the Jews' (Acts 18: 19); the reason for the emphatic 'he himself' can only be to drive home the fact that it was Paul who delivered the first missionary sermon in Ephesus.[55] Such insistence, however, hints that the real situation may have been rather more complex.

Were Luke free to create, he would have made Paul directly responsible for the evangelization of Ephesus. Luke's concern, it will be recalled, was to tie the foundation of the church there into a pattern of controlled expansion, and Paul's missionary role had been formally accepted by Jerusalem (Acts 15). It was definitely not in Luke's interest to invent the activity of people such as Prisca and Aquila. It was a fact that he had no choice but to integrate. And he did so in a way which probably corresponds to reality. The couple, he insinuates, were acting in collaboration with Paul.

[52] Juvenal, *Satires* 3. 61–78.
[53] Knibbe (1970), 275–6.
[54] See Ch. 1, 'After AD 51'.
[55] So rightly Haenchen (1971), 87, 543. The conversion of Epaenetus, 'my dear friend and the first-fruit of Asia for Christ' (Rom. 16: 5), may have been due to Paul, but not necessarily (cf. 1 Cor. 16: 15).

The letters confirm this inference. We can deduce from 1 Corinthians 16: 19, not only that Prisca and Aquila had moved from Corinth to Ephesus, but that they were very close to Paul. This verse is the most complex closing greeting in the Pauline epistles. The first of the three sources of greetings is the churches of Asia, which is perfectly appropriate in an official letter written from the first city of Asia (1 Cor. 16: 8). The second is Prisca, Aquila and their house-church, and the third is all the believers. The order is intriguing. Manifestly it was Paul's intention to mention only the churches of the province of Asia, for the reason given above. The abrupt shift from this impersonal level to the intimate level of a particular couple suggests that Aquila and/or Prisca were present as Paul ended the letter, and felt close enough to him to ask to be mentioned. The most plausible motive for such an interjection, namely a previous connection with the Corinthians (cf. Acts 18: 1–3), is confirmed by the qualification of their greeting as 'hearty'; it was not a mere ceremonial gesture.[56] The mention of their house-church in turn stimulated Paul to include a reference to 'all the believers', i.e. the 'whole' church (1 Cor. 14: 23).

Those who had received the Baptism of John

No sooner had the Christian mission in Ephesus begun than an unusual phenomenon occurred. To their immense surprise the missionaries came into contact with people who already knew Jesus! The discussion of Acts 18: 24 to 19: 7 has given rise to a lively debate,[57] which has been perfectly characterized by Käsemann in his best mordant style, 'This conspectus has brought before us every even barely conceivable variety of naivete, defeatism and fertile imagination which historical scholarship can display, from the extremely ingenuous on the one hand to the extremely arbitrary on the other.'[58]

Such confusion, I suggest, is due to a false perception of the problem. At the risk of some simplification I think it fair to say that all interpretations wrestle with the question: how could 'followers of John' be classified as 'disciples' and 'believers', i.e. as Christians? Research has been side-tracked by this formulation of the problem, which is inaccurate. Despite the title of so many articles, it is neither said nor suggested anywhere in the text that Apollos and the others were disciples of John. They had received the baptism of John, and the real question is: who administered it?

In the light of John 3: 22 the obvious answer is that it was administered by Jesus when he was preaching John's baptism of repentance in Judaea. The language of John 3. 22 contradicts R. Schnackenburg's gratuitous assumption that the period must have been very short.[59] On the contrary, the success of

---

[56] Fee (1987), 835.
[57] A survey of the opinions is given by Wolter (1987).
[58] (1964), 140.   [59] (1965), 449.

Jesus is explicitly emphasized (John 3: 26). There were many, therefore, who thought of themselves as followers of Jesus and in consequence were accepted as 'believers' at Ephesus (and elsewhere) until somehow their baptism and/or the content of their belief was questioned. They were disciples of Jesus of Nazareth (in the sense of having been converted to repentance by him), but had known him only while he was still associated with John, and had lost contact with him subsequently. They had vivid memories of Jesus as the assistant of a prophet, but knew nothing of the Passion, Resurrection, or Pentecost. It is most unlikely that they thought of Jesus as the Messiah.

Inevitably the relation of such people to Jesus would have been suspect to those who knew him as the Risen Lord. Some action was imperative if there were not to be two radically different types of followers of Jesus. The natural assumption is that Prisca and Aquila invited them to become full believers, and that those who accepted were baptized. The scenario of Acts is predictable in that it makes Paul the instrument of their conversion, but Luke adds a further dimension which suggests a new Pentecost (Acts 19: 5–7; cf. 2: 4).

One of those who had received the baptism of John and who had ended up in Ephesus, according to Luke, was Apollos, a well-educated Jew from Alexandria (Acts 18: 24). For their part, the letters reveal that an Apollos had ministered in Corinth (1 Cor. 3: 6) and that subsequently he was with Paul at Ephesus (1 Cor. 16: 8, 12). As we shall see when dealing with the Corinthian correspondence, this Apollos was a trained orator acquainted with the teaching of Philo of Alexandria. There is little doubt, therefore, that we have to do with the same person.[60]

Apart from this group, nothing specific is known about the composition of the church at Ephesus. Since the city was similar to Corinth in so many ways we can assume with some confidence that the two communities resembled each other in both size and make-up (1 Cor. 1: 26–9). Each was the city in microcosm; a few relatively wealthy members, the majority tradespeople and slaves, possibly more women than men.

MISSIONARY EXPANSION

The success of Paul and his collaborators in establishing a flourishing community in Ephesus had unexpected side-benefits in the foundation of churches elsewhere in the province. The hyperbole of Acts 19: 10 and 26 is lacking in the Pauline letters, but the existence of Christian communities outside Ephesus is attested by the greetings which 'the churches of Asia' (1 Cor. 16: 19) send to Corinth. The only names of such churches known to us from the Pauline letters

---

[60] See Ch. 11, 'The Arrival of Apollos'.

are Colossae, Laodicea, and Hierapolis (Col. 4: 13), but it would be unwise to assume that this list is exhaustive. These three are mentioned only because Paul had to ensure that neighbouring churches were not infected by the false teaching which had divided the church at Colossae.

Paul himself did not found the churches of the Lycus valley (Col. 2: 1). In the thanksgiving of Colossians, he speaks of 'the day you heard and understood the grace of God in truth as you learned it from Epaphras our beloved fellow-slave. He is a faithful minister of Christ on our/your behalf and has made known to us your love in the Spirit' (Col. 1: 6–8). In Colossians 4: 12, Epaphras is identified as 'one of yours', which is reasonably interpreted as meaning that he came from Colossae. From the compliment 'he has worked hard for you and for those in Laodicea and in Hierapolis' (Col. 4: 13), one can deduce that he was the founder of all three churches.

Was Epaphras acting on his own initiative or as Paul's agent when he evangelized the Lycus valley? The question is not really answered by a simple choice among the variants in Colossians 1: 7 on the basis of the manuscripts. Even if Epaphras were Paul's deputy, the latter could still speak of him as the representative of the Colossians in expressing their affection for the Apostle. Were this Paul's meaning, however, it does not seem likely that he would have called Epaphras 'a faithful minister *of Christ* on your behalf' (Col. 1: 7); the italicized genitive is not appropriate to a messenger from the Colossians (cf. Phil. 2: 25).[61] It rather suggests a duly authorized missionary, i.e. one sent by Paul (cf. 2 Cor. 11: 23). This is confirmed, not only by Paul's use of 'minister' and 'slave' elsewhere to identify his own role as an apostle, but particularly by the combination 'faithful minister and fellow-slave', which in this letter is applied to Tychicus (Col. 4: 7), who was certainly Paul's emissary (Col. 4: 8). The fact that Epaphras was imprisoned with Paul (Philem. 23), whereas Epaphroditus was not (Phil. 2: 25), indicates that the authorities understood the former to be Paul's agent. It is more probable, therefore, that 'on our behalf' should be read in Colossians 1: 7.[62] The warmth with which Paul speaks of Epaphras reveals his confidence in him. Epaphras could not have been responsible for whatever problems had arisen in Colossae. His relationship to Paul probably typifies that of the missionaries who were sent elsewhere in Asia.[63]

The fact that Paul himself did not go to the Lycus valley confirms what has been said above regarding his commitment to nurturing already existing communities. But if the demands on his time at Ephesus, and restrictions on his freedom of movement (see below), precluded missionary travel, he could commission others to preach in his name. Paul had at last learnt, not only that he could not do everything, but that he did not even have to try. It is unlikely that

---

[61] Abbott (1897), 200.
[62] So rightly Abbott (1897), 200; Moule (1957), 27; Lohse (1968), 53–4.
[63] See further Ch. 10, 'Missionary Strategy'.

Epaphras was the only missionary sent out from Ephesus, and it is far from impossible that most if not all of the churches in western Asia were established as part of the planned outreach of the Ephesian community guided by Paul.

If Ephesus and Laodicea, two of the seven churches of the Apocalypse (Rev. 2: 1 to 3: 22), were Pauline foundations (the latter at least indirectly through Epaphras), then there is no obstacle to attributing the creation of communities at Smyrna, Pergamum, Thyatira, Sardis, and Philadelphia to the missionary initiative of Ephesus. To these might be added Magnesia and Tralles, whose churches are known from the letters of Ignatius. All are within a 192 km. (120 mile) radius of Ephesus, and linked by major roads. Colossae, the furthest away, could be reached in a comfortable week's walk from Ephesus.

The absence of any letters from Paul to these churches cannot be construed as an objection to their Pauline origin. There are many possible reasons for his silence, but the simplest is that he had adopted a policy of delegation. He trusted the missionary responsible for a particular church to deal with whatever issues arose there. No doubt he was available for consultation, but he maintained direct contact only with the churches he had founded personally.

The letter to the Colossians is an exception to this self-imposed rule, but one which is easily explained. Epaphras had gone to Ephesus to inform Paul of the situation at Colossae and to develop a strategy for dealing with the false teaching which attracted some members of the church. There he found Paul a prisoner (Col. 4: 10, 18) and was himself held for interrogation by the Roman authorities (Philem. 23), which prevented him from returning to Colossae (Col. 4: 12–13). The defection of one of the leadership group there made it imperative to deal promptly with the situation.[64] Paul had only two alternatives, either a messenger or a letter. The former apparently was not a viable option; none of Paul's collaborators, who was free to undertake the task, had the authority that the situation demanded. A letter was the only option, and it was written by Paul to give it the greatest possible weight.

IMPRISONMENT

The scenario just outlined assumes that Paul was imprisoned in Ephesus. According to Acts, however, his sojourn in the city was entirely peaceful, with the exception of the riot of the silversmiths, which did him no damage; the only imprisonments mentioned in Acts are those at Philippi (Acts 16: 23), Caesarea (Acts 23: 23 to 26: 32), and at Rome (Acts 28: 16). It is not surprising, therefore, that from the beginning of patristic exegesis it was taken for granted that the letters in which Paul says that he is a prisoner (Eph., Phil., Col., and Philem.)

[64] See Ch. 10, 'Missionary Strategy'.

were written in Rome.[65] Only in the twentieth century did Caesarea Maritima in Palestine, and Ephesus surface as rivals to the Eternal City. Caesarea has won very little support because all the arguments invoked in its favour carry greater force when applied to Ephesus.[66] Moreover, we know from Paul himself that he experienced a life-threatening situation in Ephesus (1 Cor. 15: 32; cf. 2 Cor. 11: 23).[67] The choice, therefore, is between the capital of Asia and Rome. Unfortunately the decision must be based on vague and often ambiguous hints in the letters.

Although very different in content, Colossians and Philemon were written in identical circumstances to groups which overlapped considerably. In both letters Paul is a prisoner (Col. 4: 10–18; Philem. 1, 9, 23). In both he is accompanied by Timothy (Col. 1: 1; Philem. 1), Epaphras (Col. 4: 12; Philem. 23), Aristarchus (Col. 4: 10; Philem. 24), Mark (Col. 4: 10; Philem. 24), Luke (Col. 4: 14; Philem. 24), Demas (Col. 4: 14; Philem. 24), and Onesimus (Col. 4: 9; Philem. 10– 12).[68] In both Archippus appears among the recipients (Col. 4: 17; Philem. 2). 'These agreements do not occur in the same relationships and formulations, however, so that the thesis is unconvincing that the indubitably Pauline Philem. has been imitated by a non-Pauline writer only in these personal remarks.'[69] Three facts indicate that the house-church of Philemon was at Colossae: (1) Epaphras of Colossae knows the recipients of Philemon well enough to send greetings (Philem. 23); (2) Onesimus was from Colossae (Col. 4: 9); (3) Archippus of Colossae is among the recipients of both letters. Hence information from one letter can be used to supplement that of the other.

According to the dominant interpretation of Philemon, Onesimus, one of the bearers of Colossians, was a runaway slave who, after encountering Paul in the city in which the latter was imprisoned, was sent back to Colossae. Where had he taken refuge? It is both unreasonable and unnecessary to assume that he went all the way to Rome. The long journey involving two sea voyages was an expensive undertaking, which can only be made plausible by assumptions regarding stolen funds, or a new employer who just happened to be heading for the centre of the empire, which in turn demand other assumptions. In order to be safe Onesimus did not have to go very far. There was no police force con-

---

[65] Curran (1945).

[66] A representative voice of the Ephesian hypothesis is still Duncan (1929). For the alternative, see Johnson (1956); Robinson (1976) 60.

[67] Nothing more specific can be said because the reference to fighting with wild beasts must be taken metaphorically (cf. 2 Tim. 4: 17); so rightly Fee (1982), 770.

[68] Col. also mentions Tychicus (4: 7–8) and Jesus/Justus (4: 11) who do not appear in Philem.

[69] So rightly Kümmel (1975), 345; and most recently Knox (1990), 264–5, 'it must be recognized that the ties binding Colossians to Philemon are far stronger than any bonds with Ephesians—ties so strong and intricate, so very improbable as the inventions of even the cleverest pseudepigrapher'; against Koester (1982), 2. 267. For example, why would a forger shift the adjective 'fellow-prisoner' from Epaphras (Philem. 23) to Aristarchus (Col. 4: 10)?

stantly on the alert for fugitives.[70] Reward notices might be published,[71] but, unless the authorities were pressured by someone of irresistible influence, that was the only action they would normally take, and one cannot imagine the notices being distributed outside the immediate locality. Once in Ephesus, Onesimus would have been perfectly sure that there was only the slightest chance of being discovered. A chance encounter with an acquaintance of his master was his only danger.

Was it just bad luck that brought Onesimus into Paul's orbit? Or did he go looking for him? P. Lampe has drawn attention to a provision of Roman law which permitted a slave in danger of punishment to seek out a friend of the owner to act as an intermediary in the re-establishment of good relations.[72] Under such circumstances the slave did not become a fugitive in the legal sense. If he went to a friend of the owner, no intention to escape could be assumed. The situation is perfectly illustrated by a letter from Pliny the Younger to Sabinianus.

> The freedman of yours with whom you said you were angry has been to me, flung himself at my feet and clung to me as if I were you. He begged my help with many tears, though he left a great deal unsaid; in short he convinced me of his genuine penitence. I believe he has reformed, because he realized he did wrong. You are angry, I know, and I know too that your anger was deserved, but mercy wins most praise when there was just cause for anger.
>
> (*Letters* 9. 21; cf. 9. 24; trans. Radice)

It seems clear from Philemon 18 that Onesimus had caused some damage to Philemon.[73] It must have been rather serious, because Onesimus recognized the need for not just any advocate but one with considerable influence over his master. Although a pagan (Philem. 10), he was aware that Paul had ultimate authority over the new religious group to which his owner belonged. Hence, instead of seeking out a friend of Philemon in Colossae, he went looking for Paul.

In this scenario, which does much fuller justice to the tone and content of Philemon than the hypothesis that Onesimus was a runaway, the episode must have taken place at a time when the sense of Paul's invisible presence in the church of Colossae was strong, because he was known to be in the vicinity (cf. Col. 2: 5). This was true only when he resided in Ephesus. By the time of Paul's imprisonment in Rome he had been out of contact with the churches of Asia for several years, and it is doubtful that they even knew where he was. Moreover, in the situation envisaged, time was of the essence. The problem had to be solved

---

[70] Millar (1981), 67, 71.
[71] Two examples are given in Moule (1957), 34–7.
[72] (1985), 135–7.
[73] For a refutation of the view espoused by Goodspeed and Knox that Archippus was the owner of Onesimus, see Guthrie (1966), 247–50.

before the momentary anger of Philemon became permanent bitterness. The delay of a long journey to Rome would have made the effort of Onesimus pointless. Ephesus was at the limit of the feasible.

Only in the hypothesis of an Ephesian imprisonment does Paul's plan to visit Colossae, which was concretized in a request for lodgings to be prepared for him (Philem. 22), become intelligible. Nothing is more natural than his desire to follow up in person the impact of his letter on communities, which had been disturbed by false teaching of Jewish origin. The churches of the Lycus valley were not quite halfway between Ephesus and Galatia. If they had exhibited a partiality for Jewish-inspired doctrine might they not fall victims to the Judaizers, whom he had had to combat in Galatia? When in Rome, on the contrary, Paul's attention was focused not on the east but on the west, not on Asia but on Spain (Rom. 15: 24).

The slender clues in Philemon unambiguously point to Ephesus as its place of origin. In consequence, Colossians must have been written from the same prison. Attempts have been made to find confirmation from within the letter, but the results are unconvincing. Bowen, for example, argues that the fourteen direct allusions to the conversion of the Colossians suggests that the church had been in existence only a matter of weeks or of months at most.[74] However, one finds the same sort of allusion to the beginning of their Christian lives when Paul is speaking to the Galatians (3: 2–3; 4: 8–9), and those churches had been founded at least six years previously.

At least two of the three letters combined to create Philippians—*Letter A* (4: 10–20) and *Letter B* (1: 1 to 3: 1 and 4: 2–9)[75]—speak in favour of Ephesus as their place of origin.[76] As the seat of the proconsul, the city had a praetorium (1: 13). The great imperial estates in Asia demanded the presence of members of Caesar's household (4: 22), whose sojourn at Ephesus is confirmed by inscriptions.[77] The frequent contacts between Philippi and Paul's place of imprisonment suggest that the latter was somewhere much closer than Rome.[78]

Finally, as soon as he got his freedom Paul planned to visit Philippi (1: 26; 2: 24). Not only is this the opposite of what he tells us his plans were after Rome (Rom. 15: 24), but it can only be the visit projected in 1 Cor. 16: 5–9, which was written from Ephesus. Philippians 1: 26 and 30 give the impression that this visit will be the first since the foundation of the church at Philippi (contrast the language of Gal. 4: 13; 2 Cor. 12: 14; 13: 1), but by the time Paul got to Rome he had already visited Philippi at least twice (2 Cor. 1: 16; 2: 13).

Although not entirely free from ambiguity, the hints contained in Colossians,

---

[74] (1924), 190.

[75] On the division of Phil. into three letters, see my (1965).

[76] Gnilka (1968), 18–25; against Kümmel (1975), 324–32.

[77] Feine (1916), 94–8; Knibbe (1970), 264–5.

[78] Epaphroditus brings a gift from Philippi (4: 18) and falls sick. The Philippians hear of his illness, and he is made aware of their distress (2: 26).

Philemon, and Philippians have a cumulative force. The case for Ephesus as the prison from which they were written is much stronger than one which could be developed in favour of Rome.

Roman law at this period contained no provision for a prison sentence; detention was not used as a punishment. Individuals were removed from circulation for longer or shorter periods through being sent into exile.[79] They were held under restraint in two situations, either while under investigation[80] or after the death sentence had been passed and they were awaiting execution.[81] Other punishments (e.g. scourging, fines) were carried out immediately. In practice, of course, detention could drag on interminably.

These theoretical possibilities of the application of Roman justice are perfectly illustrated by the Acts of the Apostles, which (without assuming the historicity of details) reflects very accurately the realities of the legal situation in the provinces. Once the magistrates accepted the charges laid against the missionaries, Paul and his companions were punished without delay; they were beaten and tossed into jail for the night before being expelled from the city (Acts 16: 19–35). While awaiting judgement on accusations made against him Paul was held first in Jerusalem (Acts 21: 33 to 23: 22), then in the praetorium of Caesarea (Acts 23: 23 to 26: 32), and finally in Rome (Acts 28: 16). Peter was incarcerated while awaiting execution (Acts 12: 1–19). Access to the outside world was dependent on the whim of the official (Acts 24: 22).

There is no hint in any of the captivity letters that Paul is awaiting execution. On the contrary, he expresses his hope of being released in the near future (Phil. 2: 24; Philem. 22). We should assume, therefore, that he was being held while under investigation. If Paul offered public lectures in the hall of Tyrannus (Acts 19: 9), someone may have tried to curry favour with the authorities by drawing attention to a new religious group, which might possibly be subversive. To arrest the leader and his agents in outlying areas pending an investigation would appear a prudent decision to any administrator (cf. John 19: 12).

---

[79] 'Nothing boosts your diviner's credit so much as a lengthy spell in the glasshouse [military prison], with fetters dangling from either wrist. No one believes in his powers unless he's dodged execution by a hair's breath, and contrived to get himself deported to some Cycladic island Seriphos, and to escape after lengthy privations' (Juvenal, *Satires* 6. 561–5; trans. Green).

[80] 'Aelian next summoned Apollonius and ordered him into the prison, where the captives were not bound. "Until," he said, "the Emperor shall have leisure, for he desires to talk with you privately before taking any further steps." Apollonius accordingly left the law-court and passed into the prison, where he said, "Let us talk, Damis, with the people, for what else is there for us to do until the time comes when the despot will give me such audience as he desires?"' (Philostratus, *Life of Apollonius of Tyana* 7. 22; trans. Conybeare).

[81] Mommsen (1955), 299–305, 945–8; Humbert (1899a); Pollack (1899).

THE DATE OF GALATIANS

Before attempting to establish a chronology for Paul's stay in Ephesus, an important question must be answered: does Galatians have to be taken into account? In other words, was Galatians written during Paul's stay in Ephesus? The one clue to the date of the epistle provided by the letter itself is Galatians 2: 5 which, as we have seen, fixes a lower limit of AD 51; the letter was written after the Jerusalem Conference. The upper limit is furnished by the composition of Romans in Corinth during the winter of AD 55– 56,[82] because there is general acceptance of Lightfoot's conclusion,

> The Epistle to the Galatians stands in relation to the Roman letter, as the rough model to the finished statue. . . . The matter, which in the one epistle [Gal.] is personal and fragmentary, elicited by the needs of an individual church, is in the other generalised and arranged so as to form a comprehensive and systematic treatise.[83]

The success of his attempt to establish the priority of Galatians with respect to Romans led Lightfoot to use the same type of comparative terminological and thematic study to date Galatians after 1 and 2 Corinthians.[84] His brief treatment, however, is more impressionistic than convincing.[85] This fault was remedied by U. Borse, whose exhaustive application of the same methodology refined Lightfoot's conclusion by dating Galatians between 2 Corinthians 1–9 and 2 Corinthians 10–13, and thus he assigned Macedonia as its place of origin.[86] The strength of Borse's approach is that he uses only contacts that are unique to Galatians and the document with which he is comparing it. Its weakness is illustrated by his treatment of the formula 'another gospel' (Gal. 1: 6; 2 Cor. 11: 4). The style and construction of its context in 2 Corinthians, according to Borse, is more developed than that in Galatians; hence in this respect Galatians is more primitive, and therefore earlier, than 2 Corinthians 10–13.[87]

The underlying principle—the later is always better—is manifestly false, and the aesthetic value judgement intrinsic to the method is always debatable.[88] Moreover, there is a subtle, but unjustified, shift from priority to proximity. Borse takes it for granted that thematic and verbal agreements are always time-related, when it is much more probable in this instance that they are subject-

---

[82] See Ch. 1, 'After AD 57'.

[83] (1910), 49.

[84] Ibid. 55.

[85] Robinson (1976), 56–7, adds nothing new in his acceptance of Lightfoot's conclusion.

[86] (1972), 177. The attempt of Lüdemann (1984), 86, to reinforce the position of Borse is vitiated by the fact that his arguments are all from silence.

[87] (1972), 89.

[88] The same type of argument is used by Longenecker (1990), pp. lxxxiii–lxxxviii, to prove that Gal. antedates the Jerusalem Conference.

related. The contacts he highlights reveal only that whenever Paul dealt with the same problem he tended to express himself in a similar way. It is not at all surprising that Borse finds the greatest number of contacts between Galatians and 2 Corinthians 10–13 because in these two documents Paul is not only dealing with the same issue, namely the inroads into his communities made by Judaizers, but he does so in precisely the same bitterly disappointed frame of mind. Inevitably the same words and ideas surge to his lips.

Once the hypothesis of the proximity of Galatians to 2 Corinthians and Romans is seen to be without foundation, the opening words of the letter can be read naturally, 'I am astonished that you are so quickly deserting him who called you in the grace of Christ' (Gal. 1: 6). There is no justification for watering down the normal sense of 'quickly' (cf. 2 Thess. 2: 2; Phil. 2: 19, 24) by assuming that Paul has in mind the interval since the Galatians first became Christians. He is astounded that their resistance to the intruders was so short-lived. The brevity of the time factor is uppermost in his mind; it is not as if the endurance of the Galatians had been tested by a long period of hostile pressure. Manifestly he is contrasting previous information concerning the happy situation in Galatia with what he has now been informed is the sorry state of the community.[89] When and how did he learn that the Galatians 'were running well' (Gal. 5: 7)?

Borse, in order to maintain his Macedonian dating of Galatians, attempts to argue that 1 Corinthians 16: 1 implies that at the end of his stay in Ephesus Paul still had a good opinion of the Galatians.[90] One might agree if Paul had praised the generosity of the latter and held them up as an example. But all he in fact says is that the same administrative directive regarding the collection for the poor of Jerusalem, which he now gives the Corinthians, he had given previously to the Galatians. In any case, if the Galatians had agreed to Paul's request, he could be quite sure that they would be even more responsive to their new guides, the Judaizers, who viewed the Jerusalem church much more favourably than Paul did.

There is in fact no alternative to the only substantiated hypothesis, namely, that Paul learned of the situation of the Galatian churches when he passed through that region, and preached the collection for Jerusalem, en route to Ephesus, a journey which I have dated to the spring and summer of AD 52. It is not impossible that the Judaizers followed closely on the heels of Paul when he left Antioch, and so reached Galatia not long after he had left. If we further assume that their impact was immediate, and that Paul was warned as soon as possible, Galatians could have been dispatched before the snows closed the high country in the autumn of AD 52.

It may be doubted, however, that events moved quite so quickly, particularly

[89] These two points are well brought out by Borse (1972), 47.
[90] Ibid. 46.

since it is a question of a number of churches in Galatia. It is more reasonable to assume that the Judaizers spent the winter in Galatia, and that news of their depredations reached Paul in Ephesus only after the snows had melted and the roads were again open to traffic. We can be sure that Paul responded immediately; the use of the present participles 'who are disturbing you and who are desiring to pervert the gospel of Christ' (Gal. 1: 7) indicate that the trouble-makers are still at work as he writes, and have not entirely succeeded (cf. Gal. 5: 10).[91] Hence Galatians should be dated in the spring of AD 53.

AN EPHESIAN CHRONOLOGY

In working out the chronology of Paul's life, I argued that he arrived in Ephesus in August AD 52 and departed definitively in October 54.[92] How Paul spent the summer of AD 54 can be deduced without difficulty from 1 and 2 Corinthians. It has been touched on above and will be considered in some detail when dealing with the Corinthian correspondence. Thus the time-frame with which we are concerned is not quite two years, the twenty-two months from August 52 to May 54. The previous discussion in this chapter gave some idea of his activities during this period. The challenge now is to arrange those activities in at least a relative chronological order.

Paul's first year in Ephesus, it would appear, was trouble free. After having informed his churches in Macedonia and in Achaia where he could be reached, he was able to devote his energies to the development of the local community and to its missionary outreach into the hinterland.

The period of Paul's imprisonment must fall between the composition of Galatians, which gives no hint that Paul is a prisoner, and the writing of 1 Corinthians in May AD 54, when he is free to plan a journey through Macedonia to Corinth (1 Cor. 16: 5).

This period is further limited by two factors. Communications between Asia and Greece would have been cut from the end of the sailing season in October to the following April, and Paul is unlikely to have ventured into the interior of Anatolia in the depths of winter. If he planned to 'winter' in Corinth (1 Cor. 16: 6), it is improbable that he left Ephesus in winter. As we shall see when dealing with the Corinthian letters, all of Paul's attention in the spring and summer of AD 54 was concentrated on Corinth.

In consequence, the movements implied in Philippians, and the sending of two of its component letters, must have taken place between the spring of AD 53 and the autumn of that same year. Rather more latitude can be allowed for connections with the Lycus valley. Strong motivation and unseasonably good

---

[91] E. Burton (1921), 25.
[92] See Ch. 1, 'After AD 57'.

travel conditions would have made it feasible for Epaphras and Onesimus to get to Ephesus even in winter, and for Paul to travel to Colossae.

Let us look first at Paul's relations with Philippi. Since the prevailing wind was from the northern quadrant, one would expect travellers from Philippi to Ephesus to come all the way by boat, and Acts provides figures which are eminently plausible: Philippi to Troas five days (20: 6); Troas to Ephesus four days (20: 13–16). In total, therefore, nine days, which could be extended or diminished depending on weather conditions, and on how much time the boat spent in harbour each day. Returning home it was more sensible to travel by road to Troas, in order to avoid the delays imposed by contrary winds. Those who took ship could advance only when the wind swung briefly into the south. Troas was 350 km. (210 miles) from Ephesus, a walk of two weeks at an average of 25 km. (15 miles) per day. From there it was imperative to take a boat. Under optimum conditions the crossing to Neapolis took two days (Acts 16: 11), and the better part of another day was needed to cover the 15 km. (9 miles) from Neapolis to Philippi. The round trip could be done in twenty-six days. In order to allow time for a visit on arrival, however, a month would be a more reasonable minimum estimate.

If we assume that Galatians was written in April or May AD 53, the subsequent events of that summer with respect to Philippi can be reconstructed as follows. The church there sent Epaphroditus with gifts for Paul (Phil. 4: 18). While in Ephesus he fell ill (Phil. 2: 26), which meant that Paul had to find another messenger to carry a letter expressing his gratitude to Philippi (*Letter A*: Phil. 4: 10–20). Naturally this emissary explained what had happened to Epaphroditus, and brought back to Ephesus the news of the concern of the Philippian community for the invalid (Phil. 2: 26). It is not at all certain that Paul was under arrest at this stage. There is no necessary connection between financial aid from Philippi and imprisonment; the Philippians had previously helped him financially at Thessalonica (Phil. 4: 16), and at Corinth (2 Cor. 11: 9), when he was certainly free. At least two months should be allowed for these contacts, which brings us to July AD 53.

This date suggests the possibility that Paul's arrest was due to the zeal of the new proconsul of Asia (Phil. 1: 13). This official, who was appointed for one year, took up his duties on 1 July,[93] and may have wanted to appear decisive and energetic, when warned of a potentially subversive group. How long his investigation of the nascent church lasted we can never know, but it was not more than nine months and perhaps considerably shorter. Paul and the others may have been released before the end of the summer. The limitations on winter travel make it improbable that *Letter B* to Philippi (1: 1 to 3: 1 and 4: 2–9), Colossians, and Philemon were written after October. If Paul was released in

[93] See Ch. 1, 'Dating the Proconsulship of Gallio'.

late summer or early autumn, he could have made his promised visit to the Lycus valley. Whether he spent the winter there, or returned to Ephesus, remains an open question.

According to Acts 19: 1, Apollos had left Ephesus for Corinth prior to Paul's arrival in Ephesus. This is partially confirmed by Paul's witness that Apollos had succeeded him at Corinth (1 Cor. 3: 6), and it is unlikely to have been invented by Luke, because it is precisely the sort of the uncontrolled expansion that Luke wanted to correct and control. Apollos had certainly moved back to Ephesus prior to the writing of 1 Corinthians in May AD 54, because he is with Paul at that moment. Since the letter brought by the Corinthian delegation in the late spring of that year requested that Apollos return to Corinth with them (1 Cor. 16: 12),[94] the latter must have been in Ephesus since the previous summer. It is most economical to suppose that he brought the information which prompted Paul to write the now lost 'Previous Letter' (1 Cor. 5: 9).

In broad outline, therefore, Paul's schedule subsequent to his departure from Antioch in the spring of AD 52 was as follows:

Summer 52	Paul in Galatia.
	Apollos to Corinth.
September 52	Paul arrives in Ephesus.
	Judaizers arrive in Galatia.
Winter 52–53	Consolidation of the church in Ephesus and beginning of its missionary outreach to Asia.
Spring 53	Epaphras begins his mission in the Lycus valley.
	Bad news from Galatia.
	Paul writes *Galatians*.
Summer 53	Gift from Philippians.
	Paul writes *Letter A* to the Philippians.
	Paul and his collaborators arrested.
	Epaphras returns to Ephesus.
	From prison Paul writes *Letter B* to the Philippians, *Colossians*, and *Philemon*.
	Apollos returns to Ephesus.
	Paul writes the *Previous Letter* to Corinth.
Autumn 53	Paul travels to the Lycus valley?
Winter 53–54	Paul winters in Ephesus.
Spring–Summer 54	Intense contacts with Corinth.

[94] It is most improbable that Paul would have initiated the project of a return of Apollos to Corinth (cf. 1 Cor. 1: 12); so rightly Fee (1987), 823–24.

— 8 —

Conflict in Galatia

Two points have already been established with regard to the Galatians and the letter which Paul wrote to them. The Galatians in question were members of the Celtic tribes in the area of Pessinus (modern Balahissar),[1] and the most probable date of the letter is the spring of AD 53.[2] The task now is to explore Paul's relations with these people, and the contribution they unwittingly made to the maturation of his thought by forcing him to deal with a problem that he had not hitherto encountered. Believers with a different vision of Christianity were bidding for the allegiance of members of a community which he had founded.

GALATIA: LAND AND PEOPLE

Epigraphical studies in central Anatolia, the area settled by the Galatians, reveal that Celtic, Greek, Roman, and Phrygian types of names appear in the one family.[3] It would be difficult to find a more graphic illustration of the complex ethnic background of the members of the churches of Galatia.

The Celts or Galatians—the names are used interchangeably by the classical sources—who moved into Asia Minor in 278 BC were not the usual type of mercenaries. Warriors made up only half the 20,000 who crossed the Bosphorus; they brought their wives and children in addition to aged parents.[4] They were a nation on the move, searching for a homeland.[5] With admirable brevity Strabo summarizes,

---

[1] See Ch. 7, 'Galatia and Galatians'.
[2] See Ch. 7, 'The Date of Galatians'.
[3] S. Mitchell (1980), 1058.   [4] Livy, *History of Rome* 38. 16.
[5] 'The people who are called the Tectosages closely approach the Pyrenees. . . . It appears that at one time they were so powerful and had so large a stock of strong men, that, when a sedition broke out in their midst, they drove a considerable number of their own people out of the homeland, that other persons from other tribes made common lot with these exiles; and that among these are also those people who have taken possession of that part of Phrygia which has a common boundary with Cappadocia and the Paphlagonians' (Strabo, *Geography* 4. 1. 13; trans. Jones). The following citation may describe pure coincidence, but it should not go unremarked, 'The Christians in Gaul seem to have come from Asia Minor—certainly Irenaeus [of Lyon] did—and at any rate had close links

> The Galatians are to the south of the Paphlagonians. ... This country was
> occupied by the Galatae after they had wandered about for a long time, and
> after they'had overrun the country that was subject to the Attalic and the
> Bithynian kings, until by voluntary cession they received the present Galatia or
> Gallo-Graecia as it is called. (*Geography* 12. 5. 1; trans. Jones)

This settlement took place about 232 BC under Attalus I of Pergamum, and the
Galatians were then concentrated in the area of Ancyra (modern Ankara).
After the death of this strong king in 197 BC, the Galatians recommenced their
westward raids into Asia. They were decimated in the vicious Roman reaction
of 189 BC,[6] but their lands were not confiscated, and they subsequently became
committed allies of Rome.[7] Some 25 years later, they had peacefully penetrated
the region around the future city of Pessinus (see Fig. 4).[8]

Inevitably they brought into Asia Minor the customs which made them distinctive in the west, and which Athenaeus (*floruit c.* AD 200) found worthy of
citation from Poseidonius (135–50 BC):

> The Celts place hay on the ground when they serve their meals, which they
> take on wooden tables raised only slightly from the ground. Their food consists
> of a few loaves of bread, but of large quantities of meat prepared in water or
> roasted over coals or on spits. This they eat in a cleanly fashion, to be sure, but
> with a lion-like appetite, grasping whole joints with both hands and biting them
> off the bone. If, however, any piece proves hard to tear away, they slice it off
> with a small knife which lies at hand in its sheath in a special box. ...
>
> When several dine together, they sit in a circle; but the mightiest among
> them, distinguished above the others for skill in war, or family connections or
> wealth, sits in the middle, like a chorus-leader. Beside him is the host, and next
> on either side the others according to their respective ranks. Men-at-arms,
> carrying oblong shields, stand close behind them, while their bodyguards,
> seated in a circle directly opposite, share in the feast like their masters. The
> attendants serve the drink in vessels resembling our spouted cups, either of
> clay or of silver. Similar also are the platters which they have for serving food;
> but others use bronze platters, others still, baskets of wood or plaited
> wicker. (*Deipnosophistae* 4. 151e–152b; trans. Gulick)

> The Celts sometimes have gladiatorial contests during dinner. Having assem-
> bled under arms, they indulge in sham fights and practise feints with one
> another; sometimes they proceed even to the point of wounding each other,
> and then, exasperated by this, if the company does not intervene, they go so far
> as to kill. (*Deipnosophistae* 4. 154b; trans. Gulick)

with Asia Minor: the letter describing the course of the persecution of the Gallic martyrs was
addressed to the Christians in Asia ... and Phrygia' (Louth (1989), p. xxiv).

[6] Livy, *History of Rome* 38. 17–27.
[7] Ramsay (1900), 97, 116; S. Mitchell (1993), 27–41.
[8] Ramsay (1900), 62. A detailed map of Galatia with modern equivalents is to be found in PW 7.
530. The archaeological evidence is neatly summarized in map 4b in Mitchell (1993), 1. 53.

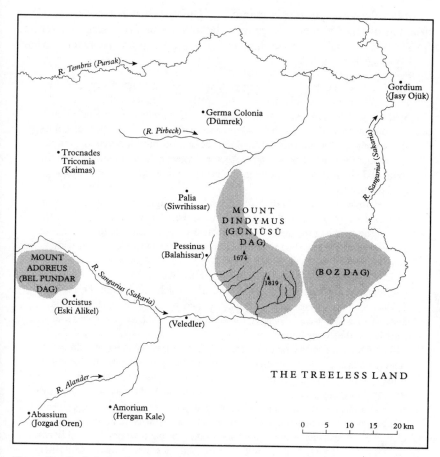

FIG. 4 Paul's Galatia (*Sources*: Richard Kiepert, *Karte von Kleinasien*, B III (Angora, 1907); PW VIII)

The Celts, even when they go to war, carry round with them living-companions whom they call hangers-on. These persons recite their praises before men when they are gathered in large companies as well as before any individual who listens to them in private. And their entertainments are furnished by the so called Bards; these are poets, as it happens, who recite praises in song.

(*Deipnosophistae* 6. 246d; trans. Gulick)

Among the barbarians, the Celts also, though they have very beautiful women, enjoy boys more; so that some of them often have two lovers to sleep with on their beds of animal skins. (*Deipnosophistae* 13. 603a; trans. Gulick)

An equally vivid portrait is painted by Didorus Siculus (80–20 BC). He is speaking of the Celts of France, but the validity of his description for the Galatians is attested by both monument[9] and text.[10] Unfortunately it is too long to be quoted in its entirety:

> The Gauls are tall of body, with rippling muscles, and white of skin, and their hair is blond, and not only naturally so, but they also make it their practice by artificial means to increase the distinguishing colour which nature has given it. ... Some of them shave the beard, but others let it grow a little; and the nobles shave their cheeks, but they let the moustache grow until it covers the mouth. Consequently, when they are eating, their moustaches become entangled in the food, and when they are drinking, the beverage passes, as it were, through a kind of strainer. ...
>
> They invite strangers to their feasts, and do not inquire until after the meal who they are and of what things they stand in need. And it is their custom, even during the course of the meal, to seize upon any trival matter as an occasion for keen disputation, and then to challenge one another to single combat without any regard for their lives. ...
>
> The clothing they wear is striking—shirts which have been dyed and embroidered in varied colours, and breeches, which they call in their tongue *bracae*, and they wear striped coats, fastened by a buckle on the shoulder, heavy for winter wear and light for summer, in which are set checks, close together and of varied hues. ...[11]
>
> The Gauls are terrifying in aspect and their voices are deep and altogether harsh; when they meet together they converse with few words and in riddles, hinting darkly at things for the most part, and using one word when they mean another; and they like to talk in superlatives, to the end they may extol themselves and depreciate all others. They are also boasters and threateners and are fond of pompous language, and yet they have sharp wits and are not without cleverness at learning. Among them are found lyric poets whom they call Bards.
>
> (*The Library of History* 5. 28–31; trans. Oldfather)

As the conquerers, the Galatians were an aristocratic caste, but this did not make them immune to their environment.[12] The extent of intermarriage with the indigenous population is underlined by Livy's characterization of the Galatians 'a mixed race'.[13] They adopted the local Phrygian religion. Not only

[9] The famous marble statue of 'The Dying Gaul' is a Roman copy (now in the Capitol in Rome) of the bronze original set up by Atallus I in Pergamum around 230 BC. It is reproduced as Fig. 7 in Mitchell (1993), I. 46.

[10] Compare the description of Livy, 'Their tall physique, their flowing red locks, their vast shields and enormous swords, together with their songs as they go into battle, their howlings and leapings and the fearful din of arms as they batter their shields following some kind of ancestral custom—all these are carefully designed to strike terror' (*History of Rome* 38: 17; trans. Bettenson).

[11] In a footnote, Oldfather comments, 'Diodorus appears to be trying to describe a kind of Scotch tartan.'

[12] See in particular Ramsay (1900), 144–5.

[13] *History of Rome* 38. 17.

was it more prudent to propitiate the local gods, but the Celtic nobility gained access to indigenous power through membership in the immensely influential priesthood of Pessinus.[14] In addition to Celtic, which continued to be spoken into the Byzantine period,[15] Greek was adopted as a second language. It was the indispensable medium of communication throughout the area, and anyone who travelled had to be bilingual.[16] None the less, the Galatians themselves were not Hellenized. 'About AD 50 Galatia was essentially un-Hellenic. Roman ideas [particularly of administration] were there super-induced directly on a Galatian system, which had passed through no intermediate stage of trans-formation to the Hellenic type.'[17] This is particularly evident in the continuing prominence in Galatia of tribal structures.

The titles of the three administrative regions set up by Augustus in 22 or 21 BC were the Sebasteni Tolistobogii Pessinuntii, the Sebasteni Textosages Ancyrani, and the Sebasteni Trocmi Taviani. The population of the three Galatian tribes was considered to be identical with the three villages trans-formed by imperial fiat into cities. The territory of Pessinus, however, with which we are concerned, stretched only from Mount Dindymus (modern Güny-üzü Dag) to the source of the river Sangarios.[18] Elsewhere in the tribal territory there were only unfortified villages.[19]

The quick-witted, enterprising Greeks of the province of Asia looked on those who dwelt in the middle of Asia Minor with contempt. The Phrygians had a reputation for being 'slow, apathetic, contented, and unutterably ignorant, incapable of being roused or excited by any cause except their vulgar and degrading superstitious rites'.[20] Understandably, then, there was nobody more despicable than a Phrygian,[21] and to be a slave among them was the nadir of human existence.[22] The Galatians for their part were considered to be large, unpredictable simpletons, ferocious and highly dangerous when angry, but

---

[14] Strabo, *Geography* 12. 5. 3; Ramsay (1900), 62; Mitchell (1993), I. 48.

[15] Pausanias, *Description of Greece* 10. 19. 11; 36. 1; Jerome, *Comm. in Ep. ad Galatas* 2. 3; Mitchell (1980), 1058.

[16] According to Mitchell (1993), I. 175, 'From these illustrations of the linguistic complexity of Roman Anatolia, it seems reasonable to conclude that between the few who spoke no Greek at all (perhaps in particular women who had less contact outside the household with commerce, official-dom, or public life) and the larger minority who had been cut off completely from their native cultural and linguistic heritage by absorption into city life, a majority of the inhabitants of Asia Minor were in some measure bilingual in Greek and an indigenous language.'

[17] Ramsay (1900), 160; cf. 142–3.

[18] Mitchell (1993), I. 87, with map 3 facing p. 40.

[19] Strabo notes that the major agglomerations were 'habitations which preserve not even traces of cities, but are only villages slightly larger than the others' (*Geography* 12. 5. 4; trans. Jones). See also Livy, *History of Rome* 38. 18–19.

[20] Ramsay (1900), 31.

[21] Dio Chrysostom, 'more despised than the Phrygians' (*Discourses* 31. 158).

[22] Dio Chrysostom, 'slaves in the interior of Phrygia' (*Discourses* 31. 113).

without stamina and easy to trick.[23] They were the archetypal barbarians. It would be hard to find a more charitable comment on the mixture of Galatians and Phrygians than that of Livy, 'a degenerate, mongrel race'.[24]

The land in which this race dwelt was hardly more interesting than its people and perhaps contributed to their lassitude. Ramsay's description has never been bettered,

> It consists of a vast series of bare, bleak up-lands and sloping hillsides. It is almost devoid of trees,[25] except, perhaps, in some places on the north frontiers; and the want of shade makes the heat of summer more trying, while the climate in winter is severe. The hills often reach a considerable altitude,[26] but have never the character of mountains. They are commonly clad with a slight growth of grass to the summit on at least one side. The scenery is uninteresting. There are hardly any striking features; and one part is singularly like another. The cities are far from one another, separated by long stretches of the same fatiguing country, dusty and hot and arid in summer, covered with snow in winter. . . . In ancient times the aspect of most of the land away from the few great cities was much the same as it is at the present day—bleak stretches of pastoral country, few villages, sparse population, little evidence of civiliza- tion.[27]

The heavy rains and snow of winter usually begin in November and last until April. The temperature can drop to -20 °C and long periods of frost are normal. The ground remains soft and muddy into June when the hot dry season begins.[28] Roman roads certainly traversed Galatia in the time of Paul, even though the great construction effort which gave Asia Minor its 9,000 km. (5,400 miles) of graded roads is dated between AD 80 and 122.[29]

The staple products of Galatia are revealed by its unique artistic creation, carved tombstones.[30] The most common motif is a distaff and spindle,[31] which

---

[23] According to Strabo, 'The whole race which is now called both "Gallic" and "Galatic" is war- mad, and both high-spirited and quick for battle, although otherwise simple and not ill-mannered' (*Geography* 4. 4. 2; trans. Jones); 'In addition to their trait of simplicity and high-spiritedness, that of witlessness and boastfulness is much in evidence, and also that of fondness for ornaments' (*Geography* 4. 4. 5; trans. Jones).

[24] *History of Rome* 38. 17.

[25] West of Abassion and south of the river Sangarios is 'the territory called Axylon ("Woodless"). This gets its name from the fact that it produces no wood at all, not even thorns or any other fuel; the people use cow dung as a substitute for wood' (Livy, *History of Rome* 38. 18; trans. Bettenson). For the localization, see the map in PW 7. 530. The lack of trees in the south of Galatia near the border with Lycaonia is noted by Strabo (*Geography* 12. 6. 1).

[26] According to Ramsay (1990), 17, the altitude of the river Sangarios is roughly 2,000 feet (615 m.) and the land rises towards the south-west to 3,600 feet (1,107 m.).

[27] (1900), 12–14.

[28] Mitchell (1993), 1. 143–5.

[29] Mitchell (1993), 1. 124–6 with maps 7 and 8.

[30] Mitchell (1980), 1070.

[31] To cite but a single volume of *MAMA*, 1. nn. 89, 93, 94, 99, 103, 104, 105, 106, 107, 113, 118, 119, 135, 219, 263, 265, 297, 300, 310, 341, 348, 353, 391, 413, 422, 423.

highlights the importance of wool in the economy of the province. Strabo noted that, 'although the country is unwatered, it is remarkably productive of sheep; but the wool is coarse, and yet some persons have acquired very great wealth from this alone. Amyntas had over three hundred flocks in this region' (*Geography* 12. 6. 1; trans. Jones). Strabo's judgement on the quality of the wool was not shared by Pliny who considered it among the finest in the world.[32] In many tombstones the distaff and spindle is associated with a mattock and pruning hook.[33] The hint that viticulture was important is confirmed by representations of a vine[34] or bunches of grapes.[35] Wine may have made life in that desolate area more bearable, but it is unlikely to have made the same contribution to the economy as the cultivation of cereals. Ears of wheat[36] and a yoke of oxen pulling a plough[37] are depicted on many tombstones. The economic situation is perfectly summarized by Mitchell, 'Grain kept the province alive, wool brought it wealth.'[38]

PAUL'S MINISTRY IN GALATIA

Jewett mentions the towns of Pessinus, Germa, and Ancyra as the area of Paul's ministry in Galatia.[39] No justification is offered, and one may presume that he simply listed the three known cities in order to justify the plural 'churches' in the address of Galatians.

Paul does use 'church' in the sense of all the believers in a town or city (1 Thess. 1: 1; 2 Thess. 1: 1; 1 Cor. 1: 2; 2 Cor. 1: 1), but he can also use it to mean a subgroup, those who assemble in a particular home (1 Cor. 16: 19; Col. 4: 15; Philem. 2; contrast Rom. 16: 23). In the latter instances 'church' is never qualified, whereas in the former it is always specified, 'in God the Father and the Lord Jesus Christ' (1 Thess. 1: 1; 2 Thess. 1: 1), 'of God' (1 Cor. 1: 2; 2 Cor. 1: 1). The fact that Galatians is addressed simply to 'the churches of Galatia' might imply that Paul is thinking of a number of house-churches within a rather restricted area. It seems more probable, however, that Paul's disappointment at the backsliding of the Galatians was so intense that he decided to treat the Galatian town communities as if they were mere secular 'assemblies', in which case we can deduce nothing about the extent of his mission field in Galatia.

If Paul reached Abassiom (modern Jüzgad Ören) on the border of Galatia,[40]

---

[32] 'The most esteemed wool is all from the neck, and that from the districts of Galatia, Tarentum, Attica, and Miletus' (*NH* 29. 33).

[33] *MAMA*, 1. nn. 126, 138, 154, 420.

[34] Ibid. 1. nn. 156, 248, 274.

[35] Ibid. 1. n. 110. Wine from the area is mentioned by Strabo (*Geography* 12. 7. 2) and Pliny (*NH* 31. 84).

[36] *MAMA*, 1. n. 5

[37] Ibid. 1. nn. 149, 293, 340.

[38] (1980), 1069; (1993), 1. 146.

[39] (1979), 60.

[40] Livy, *History of Rome* 38. 15.

he could have marched east keeping the Sangarios river on his left as did the Roman general C. Manlius. The advantage would have been to use the bridge which Manlius had built over the river. But that route would have brought him along the edge of the desolate desert of the Treeless Land,[41] and it would have been out of character for Paul to waste his energy in an area of widely scattered villages whose inhabitants—shepherds and field workers—spoke a Greek so corrupt as to be virtually unintelligible to someone like Paul.[42]

Paul, however, probably had little choice where he went. He was ill (Gal. 4: 13), and if consulted, his preference would certainly have been for the sort of urban environment in which he felt most at home and worked most effectively, and in which he could receive whatever medical care was available. The easiest way to cover the 40 km. to the Sangarios would have been by boat down the Ak Tsha'yr, which flowed into the Alander, a tributary of the Sangarios. From the latter it was only 12 km. to Pessinus. According to Strabo,

> Pessinus is the greatest commercial centre in that part of the world, containing a temple of the Mother of the gods, which is an object of great veneration. They call her Agdistis. The priests were in ancient times powerful and reaped the fruits of a great priesthood, but at present their prerogatives have been much reduced, though the market still endures. The sacred precinct was built up by the Attalid kings in a manner befitting a holy place, with a sanctuary and also with porticoes of white marble. The Romans made the temple famous when, in accordance with oracles of the Sibyl, they sent for the statue of the goddess there, just as they did in the case of that of Asclepius at Epidaurus.
>
> (*Geography* 12. 5. 3; trans. Jones)

Agdistis is better known as Cybele,[43] the Great Mother, whose cult was recognized officially by Rome in 204 BC. She was the supreme divine being of Phrygia, and the male god, Attis, was merely her inferior companion and servant.[44] Naturally the cult was especially favoured by women. Cybele was responsible for all aspects of the well-being of her people, ensuring fertility, curing disease, giving oracles, and protecting her adherents. Ecstatic states accompanied by insensibility to pain and/or the gift of prophecy were characteristic of her worship.[45]

We must presume that Paul's initial preaching in Pessinus took the form of conversations with those generous enough to give him hospitality (Gal. 4: 13–14). They were pagans; he could not have said 'formerly you did not know God' (Gal. 4: 8) to Jews. As his strength returned, he became more energetic, and his influence increased. The presence in Pessinus of pilgrims to the shrine of

[41] Ibid. 38. 18.
[42] Mitchell (1983), 1. 174.
[43] Strabo, *Geography* 10. 3. 12.
[44] Ramsay (1900), 38; Mitchell (1993), 2. 20–2.
[45] *OCD* 246.

Cybele might explain why he stayed on in Galatia after his recovery. They offered him something which he was later to see more fully realized in Corinth and Ephesus. He saw in such visitors the possibility of reaching out into the vast hinterland which he could never hope to cover. Returning to their homes they could carry his message to places to which he could not go, and whose names he may not even have known. Perhaps this is why the address of the letter is so vague.

THE CAUSE OF THE CRISIS

The problems which developed among the Christians of Galatia were not intrinsic tensions which increased until they reached flash-point. They were caused by outsiders, who attempted to persuade the Galatians to adopt a vision of Christianity which was radically different from that of Paul.

Who were the Intruders?

In the letter which he wrote to the Galatians, Paul consistently differentiates between his converts, whom he addresses as 'you' (e.g. Gal. 3: 1–5), and others to whom he disdainfully refers as 'some people' (Gal. 1: 7), 'anyone' (Gal. 1: 9), 'they' (Gal. 4: 17; 6: 13). In antiquity adversaries were never given free publicity! The clear hint that these latter were outsiders is confirmed by the verb he uses of their activity; 'to disturb, unsettle, throw into confusion' (Gal. 1: 7; 5: 10) belongs to the political language of the period and describes the work of agitators who move in to destroy a previously peaceful situation.[46]

In addition to identifying his adversaries as intruders, Paul gives us just enough information to specify further that they were not pagan philosophers (cf. Col. 2: 8) or Jews, but Christians of Jewish origin, either by birth or by conversion. They belonged to 'the circumcision' (Gal. 6: 13),[47] but considered their message to the Galatians as 'a gospel' (Gal. 1: 6).[48]

Where did these people come from? The vast majority of scholars opt for Jerusalem, but differ on the relationship of the intruders to the authorities of the church in the Holy City.[49] It is much more likely, however, that the intruders came from Antioch.[50] When Paul founded the churches of Galatia, he

[46] Lightfoot (1910), 77; Betz (1979), 49; Longenecker (1990), 16.
[47] Betz (1978), 316; Longenecker (1990), 292; against E. Burton (1921), 352–3.
[48] The fact that Paul has to correct himself and deny that there is 'another gospel' betrays his awareness that he was confronted by a different version of Christianity. So rightly Martyn (1985), 314.
[49] A good survey is provided by Longenecker (1990), p. lxxxviii–xciv.
[50] So rightly Dunn (1993), 14–17, but he mistakenly limits the sphere of Antioch's jurisdiction to the churches established during the 'first journey' of Acts.

had been acting as an agent of Antioch. Under pressure from Jerusalem, however, the Antiochean community had opted subsequently for a completely Judaized version of Christianity, which Paul could not accept (Gal. 2: 11–21). His failure to persuade those who had once sponsored his missionary drive into Europe that they were being unjust to Gentile members of the church led to a complete break.

If Paul could no longer be a member of such a community, still less could he propagate its vision of Christianity. There were many, however, who not only were prepared to do so, but believed that they had an obligation to extend to the churches founded under the aegis of Antioch the new practices adopted by the mother church.[51] The daughter communities, which naturally reflected the generous tolerance and openness of Antioch, now had to be brought into line with its new ethos. Presumably this measure was not directed exclusively at the Pauline churches; it affected those founded by Barnabas and others as well.[52]

In Jerusalem and on his return to Antioch, Paul had spoken freely of the successes with which God had blessed his ministry. This made it easy for those now sent out by Antioch to retrace his steps; they knew exactly where he had been. The group need not have been made up exclusively of Jewish Christians. The inclusion of a few Gentile Christians who had willingly accepted Judaization would have strengthened the claim that what Paul had once preached in Galatia had been superseded by subsequent developments in Christianity.

As far as Paul was concerned, the Judaizers were intruders who had no business interfering with his converts. For their part, the representatives of Antioch believed that Paul had lost his rights in the churches of Galatia. These were Antiochean foundations and the delegates felt that by repudiating his commission Paul had abdicated from any position of authority in Galatia. The delegates now wore the mantel of legitimacy; they were the official link with the authentic roots of Christianity. It is not necessary to postulate any personal animosity towards Paul. The identity of the Antiochean missionaries, as we shall see, was not rooted in an anti-Paul polemic. Perhaps they even experienced a certain sense of loss that a wonderful missionary had sidelined himself because of his tragically mistaken conviction that he alone possessed the truth. They too were convinced that they had the best interests of the Gentiles deeply at heart.

The Tactics of the Judaizers

When they arrived in Galatia, the Judaizers had two tasks. First, they had to undermine the authority of Paul. It was not enough to say that they were now

---

[51] Ibid. 13–14.
[52] Hence to speak of 'a planned and concerted anti-Pauline movement' (Barrett (1985), 22) must be done with careful nuances.

taking over. They had to discredit him. Secondly, they had to put across their version of Christianity with clarity and power. They could not simply say that Paul was wrong. They had to propose a viable alternative.

Unfortunately they did not leave the notes of their speeches, which means that their teaching has to be reconstructed from Paul's reaction. This technique of mirror reading involves rather obvious dangers.[53] The intruders, for example, may not have said exactly the opposite of Paul's response. He may have exaggerated their positions in order to facilitate his own counter-attack. Moreover, it must also be kept in mind that Paul was not confronted personally by his opponents. He had no direct knowledge of their accusations. He became aware of what was going on in Galatia only through the reports of his partisans among the Galatians.

The opacity of this filter, however, should not be exaggerated. If, as I have argued above,[54] the intruders spent the winter of AD 52–53 in Galatia, there was plenty of time for Paul's supporters to learn exactly what his rivals were saying. It is unlikely that some of those who had truly committed themselves to the Pauline gospel took very long to realize how different was the vision of Christianity now being proposed to them. Lacking the theological background necessary to develop a counter-argument, they could offer only passive resistance. Concern enhanced their concentration. They recognized the need to retain everything that they were hearing with a view to reporting to Paul in Ephesus as soon as the roads opened to travel in the spring.

Discrediting Paul

It is probable, therefore, that Paul received very accurate information regarding the tactics of the intruders, and there is in fact a rather high degree of agreement on the nature of the problems which he had to confront. F. F. Bruce vividly summarizes a large consensus when he puts the following words into the mouths of the Judaizers:

> The Jerusalem leaders are the only persons with authority to say what the true gospel is, and this authority they received direct from Christ. Paul has no comparable authority: any commission he exercises was derived by him from the Jerusalem leaders, and if he differs from them on the content or implications of the gospel, he is acting and teaching quite arbitrarily. In fact Paul went up to Jerusalem shortly after his conversion and spent some time with the apostles there. They instructed him in the first principles of the gospel and, seeing that he was a man of uncommon intellect, magnanimously wiped out from their minds his record as a persecutor and authorized him to preach to others the gospel which he had learned from them. But when he left Jerusalem for Syria

[53] Such dangers, however, have been exaggerated by Lyons (1985); see Barclay (1987), 73–93. The agnosticism of Borse (1984), 240, is entirely unjustified.

[54] See Ch. 7, 'The Date of Galatians'.

and Cilicia he began to adapt the gospel to make it palatable to Gentiles. The Jerusalem leaders practiced circumcision and observed the law and the customs, but Paul struck out on a line of his own, omitting circumcision and other ancient observances from the gospel he preached, and thus he betrayed his ancestral heritage. This law-free gospel has no authority but his own; he certainly did not receive it from the apostles, who disapproved of his course of action. Their disapproval was publicly shown on one occasion at Antioch, when there was a direct confrontation between Peter and him on the necessity of maintaining the Jewish food-laws.[55]

An Alternative Vision of Christianity

Hearing such an attack on Paul, the obvious question that his supporters asked concerned the nature of the so-called authentic gospel. In what precisely did it consist, and how was it justified? The most detailed, and carefully argued, reconstruction is that of J. Louis Martyn:

> Listen now. It all began with Abraham. He was the first human being to discern that there is but one God. Because of that perception he turned from the service of dumb idols to the worship of the true God [Jub. 12]. There God made him the father of our great nation; but that was only the beginning, for God made to Abraham a solemn utterance which through our mission has begun to find its fulfillment in the present time. Speaking through a glorious angel, God said to Abraham:
>
>> In you shall all the nations of the world be blessed [Gen. 12: 3] ... for I shall multiply your descendants as the stars of heaven [Gen. 22: 17].... Come outside, and look toward heaven, and number the stars, if you are able ... So shall your descendants be [Gen. 15: 5] ... for I speak this blessing to you and to your descendants.
>
> What is the meaning of this blessing which God gave to Abraham? Pay attention to these things: Abraham was the first proselyte. As we have said, he discerned the true God and turned to him. God therefore made an unshakable covenant with Abraham [Gen. 17: 7], and as a sign of this covenant he gave to Abraham the commandment of circumcision [Gen. 17: 10]. He also revealed to Abraham the heavenly calendar, so that in his own lifetime our father was in fact obedient to the Law, not only keeping the commandment of circumcision [Gen. 17: 23], but also observing the holy feasts on the correct days [Jub. 15: 1–2].
>
> Later, when God actually handed down the Law at Sinai, he spoke once again in the mouths of his glorious angels who pass the Law through the hand of the mediator, Moses (Galatians 3: 19). And now the Messiah has come, confirming for eternity God's blessed Law, revealed to Abraham and spoken through Moses (6: 2).
>
> And what does this mean for you Gentiles? We know from the scriptures

---

[55] (1982*b*), 26.

that Abraham had two sons: Isaac and Ishmael (4: 22). On the day of the feast of the first fruits Isaac was born of Sarah the freewoman [Gen. 21: 1–7], and through him have come we Jews, who are descendants of Abraham. Ishmael was born of Hagar the slave girl [Gen. 16], and through him have come you Gentiles. Thus you also are descendants of the patriarch. We are in fact brothers!

We also know from the scripture we have just quoted that God made his indelible promise to both Abraham and his descendants, saying 'In you shall all the nations be blessed. The inheritance of salvation is to your children's children! [Sir. 44: 21]'. That fact faces us all with the crucial question: Who is it who are the true and therefore blessed descendants of Abraham (3: 7, 29)? And the answer is equally clear from the scriptures: Abraham himself turned from idols to the observance of the Law, circumcising himself and Isaac. As we have said, he even kept the holy feasts at their precisely appointed times. And not least, by keeping God's commandments [Jub. 21. 2], he avoided walking in the power of the Evil Impulse (5: 16; cf. Genesis 6: 5). It follows that the true descendants are clearly those who are faithfully obedient to the Law with faithful Abraham (Galatians 3: 6–9). At the present holy time God has been pleased to extend this line of true descent through the community in Jerusalem, the community which lives by the Law of Christ (6: 2), the community of James, Cephas and John, *and* through the community which we represent (2: 1–10).

What are you to do, therefore, as Abraham's descendants through Ishmael, the child of Hagar, the slave-girl? The gate of conversion stands open (4: 17; 4: 9)! You are to cast off your enslavement to the Evil Impulse by turning in repentance and conversion to God's righteous Law as it is confirmed by his Christ. Follow Abraham in the holy and liberating rite of circumcision (6: 13); observe the feasts at their appointed times (4: 10); keep the sacred dietary requirements (2: 11–14); and abstain from idolatry and from the passions of the flesh (5: 19–21). Then you will be true descendants of Abraham, heirs of salvation according to the blessing which God solemnly uttered to Abraham and his descendants (3: 7, 8).

You say that you have already been converted by Paul? We say that you are still in a darkness entirely similar to the darkness in which not long ago you were serving the elements, supposing them, as Abraham once did, to be gods that rule the world (4: 3, 9). In fact the fights and contentions in your communities show that you have not really been converted, that Paul did not give you God's holy guidance. Paul left you, a group of sailors on the treacherous high seas in nothing more than a small and poorly equipped boat. He gave you no provisions for the trip, no map, no compass, no rudder, and no anchor. In a word, he failed to pass on to you God's greatest gift, the Law. But that is exactly the mission to which God has called us. Through our work the good news of God's Law is invading the world of Gentile sin. We adjure you, therefore, to claim the inheritance of the blessing of Abraham, and thus to escape the curse of the Evil Impulse and sin (5: 16). For, be assured, those who follow the path of the Evil Impulse and sin will not inherit the Kingdom of God (5: 21). It is entirely possible for you to be shut out (4: 17). You will do well to consider this possibility and to tremble with fear. For you will certainly be shut out unless you are truly incorporated into Abraham (3: 29) by observing the glorious and angelic Law of the Messiah. Turn therefore in true repentance, and come under the wings of the Divine Presence, so that with us you shall be saved as true descendants of our common father Abraham.[56]

Even though this reconstruction contains what Martyn himself terms 'a pinch of fantasy',[57] its value is to make explicit the persuasive power of the case the Judaizers were capable of making.

Against such logic, however, one must set the practical consequences for the Galatians of adopting the views of the intruders. Circumcision was an extremely painful operation for an adult, and obedience to the multifarious demands of the Law would be burdensome. Why then were the Galatians attracted to the message of the intruders? The answer must lie in their psychological make-up.

As we know from the Thessalonian correspondence, it was Paul's practice in establishing a community to give a number of general guidelines whose function was to indicate to the new believers that a different lifestyle was now expected of them (1 Thess. 4: 1–12). He expected them to work out for themselves what such incarnation of the gospel meant in practice.

The majority of his converts accepted the challenge, some with greater enthusiasm than others, and Paul intervened to refine their perception of what the following of Christ demanded only when he saw that mistakes were being made. Believers made the decisions; he acted as a sounding-board. The Galatians alone rejected the challenge. To them the few directives Paul gave (Gal. 5: 21)[58] were but feeble flickering flames which, rather than illuminating, served only to accentuate the surrounding darkness, which hid a myriad of land mines. To put a foot wrong meant death. With unusual insight Betz wrote,

> The Galatians had been given the 'Spirit' and 'freedom,' but they were left to that Spirit and freedom. There was no law to tell them what was right or wrong. There were no more rituals to correct transgressions. Under these circumstances their daily life came to be a dance on a tightrope.[59]

To those frightened by freedom, and paralysed by incertitude, the Law appeared a blessing. It was a balancing pole permitting those on the tightrope to advance confidently. Its 613 precepts were a multitude of tall steady flames, which dissipated the darkness completely. The burden and the pain, it seemed to some in Galatia, were a small price to pay for the security offered by the Law.

[56] (1985), 321–3. The references in round brackets are in Martyn's text. I have added those in square brackets. For a similar reconstruction, see Matera (1992), 7–11.
[57] The only really fantastical element is assumption that peasants dwelling in the interior of Asia Minor would be familiar with boat, compass, rudder, and anchor.
[58] See Longenecker (1990), 258.
[59] (1979), 9.

PAUL'S RESPONSE

On his long journey across Asia Minor, after his break with the church at Antioch, Paul had time to reflect on his altered situation and on the changes sweeping through the churches in Judaea. It is unlikely that he spent his time formulating contingency plans to deal with similar problems in his own communities. His Jewish converts were few in number, and it would have been reasonable to think that the further west he went the less likely they were to be influenced by the Jewish nationalism which had worked to his advantage in Jerusalem, but to his disadvantage in Antioch. The east had surrendered to those with a different vision of Christianity, but the west was his. It is unlikely, therefore, that on his second visit he said anything to the Galatians about what had taken place since his previous stay among them. On the contrary, he probably insisted on the unity of the Christian movement and the reciprocity of its various parts in order to dispose the Galatians to contribute to the collection for the poor of Jerusalem (1 Cor. 16: 1).

It is easy to imagine the shock produced on Paul by the totally unexpected information that people from Antioch were bent on taking over his foundations in Galatia, and that his converts were proving receptive to a totally different vision of Christianity. The sense of bewilderment (Gal. 5: 7), almost of despair (Gal. 4: 11), comes through very clearly in his response, but the dominant emotion is restrained anger. The contrast between the venemous bluster of 2 Corinthians 10–13 and the cold fury of Galatians is striking. In both cases Judaizers are the problem, but Paul's bitter disappointment with the Corinthians finds relief in recrimination, whereas in Galatians it is channelled into an argument of icy precision.

One might have expected the opposite to be the case, since control comes with age and experience. The Corinthians, however, had disappointed Paul many times. Whatever provoked the outburst of 2 Corinthians 10–13 was the last straw. Moreover, it came just at the moment when the Apostle thought that he was free of maintenance, and could return to founding new churches. When he wrote Galatians, Paul had experienced misunderstanding on the part of the Thessalonians, but that was an accident, and he was not the target of any personal animosity. The deliberate opposition revealed by the report from Galatia was an entirely different matter. And there was the very real possibility that it would be systematically extended to the other churches he had founded under the aegis of Antioch. Paul realized that his credibility was at stake. His gospel was threatened. His whole future was endangered.

A Complex Strategy

It was clear to Paul that the situation was much too serious for an outburst which would only serve to relieve his feelings. To vent his spleen on already bewildered Galatians would play into the hands of his adversaries; expostulation is often a sign of guilt (cf. 2 Cor. 11: 7–11). He realized that he had to produce a carefully crafted response to each detail of the arguments urged against him. It was not merely a question of reassuring the Galatians that his gospel was the truth. The intruders were still in Galatia (Gal. 1: 7; 5: 10),[60] and it was much more important to persuade them that their perspective on the gospel was not at all as well-founded as they imagined.

Although addressed to the Galatians (Gal. 1: 2; 3: 1), the letter could not be kept from the intruders, particularly if it was read in public (cf. Col. 4: 16), and Paul was certainly aware of this. In fact it became the basis of his strategy. Inevitably he speaks directly to the Galatians, but the intruders are its real audience.[61] If their presence in Galatia was but the first step in an effort by Antioch to recover what it considered its daughter churches, Paul could not content himself with dealing with the symptoms by detaching the Galatians from the Judaizers. He had to go to the root of the problem by developing a long-term solution. The only way to deter any further advance into his territory, and to secure permanently the future of the Galatians was to undermine the convictions of the Judaizers. Thus he made the crucial decision to focus on the Judaizers, leaving the Galatians in the background. The recovery of the latter was to be a by-product of the defeat of the former.

Paul could not have expected the Galatians, who were converts from paganism (Gal. 4: 8), to grasp the force of arguments which depended on a detailed knowledge of Jewish tradition.[62] Such carefully calculated thrusts were designed to throw the intruders into disarray. The ensuing consternation, Paul hoped, would be the most persuasive argument as far as the Galatians were concerned. He counted on re-establishing his authority among them by reducing the Judaizers to silence. Of course, if the Galatians caught the drift of his arguments, so much the better. Moreover, his evocation of their conversion experience kept them in the picture. They could understand the thrust of such

---

[60] E. Burton (1921), 25.

[61] This point is generally ignored in treatments of Gal., and means that no conclusions can be drawn from the letter regarding the culture of the Galatians (against Betz (1979), 2), which is sometimes invoked by partisans of the South Galatia hypothesis.

[62] To those who gratuitously assume that the Galatians were God-fearers, I can only recommend reading the desperate effort of Lightfoot (1910), 9–11, to prove that there were Jews in northern Galatia. Betz (1979), 5, refers to Jewish tombstones in 'the inner parts of Anatolia', but gives neither date nor location. Mitchell (1993), 2. 31–7, shows that Jewish communities existed on the western and southern fringes of Phrygia, but not in Galatia.

an appeal, but so could the intruders, whose conversion to Christianity was in no way related to the Law.

The sophistication of this approach to a dangerously volatile situation both confirms what was said above regarding the detailed information that Paul had of his opponents' arguments, and at the same time underlines his mental capacity and intellectual formation. Only someone totally convinced of the quality of his rhetorical ability, and literary skill, would have attempted to carry out such a delicate strategy by letter. It would have been much easier in person. If Paul did not take this latter option, it can only be because something made a visit to Galatia impossible (Gal. 4: 20). Perhaps the moves which led to his imprisonment in Ephesus had already begun; in which case flight might be taken as evidence of guilt. Or there may have been sensitive problems within the Ephesian community, about which we have hints in Philippians,[63] and which made it necessary for him to stay there.

A Divine Commission

The attack on his personal status as a missionary forced Paul to give form to reflections, which must have been maturing since he left Antioch. He knew the strength of his opponents' case. He had in fact accepted a commission from the Antiochean church, and his participation in its delegation to Jerusalem (Gal. 2: 1–10) was at least implicit recognition of the authority of the mother church. Now it became imperative for him to justify the independence, which had been thrust upon him by the changes at Antioch.

A tactical mistake on the part of the intruders made Paul's about-face a little easier. In order to strengthen their position they had insisted on Paul's dependence on Jerusalem which, they claimed, was the source of the authentic gospel. Things would have been very different had they dwelt on the long association of Paul with the church of Antioch. Not only had he lived there for considerable periods, but he had been co-opted into missionary work by Barnabas. It would have been impossible for Paul, who insisted so strongly on the importance of community, to deny having belonged to Antioch. And such belonging, from his perspective, implied dependence.

It was relatively easy, however, for Paul to document how little time he had in fact spent in Jerusalem *as a Christian*. The italicized words are important because, of course, he had spent some ten to fifteen years there as a Pharisee.[64] If the Judaizers had not mentioned this, Paul was not going to complicate matters by bringing it up. The situation demanded a certain economy with the truth. And it was a basic rhetorical rule, with respect to the statement of facts in a speech for the defence, that while anything that might be disadvantageous to

---

[63] See Ch. 9, 'Opposition at Ephesus'.
[64] See Ch. 3, 'Pharisaic Studies'.

the defendant should not be omitted, it did not have to be emphasized.[65] Thus as regards his first contacts with Christians Paul speaks only of having persecuted 'the church of God' (Gal. 1: 13) and 'the churches of Christ in Judaea' (Gal. 1: 22). The Holy City is not mentioned.

Subsequent to his conversion Paul had made only two visits to Jerusalem, both very brief. Some three years after his conversion, he had spent fifteen days in Jerusalem, and his contacts had been limited to Cephas and James (Gal. 1: 17– 19). The second was fourteen years later (Gal. 2: 1). Syria and Cilicia (Gal. 1: 21) are mentioned as his mission fields in the interval, but we know that he went much further.[66] In this silence we catch a glimpse of Paul's rhetorical skill. Quintilian had advised orators, 'Whenever a conclusion gives a sufficiently clear idea of the premises, we must be content with having given a hint which will enable our audience to understand what we have left unsaid.'[67] It is precisely because Paul had been to the Galatians, and was heading west when he left them, that he could afford not to mention his movements. Such discretion would have made his presentation all the more convincing, because it betrayed a confidence that carried its own persuasive power. To have given details which, from the point of view of the Galatians, were unnecessary, might have created an impression of anxiety. The assumption of shared knowledge flattered his readers.

Although nothing had been said about his relationship to the church of Antioch, Paul had to pre-empt the option by asserting from the outset that his apostolic mandate did not come 'from men or through a man' (Gal. 1: 1). His commission did not derive from any community, nor from any church leader, but came directly from Jesus Christ, whose authority was guaranteed by his resurrection. The unstated implication was that Jesus alone had the right to judge whether Paul was a faithful envoy. The miraculous character of Paul's conversion was not something that his opponents could deny; the Galatians were aware that his first contacts with Christians had been as a persecutor (Gal. 1: 13).

But his opponents could assert that for the content of his gospel he was dependent on the Christian tradition most authoritatively represented by Jerusalem. Foreseeing this objection, Paul insisted that 'the gospel preached by me is not worked out by man;[68] for I did not receive it from anyone nor was I taught it, but it came through a revelation of Jesus Christ' (Gal. 1: 11). Here it would be easy to charge Paul with being somewhat less than honest, because he had learnt much from the Christian communities of Damascus, Jerusalem, and

---

[65] Quintilian, *Institutio Oratoria* 4. 2. 101.
[66] See Ch. 1, 'Galatia and a Journey into Europe'.
[67] *Institutio Oratoria* 4. 2. 41.
[68] In Paul's lexicon 'according to man' (Rom. 3: 5; 1 Cor. 3: 3; 9: 8; 15: 32; Gal. 3: 15) is the antithesis of 'according to God' (Rom. 8: 27; 2 Cor. 7: 9–11; Eph. 4: 24), and evokes fallen humanity.

Antioch in which he had lived. He was thinking, however, of the core of his law-free gospel which, as we have seen, flowed directly from the rearrangement of his ideas caused by his encounter with the Risen Lord.[69] What he absorbed from believers in Damascus, Jerusalem, and Antioch was so thoroughly sifted through his mental filters that it became merely the confirmation and elaboration of his intensely personal fundamental insight.

It is doubtful that Paul was conscious of the selectivity operative in his appropriation of the embryonic Christian tradition. That which harmonized with his perspective was integrated, but that which did not fit was ignored without being repudiated. Thus, for example, at the beginning, he had no objection to Jewish converts continuing to observe the Law even in his own communities. He was content as long as they did not impose it on converts from paganism. It was much later when he recognized the dangers of this unreflecting concession, and only then did he repudiate it.

Not unnaturally Paul got a reputation for being erratic, which surprised and angered him. He was consistent, however, only in what he positively chose from the Christian tradition; what he accepted or permitted, however important it might be to others, was to him irrelevant and implied no commitment on his part. His focus on what he considered the essential was from another angle tunnel vision. What he saw was clear but severely limited. The obscure periphery ever retained its capacity to surprise him.

The Living Christ

In the Thessalonian correspondence Paul had not gone beyond the traditional formulations concerning the saving death of Jesus and his anticipated return in glory. In Galatians, on the contrary, we find the beginnings of his distinctive Christology. His need to develop a response to the prominence given to Abraham by the Judaizers forced him to reflect more deeply on Jesus Christ. Paul had always recognized him as the divinely appointed agent of salvation, but now he is forced to explore Jesus' individual humanity, a quest which led him to unexpected insights regarding the nature of the Christian community.

The intruders had conceded the Messianic character of Jesus. His inauguration of the eschaton was the justification of their mission to Gentiles. Although Jews took it for granted that the Messiah would be a human being,[70] his attributes and achievements were not those of an ordinary man.[71] The stress on universal dominion, for example, could easily be misunderstood as according him divine stature. Paul, in consequence, had to ensure that the Galatians

---

[69] See Ch. 4, 'Recognizing the Risen Lord'.

[70] Mowinckel (1959), 285, 323.

[71] See in particular the *Psalms of Solomon* 17 and 18, and the comments of Mowinckel (1959), 311–21.

understood, and that the intruders were reminded, that Jesus was 'born of woman' (Gal. 4: 4), the standard Jewish expression to indicate a normal member of the human race.[72]

The dimension of Jesus' life that Paul highlights is his *pistis*. In this epistle mention is made of *pistis Iêsou Christou* (2: 16; 3: 22; cf. Rom. 3: 22), of *pistis Christou* (2: 16; cf. Phil. 3: 9), and of *pistis tou hyiou tou theou* (2: 20); in Romans we find *pistis Iêsou* (3: 26). The problems of interpretation are manifest because *pistis* can mean 'belief' or 'fidelity/faithfulness', and the genitive 'of Jesus' can be either objective or subjective. The older commentators took it for granted that Paul intended 'belief' coupled with an objective genitive ('faith in Christ') but from the beginning of the twentieth century eminent voices have been raised in favour of the subjective genitive ('Christ's faith').[73]

The polarized debate has been given a new dimension by the observations of S. K. Williams,

> First, in these four texts (Rom. 3: 22; Gal. 2: 16; 3 : 22; Phil. 3: 9), each instance of *pistis Christou* occurs in a prepositional phrase indicating means or basis. Second, each of these prepositional phrases expresses the means by which *God* effects salvation. Third, it is striking that in each case we find, in addition to *ek* (or *dia*) *pisteôs* [*Iêsou*] *Christou*, another word or phrase which refers explicitly to the *believer's* faith.[74]

The implication of these observations is that the faith/fidelity of Christ[75] is evoked, not in and for itself, but because it is both the cause and exemplar of the the faith/fidelity of Christians. Their active commitment is both enabled by, and modelled on, that of Christ.

Explicit confirmation of this interpretation is furnished by a literal translation of Galatians 2: 20, 'I have been crucified with Christ. It is no longer I who live, but Christ who lives in me, and the life I live in the flesh *I live by faith, that of the Son of God*, who loved me and gave himself for me.' This text also allows us to go a step further because it identifies the *pistis* of Christ as love expressed in self-sacrifice for others;[76] this is 'faith working through love' (Gal. 5: 6). The magnitude of the love is revealed by the form of the self-sacrifice, the horrible death by crucifixion. That love, however, is not merely a fact; it is the power whereby Paul has been raised from 'death' to 'life'.

Christ's self-giving is the creative act, which is the essence of authentic humanity. Paul already has in mind a thought he will formulate only sometime

[72] Job 14: 1; 15: 14; 25: 4; Matt. 11: 11; 1QH 13: 14; 1QS 11: 21. There is no evidence that Paul ever envisaged the divinity of Christ; see my (1982b), 58–69.

[73] When combined, the bibliographies of Longenecker (1990), 87, and Matera (1992), 104, cover the significant works. [74] (1987), 443.

[75] Matera (1992), 100–1, has acutely observed that the genitives in Gal. 2: 16 are pointless unless they are intended to be subjective, and noted the strict parallelism between *ek pisteôs Christou* (Gal. 2: 16; 3: 22; Rom. 3: 26) and *ek pisteôs Abraam* (Rom. 4: 16) which is certainly subjective.

[76] The 'and' in 'who loved me and gave himself for me' is explanatory (BDF §442[9]).

later, 'without love I am nothing' (1 Cor. 13: 2). Creative love is what makes a person both human and Christian; it is the law of the believer's being. Thence Paul is led to the conclusion that the law is Christ, 'Bear one another's burdens and so fulfil the law of Christ' (Gal. 6: 2).[77] Christ's comportment exemplifies authentic behaviour. This is the true answer to the Galatians' question: how do we know what to do?

The life of Christ revealed to Paul that the one essential command of the Law is to love one's neighbour (Gal. 5: 14). Thus, in opposition to the intruders, who saw the Messiah as affirming and interpreting the Law, and thereby subordinating himself to it, Paul saw the Law as subsumed in Christ. The perfection of his love (Gal. 2: 20) was all that the Law could possibly demand. Christ, then, was the Law in the most radical sense. At one stroke Paul replaces obedience by faith/fidelity, and instead of describing faith/fidelity, which risked creating a new Law, he illustrated it by the behaviour of Christ. It was up to each believer to discern how in any given set of circumstances the creative, self-sacrificing love demonstrated by Christ should be given reality.

Only in this perspective can we understand what Paul means when he says 'I live now, not I, but Christ lives in me' (Gal. 2: 20). The egocentricity of 'death' has been replaced by the creative altruism of 'life'. Not only is his new being created by Christ, but his comportment is modelled on that of Christ. In the act of loving, Paul is Christ, in so far as he makes present in the world the essence of Christ's being. But this is true of all committed believers. Hence, they are together Christ. They have 'put on Christ' and are 'one person in Christ Jesus' (Gal. 3: 27–8). In opposition to those under the Law, who acquire a functional unity through obedience to commandments, the Christian community is an organic unity; its members are the integral parts of a living being (Gal. 5: 4).

In this insight we have the seeds of two further developments in Paul's Christology, the giving of the name 'Christ' to this new reality the believing community (e.g. 1 Cor. 6: 15), and the clarification of its nature as 'the body of Christ' (e.g. 1 Cor. 12: 12). It will take another crisis, however, to force them to the surface of his mind. In Paul's character a certain intellectual lethargy was the enemy of progressive logic. He never pursued a line of thought for its own sake. He functioned most effectively in reaction, but only to the limit of the concrete problem. He had a tenacious mind, however, and was instinctively consistent. Each new problem, in consequence, stimulated greater profundity; it did not lead to fragmentation. His Christology grew as a coherent whole, and what at first might appear to be *ad hoc* solutions can always be traced back to basic interrelated insights.[78]

---

[77] On the problems of this text, see Ch. 6, 'Pastoral Instruction'.

[78] Barrett (1994), 56, asserts: 'Beyond the occasionalism of Paul's theology there is a real unity; he reacts to circumstances spontaneously, but he does not react at random; he reacts in accordance with principles, seldom stated as such but detectable.'

Faith and Law

As we have seen, the argument of the Judaizing intruders was essentially a review of the history of salvation starting with Abraham and moving via his covenant and circumcision to the centrality of the Law. The story was not only familiar to Paul, it was the founding narrative of his people. His acceptance of Christ, however, enabled him to see it from a new and different perspective. He now knew that the true story was the life of Christ. All that had gone before was merely a preface. Given the directness of Paul's character, we can be sure that in his preaching he ignored the preparatory stages in order to concentrate on the essential, the revelation brought by Christ. It is most unlikely, therefore, that he had worked out an alternative version of the history of salvation, which he could now produce in order to counter the intruders.

The fundamental insight which enabled Paul to begin to unravel the apparently seamless argument of his adversaries came from the observation that the Galatians had been graced by the Spirit and had experienced the power of God simply because they had accepted Paul's preaching.[79] This demonstrated the irrelevance of the Law, of whose demands the Galatians heard only long after their conversion (Gal. 3: 1–5). It also directed Paul's attention to the fact that, in precisely the same way (Gal. 3: 6), Abraham had responded to God's word and had been blessed for it (Gen. 12: 1–2; 15: 1). This act of faith was the basis of all that followed, first the covenant (Gen. 15: 18) and then the additional requirement of circumcision (Gen. 17: 10). Faith, therefore, is fundamental, and all else is secondary (Gal. 3: 11).

Having established faith, not obedience to the Law, as the essential characteristic of Abraham, Paul takes up the question of his descendants, on which the intruders had laid so much weight. Capitalizing on the fact that the singular 'to the seed' is used in the promise to Abraham (Gen. 13: 15; 17: 7–8), Paul identifies Christ as *the* descendant of Abraham (Gal. 3: 16). Hence, it is those who are 'of Christ'—who are brought into being by him, and who reproduce his faith/fidelity in their comportment—who are the genuine descendants of Abraham (Gal. 3: 26–9). The intruders, of course, correctly understood 'seed' as a collective, but Paul's bold and unprecedented insistence on taking the singular literally cut the ground from under them. It is a perfect debater's argument, simple, unambiguous, impossible to refute. And if pressed Paul could always allow the collective sense, because he knew it to be verified by those who had put on Christ!

Such legalistic aggressivity, which unconsiously reveals something of the quality of Paul's education, is accentuated by his treatment of circumcision.

---

[79] The enigmatic 'hearing of faith' is primarily a hearing which culminates in faith. While in Antioch the intruders had listened to the same words that the Galatians heard but reacted very differently. See my (1964), 217–22, and Williams (1989).

With magnificent aplomb he simply ignores the fact that Abraham had been circumcised (Gen. 17: 23), and deflects attention from the problem by speaking of 'the circumcision' (Gal. 2: 7–9). This unprecedented way of referring to the Jewish people[80] had two advantages. It caused the recipients to focus on the present rather than on the past, and capitalized on the repugnance with which the Greco-Roman world viewed circumcision.

The notion of covenant could not be dismissed quite so easily, not least because the words 'this cup is the new covenant in my blood' (1 Cor. 11: 25) were part of the eucharistic liturgy, which Paul had inherited from Antioch, and which he had passed on to the Galatians. The intruders exploited the intrinsic connection between covenant and Law, and insisted that a new covenant carried the connotation of a new law. They could claim the support of Jeremiah, 'This is the covenant which I will make with the house of Israel after those days, says the Lord: I will put my law within them, and I will write it upon their hearts, and I will be their God and they will be my people' (31: 33). The internalization of the Law, in the view of the intruders, implied its continuing validity.

Since Paul could not reject outright the concept of a new covenant he had to ensure that it could not be used as a premiss in the way the intruders found so convenient. What he did was to divorce law from covenant in an intellectual tour de force, which highlights the extraordinary flexibility and power of his mind.

First, he associates covenant with freedom. Christians are the children of the Jerusalem above (Gal. 4: 26), which is identified with Sara, and thus belong to the covenant of the free woman (Gal. 4: 31). This is the antithesis of the covenant of the slave woman associated with Mount Sinai (immediately evocative of the Law), and with the present Jerusalem (Gal. 4: 22–5). This covenant of freedom is the covenant of Abraham (Gal. 3: 17) which, Paul insists, is essentially promise (Gal. 3: 16–18, 21; 4: 28) not legislation.

Secondly, Paul points out that covenant and Law are not indissolubly linked. The Law cannot have been part of God's original plan because it appeared only 430 years after the covenant/promise (Gal. 3: 17). Moreover, the Law was ordained, not by God, but by angels (Gal. 3: 19). Hence it cannot modify in any way the covenant/testament drawn up by God. Finally, the Law does not enjoy the permanency of the covenant/promise, because it was given only for a limited time (Gal. 3: 19). Its role ceased once the promise to Abraham had been fulfilled in Christ (Gal. 3: 14). Believers, in consequence, could see themselves as partners in a new covenant without in any way being bound by the Law.[81]

---

[80] Outside the Pauline corpus it is found only in Acts 10: 45 and 11: 2.

[81] Paul's approach to the same problem of covenant and Law in 2 Cor. is at once simpler and more elegant. With the maturity of greater experience he makes a neat distinction. 'God qualified us to be ministers of a new covenant not of the letter but of the spirit' (2 Cor. 3: 6). Most commenta-

Set Free for Freedom

A feature of the vocabulary of Galatians is the frequency of 'slave' (1: 10; 3: 28; 4: 1) and its cognates 'to perform the duties of a slave' (4: 8, 9, 25; 5: 13), 'to enslave' (4: 3), and 'slavery' (4: 24; 5: 1). Previously Paul had used only the first verb, and then in the sense of freely chosen service of God (1 Thess. 1: 9). The irruption of slave language into his lexicon when dealing with the crisis in Galatia is related no doubt to his awareness of the constraint imposed by the Law (Gal. 2: 23), but this is not the whole explanation. The Law was not the only slave-master. It served as the trigger which brought into focus Paul's experience as a traveller.[82] Society itself imposed a certain comportment on its members, which Paul knew to be inimical to authentic human development. He recalled the times when he was obliged to behave selfishly merely in order to survive and realized that the Galatians had been subject to the same pressures.

In order to convey this idea to them as economically as possible, he said that they 'had been enslaved to *ta stoicheia tou kosmou*' (Gal. 4: 3, 9). Unfortunately he was too sparing with words. The variety of translations of the Greek formula—'the elemental spirits of the world' (*NRSV*), 'the elemental principles of this world' (*NJB*), 'the elements of the world' (*NAB*)—highlights the inconclusive character of the debate, which is due to the fact that nine diverse meanings are attested for *stoicheon*.[83] Dunn is probably correct in seeing the phrase as Paul's 'way of referring to the common understanding of the time that human beings lived their lives under the influence or sway of primal and cosmic forces'.[84] The real question however is: how did they experience such pressure? Evidently, through the circumstances of their daily lives. What Paul wanted to get across was that society in its most basic elements, the very structure of society, was oppressive.

He uses the same enigmatic formula twice in Colossians, and there the allusion to society is made unambiguous by the apposition of 'human tradition' (Col. 2: 8) and 'belonging to the world' (Col. 2: 20). By the time he comes to write to the Romans he will express the same idea by saying that 'all, both Jews and Greeks, are under the power of Sin' (Rom. 3: 9). In all cases he is thinking of

tors, perhaps under the influence of Rom. 7: 6, understand him to be referring to a 'old covenant of the letter' and a 'new covenant of the spirit' (e.g. Furnish (1984), 199; Klauck (1986), 37; Westerholm (1988), 212). Grammatically, however, 'not of the letter but of the spirit' qualifies 'new covenant' (so rightly Windisch (1924), 110). Paul is not distinguishing between an old and a new covenant, but between *two types of new covenant*, one characterized by 'letter', the other by 'spirit'. In effect he is saying 'we are not letter-ministers but spirit-ministers of the new covenant' (Plummer (1915), 88). 'Letter-ministers', of course, are the Judaizers, who in practice reduce the new covenant to the old covenant.

[82] See Ch. 4, 'Dangers on the Road'.
[83] Blinzler (1963).
[84] (1993), 213.

the total control over the individual exercised by the false value system of society.[85]

One such false value was the blind obedience the Jews gave the Law. Just as those living in polluted environments have no alternative but to breathe in toxins, so those born into the world are automatically infected by its attitudes and standards, its root principles. They can no more offer opposition than wood chips tossed into a fast flowing river. Paul deliberately evokes enslavement in order to underline that no resistance is possible. The echos of his own experience, both religious and secular, are unmistakable.

Freedom becomes a reality only 'in Christ', namely, in and through the Christian community. But Paul was not so naïve as to believe that the deeply ingrained habits of a lifetime were automatically eradicated by the act of conversion. Much more than nominal membership was necessary. Thus he warns the Galatians that if the victory of 'the desires of the spirit' over 'the desires of the flesh' (Gal. 5: 17) is a victory only in principle, then their freedom will exist solely in theory. Only those who have in fact 'crucified the flesh with its passions and desires belong to Christ Jesus' (Gal. 5: 24). In other words, the Galatians have only been 'set free for freedom' (Gal. 5: 1).[86] A possibility has been offered them; it is up to them to make it real. How?

Paul's answer is unambiguous. 'You have been called to freedom; only do not use your freedom as an opportunity for the flesh, but through love be servants of one another' (Gal. 5: 13). The choice is theirs. If their behaviour reflects the egocentricity of the 'works of the flesh' (Gal. 5: 19–21), they accept again 'the yoke of slavery' (Gal. 5: 1). Free service of others in love, on the contrary, is the only authentic response to the summons to freedom; what this involved in practice is outlined in Galatians 5: 22–3.

Here we catch a glimpse of the principle underlying Paul's understanding of enslavement and freedom. What promotes community generates freedom, whereas what militates against community destroys freedom. In the last analysis, therefore, freedom is a property of the community, not a possession of the individual. Only those believers who belong to an authentic community are free. Those who belong to a nominal community are not. The only protection against the all- pervasive power of the false value system of the world is afforded by the support and inspiration offered by the lived authentic values of fellow-Christians.

*

Painful as the experience must have been, the crisis provoked by the Judaizing intruders in Galatia was of crucial significance in Paul's intellectual

---

[85] For more detail, see Ch. 13, 'Sin, Law and Death'.

[86] E. Burton (1921), 271, translates 'With this freedom Christ set us free' thereby betraying his desire to see freedom as a gift rather than as an achievement. Acts of sacral manumission reveal the dative to be purposeful, 'for freedom'; so Betz (1979), 255–6; Barrett (1985), 55; Longenecker (1990), 224.

development. In reaction to the prominence given Abraham, Paul was obliged to explore more profoundly than ever before the faith/fidelity of Christ, and its relation to that of Christians. This led him to a critical new insight into the relationship between Christ and the believing community. Reflection on the constraint imposed by the Law brought to mind the pressures of society and obliged him to define more clearly than hitherto the difference between life in the world and life in the Christian church. For the first time he grasped the nature of authentic freedom.

In Galatians these seminal ideas appear in their embryonic form. Their formulation is not as clear as one would wish, and it appears that Paul is not fully aware of their implications. Fortunately there would be other crises to stimulate their exploitation.

Partnership at Philippi

Philippi was the first European city to be evangelized by Paul.[1] He arrived there in the late summer or early autumn of AD 48, having tramped across western Turkey from Galatia.[2] His ship from Troas docked at Neapolis, modern Kavalla (Acts 16: 11). According to Luke, Paul did not spend any time in this port city, but continued inland to Philippi (Acts 16: 12).[3] This is confirmed by Paul's hint that his first converts came from Philippi (Phil. 4: 15).

Given Paul's subsequent preference for coastal cities, notably Corinth and Ephesus, his haste to move inland is surprising. At this point in his career, however, he had not realized that he would have to keep in touch with his foundations. He understood his mission as simple evangelization, to plant the gospel and march on; the watering of the seed was not his responsibility (1 Cor. 1: 17a). It was only two years later, when he arrived in Corinth and was forced to concern himself with the affairs of the church at Thessalonica,[4] that he became aware that facility of communications had to be a critical factor in the choice of a missionary base.

PHILIPPI

The most detailed ancient description of Philippi is that of Appian:

> Philippi is a city that was formerly called Datus, and before that Crenides, because there are many springs bubbling around a hill there. Philip [II of

---

[1] The word 'Europe' is first attested in the 7th cent. BC Homeric Hymn to Apollo, and by the 5th cent. BC it had become the name of one of the three great territorial divisions of the ancient world, the others being Asia and Libya (= Africa). According to Herodotus, 'The Persians claim Asia for their own, and the foreign nations that dwell in it. Europe and the Greek race they hold to be separate from them' (1. 4). Appian places 'the main pass from Europe to Asia' 4 kms north of Philippi (*Civil Wars* 4. 106; trans. White), but the water barrier of the Hellespont and Bosphorus is a much more natural line of division (Pliny, *NH* 5. 141; Philo, *Leg. ad Gaium* 281).

[2] See Ch. 1, 'Prior to AD 51'.

[3] The distance is given as 12 Roman miles by the *Antonini Itinerarium* and as 10 Roman miles by the Bordeaux Pilgrim.

[4] See above, Ch. 5.

Macedon in 356 BC] fortified it because he considered it an excellent stronghold against the Thracians, and named it from himself, Philippi. It is situated on a precipitous hill and its size is exactly that of the summit of the hill. There are woods on the north through which Rhascupolis led the army of Brutus and Cassius. On the south is a marsh extending to the sea. On the east are the gorges of the Sapaeans and Corpileans, and the west a very fertile and beautiful plain. . . . The plain sloped downward so that movement is easy to those descending from Philippi, but toilsome to those going up from Amphipolis. There is another hill not far from Philippi which is called the Hill of Dionysus, in which are gold mines called the Asyla. (*Civil Wars* 4. 105–6; trans. White)[5]

This description sets the scene for the battle of Philippi in 42 BC, when Octavian and Antony defeated Brutus and Cassius, and accurately reflects the excellent quality of Appian's sources. The construction *c.* 130 BC of the Via Egnatia, the great Roman road running across northern Greece from the Adriatic Sea to Neapolis, had brought no great prosperity to Philippi. Things changed, however, after the battle. More space was necessary to accommodate the Roman veterans settled there by Mark Antony. The town spilled down the mountainside towards the swampy land surrounding the lake.[6] Coins attest its status as a colony with the title *Antoni Iussu Colonia Victrix Philippensium.*[7] A further influx took place after Octavian's defeat of Antony in 31 BC. According to Dio Cassius, 'By evicting those communities in Italy which had taken Antony's side, Octavian was able to settle his soldiers both in their cities and on the lands of his opponents. He compensated most of those who had been penalized in this way by allowing them to settle in Dyrrachium, Philippi, and elsewhere' (*Roman History* 51. 4. 6; trans. Scott-Kilvert).

Thereafter the official title of the colony became *Colonia Iulia Augusta Philipp(i)ensis.*[8] The settlers naturally retained their privileges as Roman citizens, and Philippi enjoyed the Ius Italicum.[9] It was as if the city had been transferred to the soil of Italy; its residents were not subject to provincial land and personal taxes, and in theory at least were independent of the governor of the province of Macedonia.[10]

The city was the civic and administrative centre of an area of some 2,100 sq. km. (730 sq. miles).[11] The vast majority of the settlers lived on their land, but the city was the market for their produce, and the source of services and manu-

---

[5] Good maps are given in Schmidt (1938), 2210, 2218.

[6] See the plan in Finley (1977), 176.

[7] Schmidt (1938), 2233.

[8] Ibid. 2234.

[9] Justinian, *Digest of Roman Law*, 1. 16. 6–7.

[10] von Premerstein (1917).

[11] Papazoglou (1982) with map 2 in Portefaix (1988). The only variant of Acts 16: 12 which corresponds to historical reality is the Western Text, which identifies Philippi as 'a city of the first district of Macedonia, a colony'. The capital of the first district was Amphipolis (Livy, *History of Rome* 45. 29. 9). See J. J. Taylor (1994b), 244–5.

factured goods. As it expanded, facilities increased, and the city became a more attractive stopping-place for those travelling on the Via Egnatia. Increased opportunities drew in non-Roman immigrants from Greece and further east.

As the tongue of the dominant class, Latin was the official language throughout the city and colony. Greek, of course, continued to be used by the indigenous population. The relationship of the two is very aptly illustrated by Paul's use of *Philippêsioi* (Phil. 4: 15), which is derived from the Latin *Philippenses*, rather than the more authentically Greek forms *Philippeis* or *Philippênoi*.[12] A Roman veneer had been applied to a population that remained essentially eastern. Moreover, many of the colonists, and certainly their descendants, would have known at least some Greek.[13]

The great preponderance of Latin in official inscriptions only serves to highlight the fact that Greek is the language of half of the inscriptions pertaining to the worship of the Egyptian gods, who enjoyed the only real temple on the slopes of the acropolis. Its devotees were better off than others, who could not afford the same quality of construction. A quarry near the base of the hill housed wooden sanctuaries of the Roman deities Silvanus and Diana, and the eastern Magna Mater. Scattered over the acropolis are 187 rock-cut reliefs of rather poor workmanship. The figure of Diana predominates, but the Thracian rider, Magna Mater, Isis, Jupiter and Minerva are also represented. The Roman state religion dominated the area south of the Via Egnatia. The emperor and the Capitoline triad were venerated there in addition to Diana and Mercury.[14]

The Founding of the Church

The abundance of evidence for the religious preferences of the pagan population of Philippi makes the absence of any archaeological or epigraphic hint of a Jewish presence significant. Luke's source evokes a 'place of prayer' outside the city near a river to which Paul and his companions went on the sabbath (Acts 16: 13). The implication that it was a Jewish place of worship cannot be denied. The term 'place of prayer' does not exclude a building,[15] but neither does it necessarily imply one. The known first-century Diaspora synagogues, however, are all within cities,[16] as one might have expected, since Jews had a legal right to a place of worship. If they were too few to build a synagogue, a room in one of their houses would be the obvious place to meet for

[12] Ramsay (1899), 116.
[13] It should also be kept in mind that, if Roman boys studied Homer at school (Pliny, *Letters* 2. 14. 2; Quintilian, *Institutio Oratoria* 1. 8. 5), their sisters felt that a knowledge of Greek added to their attractiveness (Juvenal, *Satires* 6. 184–99),
[14] Excellent surveys of the archaeological and epigraphic data, with full references to the sources, are provided by Portefaix (1988), 70–2, and Hendrix (1992a), 316.
[15] So J. J. Taylor (1994b), 247.
[16] Kraabel (1979).

study and prayer. Moreover, no Jews are mentioned among Paul's converts. Lydia is explicitly identified as a Gentile (Acts 16: 14) and the same must be said of the jailer (Acts 16: 30–1). Finally, the group whom Paul found assembled did not contain any men. Luke's source mentions only women (Acts 16: 13), one of whom, Lydia, is identified as a 'worshipper of God'. In other words, she was a God-fearer, a pagan who associated herself with Judaism but without becoming a formal convert.[17] Since a Jewish presence in her home town, Thyatira in Asia, is apparently attested,[18] it is not necessary to assume that she was attracted by a Jewish community in Philippi.

In order to account for these data, it is necessary only to assume that the place of prayer served, not Jewish residents of Philippi, but Jewish travellers on the Via Egnatia, who happened to be in the city on the sabbath. Lydia herself had come from afar. In the absence of transient Jews, local God-fearers gathered there, and perhaps Jewish women married to pagans. According to Luke's source, it was among this group that Paul made his first converts. Some confirmation is provided by the letters which attest the prominence of two pagan ladies, Euodia and Syntyche, who expected recognition for their contribution to the evangelization of the city (Phil. 4: 2–3; see below).

The tradition history of the material contained in Luke's detailed account of Paul's experiences in Philippi (Acts 16: 13–40) is complex.[19] The narrative of the encounter with the magistrates, however, and its consequences (beating, imprisonment, apology), 'belongs to a first-class source, indeed an eye-witness account [whose] details are historically exact'.[20] Moreover, it is confirmed by the letters. In writing to the Thessalonians Paul mentions that he and his companions had 'already suffered and been shamefully treated (*hybristhentes*) at Philippi' (1 Thess. 2: 2). The verb *hybrizō* is perfectly apt to describe the punishment of a Roman citizen without even the semblance of a trial. There can be no serious doubt that Philippi was one of the places where Paul was imprisoned and beaten with rods (2 Cor. 11: 23–5).

The disagreeable episode ends with the departure of Paul from Philippi (Acts 16: 40). How long had he spent there? This is one question which Luke does not answer. Haenchen rightly refuses any real value to the two chronological indications in Acts.[21] The initial allusion to 'some days' (16: 12) probably refers to the time between the arrival of the missionaries and the sabbath. The subsequent mention of 'many days' (16: 18) is merely an ingredient in Luke's story-telling technique. There is something, however, in Luke's account which suggests a more realistic solution.

What concerned those who dragged Paul before the magistrates was the loss

---

[17] See my (1992c).
[18] Schürer (1973–87), 3. 19.
[19] See Boismard and Lamouille (1990), 2. 288–93; 3. 214–23.
[20] J. J. Taylor (1994b), 253.   [21] (1971), 494–5.

of their livelihood, but what they said in court was 'These men are disturbing our city; they are Jews and are advocating customs that are not lawful for us as Romans to adopt or observe' (Acts 16: 20–1). The discrepancy permits us to separate the the occasion from the charge. It is possible to refrain from judgement on the exorcism, while at the same time according the charge serious historical probability.[22]

Loisy's interpretation of the charge,[23] as implying a missionary effort of considerable duration and success, is confirmed by Paul's correspondence with Philippi. The letters reveal a well-organized, generous community, with the energy to support Paul's missionary endeavours elsewhere (Phil. 4: 15–16). In no other letter does Paul single out women 'who have laboured side by side with me in the gospel' (Phil. 4: 3). Nowhere else does he thank a church, whose very existence is a 'holding forth of the word of life' (Phil. 2: 16), for its 'partnership in the gospel' (Phil. 1: 5). What these allusions imply about the relationship of the believers to the Apostle, and their lived embodiment of authentically Christian values, could not have been achieved in a brief visit.[24]

We must assume, in consequence, that Paul spent at least the winter of AD 48–49 in Philippi where he made converts among pagans. It is entirely possible that his stay there was cut short by the sort of event reported by Luke's source. Even though there was no organized Jewish proselytization in the first century,[25] sufficient Romans had been attracted to Judaism that Tiberius in AD 19 felt himself obliged to react against the phenomenon by expelling the majority of Jews from Rome.[26] The example of the Eternal City would carry weight in a Roman colony.

A SERIES OF LETTERS

The New Testament contains only one canonical letter to Philippi, but from the beginnings of critical study of the New Testament, serious doubts about its integrity have been voiced. Some commentators distinguish two letters. The majority detect three letters. But there always have been those who maintain the unity of the epistle. The history of the debate has been summarized at length by B. Mengel,[27] but more thoroughly by D. E. Garland.[28] The only new argument to appear subsequently has been the thesis that Philippians exhibits the rhetorical schema and so must be a literary unity.[29] Obviously the historical

---

[22] So rightly Haenchen (1971), 496 n. 5.
[23] (1920), 639.
[24] Similarly Hawthorn (1983), p. xxxv.
[25] See my (1992a) and Will and Orrieux (1992).
[26] For details, see Smallwood (1981), 202–9.
[27] (1982).   [28] (1985).
[29] D. F. Watson (1988); Bloomquist (1993).

reconstruction of Paul's relations with Philippi changes radically if there is a series of letters rather than a single communication. Hence, some attention must be devoted to this problem.

Methodologically, literary unity is a presumption. It cannot be proved without the direct witness of the author. The presumption, however, can be overturned by arguments whose effect is to show that particular combinations are in themselves improbable or incompatible with a given author's style and approach. In his letters Paul regularly begins with what is uppermost in his mind, e.g. the backsliding of the Galatians (Gal. 1: 6), the factions at Corinth (1 Cor. 1: 10). In Philippians, however, we encounter the exact opposite.

An Ambivalent Expression of Gratitude

Paul's gratitude for the financial assistance of the Philippians appears only in 4: 10–20, at the very end of the letter. Efforts have been made to interpret 1: 5 and 2: 30 as expressions of thanks.[30] It is clear to any sensitive reader, however, that these allusions rather presume that the precise nature of the service has already been acknowledged. If the Philippians had not been thanked previously, it is inconceivable that their financial aid should not have been mentioned in 2: 25–30. Moreover, the hypothesis that 4: 10–20 belongs to the letter carried by Epaphroditus on his return to Philippi involves the unacceptable assumption that Paul did not avail himself of the messengers, who brought the news of Epaphroditus' illness back to his community (2: 26), to thank the Philippians for their gift. All of these difficulties disappear if Philippians 4: 10–20 was originally an independent letter, and the first addressed by Paul to the Philippians. For this reason I call it *Letter A*.

The one clue to the dating of *Letter A* is its self-conscious, defensive tone. At first sight this is surprising because Paul was used to receiving aid from the Philippians; they had assisted him financially more than once at Thessalonica (Phil. 4: 16), and subsequently at Corinth (2 Cor. 11: 9). If he found the gesture offensive, he had had many opportunities to ensure that it was not repeated. Paul's embarassment becomes understandable if the gift came at a time when there was danger that it might be misinterpreted. This condition was verified only after he had begun to preach the collection for the poor of Jerusalem. At that point acceptance of a personal gift could appear as if he were appropriating funds given for another purpose. It became important to emphasize that he had not solicited funds from the Philippians (Phil. 4: 17), and that he needed nothing more (4: 18). *Letter A*, in consequence, must be dated after the Jerusalem assembly.

There is no need to assume that Paul was under arrest when he wrote *Letter*

---

[30] Garland (1985), 153. Contrast the use of the technical formula 'here is my receipt for everything' in 4: 18 (Hawthorn (1983), 206).

A.[31] He had not been imprisoned when he benefited by previous subsidies from Philippi. Of all his foundations the Philippians alone had the insight to recognize that Paul's efforts to be financially independent were not entirely successful and, when they were in a position to assemble some surplus cash, they sent it to him no matter where he was. In the present instance we must assume that Paul had informed Philippi, and the other European churches, where he was to be found in case he was needed.

The fact that Paul received money on a fairly regular basis from Philippi implies some organization. The church there must have delegated responsibility for the collection and transmission of funds to certain members of the community. In all probability these individuals were the *episkopoi kai diakonoi* 'supervisors and assistants' who are mentioned in the address (Phil. 1: 1), which may have belonged to any or all of the three letters.[32] With regard to its leadership structure Philippi was exactly the same as the other Pauline churches. Paul did not select leaders. He expected them to emerge from the community as their gifts were expressed in service. What he says to the Philippians, 'Mark those who so live as you have an example in us' (3: 17), echoes what he had written to the Thessalonians (1 Thess. 5: 12), and anticipates what he would direct the Corinthians to do (1 Cor. 16: 15–18).

Other churches looked up to Paul as their founder, and treasured his letters, but did not send him financial assistance. This was not because they lacked resources. The adjective all antiquity applied to Corinth was 'wealthy',[33] and Ephesus was not far behind. Unless we are to assume that such communities were not animated by a Christian spirit, the generosity of the Philippians cannot be explained merely by fraternal charity. If they gave despite their poverty (2 Cor. 8: 1), it must have been for something over and above their affection for Paul. The reason suggested by Philippians is the apostolic spirit of the church at Philippi. If Paul gives thanks for their 'partnership in the gospel' (Phil. 1: 5; cf. 1: 7), it must be because the Philippians actively participated in the evangelization of their city (cf. Phil. 4: 3).

Very quickly they became aware of the drain on their time and energy; they still had to earn a living. Yet they were in a much better position than a missionary like Paul. They had remunerative occupations with an established clientele and a stable network of family and friends. A new city offered Paul no guarantee of employment. He was always the vulnerable outsider, operating

---

[31] A mistake made by Garland (1985), 152, in order to create a spurious objection, 'Why would Paul spin off a thank-you note, a dankelose Dank at that, and not explain his personal situation that had aroused the Philippians' concern in the first place and had prompted them to dispatch Epaphroditus with their gift?'

[32] *Episcopos* is used of those entrusted with fiscal supervision; see LSJ. In the Essene writings one of the responsibilities of the 'overseer' was to collect and distribute charitable funds (*CD* 14. 12–16).

[33] Homer, *Iliad* 2. 570; Strabo, *Geography* 8. 6. 20.

218   *Partnership at Philippi*

without any cushion of connections. On the basis of their own experience, the Philippians recognized that if Paul was to live as an apostle he needed to be subsidized.

Once he had the opportunity to reflect on the implications of his generous gesture at the meeting in Jerusalem (Gal. 2: 10), Paul quickly realized that a life which had never been easy was going to become much more difficult. The experience of his first journey into Europe had taught him that, as the demands of his ministry became more pressing, the less time he had to earn his living, and the more dependent he would become on gifts from others. Now that he was committed to requesting funds to be held in trust for Jerusalem, it became imperative for him to devise a way which would make clear that he was not using for his own needs money given for the poor of the Holy City. Paul, it will be recalled, lived in a world in which every official stole from the public purse; questions were raised only when they took too much.[34] Tax collectors were hated because only a percentage of what they exacted went for its ostensible purpose (cf. Luke 3: 12–13).

No doubt Paul considered and rejected a number of different plans as he plodded across Asia Minor. By the time he reached Galatia, however, he had a satisfactory answer, which he subsequently repeated to the Corinthians:

> Now concerning the contribution for the saints. As I directed the churches of Galatia, so you also are to do. On the first day of every week, each of you is to put something aside and store it up, as he may prosper, so that contributions need not be made when I come. And when I arrive, I will send those whom you accredit by letter to carry your gift to Jerusalem. If it seems advisable that I should go also, they will accompany me.   (1 Cor. 16: 1–4)

This strategy had several advantages. Weekly savings were certain to produce a greater sum than anyone could contribute at short notice. Paul was not responsible for safeguarding funds entrusted to him, and so his mobility was not impaired. Once the contributions were assembled, they were the responsibility of representatives of the donors. Paul himself was involved with the transmission of the gift to Jerusalem only to the extent that his accompanying letter identified the gift as the fulfilment of his promise to the three Pillars (Gal. 2: 10).

Despite such precautions, however, within a year or so rumours spread by Paul's enemies smeared his reputation at Corinth.[35] The unsolicited personal gift from Philippi may have provided them with the opportunity to inject a note of distrust into Paul's monetary arrangements.

---

[34] Narcissus, the Secretary, and Pallas, the Treasurer, 'were able to acquire such riches, by illegitimate means, that when one day Claudius complained how little cash was left in the imperial treasury, someone answered neatly that he would have heaps of pocket money if only his two freedmen took him into partnership' (Suetonius, *Claudius* 28; trans. Graves).

[35] See Ch. 12, 'Financial Assistance'.

Two Further Letters

The material remaining after the abstraction of *Letter A*, namely, Philippians 1: 1–4: 9, is not a literary unity. It falls into two parts each with a different atmosphere and concern. In 1: 28 Paul's attitude towards a threat to the community at Philippi is one of calm superiority. But the sneering tone of the deliberately insulting comments in 3: 2 and 19 conveys a hint of desperation. Manifestly one danger is much more serious than the other.

In the latter instance the community is menaced by Judaization; the allusions to mutilation (3: 2) and to the stomach as the matter of ultimate concern ('their god is their belly', 3: 19) unambiguously evoke circumcision and Jewish dietary laws, respectively. When confronted by Judaizers, here as elsewhere (e.g. Gal. 3: 1), Paul's fear is that his converts will be seduced from the true faith. The method of alienation evoked in 1: 28, however, involves threats, and or force, or both, an approach which might 'frighten' the Philippians. Paul evidently is thinking in terms of persecution by pagans.

When associated with these two contrasts, an argument from silence gains weight. Paul mentions his imprisonment in 1: 7, 13, and a further reference would be perfectly in place in the evocation of his sufferings in 3: 8–11. The absence of any hint might suggest that ch. 3 was written after Paul's release.

If we call *Letter B* the section of Philippians in which an imprisoned Paul is complacent regarding the effect of pagan persecution, and *Letter C* the section in which he fears Judaizing infiltration, where do both begin and end?

Philippians 3: 1 is the key to the answer. The words 'For the rest, my brethren, farewell in the Lord' in the first part of the verse are closely paralleled by 2 Corinthians 13: 11, 'For the rest, brethren, farewell', which is the conclusion to 2 Corinthians 10–13. The second part of the verse reads, 'To write the same things to you is not irksome to me, and is safe for you' (3: 1b). Some commentators understand 'the same things' as an allusion to Paul's repetition of the preceding exhortation.[36] But in how could such reiteration contribute to the Philippians safety? Hence 'the same things' must refer to what Paul is going to write. In no way, however, does 3: 2 ff. repeat anything in chs. 1–2. The only alternative to uncontrollable hypotheses of oral or lost instructions, is to look forward to the admonition addressed to Euodia and Syntyche in 4: 2, for there Paul repeats in a more specific form the pleas for unity in 1: 27 and 2: 2–4.[37] The partisanship and vain ambition which Paul deprecates in 2: 3 have a specific application in the case of these two women.[38] He may have hoped that they would recognize the general pleas as directed to them, but at the last moment

[36] e.g. Hawthorn (1983), 124.
[37] Beare (1969), 102.
[38] So rightly Beare (1969), 143.

he decided that it would be safer to be explicit.[39] Hence, *Letter B* consists of Philippians 1: 1–3: 1 and 4: 2–9, while *Letter C* is made up of 3: 2 to 4: 1.

THE LETTER FROM PRISON

Most unusually, *Letter B* tells us more about Paul's situation in Ephesus than it does about the Philippians. This is certainly a reflection of the closeness of Paul's relationship with the church at Philippi and the quality of its community life. It suffered the minor crises typical of a growing, vital community, but there were no serious problems, and in great part he could write for the pleasure of maintaining contact.

From the way Paul introduces the topic (1: 7, 12), it would appear that the Philippians already knew that he had been imprisoned. His focus is on the impact of his incarceration, both as regards the Christian community and those with whom he came into contact while in prison. He is held while under investigation in 'the praetorium' (1: 13), the official residence of the governor of Asia, and in which he also exercised his juridical functions.[40] Since the capital of the Attalid kings had been Pergamum, it seems likely that this was one of the new edifices erected by Augustus.[41]

The form of detention was entirely at the discretion of the magistrate, whose decision was determined not only by his own personality and the nature of the case, but particularly by the degree of influence the prisoner and his friends could bring to bear. The treatment accorded the rich, particularly in their own city, differed significantly from that meted out to the poor and strangers.[42] His place of detention identifies Paul as one of the latter. Even though Philippians 1: 13 uses 'bonds' in the sense of 'imprisonment', which conformed to contemporary usage,[43] it is virtually certain that the expression should be taken literally.[44] What precisely this involved is a matter of speculation. Paul may have been chained to a soldier (cf. Acts 28: 16, 20), or to the wall of his cell, or he may have been forced to wear handcuffs or leg-irons.

Although his movements were hampered, the conditions under which Paul

[39] This hypothesis is an adaptation of the hypothesis of Furnish (1963) that Paul originally intended the bearer of the letter to deliver the reprimand orally.

[40] On the different senses of *praitorion*, see in particular Benoit (1952), 532–6.

[41] See above, Ch. 7, 'The City'.

[42] Powerful friends of the future Agrippa I managed to ensure that 'the soldiers who kept him should be of a gentle nature, and that the centurion who was over them, and was to eat with him, should be of the same disposition, and that he might have leave to bathe himself every day, and that his freedmen and friends might come to him, and that other things that tended to ease him might be indulged him' (Josephus, *AJ* 18. 203; trans. Whiston and Margoliouth).

[43] Mommsen (1955), 300. 'Bonds' covered both 'chains' and 'fetters' (cf. Mark 5: 4; Luke 8: 29). The cognate means 'prisoner' (Philem. 9). Paul's situation cannot be identified as a form of house arrest; see Mommsen (1955), 217.

[44] Ibid. 301.

was imprisoned in Ephesus cannot have been too severe. He was not placed in solitary confinement. He could communicate with his collaborators, who were held with him (Phil. 2: 19; Philem. 23; Col. 4: 10). One of them may have served as the secretary he needed to write the letters to Philippi, Colossae, and Philemon, but the fact that outsiders such as Epaphroditus were able to visit him and receive commissions from him (Phil. 2: 25) leaves open the possibility that he had access to a professional secretary.

Dealing with the Possibility of being Executed

Paul admits, however, that at one stage he had to face the possibility that he would be executed (Phil. 1: 20–5). Like all his contemporaries he knew that the arbitrary abuse of authority was restrained only by the fear of reprisals. As an outsider lacking any high-level local support in Ephesus, there was no way he could create difficulties for the proconsul of Asia. Paul confesses that death greatly appealed to him, not because he was tired of life or afraid of suffering, but because it would mean union with Christ.

Quite clearly Paul is thinking in terms of conscious personal fellowship with Christ. In 1 Thessalonians 4: 16–17, however, he had said that the dead would be restored to life, and would be with the Lord, only at the Second Coming. If he now thought that there would be no delay in full communion with Christ, would not the resurrection be superfluous? Paul would have been shocked at such a conclusion (Phil. 3: 20–1). The inconsistency derives, not from a change of position, but from the fact that, while the human mind can envisage its own annihilation, it cannot conceive of an interruption in its existence. Hence the paradox that in order to benefit by resurrection one must continue to exist; otherwise it would be the creation of an entirely new being.[45]

Speculation on how Paul conceived the so-called 'intermediate state' is pointless. In terms of understanding his personality, it is much more revealing to note the thrust of his internal debate, which pivots on the conviction that what is best in theory is not always to be chosen in practice. To die and be with Christ is the best option absolutely speaking, but that is not an adequate basis for a decision. The needs of the Philippians and others make it imperative to choose life and struggle; 'to remain in the flesh is more necessary on your account' (Phil. 1: 24). In other words, the decisive criterion in Paul's moral judgement is not whether a course of action is good or bad in itself, but whether it will empower or injure one's neighbour. This key insight, derived from a tension-filled experience, will play a critical role in Paul's correspondence with the Corinthians (cf. 1 Cor. 8).

---

[45] This is why even those Jewish texts which speak most explicitly of resurrection contain statements which seem to imply the contradictory belief in immortality of the soul; see Cavallin (1974), 199.

The recognition that he was still needed perhaps contributed to Paul's conviction that divine providence would ensure release in the not too distant future (Phil. 2: 24). His hope was also fed by the awareness that all those in the praetorium with whom he came in contact were convinced that he was neither a revolutionary nor a criminal (Phil. 1: 13). It would not be long, he imagined, before word of his innocence filtered up to those responsible for the disposition of his case.

As soon as he regained his freedom, Paul planned to make a visit to Philippi (Phil. 2: 23–4). This may have been conceived as a tactful gesture, or there may have been pastoral reasons. In any case, he had to prepare for the eventuality that even if he were exonerated, he might be expelled from Ephesus, as he had been from Philippi. He was not a citizen with legally guaranteed rights. As a Jew he had an indirect legal status in so far as he was accepted by the *politeuma*, the official corporation representing the Jewish community *vis-à-vis* the civil authorities.[46] A word from the latter, of course, would make him unwelcome among his own people.

Opposition at Ephesus

As things turned out, once he was freed Paul was permitted to remain in Ephesus (1 Corinthians 16: 8). He did not carry out his plan to visit Philippi. It was still on his calendar as his next stop when he wrote 1 Cor. 16: 5, but a sudden deterioration in the situation in Corinth demanded his presence there, and it was only on his way back through Macedonia that he finally revisited Philippi (2 Cor. 1: 16).

The reason why Paul stayed on in Ephesus was the unhappy situation of the community there. His imprisonment had split the church into three factions (Phil. 1: 14–15). One group was frightened into silence. Its members presumably were considering whether it was wise to remain Christians. The majority, however, became even more active missionaries. This group was not homogeneous. The preaching of one faction, in Paul's judgement, was inspired by envy, rivalry, selfish ambition, insincerity, a desire to injure Paul, and embodied an element of pretence. It would be difficult to find a harsher catalogue. Yet he does not accuse them of preaching another Jesus (cf. 2 Cor. 11: 4). The proclamation of the others, on the contrary, was rooted in good will and love, and was characterized by truth.

The latter makes it clear that the distinction between the two groups lay in their relationship to Paul; no doctrinal difference is even hinted at. He liked one party and reciprocated the dislike of the other. Commentators have made no plausible suggestions as to the identity of the factions, or the roots of the per-

---

[46] See Smallwood (1981), 225–6.

sonality conflict.[47] One becomes apparent, however, once it is recalled that Prisca and Aquila had been at work in Ephesus for a year before Paul's arrival.[48] It is not unknown for members of a community to resent the assumption of authority by a latecomer.[49] The history of the Essenes provides an instructive parallel. The movement split, when the Sadokite high priest dispossessed by Jonathan became a member of the sect and tried to assume control.[50] Similarly when Paul arrived in Ephesus, there must have been some who did not welcome him with open arms. Even though he presumably had been warmly recommended by Prisca and Aquila, certain believers saw him as an intruder. When he landed in gaol, they were delighted to be in a position to show him that he was in no way necessary to the life and mission of the church. It had grown without him in the past and could expand without him in the future.

Given the extremely positive way he speaks of Prisca and Aquila subsequently (Rom. 16: 3–4), it seems improbable that they took any part in the opposition to Paul.[51] Nor are they covered by the blanket criticism of the whole community at Ephesus (with the exception of Timothy) as fundamentally selfish. 'They all look after their own interests not those of Jesus Christ' (Phil. 2: 21). The context limits the applicability of this apparently universal criticism to the issue of going to Philippi.[52]

His forced inactivity gave Paul the leisure to worry about the fate of the Philippians, who were suffering persecution (Phil. 1: 28). He seethed with anxiety, and desperately needed someone to go to Macedonia, and to bring back word of the state of the community. Timothy was prepared to undertake the task, but Paul preferred to keep his closest collaborator with him until his fate should be decided (Phil. 2: 23). The refusal of others was perhaps motivated by the realization that Paul could communicate with Philippi via the letter sent with Epaphroditus (Phil. 1: 25), and that the problems there were not so severe as to need the additional presence of a trouble-shooter. Why, argued Ephesian believers, should they interrupt a fruitful missionary effort in their own city simply in order to gratify Paul's desire for information? On the contrary, was it not selfish of him to prefer his own consolation to the spread of the gospel? Not unnaturally, Paul did not see the matter in this light!

[47] Hawthorn (1983), 37–8, noncommittally notes a number of far-fetched hypotheses.
[48] See Ch. 7, 'The Founding of the Church'.
[49] With his customary insight, Lohmeyer (1974), 47, noted, 'Von neuem bestätigt sich dann hier, das Paulus an dem Orte seiner Haft nicht die Autorität eines Begründers und Leiters beanspruchen, sondern nur die Distanz eines Zuschauers wahren kann.' Unfortunately, he saw it as an argument in favour of Caesarea.
[50] See my (1985), 239–41.
[51] Something similar happened at Corinth later, where Paul had problems, not with Apollos (1 Cor. 16: 12), but with those who attached themselves to him (1 Cor. 1: 12).
[52] So rightly Beare (1969), 97, followed by Hawthorn (1983), 111.

Paul's tantrum betrays a wilfulness that could not bear to be thwarted.[53] The childishness of the identification of his needs with those of Christ needs no emphasis. Were there other outbursts of this type, as he tried to establish his authority at Ephesus, the natural reluctance of the community to accept a new-comer would be intensified, and the opposition discussed above becomes more explicable. The hostility which Paul attracted was not entirely due to his theological positions. His own character traits were also a significant factor.

Tensions at Philippi

The degree of Paul's self-absorption at this point in his career is remarkable. That he should reveal his feelings so frankly to friends is understandable. But *Letter B* was addressed to a church that was itself bedeviled by a clash of personalities!

I noted above that the directive 'Do not act out of a spirit of rivalry, nor out of vain ambition, but in humility count others better than yourselves' (Phil. 2: 3) has a specific application to the dispute between Euodia and Syntyche (Phil. 4: 2). Both of these ladies had participated in the spread of the gospel, 'they fought at my side for the gospel' (Phil. 4: 3), and evidently felt that their talents and devotion had earned them an authoritative role in the nascent church. Such ambition would be irrelevant unless they both had supporters. It is natural, therefore, to think that each headed a house-church, as did Phoebe at Cenchreae (Rom. 16: 1–2). Their competitive attitude engendered a disruptive spirit, which endangered the future of the community.

One wonders what the Philippians made of Paul's call for unity and reconciliation, when he exhibited nothing but contempt for those at Ephesus who disagreed with him? Did he perceive that he was sending contradictory messages when he told them 'Do what you have heard and seen in me' (Phil. 4: 9)? Even when he recognized that his duty was to rise above hurt feelings, he could not resist a mean aside, 'What then? Only that in every way, whether *in pretence* or in truth, Christ is proclaimed. And in that I rejoice' (Phil. 1: 18). The sincerity of his pleasure is at least open to question. If he recognized that the power of the gospel was derived from its effective incarnation in those who preached it (1 Thess. 1: 6–8; Phil. 2: 14–16), how could he even admit the possibility that it could be proclaimed with false motives, as part of a plan to hurt a fellow-believer?

---

[53] Although written from a perspective with which I do not wholly agree, there is much truth in Fortna (1990).

A Liturgical Hymn

A further indication of Paul's self-absorption is his citation of a magnificent Christological hymn, which is perhaps the most damning condemnation, albeit implicit, of his egocentric attitude.

Since the beginning of the twentieth century it has been recognized that the rhythm and formulation of Philippians 2: 6–11 make it stand out from its present context in the letter. Since Paul does not craft his paragraphs with the great care displayed in these verses, there is a wide consensus that he is quoting a pre-existent document. Its identification as a hymn is due to the detection of different strophes. There is a great deal of disagreement on the number of strophes,[54] but I remain convinced that a three-strophe arrangement best respects the formal elements in the text.[55]

I	v. 6a	Who being in the form of God
	v. 6b	Did not claim godly treatment
	v. 7a	But he emptied himself
	v. 7b	Taking the form of a servant.
II	v. 7c	Being born in the likeness of men
	v. 8a	And being found in shape as a man
	v. 8b	He humbled himself
	v. 8c	Becoming obedient unto death.
III	v. 9a	Therefore God super-exalted him
	v. 9b	And gave him the supreme name
	v. 10a	So that at Jesus' name every knee should bow
	v. 11a	And every tongue confess 'Jesus Christ is Lord'.

The clarity of the pattern is its own justification. Something so perfect did not happen accidentally. Only deliberate intention explains the structural balance of the first two strophes. The double mention of 'God' in the first strophe matches the repeated reference to 'man' in the second. In both strophes the third line contains the verb followed by a reflexive pronoun, whose meaning is explained in the fourth line. The humiliation of the first two strophes gives way to exaltation in the third. To the elevation of Jesus in line one of the third strophe corresponds the submission of humanity in line three. The name conferred on Jesus in line two is proclaimed in line four.

No one who goes to the trouble of creating such a perfect arrangement will destroy it. Hence, the extra words which appear in the letter, namely, 'death on a cross' (v. 8c), 'in heaven, on earth, and under the earth' (v. 10b), and 'to the glory of God the Father' (v. 11b), must have been added by a hand other than

---

[54] A brief survey is given by Hawthorn (1983), 76–7.
[55] Jeremias (1953) and (1963).

that of the original composer. Whose was it? The insistence on highlighting the brutal modality of Christ's death points to Paul.[56] Paul, therefore, not only quotes a hymn, but adapts it to his own theological perspective. Originally the hymn must have been the inspired composition of a charismatic believer (1 Cor. 14: 26; Col. 3: 16), which Paul saw as reflecting to a great extent his vision of Christ. He accepted what it said, but made explicit what he felt was lacking.

We do not know in which community the hymn originated, but in all probability it was one which had been founded by Paul. The strong emphasis on the deliberate choice involved in the self-sacrifice of Christ—'he emptied/humbled himself'—reflects the perspective of Galatians 2: 20, 'he loved me and gave himself for me' (cf. Gal. 1: 4). The insistence that Christ became Lord is echoed in 1 Corinthians 15: 45; Romans 1: 3– 4 and 14: 9. The hymn grew out of Pauline teaching.

In biographical terms the importance of the hymn is twofold. It tells us something about the way Paul interacted with his communities, and it reveals a critical development in his understanding of the person of Jesus Christ.

The Teacher Learns

Paul's message was always very simple. This caused problems in Galatia, and would again at Corinth. He did not believe in a speculative theology. All that was necessary, in his eyes, was to understand what Christ had done for us and to act accordingly. What this meant in practice was a matter for each community to decide. He had made this clear to the Galatians, and says the same thing to the Philippians; 'work out your salvation in fear and trembling, for God is at work in you, both to will and to work for his good pleasure' (Phil. 2: 12–13). The depth of Paul's conviction that the local church should be autonomous in its development is underlined by his willingness to learn from it, not only by way of challenge but, as in the case of the hymn, by way of formulation. The hymn gave dramatic, memorable formulation to his thought, and he acknowledged it publicly by citation.

Paul's reflection on the tangible evidence of the action of the Spirit in such 'spiritual songs' (Col. 3: 16) culminated ultimately in his vision of the community as a spiritual temple (1 Cor. 3: 16–17; 6: 19), in which the presence of God made itself effective through a variety of gifts (1 Cor. 12–14).[57]

---

[56] The traditional material discerned in the Pauline letters mentions only the fact of the death of Jesus without specifying its modality; see Rom. 1: 3–4; 4: 25; 8: 34; 1 Cor. 15: 2–7; Gal. 1: 3–4; 1 Thess. 1: 10.

[57] A number of scholars have argued that Paul borrowed the concept of the community as a spiritual temple from Qumran. Earlier studies are reviewed and expanded by Klinzing (1971), 210. The Essenes, however, derived their concept of a spiritual temple from that of spiritual sacrifice (1QS 8. 5–9; 9. 3–5), whereas for Paul it was a deduction from the indwelling of the Spirit.

Paul's Adoption of an Adamic Christology

The Christology of the hymn is fiercely debated, and it has been made to say many different things about the person and role of Jesus.[58] There is little doubt in my mind, however, that it was intended to be read against the background of the story of Adam as filtered through the sapiential literature.[59]

In Galalatians the Judaizers' stress on the figure of Abraham forced Paul to penetrate more deeply than hitherto into the mystery of the person of Jesus Christ. The hymn took him even further back into the history of salvation. Paul's adoption of its Adamic perspective on Christ proved to be a decisive development which would influence all his subsequent soteriological teaching.[60]

If one is prepared to cleave to the essential, the insight of the hymn can be summarized without great difficulty. As the righteous person par excellence, Christ was the perfect image of God. He was what Adam should have been, the inspiring illustration of what God intended a human being to be. Christ's sinlessness gave him the right to be treated as if he were a god, that is, to enjoy the incorruptibility in which Adam was created. This right, however, he did not use to his own advantage. On the contrary he gave himself over to the consequences of a mode of existence inaugurated by fallen Adam. He freely chose the life of a slave which involved suffering and death, the state which Adam experienced as punishment. Although in his human nature Christ was identical with other members of the human race, he in fact differed from them because he had no need to be reconciled with God. It was this which enabled him to become their saviour through obedience and death. Therefore God exalted him above all the just who were promised a kingdom, and transferred to him the title and authority which previously had been God's alone. He became the Lord, whom every voice must confess and to whom every knee must bow.

This is not the place to detail the light these insights throw on the human condition, on the nature of salvation, and on Christ's salvific role. Here it must suffice to indicate the broad outlines of ideas which Paul will develop in subsequent letters.[61] The state of humanity was not simply a given, but was a living out of the consequences of Adam's sin; it was a radically unnatural way of being. Salvation was the reacquisition of Adamic identity as portrayed by Jesus in his revelation of love as the essential constituent of authentic humanity.

[58] See Martin (1983).
[59] In addition to my (1976), see Dunn (1980), 113–21.
[60] It is possible to read Gal. 4: 4 as indicating that Paul was already thinking of Christ in Adamic terms (so Dunn (1980), 41), but I doubt that anyone would think of doing so without the much clearer indications in Phil., 1 Cor., and Rom.
[61] See Dunn (1980), 98–128.

THE LETTER OF WARNING

Letter C (Phil. 3: 2 to 4: 1) begins with a triple imperative *blepete*. It is usually translated as 'look out for' (*RSV*), 'beware of' (*NRSV, NAB*), 'be on your guard against' (Phillips). G. D. Kilpatrick, however, has argued that when *blepein* is used in this sense it is followed either by *mê* with the subjunctive or by *apo* with the genitive. Here we have the direct object and so, he claims, it should be translated 'consider, take note of'.[62] In consequence, a number of commentators have insisted that the function of 3: 2 to 4: 1 is not to warn the Philippians against any particular group, but to hold up the Jews as a cautionary example.[63] The church was not menaced by intruders, but by attitudes among its members which Paul desired to correct by illustrating their effect among Jews.

Despite its apparently sound grammatical base, this interpretation fails to do justice to either the tone or content of *Letter C*. It is implausible that Paul would use the Jews as a cautionary model for a church whose members had come entirely from paganism. If his intention was to contrast reliance on self with reliance on divine help, there were many examples from Greek history which would speak directly to Philippians. If Judaism did enter his mind, he had only to mention his own experience. There was no need to refer to dogs, evil doers, and mutilators, whose god was their belly (Philippians 3: 2, 19). The viciousness of such invective betrays the depth of Paul's fear for the future of the Philippians. The tone evokes, not a remote possibility, but an imminent danger. It is not surprising, therefore, that even those who acknowledge the accuracy of Kilpatrick's observations do not always endorse his interpretation, and continue to read *Letter C* as a warning.[64]

There is a wide variety of opinions regarding the identity of those against whom the Philippians are warned.[65] Most have only a tenuous basis, if any, in the letter. The first element in Phil. 3: 2 'Beware of the dogs' would make one think of Gentiles who, because they did not discriminate in what they ate, were considered 'dogs' by Jews.[66] The third element, however, 'Beware of the mutilators. For we are the true circumcision' clearly refers to Jews, or Christians of Jewish origin. The possibility that Paul has two distinct groups in mind is excluded by the Gentile character of the church. Paul would not use 'dogs' in a sense applicable to the Philippians. Hence we must assume that he is

---

[62] (1968), 146–8.
[63] e.g. Caird (1976), 131; Garland (1985), 165–6; Hawthorn (1983), 125.
[64] e.g. Martin (1976), 124; Bruce (1983), 80.
[65] Hawthorn (1983), 163, lists Judaizers, libertines, gnosticizing believers, faithful who feared persecution or who did not recognize the decisive eschatological significance of the Cross and its imperative of self-sacrifice, and proselytizing Jews.
[66] Billerbeck (1922–8), 1. 724–5; 3. 621–2.

turning back on Jews one of their most vicious slurs. There can be little doubt that Paul intends to evoke circumcision and the dietary laws.

Those from whom danger comes are also alluded to in 3: 19, which when translated in such a way as to respect its structure—'they have made their stomach and their glory in their shame their god'[67]—reveals that Paul has in mind two matters of ultimate concern, one is 'stomach' and the other 'shame'. The relationship between stomach and diet needs no emphasis. That between shame and circumcision becomes evident only when it is recalled that Greeks and Romans heaped scorn and ridicule on circumcision,[68] and that out of shame some Jews underwent an operation to restore the foreskin.[69]

It is difficult to imagine that the second element in Philippians 3: 2, 'beware the evil workers', is directed against all Jews. If Paul had suffered at the hands of some, most were totally ignorant of his existence. The substantive implies energetic effort. Since it cannot refer to Jewish proselytization,[70] it must allude to the phenomenon with which Paul had to deal in Galatians. The parallel 'deceitful workers' (2 Cor. 11: 13) confirms that what Paul feared was that the Philippians would come under pressure from Jewish Christians to adopt circumcision and the dietary laws. For Paul Judaizers were the real enemies of the cross of Christ (Phil. 3: 18), because they denied its salvific value. For them Christ simply inaugurated the eschaton; salvation was still conditional on observance of the Law. Those who thought of a crucified saviour as folly (1 Cor. 1: 23) were less dangerous.

This interpretation is reinforced by the parallels with Galatians, whose broad outline *Letter C* reproduces. The autobiographical material (Phil. 3: 4–8) is reminiscent of Galatians 1, but it is not used in precisely the same way. In Galatians Paul was concerned to demonstrate his independence of Jerusalem, and thus indirectly of Antioch, whereas here his point is to show that he had once been a strictly observant Jew, but had found something better. The contrast between righteousness acquired by obedience to the Law and righteousness given by God 'through the faith/fidelity of Christ' (Phil. 3: 9 = Gal. 2: 16) is evocative of Galatians 3–4. The admonition that salvation is not an immutable given, but an ongoing struggle towards a future prize (Phil. 3: 10–16) could serve as an accurate summary of Galatians 5–6. Note in particular the parallel between Paul's bearing the stigmata of Jesus (Gal. 6: 17) and his sharing in the fellowship of Christ's sufferings (Phil. 3: 10). The concluding exhortation to imitate Paul (Phil. 3: 17) echoes Galatians 4: 12, but in view of the divisions within the church at Philippi the Apostle creates the word 'fellow-imitators' to underline the corporate dimension of the believers' existence.[71]

---

[67] Hawthorn (1983), 166.   [68] Martial, *Epigrams*, 7. 35 and 82.
[69] 1 Macc. 1: 15; Josephus, *AJ* 12. 241. See Hall (1988).
[70] See my (1992a) and Will and Orrieux (1992).
[71] See in particular Hawthorn(1983), 160.

Where and how did Paul learn that Philippi was menaced by Judaizers? I have argued above that the Judaizers who troubled the churches of Galatia were sent by Antioch to reform the churches founded by Paul.[72] Their natural course subsequently would have been to follow his tracks to Europe, where Philippi was his first foundation. Once he realized what was going on in Galatia, Paul should have anticipated this danger, but there is no hint in either *Letter A* or *Letter B* that he did. Such carelessness may be another aspect of the self-absorption so evident in *Letter B*.

It is not impossible that Paul suddenly woke up to the potential threat to Philippi, and dashed off *Letter C* in fulfilment of his responsibility. In this case, however, one might have expected a hint of self-reproach. The note of urgency, which penetrates the letter, suggests rather that it was a reaction to precise information. The simplest hypothesis is also the most probable. One of Paul's supporters came from Galatia to inform him of the plans of the Judaizers to move against Philippi, and the other Pauline churches. It would be most surprising, if those who had alerted him to the presence of the latter should not have kept him in touch with the evolution of the situation in Galatia. A more complex hypothesis might claim that the Judaizers knew that Paul was headed for Ephesus when he left Galatia, and followed him there. When they failed to convince him and his converts, they headed north for Philippi. Not only is this hypothesis more complicated than the data demands, but there is no hint in Colossians or Philemon that Judaizers of the Galatian type had come west into Asia from Galatia.

---

[72] See Ch. 8, 'Who Were the Intruders?'.

— 10 —

Contemplation at Colossae

PAUL's strategy during his two years and three months' residence in Ephesus (Acts 19: 8–10) had two facets. He stayed in the city dedicating himself to the formation of the community and to maintaining contact with his other foundations. The church, however, had to be apostolic. Hence, he commissioned others to spread the gospel outside the urban area, following the pattern dictated by the Roman roads radiating out from the capital of Asia.[1] As we have seen, some went north to Smyrna and Pergamum. Others took a road angling off to the north-east, and evangelized Philadelphia, Sardis, and Thyatira. Still others took the great common highway to the east and brought the gospel to Magnesia and Tralles.[2] One went much further, into the Lycus valley on the fringes of the province of Asia. It was his homeland (Col. 4: 12).

The Roman road, which Epaphras followed, was constructed by Manius Aquillius, who was proconsul of Asia 129–126 BC. For the first 80 miles out of Ephesus it followed the north bank of the river Meander, which it crossed on a bridge at Antioch-on-the-Meander, and continued along the south bank until it was blocked by a tributary, the Lycus (modern Çürük-su), coming in from the south-east.[3] Turning to stay on the west side of this considerable river, the road first reached Laodicea, and then Colossae (192 km. or 120 miles from Ephesus), after which there was a bifurcation. The road of Manius Aquillius swung south to the coast. The Cilician road curved to the north to the cities of Paul's 'first journey' (Acts 13–14).[4]

THE LYCUS VALLEY

The eye of anyone entering the valley from the west is caught by a dazzling blaze of white against the brown of the cliff across the river. For millennia, mineral-saturated hot water has poured down the slope gradually building up a

---

[1] J. J. Taylor (1994*b*), 195.
[2] See Ch. 7, 'Missionary Expansion'.
[3] Strabo, *Geography* 14. 2. 29.
[4] French (1980), 707; map 3 in S. Mitchell (1993), 1. 40.

deposit so that today it looks like 'foaming cataracts frozen in the fall'.[5] The phenomenon was known to Strabo, who noted the ingenious use the natives made of it, 'the water of the hot springs so easily congeals and changes into stone that people conduct streams of it through ditches and thus make continuous stone fences'.[6] Today the site has the entirely appropriate name of Pamukkale, 'Cotton Castle', but in the first century it was known as Hierapolis.

It would be most unusual if the unique properties of the waters had not attracted settlers to Hierapolis from remote antiquity, but it, and its neighbour Laodicea (6 miles away across the river) appear on the stage of history only in the Hellenistic period. The oldest documented town in the valley is Colossae, which is 11 miles upstream from Laodicea. The double 'ss' in its name is thought to be a relic of a pre-Greek language,[7] and it is mentioned in the fifth century BC by Herodotus (7. 30) and Xenophon (*Anabasis*, 1. 2. 6) as a large and prosperous city.

The virtual monopoly that Colossae enjoyed in the exploitation of the natural resources of the valley came under threat in the third century BC, when Seleucid monarchs intervened to create new commercial centres. Antiochus I Soter (281–261 BC) raised Hierapolis to the status of a city,[8] and his son Antiochus II Theos (261–246 BC) conferred the same favour on a settlement called Diospolis/Rhoas, whose name he changed to Laodicea to honour his wife.[9] In 220 BC a certain Achaeus raised the standard of rebellion in Laodicea against Antiochus III the Great (233–187 BC).

The rising was abortive, but in order to guarantee that it could not happen again, Antiochus III, around 213 BC, settled 2,000 Jewish families from Babylon and its environs in Phrygia and Lydia.[10] It would be most surprising if a significant number of these colonists did not end up in the Lycus valley.[11] A century and a half later, the Jewish population was considerable. In 62 BC the district of which Laodicea was the capital had at least 11,000 adult male Jews.[12] Some twenty years later, the authorities of Laodicea assured the Roman authorities that Jews would not be hindered in the practice of their religion.[13] The presence of a Jewish community with its roots in Babylon, is crucial for an understanding of the problems that Paul and Epaphras had to confront.

Despite the seniority implicit in giving its name to a particular colour (see below), Colossae lacked certain advantages enjoyed by its younger rivals.

---

[5] Lightfoot (1904), 10.

[6] *Geography* 13. 4. 14. Similarly Vitrivius, *De Architectura* 8. 3. 10.

[7] McDonagh (1989), 370.

[8] Kolb (1974); Bruce (1992*a*), 195.

[9] Pliny, *NH* 5. 105.

[10] Josephus, *AJ* 12. 148–53; see Schürer (1973–87), I. 17 n. 38.

[11] Lightfoot (1904), 19.

[12] This is the calculation of Lightfoot (1904), 20, based on the twenty pounds of gold confiscated by Flaccus; see Cicero, *Pro Flacco* 68.

[13] Josephus, *AJ* 14. 241–3.

Laodicea was the capital of the district. The courts of the proconsul of Asia might be infrequent, but its role as the financial and tax centre gave it a latent power, which proved attractive to those interested in policy and business.[14] Inevitably leisure facilities would be better than elsewhere in the vicinity; gladiatorial shows are attested.[15] Hierapolis no doubt enjoyed a share of this tourist market. The pleasures of natural hot baths were intensified by the medicinal properties of the waters and drew seekers of luxury and health from a wide area. The merely curious no doubt flocked to inspect the Plutonium, a cave whose poisonous vapours slew animals.[16]

The extent to which Colossae lost out in the prosperity stakes is graphically illustrated by the dearth of visible remains when compared with the extensive ruins of Laodicea and Hierapolis.[17] It cannot even boast a famous name, whereas Hierapolis could claim the Stoic philosopher Epictetus (AD 55–135), and Laodicea the rhetorician Zeno, the bravery of whose son, Polemon, when the city was attacked by the Parthians in 40–39 BC, won him the kingdom of Cilicia Tracheia.[18]

The volcanic springs and underground rivers alerted Strabo to the unstable character of the land in the Lycus valley, 'if any country is subject to earthquakes, Laodicea is' (*Geography* 12. 8. 16). Many went unrecorded, but major earthquakes hit in the reign of Augustus,[19] and again in AD 60, as Tacitus reports, 'In the Asian province one of its famous cities, Laodicea, was destroyed by an earthquake in this year, and rebuilt from its own resources without any subvention from Rome.'[20] No earthquake that devastated Laodicea would have spared its neighbours. The recovery of Hierapolis is guaranteed by the existence of a bishopric there at the beginning of the second century AD, headed by Papias.[21] Colossae, on the contrary, sinks into oblivion.[22]

In their heyday these cities lived from wool. The Lycus valley was a vast pasture in which numerous flocks wandered. In this it was no different from much of Anatolia. Yet the inhabitants managed to carve out a unique niche in the textile market by the quality of their products. According to Strabo, 'The country around Laodicea produces sheep that are excellent, not only for the softness of their wool, in which they surpass even the Milesian wool, but also for its raven-black colour, so that the Laodiceans derive splendid revenue from it, as do also the neighbouring Colossians from the colour which bears the same name' (*Geography* 12. 8. 6; trans. Jones).

It would appear that the glossy black fleeces associated with Laodicea were natural. Certainly this is the interpretation of Vitrivius, for whom it was

---

[14] Cicero, *Att.* 5. 15; *Fam.* 3. 5.
[15] Cicero, *Att.* 6. 3.
[16] Strabo, *Geography* 13. 4. 14.
[17] McDonagh (1989), 370–81.
[18] Strabo, *Geography* 12. 8. 16.
[19] Suetonius, *Tiberius* 8.
[20] *Annals* 14. 27. 1; trans. Grant.
[21] Eusebius, *Church History* 2. 15; 3. 36–9.
[22] The silence of Pliny in *NH* 5. 105 is given significance by Aletti (1993), 11 n. 3.

explained by the water of certain springs from which the sheep drank (*De Architectura* 8. 3. 14). Strabo's failure to specify the precise colour associated with Colossae is remedied by Pliny, who tells us that *colossinus* is a purple resembling that of the cyclamen blossom (*NH* 21. 51; cf. 25. 114). That the unusual characteristics of the water of the region contributed to the distinctive colour is suggested by a note of Strabo apropos of a neighbouring city, 'The water at Hierapolis is remarkably adapted also to the dying of wool, so that wool dyed with the roots [madder-root] rivals that dyed with the cossus [kermes-berries] or with the marine purple' (*Geography* 13. 4. 14; trans. Jones).

MISSIONARY STRATEGY

When Paul marched across Asia Minor for the second time, his goal was Ephesus, and he did not attempt to found new communities.[23] Where then did he meet Epaphras, a native of Colossae (Col. 4: 12)? The encounter could have taken place on the road. Paul would have been glad of a companion, a potential convert, whose presence enhanced his security. Or it might have been in Ephesus. But it could have been much further afield. The probability is that Epaphras was in some way associated with the export of textiles from the Lycus valley, and if Lydia from Thyatira was selling in Philippi (Acts 16: 14), it is not at all impossible that the superior product of the Lycus valley was being marketed by Epaphras in Macedonia or Achaia. This latter hypothesis, however, is not really plausible. If Epaphras had been commissioned by Paul in Greece to plant the gospel in his home valley when he returned, it is rather improbable that Paul would not have turned aside for a few days, after having visited the Galatians, in order to check on how things were going. If he did not do so (Col. 2: 1), it can only be because the churches in the Lycus valley did not yet exist.

Wherever they met, Epaphras was formed as a missionary by Paul in Ephesus, and he must have been typical of those whom Paul chose to fan out to found other churches. From personal experience in Asia Minor and Macedonia, Paul knew the difficulty of coming into a strange city in which he knew no one. He had to find work in a congenial situation which would permit him to preach. Where was he to begin? He must have recognized immediately how much easier his task became when he linked up with Prisca and Aquila in Corinth. They provided a base and a ready-made set of contacts. In any case, thereafter he built it into his missionary strategy. He left the couple in Ephesus in order to have everything in readiness for his return from Jerusalem,[24] and later would

---

[23] See Ch. 7, 'Maintenance not Mission'.
[24] See Ch. 7, 'The Founding of the Church'.

send them to Rome to prepare for his arrival there.[25] They now carried the burden of loneliness and alienation, but they were strengthened by the confidence that Paul would soon arrive to share the responsibility.

Paul could have demanded of others, and probably did, the sacrifices he demanded of Prisca and Aquila. There are always those willing to strike out into unknown territory. But would it not have been much more efficient to select as missionaries those who started with a built-in advantage? The prime candidates were the energetic and enterprising women and men, like Epaphras, who came to the capital of Asia on business. It did not matter whether they were acting as principals or agents, they returned home to a network of acquaintances rooted in long-standing family, social, and business contacts. They did not have to look for work. They were known and trusted. The respect they had earned guaranteed that there were always at least some sympathetic ears to hear their first stumbling sermons.

The freedom of Epaphras to make a trip to Ephesus in order to seek Paul's advice when problems developed at Colossae suggests that he was in business on his own account. The alternative is to suppose that he converted his employer, who proved to be most sympathetic in terms of time off in order to permit Epaphras to discharge his duties as founder of the church. The hypothesis is not impossible, but it is more complicated, and even Paul did not take the Christianity of employers/owners for granted. The relationship of Philemon and Onesimus is a case in point.

The legal aspect of this dispute has already been dealt with.[26] Here we must confront much simpler questions, which lead us into unexplored aspects of the evangelization of the Lycus valley. How did Onesimus, a pagan (Philem. 10), know of Paul's influence on his master, and how did he know where to find Paul in far-away Ephesus? In the light of the preceding discussion, one is immediately inclined to consider Epaphras as the source of this information. In Philemon 19, however, Paul takes the pen from the hand of his secretary to guarantee the repayment of whatever damage Onesimus had caused, and underlines his creditworthiness by pointing out that Philemon is in the Apostle's debt, 'you owe yourself to me'.

The natural interpretation of this phrase is that Philemon had been converted by Paul personally, presumably in Ephesus.[27] It would have been natural of him to speak to his household of the importance of Paul in his life. Acceptance of this interpretation, however, leads to unacceptable consequences. It means that Paul had sent two apostles to the Lycus valley, but gives the credit for the establishment of the churches of Colossae, Laodicea, and Hierapolis to only one, Epaphras (Col. 4: 13). Had Philemon made no contribution, Paul's

[25] See Ch. 13, 'Planning for the Future'.
[26] See Ch. 7, 'Imprisonment'.
[27] So Lightfoot (1904), 303, 342; Lohse (1968), 284–5.

compliments in Philemon 6–7 appear as condemnation by faint praise. Would Paul have slighted Philemon just at the moment he wanted something from him? The difficulty of admitting that Paul would have acted so stupidly forces us to consider the possibility that, in writing Philemon 1 and 19, he was acting on the principle that masters are responsible for the actions of their agents. If those who command are liable for damages, they can also claim the credit for success. In other words, Philemon was converted by Epaphras as Paul's agent.[28]

Philemon was followed into the faith by his wife Appia,[29] and by Archippus (Philem. 1). Together they became the nucleus of a house-church (Philem. 2), which may have been the first of a number of such sub-units which together made up the 'whole church' of Colossae. The social status of Philemon can be deduced from his ownership of at least one slave; it is confirmed by his possession of a house large enough to contain a guest-room (Philem. 22). Had Paul impressed on Epaphras the strategy, which he himself was to employ so successfully at Corinth? It was important to recruit quickly one or two people who could provide a centre for the nascent community.[30] Nympha may have played this role at Laodicea, where she became responsible for a house-church (Col. 4: 15). Philemon is called 'fellow-worker' and Archippus 'fellow-soldier' (cf. Phil. 2: 25). The implication that both were active in the development of Christianity in the Lycus valley is confirmed by Paul's treatment of the former as 'a partner' (Philem. 17; cf. Phil. 1: 5) and by the words addressed to the latter in Colossians 4: 17, 'see that you fulfil the ministry which you have received in the Lord'.

The curious form of this admonition—it is introduced by 'Tell Archippus'—and the contrast with the complimentary epithet in Philemon 2, indicate that the status of Archippus had changed between the writing of Philemon and Colossians.[31] What might have happened? One scenario which deals adequately with the data runs as follows. Onesimus had injured his master Philemon in a serious way. Epaphras sent Onesimus to Ephesus to beseech Paul's mediation in the dispute with Philemon. Although not a Christian, Onesimus, like any of the servants, was fully aware of the composition of the little community that met in his master's house. Once Onesimus had been baptized (Philem. 10; cf. 1 Cor. 4: 15), the significance of the active missionary role of Philemon and Archippus became evident to him. And he told Paul, who needed to flatter Philemon in order to win a favour from him. Paul had never written this sort of letter before, and it demanded serious reflection. Before the missive was finished, Epaphras arrived and was arrested by the Romans because of his

---

[28] Some scholars have read Philem. 5 as implying that Paul did not know Philemon personally; see Lohse (1968), 270 n. 4.

[29] So rightly Lightfoot (1904), 304; Lohse (1968), 267.

[30] See Ch. 11, 'The First Converts'.

[31] The anteriority of Philem. is defended by Lohse (1968), 247.

official association with Paul. In addition to informing Paul about the false teaching that circulated at Colossae, he spoke sadly about Archippus. Given the gravity of the theological situation, it does not seem adequate to postulate merely that Archippus was somehow less active than hitherto. 'Tell Archippus' (Col. 4: 17) makes sense only if he had left the community and would not hear the letter when it was read publicly (Col. 4: 16). Had he become involved with the false teachers to the point that he no longer found the liturgy of the church satisfying? Only an affirmative answer explains the urgency of Colossians. If a leader of Archippus' quality had been seduced by esoteric teaching, the danger for others in the community was very real. A response could not wait until Paul or Epaphras was released from prison.

PAUL'S APOSTOLIC OFFICE

No consensus exists regarding the authenticity of Colossians. The scholarly community is split down the middle. Those who affirm Pauline authorship, however, are rather more hesitant than those who deny it.[32] None the less, the conclusions of the latter are not always as well founded as the force with which they are articulated would appear to indicate.

The stylistic argument, which has always been considered the most objective, must be set aside.[33] Paul's use of co-authors and secretaries precludes the establishment of a writing style exclusive to the Apostle against which letters can be measured.[34] Equally, without evidence that it was a standard pseudepigraphic technique, the names and personal notices (Col. 1: 7–8; 2: 1; 4: 7–18) cannot be dismissed as an artificial attempt to give Colossians a place in Paul's ministry. Those who maintain the inauthenticity of 2 Thessalonians and Ephesians find no reason to postulate such a requirement. If, as I have argued above,[35] such personal references are taken seriously, then Colossians must be dated to the summer of AD 53, during Paul's imprisonment at Ephesus. The critical questions, then, are: (1) are the differences between Colossians and the other letters written during this period as great as have been thought? (2) if so, can they be explained as due to the particular circumstances under which this letter was written?

Opponents of the authenticity of Colossians find justification for their position in its view of Paul's apostolic office and of the value of his sufferings. The latter, we are told, are understood to have a vicarious value, and Paul is

---

[32] Compare Aletti (1993), 280, with Furnish (1992), 1094.

[33] The observations of Bujard (1973) are negated by the more sophisticated studies of Kenny (1986) and Neumann (1990). See also Sappington (1991), 23.

[34] See my (1995), ch. 1.

[35] Ch. 7, 'Imprisonment'.

presented as the universal, unique apostle. Such inflation of his position is incompatible with his historical role, and thus points to the artificiality of Colossians.[36]

Universalistic language is certainly not lacking in Colossians, but when looked at closely, it does not confirm the interpretation forced upon it. The note that the gospel is bearing fruit and growing 'in the whole world' (Col. 1:6) is a simple reflection on the success of the ministry to the Gentiles, and in no way implies that Paul alone was responsible. On the contrary, the intrinsic power of the word of God is an authentically Pauline theme (cf. 1 Thess. 1: 5; 2: 13). Later Paul speaks of 'the gospel which you heard, which has been preached to every creature under heaven, and of which I Paul became a minister' (Col. 1: 23). The lack of the definite article before 'minister' underlines that Paul is not the exclusive agent of propagation, and the hyperbole is precisely paralleled in the Apostle's very first letter, both with respect to the past tense and to the universal extension, 'your faith in God has gone forth everywhere' (1 Thess. 1: 8). Finally, the sole implication of 'teaching everyone in all wisdom that we may present everyone mature in Christ' (Col. 1: 28) is that Paul's message is for all without exception.

Colossians differs from the other Ephesian letters in that it is written to a church that Paul did not found. It would not have been written, as we have seen,[37] had Epaphras been free to return to the Lycus valley after having consulted Paul in Ephesus. Only when he too landed in prison did it become imperative to devise another way of dealing with the situation. Apparently no competent emissary was available and that left a letter as the only option. Who was to write it? Epaphras was the obvious candidate, since the Colossians were his people, and his problem.[38] Could he not say on paper what he planned to say verbally? Some in Paul's entourage were less sanguine. To express oneself adequately in writing in circumstances where a false word could be disastrous demands a very special skill. Epaphras had given no evidence of this talent, which Paul had demonstrated in his letter to the Galatians. In any event Paul accepted the responsibility. Epaphras might then appear to be the obvious choice as co-author; he was the local expert. The same could be said of Apollos with respect to the developing situation in Corinth,[39] but, as in the present case, Paul preferred to rely on Timothy.

Paul's lack of personal involvement with the Christians in Colossae and his sense of the autonomy of the local church explains the universalism of the texts just discussed. On the formal level, he could and did point out that Epaphras had been acting as his agent (Col. 1: 7), but on a a more profound level, he felt

---

[36] Furnish (1992), 1094.
[37] Cf. Ch 7, 'Missionary Expansion'.
[38] There are those who claim that he in fact wrote the letter.
[39] See Ch. 12, 'Weaning'.

the need to evoke the world-wide scope of his apostolic responsibility in order to justify his concern for the Colossians (Col. 1: 25). The context in which this must be understood is the Jerusalem agreement (Gal. 2: 9), which authorized all missionaries to go everywhere.[40] In Paul's mind this meant that he could not exempt himself from any effort that might draw people to Christ. In no way does it imply that he felt that everything had to be done under his aegis. He took it entirely for granted that other missionaries would work in parallel with him and made it a principle not to duplicate their efforts (Rom. 15: 20).

Paul's interpretation of his sufferings in Colossians 1: 24 has caused much ink to be spilled. The *NRSV* reflects the common translation of this verse, 'Now I am rejoicing in my sufferings for your sake, and in my flesh I am completing what is lacking in Christ's afflictions for the sake of his body, that is, the church.' Questions immediately arise: was Christ's sacrifice somehow imperfect? is the genitive 'of Christ' to be understood as subjective, as objective, as qualitative, etc.? how can Paul's sufferings be added to those of someone else? is there a quota of sufferings that must be endured before the Parousia? The mind reels before the permutations and combinations of the possible answers, each of which has found an advocate.[41]

Such complications, however, arise only because the translation is faulty. The interpretation of the verse is greatly simplified if the order of the key Greek words is respected,[42] and hyphens are added for clarification, 'I am completing what is lacking in Christ's-afflictions-in-my-flesh for the sake of his body.' Paul is not speaking of the sufferings of Christ in themselves, but of his own sufferings, which in a certain sense are those of Christ. In a previous letter he had written, 'I live now, not I, but Christ lives in me' (Gal. 2: 20). Subsequently he would try to get the same idea across by writing 'always carrying in the body the dying of Jesus' (2 Cor. 4: 10). Paul did not need great insight to know that his pain would be prolonged, and he was fully aware of precisely how it benefited the church. His imprisonment dramatized his commitment to Christ, which both impressed pagans and fortified believers (Phil. 1: 13–14). Colossians 1: 24 is nothing more than a typically Pauline Christologization of this theme.

The intensification of Paul's functional identification with Christ in Colossians can be seen as simply the logical consequence of the insight of Galatians 2: 20, but it was not Paul's style to develop methodically the ramifications of an idea. There must have been something in the attitude of the Colossians towards Christ which stimulated his reflection.

---

[40] See Ch. 6, 'The Agreement'.
[41] They are documented by Kremer (1956).
[42] Aletti (1993), 135.

THE COSMIC CHRIST

The difference between the Christology of Colossians and that of the other letters is usually explained in one of two ways. Defenders of the authenticity of Colossians date it at the very end of Paul's life in order to give enough time for such a radical development of his thought. Those who refuse the authenticity of Colossians find the difference so great as to make Pauline authorship inconceivable. A third approach, which does more justice to the data of the letter, has been espoused by C. K. Barrett, 'It seems rather that the Colossians ... had done their best to give Christ a prominent place in the realm of cosmic speculation. What they had not done, and the editor now proceeds to do, is to recognize his earthly activity.'[43] In other words, the concern of Colossians is not to lift its readers into the cosmic sphere, but to ensure that they do not lose contact with the mundane. The Saviour must stand on terra firma. His disciples must not retreat into ascetic isolation.

As Barrett perceived, the clearest illustration of what is actually going on in the letter is found in Colossians 1: 15–20. Didactic hymns were part of the liturgy at Colossae (Col. 3: 16), and it is generally recognized that Paul is here quoting one of these hymns. Its precise extent and structure has been the subject of intense debate.[44] This is not the place to enter into dialogue with the wide variety of views which have already been expressed. The justification of my position will emerge, I hope, from the coherence of what follows.

The original hymn was made up of two four-line strophes, which are identical in structure:

I	v. 15a	Who is (the) image of the invisible God
	v. 15b	First-born of all creation
	v. 16a	For in him were created all things
	v. 16f	All things through him and to him were created.
II	v. 18b	Who is (the) beginning
	v. 18c	First-born from the dead
	v. 19	For in him all the Fullness was pleased to dwell
	v. 20a	And through him to reconcile all things to him.

The repetition of key terms in the same order in each strophe reinforces the structure. The first two lines of each strophe are affirmations which are subsequently justified in the last two lines. Such perfection of balance betrays a deliberate creative effort. No artist who had invested so much would destroy the elegance of his work. The elements in the existent text which disturb the balance must have been added by a later hand, more concerned with content

---

[43] (1994), 146.
[44] The most detailed survey is that of Benoit (1975).

than with form. The same phenomenon was noted apropos of the hymn cited by Paul in Philippians 2: 6–11.[45] Such similarity greatly reduces the subjective factor which is integral to every literary judgement. It is instructive to put the two sets of additions in parallel:

Colossians 1	*Philippians 2*
v. 16b in heaven and on earth	v. 10b in heaven, on earth, and
v. 16c visible and invisible	under the earth
v. 16d whether thrones or dominations	
v. 16e or principalities or powers	
v. 17 And he is before all things and all things in him hold together	
v. 18a And he is the head of the body, the church	
v. 18d that he might in everything become pre-eminent	
v. 20b making peace by the blood of his cross through him	v. 8c even death on a cross
v. 20c whether those on earth or those in heaven	v. 10b in heaven, on earth, and under the earth
	v. 11b to the glory of God the Father

The similarities are so obvious as to hardly need pointing out. In both instances the redactor is concerned (1) to insist on the modality of the death of Christ, and (2) to restrict the meaning of 'all things' to intelligent beings. In the case of the Philippian hymn there is no doubt that the redactor was Paul; the authenticity of that letter is unquestioned. The language furnishes confirmation. If we leave aside the work of the evangelists, 'cross' and 'to crucify' are virtually exclusively Pauline terms in New Testament usage.[46] This is all the more significant in that the traditional material, which Paul incorporates into his letters, mentions only the fact of the death of Christ without specifying its manner.[47] The parallels create a *prima facie* case that the redactor of the Colossian hymn was also Paul. It is typical of him to emphasize the 'blood' of Christ (v. 20b).[48] It is also characteristic of Paul to stress that Christ gained something by the resurrection (v. 18c).[49]

There is an obvious quantitative difference between the retouches of the two hymns. Those in Colossians 1: 15–20 are much more extensive than those in

[45] See Ch. 9, 'A Liturgical Hymn'.
[46] The exceptions are Heb. 6: 6; 12: 2; Rev. 11: 8.
[47] The texts commonly cited are Rom. 1: 3–4; 4: 25; 8: 34; 10: 8–9; 1 Cor. 15: 2–7; Gal. 1: 3–4; 1 Thess. 1: 10.
[48] Rom. 3: 25; 5: 9; 1 Cor. 10: 16, 11: 25, 27.
[49] Rom. 1: 3–4; 14: 9; 1 Cor. 15: 45.

Philippians 2: 6–11. The natural inference is that the original Philippian hymn was closer to Paul's theological perspective than the hymn which Epaphras brought from Colossae. This in turn opens the possibility that Paul retained the hymn for a specific purpose without accepting all its dimensions.

The distinction of two literary levels permits us to develop two readings, namely, the meaning of the original hymn, and the meaning Paul gave it by means of his additions.

The Original Hymn

The basic theme of the original hymn is obviously the mediation of Christ, first in creation and then in reconciliation. God is mentioned explicitly only as the referent of 'image', but he is certainly evoked by the passive verbs in verse 16a and 16 f, and possibly may be the subject of 'was pleased' (v. 19). The creative power of God is revealed in the action of his chosen instrument, and thereby Christ is exalted above all other beings.

In the first strophe there is no real difference between the formulae 'in him' and 'through him'; the former can be instrumental, and is to balance 'in him' in the second strophe (v. 19) where, however, the meaning is different. On his first reading Paul may have understood 'image of God' in the light of his Adamic Christology based on Genesis 1: 27, but that would have quickly been corrected.[50] The combination of 'image' with 'first-born of all creation' is more likely to have evoked the figure of Wisdom in the sapiential writings, notably Wisdom 7: 22–6; 8: 6; 9: 9. Paul after all was a child of the Hellenistic synagogue. The ambiguity of 'first-born of all creation' is remarkable: is he *of* or *above* creation? The emphasis on 'all creation' and 'all things' makes the cosmic dimension unambiguous. The participation of Christ in the act of creation extended to the totality of being. But in what sense? The context is of no help, no more than it is in answering the question arising from the one element which does not fit the sapiential background, namely, the presentation of all reality as directed, 'to him'. What precisely does this mean? The failure of exegetes to reach a consensus on the answers to these questions suggests that obscurity was intended by the author(s).

In the second strophe 'beginning' again evokes Wisdom (Prov. 8: 22), but leaves vague the sense in which Christ vanquished death. The original author may have thought in terms of immortality, in keeping with the sapiential inspiration of his approach, but the formulation does not exclude resurrection. Christ's being the first to experience life after death is due to a divine gift. The formula 'to be pleased to dwell' occurs regularly in the Old Testament with God as subject, e.g. 'the mountain in which God was pleased to dwell' (LXX Ps.

---

[50] Dunn (1980), 188.

69: 17).[51] Why the surrogate 'Fullness' should have been used here is unclear. Perhaps the author felt it would be more appropriate to the cosmic dimension of his theme, or feared that the Old Testament formula would be misunderstood. Those coming from a Jewish background would never have taken divine indwelling to mean that the person or place was divinized. A pantheist from Phrygia, however, would have gone away with a very different impression. Christ's salvific mission is evoked only in the last line and by the verb 'to reconcile', which for the first time introduces a hint of tension within creation.

The hymn is a perfect example of what Paul calls 'beguiling, persuasive speech' (Col. 2: 4). Formal beauty clothes an abstract vision of Christ, which is allusive rather than explicit. The lapidary phrases are redolent of profundity, but yield no clear understanding. The pervasive ambivalence indicates that a univocal meaning was not intended. The hymn could be sung or recited by all believers (Col. 3: 16) in the belief that they were articulating a mystery beyond them. The initiated could debate the questions that still test the ingenuity of exegetes. In opposition to the hymn cited in Philippians 2: 6–11, nothing in the original hymn betrays Pauline roots. The preaching of Epaphras has been divested of realism by being transposed into a loftier and colder dimension.

Paul's Revisions

The truth of the titles given to Christ meant that Paul could not reject the hymn out of hand. It evoked aspects of Christ that he would not have chosen to emphasize, but they were rooted in the revelation accorded to his people. To accept them was the price he had to pay for the lapidary formulae of the hymn which he realized he could turn against its originators.[52]

Before discussing this point, it is important to note the flexibility of Paul's mind. He did not disdain to take over a key concept of the hymn. He had already used 'fullness' in the phrase 'in the fullness of time' (Gal. 4: 4). He now adopts the personal dimension of 'Fullness', and integrates it into his own thought. Later in the letter it appears in a formula where reality replaces mystification, 'in Christ the whole fullness of deity dwells bodily' (Col. 2: 9). By using the explicative genitive, 'of deity', Paul demonstrates that he understood correctly the role of the term in the hymn and removes any possible ambiguity. By the introduction of 'bodily' he directs the readers attention to the physical existence of him who is now the Risen Lord (cf. Col. 1: 22). Paul's concern is to block any tendency to disassociate Jesus and Christ; 'you received Christ (as) Jesus the Lord' (Col. 2: 6).[53] As in the original hymn, the presence of the verb 'to

---

[51] For other references, see Aletti (1993), 110 n. 81.

[52] This is the answer to those who object that, if Paul did not agree with part of the contents of the original hymn, he would have omitted the ideas which did not harmonize with his own theology, e.g. O'Brien (1982), 56. [53] See in particular Lightfoot (1904), 110, 174.

dwell' in 2: 9 means that Paul is thinking in terms of his Jewish formation,[54] and the statement cannot be read as if it were the Pauline version of the prologue to John's gospel.[55]

In his redactional additions Paul's concerns are both negative and positive. He has to reduce the spirit world to its proper proportions and to replace Christ in his essential role.

By inserting 1: 16b–e, 20c Paul restricts the meaning of 'all things' to angelic and human beings. The prominence given to the angelic powers by listing their names is striking,[56] and must be understood in the light of the reference to 'the worship of angels' (2: 18). The meaning of this cryptic phrase is disputed, but it seems most probable to understand it as Paul's way of asserting that certain Colossians were being encouraged to give too much importance to visions of the throne of God surrounded by adoring angels.[57] From Paul's perspective such an invitation into a totally unattainable world compromised the primacy and centrality of Christ in the real world. Paul cleverly turned the tables on the teachers at Colossae, by using the creation dimension of their hymn to underline that, as the one responsible for the coming into being of the spirit powers, Christ was infinitely superior to them (Col. 2: 10).

The addition of 'those on earth or in heaven' to the last line of the second strophe (1: 20c) parallels that in 1: 16b ('those in heaven or on earth') but chiastically reverses the order, so that 'those in heaven' occupies the dramatic final place. Paul thereby again uses the original hymn to create a new argument against its writer(s). The need of human beings for reconciliation needs no emphasis, and the point is made several times during the letter (1: 21; 2: 13; 3: 7, 13b). Paul's insertion, however, insinuates that the spiritual powers also need reconciliation.[58] The angelic world, therefore, cannot be viewed uncritically. Wicked angels are unlikely to be satisfactory mediators between humanity and God. But how are terrestrial beings to judge their celestial counterparts? Paul does not need to make explicit the futility of the exercise. His rhetorical training had made him aware that conclusions are more convincing when drawn by the audience.

To the intellectual pleasure of seeing the Colossian teachers hoist with their

---

[54] 'In the last days God will send his compassion on earth and, wherever he finds bowels of mercy, he dwells in him' (*Testament of Zebulun* 8. 2; cf. *Jubilees* 1. 17; *1 Enoch* 49. 2–3).

[55] See particularly Wright (1990), 462–3.

[56] Aletti (1993), 102, notes that the heaping up of terms is a Pauline trait (1 Cor. 15: 24; Rom. 8: 38). These two letters, however, are subsequent to Col. and the technique probably owes its origin to the line of thought into which the crisis at Colossae forced him.

[57] O'Brien (1982), 142–3, and Aletti (1993), 196, against Lohse (1968), 174. The Jewish texts documenting apocalyptic visions, which find their climax in a glimpse of the heavenly throne praised by the angelic hosts, are best assembled and discussed by Sappington (1991), 55–111.

[58] Manifestly Paul is thinking of the fallen angels; 'some of the angels of heaven transgressed the word of the Lord, and behold they commit sin and transgress the law' (*1 Enoch* 106. 13–14; cf. 6. 1–8; 15. 1–12; 69. 3–16; 86. 1–88. 3; *2 Bar.* 56. 11–13; *Sir.* 16: 7).

own petard, Paul adds the satisfaction of directing their attention to the precise modality of Christ's achievement by inserting the phrase, 'making peace by the blood of his cross' (1: 20b).[59] The amendment evokes that of the Philippian hymn, 'even death on a cross' (Phil. 2: 8c), but the formulation is infinitely more dramatic. The graphic imagery will be intensified subsequently by the mention of 'nailing' (2: 14). Paul will never let anyone forget that redemption has been achieved within history through agonizing suffering. His choice of the verb 'to make peace' probably has less to do with any supposed animosity between heavenly beings, or between celestials and terrestrials, than with the tensions within the Colossian church. The letter contains evidence that the false teachers did not have it all their own way; 'let the peace of Christ rule in your hearts' (3: 15). Note also the letter's emphasis on mutual forgiveness (3: 13), and on unity, 'knit together in love' (2: 2), 'put on love which binds everything together in perfect harmony' (3: 14).

The theme of unity also appears in the second part of each of the two-line insertion which Paul places between the strophes of the original hymn (1: 17–18a). The first part of each simply marks the supremacy of Christ. 'All things in him hold together' is another matter. One has only to reflect on how this worked in practice in order to realize that Paul in verse 17 is parodying, not only the tone, but the enigmatic character of his source.[60] This ability to find a verb whose ambivalence fits so perfectly into its context that very astute exegetes have taken it seriously once again reveals the quality of Paul's education. What Paul really wants to convey is clearly expressed in the second line, where the unity of the church is defined as that of a 'body' (v. 18a). This is but the logical extension of the insight of Galatians, 'you are all one in Christ Jesus' (Gal. 3: 28). That Paul is here thinking along the same lines is clear from the extremely close parallel between this verse and Colossians 3: 11, which speaks of the new man 'where there cannot be Greek and Jew, circumcized and un-circumcized, barbarian, Scythian, but Christ is all, and in all'.

The sources of Paul's vision of the church as a 'body' have been long debated. The predominant view that he drew on Greek philosophical reflections on the body politic is also the most implausible.[61] It is psychologically impossible that Paul should have taken over to describe the church a term used to characterize society. The latter appeared to him as riven by divisions (Gal. 3: 28; 5: 19–20), whereas the basic quality of the church was unity rooted in love

---

[59] Barrett (1994), 149, rightly underlines that the Christology of the Colossians was 'deficient in emphasis upon the historic acts of the man Jesus. It was static and ontological rather than eschatological.'

[60] The commentary of Lightfoot (1904), 154, is typical, 'He impresses upon creation that unity and solidarity which makes it a cosmos instead of a chaos.' This is merely to reformulate Paul, not to explain him. Dunn (1980), 191, is much closer to the mark in asking 'is deliberate ambiguity intended?'

[61] Arrianus, *Dissertationes* 2. 5. 24–7; 2. 10. 3–4.

(1 Thess. 4: 9). It is much more likely that he was jolted into thinking of the church as a 'body' by reflecting on the most memorable feature of the temples of Asclepius scattered throughout the eastern Mediterranean, namely, the ceramic representations of parts of the body which had been cured.[62] The recommendation of Vitruvius that such temples be sited only in areas with clean air and pure water made them favourite places of recreation,[63] and there is no reason to think that Paul did not frequent them on occasion. The sight of legs which were not legs, brought Paul to the realization that a leg was truly a leg only when part of a body. Believers, he inferred, were truly 'alive' only when they 'belonged' to Christ as his members (Col. 2: 6, 13; 3: 4). The 'death' of egocentric isolation has been replaced by the 'life' of shared existence.

When prolonged, this same line of thought gives us, 'holding fast to the head, from whom the whole body, nourished and knit together through its joints and ligaments, grows with a growth that is from God' (Col. 2: 19). This use of 'head' in the sense of 'source' is better attested than the alternative meaning 'superior', which is certainly the sense in Colossians 2: 10.[64] The vision of the church as the Body of Christ also appears in later letters.[65] It is not surprising that they do not take up the distinction of 'head' and 'body', which was dictated here by the theological climate of the Colossian church. In neither Corinth nor Rome was the supremacy of Christ questioned.

Paul's insistence that Christ is present in him, and in all members of the Church draws the cosmic dimension of the Christological reflection of the Colossians down into ecclesiology. Paradoxically this point is further under-lined by another specific feature of Colossians, namely, its identification of the gospel (1: 5, 23) as 'the mystery' (1: 26, 27; 2: 2; 4: 3), a theme that will appear later in 1 Corinthians 2: 6–9. This shift was no doubt inspired by the philo-sophical approach to religion that had become fashionable at Colossae (2: 4, 8, 18). There is no intention to exalt the gospel to a level that it did not previously enjoy. What Paul wants to get across is that 'the mystery' is no longer a mystery! The revelation which Jews and Gentiles struggled to find is no longer a secret. It has now been revealed by Christ and in Christ (1: 27; 2: 2; 4: 3). The riches of assured understanding, wisdom, and knowledge are achieved, not by contemplation of a heavenly, spirit-filled dream world, but by reflection on, and commitment to, Christ (2: 2–3). Sometime later Paul will express the same idea by presenting Christ as 'the power of God and the wisdom of God' (1 Cor. 1: 24).

---

[62] Hill (1980).
[63] *De Architectura* 1. 2. 7.
[64] Fitzmyer (1989).
[65] See Ch. 11, 'The Body of Christ'.

ESCHATOLOGY AND ETHICS

Opponents of the authenticity of Colossians give great importance to its realized eschatology, which they claim is incompatible with the futurist eschatology characteristic of the genuine letters. A close reading, however, reveals that future statements predominate in Collosians. Christ is 'the hope of glory' (1: 27) because 'when he appears, you will appear with him in glory' (3: 4; cf. 1 Thess. 4: 17). There will be a final judgement (3: 6, 15–16), at which both the good and the bad will be assessed. The good will be presented holy and blameless, only if they continue in the faith (1: 22; cf. 1 Thess. 3: 13). It is within this clearly defined context that the two statements of realized eschatology— 'you were raised with him' (2: 12); 'if you have been raised with Christ' (3: 1); 'you died and your life is hidden with Christ in God' (3: 3)—must be understood. Manifestly they cannot mean that the Colossians have already been physically raised from the dead. They are simply an alternative, and more vivid, expression of the body theme, 'you who were dead God has made alive together with him' (2: 13; cf. 3: 3). Grace has brought about a fundamental change. For Paul it was imperative to make sure that the Colossians understood that Christ had done everything essential. His plenitude meant that there was nothing that the spirit powers could add.[66]

In 1 Thessalonians 4: 1–12 Paul found it helpful to remind his converts that their life-style as Christians must be radically different from their previous comportment. In Colossians he does the same, but more explicitly, and at greater length (3: 1 to 4: 6). No doubt as his experience as a pastor increased, the generosity of his assumptions regarding human nature diminished. No longer did he take it for granted that his converts would be as whole-hearted as he had been in working out what behaviour was appropriate to life in Christ. In addition the situation at Colossae was complicated by the fact that some members of the church were attracted to excessive ritualism and ascetic rigour (2: 16–23), and looked down on others who did not share their views.

The insistence on Jewish observances—'matters of food and drink, a festival, a new moon, or a sabbath' (2: 16)—reveals that the situation at Colossae was analogous to that which obtained at Galatia, yet Paul's reaction is completely different. Nowhere in Colossians does he evoke faith or the works of the Law. This has led some to disqualify him as the author of Colossians.[67] They should rather have questioned just how similar the two situations were.

In Galatia, as we have seen, the churchs were troubled by intruders insisting

---

[66] Aletti (1993), 208, is entirely correct in writing, 'le glissement des catégories est davantage dû à la logique interne de l'épitre qu'à un changement radical d'eschatologie'. Similarly Sappington (1991), 226–7.
[67] Furnish (1992), 1093.

that they had a mandate from the mother church in Antioch systematically to correct the gospel, which Paul had preached, by imposing full observance of the Law.[68] It was a frontal attack on Paul personally, and on all that he believed. Nothing of the sort occurred in the Lycus valley. The churches there were not founded under the aegis of Antioch. There is not the slightest hint that Paul's authority was questioned. There is no conclusive evidence that the false teachers were fellow-Christians, or that they were active proselytizers. Some believers may have been attracted to a form of esoteric Jewish teaching which circulated at Colossae and wanted to share their new insights with others. They did not reduce Christ to irrelevance as did the intruders in Galatia, but rather exalted his mediatory role. The problem with which Paul had to deal was not a doctrinaire attitude towards the Law, but the ascetic-mystical piety of Jewish apocalypticism,[69] whose roots were more emotional than theological. In Galatia Paul had to counter a really serious threat, which was being pushed home as a matter of principle. At Colossae the issue was a fashionable fad, whose followers sought 'heavenly ascents by means of various ascetic practices involving abstinence from eating and drinking, as well as careful observance of the Jewish festivals. These experiences of heavenly ascent climaxed in a vision of the throne and in worship offered by the angelic hosts surrounding it.'[70] Jewish observances were important, not in themselves, but as the means to an end.

   Given such differences, it would have been surprising to find Paul using at Colossae the tactics which had suited the situation in Galatia. There he had to demolish a thoroughly worked out vision of Christianity, whose coherent arguments were rooted in revelation. His opponents at Colossae, on the contrary, had no such intellectual depth. They described mysteries, apocalyptic visions, whose reality no one could verify. In contrast to the well-rooted epresentatives of Antioch, they floated in a fantasy world. Paul's concern was to restore a sense of reality, to set the feet of the misguided on solid ground. They grasped at shadows; he had to show them that Christ was substance (2: 17). The most effective tactic was not to challenge the mystics head on, but to consistently introduce discreet modifications, whose cumulative impact would subvert their teaching completely. He relied on the sobering effect of the calm assumption of authority.

THE HOUSEHOLD CODE

It is in this perspective that we must approach Paul's use of a household code (3: 18 to 4: 1) which, if he were true to himself, he should never have employed.

---

[68] See Ch. 8, 'The Cause of the Crisis'.
[69] This has been most thoroughly argued by Sappington (1991), 150–70.
[70] Ibid. 170.

Nothing similar appears in any previous or subsequent letter of his. This series of paired injunctions (wives–husbands / children–parents / slaves–masters), not only represented the conventional morality of society, a social grouping that for Paul was the antithesis of the church, but it flatly contradicts the structure of the church 'where there cannot be Greek and Jew, circumcised and uncircumcised, slave and freeman' (3: 11). The use of the code here makes sense only as an invitation to the Colossians to leave the mystical realm of the angels and to return to the real world, where the fabric of daily life was woven from a multitude of interpersonal relations, of which the most basic were the three pairs listed here.

What is said to slaves stands out from the others both quantitatively and qualitatively (Col. 3: 22–5). Inevitably commentators have seen a relationship to the situation of Onesimus, who was returning to confront his injured master (Col. 4: 9). In that case, however, one would have expected either a warning to slaves not to imitate the dishonesty of Onesimus, or an expanded monition to owners on how slaves should be treated. Neither appears here. It might be thought that the directive was made necessary by agitation among Christian slaves at Colossae to be given the equality they theoretically enjoyed as members of the same Body, but the formulation militates against this interpretation.[71]

We are forced to conclude that Colossians 3: 22–25 reflects Paul's habitual attitude towards slaves who accepted Christianity. Contrary to what one might have expected, he was not concerned with their liberation. Within the community he took it for granted that they would show and share the love that was its most distinctive feature, but there is no hint that he did anything to change the social order. This is well illustrated by the case of Onesimus. Paul's request was that he should not be treated 'as a slave' when he returned to Colossae. There is no demand that he be manumitted (Philem. 16–17).[72] When sometime later Paul was forced to confront the issue of slavery by the Corinthians, who believed that their relationship with God could be improved by a change in social status, he responded (a) that no change should be initiated for the sake of principle; (b) that a social change could be initiated to compensate for a human weakness; and (c) that a social change initiated by factors outside one's control could be accepted (1 Cor. 7: 17–24).

In Colossians 3: 22–5 all Paul's attention is focused on the authenticity of a slave's life. Of the six relationships dealt with in the household code, only two were likely to be marked by deception. In many ways the position of children was parallel to that of slaves; they did not control their lives. The former, however, are minors, and it is a father's duty not to create the conditions for deception. Even if ill-treated, slaves are adults, and responsible for their

---

[71] So rightly Aletti (1993), 254–5.
[72] Against Bartchy (1992), 71.

attitudes. What Paul does not want them to do is to obey orders to the letter, while the heart raged, and hate corroded the spirit. The reason behind this position can be deduced from Colossians 4: 5–6, where Paul stresses the witness value of the comportment of Christians (cf. 1 Thess. 4: 12). The internal tension, which was the occupational disease of slavery, had to be resolved in order to permit the transforming effect of grace become visible.[73]

DID PAUL VISIT COLOSSAE?

While in prison in Ephesus Paul planned two visits as soon as he was released, one to Philippi (Phil. 1: 26; 2: 24) and the other to Colossae (Philem. 22). He did not make the visit to Philippi. It was still on his agenda in the early summer of AD 54 when he wrote 1 Corinthians 16: 5. Was the visit to Colossae also aborted?

Paul was probably released in the latter part of the summer of AD 53. We have seen that the likely reason why he did not go to Philippi was the situation in the church at Ephesus, where there was significant opposition to his leadership.[74] The round trip would have taken the minimum of a month[75] and brought him dangerously close to the moment when normal sea travel ceased. If ships no longer sailed from Neapolis, he would be trapped in Macedonia for the winter, with unacceptable consequences. A prolonged absence might guarantee the success of a different vision of Christianity at Ephesus. Moreover, to leave the city at that crucial moment might be interpreted as the flight of a coward.

These reflections do not militate with the same force against a visit to Colossae. Paul's reason for going to Philippi was essentially for the pleasure of seeing believers who had always been loyal and co-operative. It was intended to refresh his spirit after a tense time under investigation. Apart from the tension generated by the personal competition of Euodia and Syntyche (Phil. 4: 2), there were no problems that imperatively demanded his presence. It would not have been difficult for Paul to rationalize his failure to keep his promise to the Philippians as the repudiation of a selfish decision made in a moment of weakness.

The planned visit to Colossae could not be avoided so easily. An important doctrinal point was at stake, and, Paul had accepted responsibility for the work of his agent, Epaphras, by writing letters to the Colossians and to the Laodiceans. He needed to know whether the way he had dealt with the false teachers had been successful. If not, it was imperative to make a further effort. Paul, however, did not have to go to Colossae himself. There were at least two other sources of information available. Tychicus was a permanent member of

---

[73] See Crouch (1972), 160.
[74] See Ch. 9, 'Opposition at Ephesus'
[75] See Ch. 7, 'An Ephesian Chronology'.

Paul's entourage (2 Tim. 4: 12). Although it is not stated, it must be presumed Paul expected him to return to Ephesus with a report on the situation at Colossae.[76] Epaphras would certainly have returned to Colossae the moment he was released, and could be back in Ephesus within two weeks, if the situation warranted it.

Paul must have wondered whether he should rely on second-hand information or go and see for himself? A decisive factor in his internal debate was his conviction of the autonomy of the local community. Whatever needed to be done as the result of his letter would have to be accomplished from within, as a member of the community. In many other churches this would mean no more than slipping back into the niche which he had occupied for a year or more. At Colossae he knew people only by reputation. He had met none of them personally (Col. 2: 1). To know them, and to be known by them, would take time. The more he thought about it, the more Paul became convinced that, if he went to the Lycus valley, he would be obliged to spend the winter there.

Could he afford so much time away from Ephesus? On balance Paul thought not. His personal position there was in danger, and it was the place which he had selected as his base for contacts with other churches. He had just had to deal with a problem in Corinth via the Previous Letter (1 Cor. 5: 9). Moreover, he had not formally promised the Colossians that he would visit them. Nothing about a visit is mentioned in Colossians and he did not plan to stay with Epaphras, the leader of the community, with whom he had shared a prison cell. The request for a guest-room was addressed to Philemon (Philem. 22), and could be considered a purely private matter. It would be more prudent, Paul decided, to wait for the reports of Tychicus and Epaphras.

What news they brought him we shall never know. From the following spring Paul was completely absorbed by the problems of the church in Corinth, and left Colossae to the care of Epaphras.

---

[76] Some manuscripts of Col. 4: 8 make this explicit by transforming 'in order that you may know the things concerning us' into 'in order that I may know the things concerning you'; see Lightfoot (1904), 233, 253.

— 11 —

Confusion at Corinth

AFTER leaving Philippi, Paul went to Thessalonica, and thence via Athens to Corinth, where he arrived sometime in the early spring of AD 50. The choice of Corinth as a missionary centre was well motivated,[1] and the community he founded there reflected the best features of the dynamic trading city. Its members were committed and enthusiastic, and did not hesitate to accept the responsibility of working out what Christianity meant for them. But they proved to be the most exasperating church with which Paul had to deal. The imprecision of his preaching exacerbated a positive genius on their part for mis-understanding him. Virtually every statement he made took root in their minds in a slightly distorted form, and from this defective seed flowered bizarre approaches to different aspects of the Christian life.

Paul, in consequence, found himself obliged to think much more deeply about a whole array of issues. More importantly for our purposes, the pressures of an extremely turbulent relationship forced to the surface aspects of his per-sonality which are not perceptible elsewhere. The intense emotion which imbued all his dealings with the Corinthians acts as a prism through which facets of his character are refracted in vivid colours. Although Galatians is ostensibly the most autobiographical letter, it remains very much on the surface of things. The external events listed in Galatians 1–2 disclose virtually nothing of the complex nature of the Apostle. The Corinthian correspondence is much more self-revelatory. In it the Apostle unwittingly lays bare his soul.

LETTERS TO CORINTH

The intensity of Paul's relationship with the Corinthians is illustrated by the fact that he wrote more letters to them than to any other church. The New Testament contains only two letters, but these mention two others, the Previ-ous Letter (1 Cor. 5: 9) and the Painful Letter (2 Cor. 2: 4). Hence, four in all. From the end of the eighteenth century, however, doubts have been raised

---

[1] See above, Ch. 5, 'The Move to Corinth'.

regarding the integrity of both 1 and 2 Corinthians.[2] The division of 2 Corinthians into two originally independent letters was postulated in 1776. It took a hundred years for the integrity of 1 Corinthians to be called into question. From that moment hypotheses became ever more complex as fragments from one letter were associated with those from another. This trend in New Testament research reached its climax with the thesis that originally the Corinthian correspondence consisted of nine distinct letters.[3]

It is easy to mock the arbitrariness of such theories. The exaggerations of some practitioners, however, do not invalidate the method. Partition theories are never developed for their own sake; they are designed to account for observations that are made in good faith, and so they deserve to be taken seriously, even if ultimately they do not command assent. The litmus test is: are the internal tensions so great as to destroy the methodological assumption of literary unity? Answers will vary because complete objectivity is impossible; every literary judgement necessarily embodies a subjective element. A detailed evaluation of all the proposals would take us too far afield. None the less, something must be said because the number and order of the letters is obviously fundamental to any reconstruction of Paul's relations with Corinth.

1 Corinthians

1 Corinthians is relatively easy to deal with because the principle on which the partition theories are based can be discerned without difficulty. The salient feature of 1 Corinthians is the absence of any detectable logic in the arrangement of its contents. In some minds this produced an impression of disorder, which was explained by postulating an inept conflation of a number of letters. The key used by such scholars to unlock the secret of the original documents has been the various sources of information which Paul had available to him, namely, Chloe's people (1 Cor. 1: 11), the letter sent by the Corinthians (1 Cor. 7: 1), and the delegation comprising Stephanas, Fortunatus, and Achaicus (1 Cor. 16: 17). Different parts of 1 Corinthians are combined to create what are considered to be appropriate and internally consistent responses to each one taken singly or in combination.[4]

Once the underlying assumptions are brought out clearly, the fundamental flaw of this methodology becomes apparent. It assumes knowledge superior to that of the author by dictating what he should have said. The resultant letters owe more to the aesthetic sense of the scholar than to any objective factors.

---

[2] A survey of the history of research into the literary unity of the Corinthian letters is provided by Betz (1985), 3– 36.

[3] Schmithals (1973). For a survey of the various modern partition theories, see Sellin (1987), 2965–8.

[4] In addition to Schmithals, see the much more sober Héring (1959), 10–12; Schenke (1969); Senft (1979), 17–25.

Moreover, as G. D. Fee has rightly pointed out, all the so-called internal con-tradictions in 1 Corinthians can be resolved by a more exacting exegesis.[5] Finally, no satisfactory explanation is ever provided for the procedure, and no justification for the intention of the redactor(s), who gave 1 Corinthians its present shape. It is not surprising, therefore, that the major commentators have been firmly in favour of the unity of 1 Corinthians.[6]

2 Corinthians

Those who maintain the unity of 2 Corinthians are much fewer than those who insist on the integrity of 1 Corinthians. The reason is the radical break between chs. 1–9 and chs. 10–13. It is impossible that Paul should have followed his cele-bration of his reconciliation with the church of Corinth (chs. 1–9) by a torrent of reproaches and sarcastic self-vindication (chs. 10–13).[7] Even the most committed defenders of the unity of 2 Corinthians have to recognize the force of such observations. How they deal with them is another matter, and the approach of E.-B. Allo is typical.[8] Sometime after finishing chs. 1–9, he claims, Paul received news that intruders at Corinth had spread rumours that he was appropriating funds for Jerusalem for his personal use, and that the community there had not come to his defence. In reaction, we are told, he wrote the blister-ing attack in chs. 10–13 and attached it to chs. 1–9.

This hypothesis redefines the concept of literary unity in such a way as to make it meaningless. If chs. 10–13 werr written after a certain interval, and motivated by a concern other than that animating chs. 1–9, it is a separate letter by normal standards. Were chs. 1–9 still in Paul's possession, when the informa-tion arrived, the anger bubbling to the surface in chs. 10–13 makes it more likely that he would have torn up chs. 1–9, and sent only chs. 10–13. Allo's reconstruction is psychologically impossible.[9]

The view that chs. 1–7 exhibit a complete rhetorical argument, and that, in consequence, chs. 8–9 are superfluous and must be treated as a later addition, since they are too long to be a postscript,[10] is contradicted by a letter of Cicero to Atticus (*Att.* 12. 28–9), in which the postscript is proportionately much longer than 2 Corinthians 8–9 relative to chs. 1–7.[11] Rhetorical criticism is also

---

[5] (1987), 15–16. For greater detail, see Merklein (1984).
[6] e.g. Barrett (1968), 15; Conzelmann (1975), 3–4; Merklein (1992), 46–8.
[7] The best description of the difference remains that of Plummer (1915), pp. xxix–xxx.
[8] (1956b), 267–8.
[9] Other defences are even more unsatisfactory. Bates (1965) treats chs. 10–13 as a 'recapitula-tion' of chs. 1–9, a hypothesis which is not justified by the content and which fails to explain the change of tone. Hyldahl (1973) forces the exegesis of key texts in an attempt to prove that there was no time for chs. 10–13 to be written except as the continuation of chs. 1–9.
[10] Kennedy (1984), 92.
[11] A rigid application of their own rhetorical rules is forcefully opposed by the *Rhetorica at Herennium* 3. 16, and Quintillian, *Institutio Oratoria* 2. 13. 1–7.

invoked to prove that 2 Corinthians 8 and 9 were originally independent letters by claiming that each exhibits the rhetorical schema.[12] This is not in fact the case.[13]

The overlap between the 2 Corinthians 8 and 9 does, however, require an explanation. The shift into the first-person singular at the beginning of ch. 9, when coupled with the parallel in structure with Galatians 6: 11–18, where a personal appeal also shades into a profound theological argument, identifies ch. 9 as a personal postscript authenticating the letter. This was as much part of Paul's epistolary technique as it was of that of his contemporaries.[14]

If 2 Corinthians 1–9 and 2 Corinthians 10–13 are two letters, which was written first? A significant number of scholars follow A. Hausrath, who in 1870 argued that chs. 10–13 should be identified with the Painful Letter mentioned in 2 Corinthians 2: 4. The most able defence of this hypothesis is that of F. Watson,[15] but his arguments ultimately fail to carry conviction.[16]

2 Corinthians 10–13 was occasioned by an attack on Paul's apostolic authority by Judaizing intruders, a subject which is never evoked in what Paul says of the Painful Letter. Equally, the issue which gave rise to the Painful Letter was an insult to Paul by an individual; this is not even hinted at in chs. 10–13. Finally, 2 Corinthians 10–13 was written in preparation for a visit to Corinth in the immediate future, Paul's third (12: 14; 13: 1–2), whereas the Painful Letter was written as a substitute for a visit, which Paul had promised on his second visit but then refused to make (2 Cor. 2: 1–4).

The other missing letter is that mentioned in 1 Corinthians 5: 9. Héring identified 2 Corinthians 6: 14 to 7: 1 as part of this epistle,[17] and he has been followed by others, who expand it with material drawn from 1 Corinthians.[18] This thesis has been facilitated by the impression that this block of material breaks the connection between 2 Corinthians 6: 13 and 7: 2. Many in fact treat it as a post-Pauline interpolation of Essene inspiration.[19] In fact, however, the so-called Essene linguistic parallels are much inferior to those drawn from the language of the Diaspora synagogue which is best represented by Philo. When understood in this sense, 6: 14 to 7: 1 fits perfectly into the argument of 2 Corinthians 1–9.[20]

To draw together the strands of this discussion, Paul wrote five letters to Corinth: (1) the lost Previous Letter; (2) 1 Corinthians; (3) the lost Painful Letter; (4) 2 Corinthians 1–9; and (5) 2 Corinthians 10–13. The relative order of nos. 1 and 2 has been established, as has that of nos. 3–5. The one question remaining open is the relationship of 1 Corinthians to letters 3–5.

[12] Betz (1985), pp. vii–viii.
[13] Stowers (1987), 727–30.
[14] Richards (1991), 80–90, 176–82.
[15] (1984), 324–46.
[16] For a more detailed response, see my (1991b).
[17] (1959), 10–12.
[18] Schmithals (1973), 276; Hurd (1965), 238.
[19] e.g. most recently Klauck, (1986), 60–1.
[20] See my (1988a).

If we abstract from theories which completely dismember 1 and 2 Corinthians, no one (to the best of my knowledge) has suggested that 1 Corinthians should be dated after either of the component elements of 2 Corinthians The principal argument for the traditional arrangement is the references to the collection for the poor of Jerusalem. In 1 Corinthians 16: 1–4, in response to a request from the Corinthians, Paul gives practical directives as to how the money at Corinth should be assembled. From 2 Corinthians 8–9, however, it is clear nothing had in fact been done even though a year had elapsed; their commitment had to be revived (9: 2).[21]

Moreover, 1 Corinthians gives the impression that it is Paul's first contact with Corinth since his founding visit. Many problems have accumulated. All his information is second-hand. And he has had to send Timothy, who is not listed in the address of the letter (1: 1), to provide an independent assessment of the situation there (4: 17; 16: 10–11). From 2 Corinthians 1–9, on the contrary, we learn that Timothy is again with Paul (1: 1), and that the latter has been recently in Corinth, a visit whose bitter consequences totally exclude its identification with the founding visit (2: 1–4). Moreover, Paul's planned visit to Macedonia (1 Cor. 16: 5) from Ephesus was actually made via Corinth (2 Cor. 1: 16).

THE ROAD FROM ATHENS TO THE ISTHMUS

The reasons for Paul's move from Athens to Corinth have been discussed already in the context of his correspondence with Thessalonica.[22] Now we must look at his first visit there in some detail.

In order to get from Athens to Corinth, Paul and his companions, Timothy and Silas (2 Cor. 1: 19) had two options. They could go by land or by sea. Ships plied regularly between Piraeus and Schoenus or Cenchreae via the island of Salamis. Delays, however, were frequent. Bad weather, adverse winds, or simply bad omens impeded departure. The sea route, therefore, could take as long as the two-day land route via Megara.[23] Since Paul's contemporaries undertook a sea voyage solely when it was the only way to get from one place to another, or when it offered immense savings of time and energy,[24] there is little doubt that the Apostle opted for the overland route.

---

[21] e.g. Lüdemann (1984), 87; Wolff (1989), 3.

[22] See Ch. 5, 'The Move to Corinth'.

[23] 'Diognetus by spending the night near Megara could very easily be in Athens on the following day. Or else, if he preferred, at Eleusis. Otherwise he could take a shorter way through Salamis, without passing through any deserts' (Dio Chrysostom, *Discourses*, 6. 6; trans. Cohoon).

[24] Everyone was conscious of the unpredictable dangers of the sea. Horace evoked the invention of the boat with the words, 'Oak and triple bronze were about his breast who first committed his fragile boat to the surly sea' and so 'impious boats traverse the sounds that ought to remain

When Paul set out from Athens he had a walk of some 80 km. (50 miles) ahead of him, a route rich in religious associations, but whose danger was underlined by the epic deeds of Theseus. At every step of the way there was something to remind the Apostle of the religious and political history of Greece.[25] For the first 22 km. (14 miles) his path followed the Sacred Way to Eleusis, along which passed the great procession each autumn to honour Demeter. Once he reached the shore of the Eleusinian Gulf he had on his right the salt-water fish-ponds sacred to the Maid and Demeter, and the Rharian meadow, the first place ever sown or cropped, according to Greek legend. To balance such tranquil scenery memory carried the story of Procrustes, a brigand who made his victims fit his bed by racking the short and amputating the long. One may doubt that Paul wasted time admiring the great sanctuary at Eleusis. His concern must have been to reach Megara, 19 km. (12 miles) further on, before nightfall. It would have been an exceptionally long day's walk.

That night Paul's thoughts would have been concentrated, not on the problems that would face him on arrival at Corinth, but on the first part of his journey next day, which presented a more immediate threat. The 8 km. (5 miles) section of the road known as the Sceironian Rocks is described by Strabo:

> They leave no room for a road along the sea, but the road from the Isthmus to Megara and Attica passes above them. However, the road approaches so close to the rocks that it many places it passes along the edge of precipices, because the mountain situated above them is both lofty and impracticable for roads. (*Geography* 9. 1. 4; trans. Jones)

According to Pausanias the track was hewn for active men and was enlarged into a real road only by Hadrian.[26] Its association with bandits did not end when Theseus threw Sceiron into the sea. He had been a robber who forced travellers to wash his feet and, as they finished, kicked them over the cliff.[27] In Paul's day it was a place in which he certainly experienced 'danger from robbers ... dangers in the wilderness' (2 Cor. 11: 26).

Once Paul reached Crommyon (today Hagios Theodoros) he was in Corinthian territory, in an area associated with two further exploits of Theseus, the slaughter of the sow Phaea and the execution of the bandit Sinis by the technique he himself used to dispose of his victims; he was torn in two when the bent pine trees to which his legs were tied were allowed to spring up.[28]

---

unstained' (*Odes* 1. 3. 9 and 23; trans. Shepherd). The ship, in other words, was conceived by a sadistic degenerate committed to the destruction of humanity.

[25] Two ancient descriptions are extant. Pausanius (*Description of Greece* 1. 36. 3 to 44. 10) goes from Athens to Corinth, as did Paul. Strabo travels in the reverse direction (*Geography* 9. 1. 1–16).

[26] *Description of Greece* 1. 44. 6.

[27] Ibid. 1. 44. 8; Strabo, *Geography* 9. 1. 4.

[28] Pausanias, *Description of Greece* 2. 1. 3.

At Schoenus Paul had his first experience of the dynamism of Corinth.[29] If the isthmus was a land-bridge, permitting trade to flow easily between the Peloponnese and the Greek mainland, it was a barrier to east–west shipping, and mariners needed an alternative to the long route around the Peloponnese.[30] As early as the sixth century BC, the Corinthians thought of cutting a canal.[31] Like later plans, this project came to nothing,[32] and an ingenious, provisional solution remained in place for 1,300 years. The Corinthians laid a paved road, the *diolkos*, to join the Corinthian and Saronic gulfs. About 400 metres of the road have been excavated on the west side of the Isthmus. The width varies from 3.4 to 6 metres. Grooves cut in the paving 1.5 metres apart guided the wheels of the wooden platform (the *holkos*), on which small boats and goods were hauled across the isthmus. Earthen tracks on either side were used by pack animals.[33]

The jostling, shouting multitude of labourers along this road, through whom Paul had to push his way, would have been his first concrete perception of what life at Corinth was going to be like. Thus far he had encountered nothing similar. The provincial towns of Asia and Macedonia were, by comparison, sleepy oases of leisure, in which his mission would have been an agreeable distraction. Corinth had more business than it could comfortably handle. The immense volume of trade was augmented by huge numbers of travellers. Profit came easily to those prepared to work hard, and cut-throat competition ensured that only the committed survived.[34] Would people so busy and preoccupied, so eager in their pursuit of gain, have any time to listen to his message? The obstacles appeared greater than those he encountered at Athens, where at least he was given a hearing (Acts 17: 16–34).

Once through the crowds, Paul found himself in a different world. The road led to the sanctuary of Poseidon at Isthmia with its temple, theatre, and stadium.[35] Paul cannot have been unaware that this was the scene of the Isthmian Games, one of the four great pan-Hellenic festivals, which was cele-

---

[29] On Corinth, see my (1992*b*) and (1992*e*), and Engels (1990).

[30] The perils of the 8 km. (5 miles) wide channel between Cape Malea, the south-eastern tip of Greece, and the island of Cythera were emphasized by Homer (*Odyssey* 9. 80; cf. 3. 286; 4. 514; 19. 186). Thereafter, it became part of the literary tradition, which it was in the interest of Corinth to reinforce by repetition of the proverb, 'When you double Cape Malea forget your home' (Strabo, *Geography* 8. 6. 20). The strong element of hyperbole is revealed by the mention on the tombstone of Flavius Zeuxis of Hierapolis in Asia Minor that 'as a merchant he rounded Cape Malea 72 times on voyages to Italy' (*CIG* 3920). The channel was clear of rocks and all the difficulties were caused by the 50% chance of contrary winds, which meant delays that merchants could not afford.

[31] Diogenes Laertius (1. 99) gives the credit to Periander (625–585 BC).

[32] Pliny, *NH* 4. 9–11.

[33] Wiseman (1978), 45.

[34] This is the original sense of the proverb, 'Not for everyone is the voyage to Corinth' (Horace, *Epistles* 1. 17. 36), although both Strabo (*Geography* 8. 6. 20) and Gellius (*Attic Nights* 1. 8. 4) give it an exclusively sexual connotation.

[35] Pausanias, *Description of Greece* 2. 1. 7 to 2. 2.

brated every two years in the late spring. Perhaps there were still traces of the games of AD 49 celebrated some nine months earlier. He was too mature for his blood to stir at the thought of the dry celery crowns awarded the victors.[36] If anything, he thought of the unity that the festival achieved among Greeks from all over the known world. As he marvelled at what the games did for Greek identity, did he pray that the members of his far-flung churches would feel united by a bond equally vivid and secure?

Such spiritual thoughts would have been complemented by the grateful realization that he should have little difficulty finding work in Corinth. During the week of the festival visitors thronged the area and all Corinth went out to serve and to celebrate with them.[37] The former needed tents in which to stay, and the latter brought booths in which to display their wares. A good tent-maker would find plenty to do. Repairs were as necessary as the manufacture of new tents, and the next Isthmian Games were only fifteen months away. The lift of the heart caused by the solution to one of his many problems carried him easily up the final 10 km. (6 miles) to the city (see Fig. 5).

THE NARRATIVE OF ACTS

Paul's sojourn in Corinth is recounted in some detail in Acts 18: 1–18. Unfortunately, this account cannot be accepted at face value. Not only do we have the problem of two textual traditions (the Western and the Alexandrian), but the narrative abounds in hints of redactional activity.[38] Luke's account of Paul's ministry at Corinth is in fact a many-layered text.[39]

The most primitive story[40] narrated only an abortive attempt to convert Jews, after which Paul was consoled by a vision of Christ, whose efficacious protection was immediately demonstrated by the refusal of Gallio to hear the charge laid against Paul by the Jews. I see no reason to refuse the historicity of the events narrated in this document.[41] Manifestly, however, it is not a complete account of Paul's founding visit to Corinth. That much has been omitted is indicated by the fact that Paul both stayed a considerable time and made many converts (18: 18).

---

[36] Dio Chrysostom, *Discourses* 8. 15; Plutarch, *Quaestiones conviviales* 5. 3. 1–3 (675D–677B); on which, see Broneer (1962).

[37] A vivid description is given by Dio Chrysostom, *Discourses* 8. 6–10.

[38] Boismard and Lamouille (1990), 3. 228–32. Their arguments are of a completely different order from those put forward by Lüdemann (1984), 157–62.

[39] For the details see Boismard and Lamouille (1990), 2. 247–9, 300–3, 366.

[40] Consisting of verses 1, 4a, 5b, 6, 9, 10, 12–14a, 15b, 16–18.

[41] The chronological implications of Paul's encounter with Gallio have been worked out in Ch. 1, 'Paul's Encounter with Gallio'.

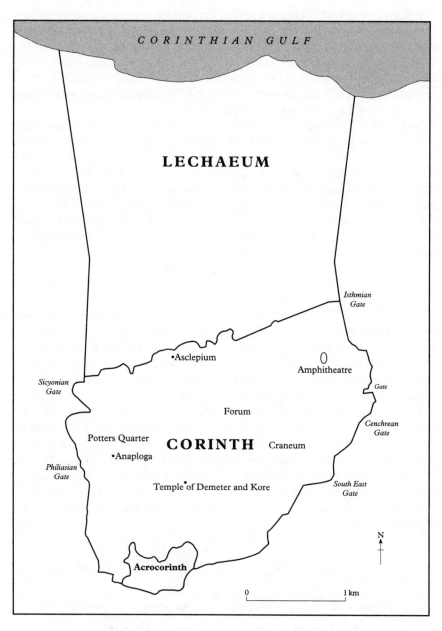

FIG. 5 Corinth: The Walls (*Source*: David Gilman Romano, *The Corinthia in the Roman Period*, JRA Sup. 8, ed. T. E. Gregory (Ann Arbor, 1993))

Working with Prisca and Aquila

Redactional additions fill out the story. The first introduces Prisca and Aquila as Paul's hosts (18: 2–3). This should be accepted as historical fact because, as we have seen, 1 Corinthians 16: 19 demonstrates that they had been in Corinth and that they were particularly close to Paul.[42] According to the Western text, however, the reason why Paul stayed with them was because he had previously known Aquila, since they belonged to the same tribe. This is corrected in the Alexandrian text, which presents Paul as joining them because they were of the same trade.

The former is suspiciously like a deduction from a coincidence. Both were of the tribe of Benjamin (or simply Jews; cf. Acts 10: 28), hence they must have known each other! Where they might have made each other's acquaintance is conveniently passed over in silence. As far as our sources go, their paths had never crossed. Recognition of this difficulty explains the Alexandrian version; 'he came to them because he was of the same trade and he remained with them and he/they worked, for they were tentmakers by trade' (18: 3).

The last clause apparently has only Prisca and Aquila in view; its function is to explain 'same trade'. J. J. Taylor argues that logic dictates that the same couple are the subject in 'they worked'.[43] In which case there would be no mention of Paul's manual labour; 'he lodged with them' but 'they worked'. The continuation of the narrative makes it certain that the redactor did not intend this contrast. The money brought by Silas and Timothy, the editor lets us understand, made it possible for Paul *to give up manual labour* and to give himself full time to the ministry of the word (18: 5); previously he had preached only on the sabbath (18: 4).[44] The scribe who transformed the plural into 'he worked' realized the necessity of avoiding a false impression. Common sense militates against strict grammatical logic in such constructions. Paul worked side by side with Prisca and her husband as a tent-maker.

Can any confidence be placed in this information, which entered the narrative so late? What is secondary from a literary point of view does not necessarily imply fabrication. In this case invention can be excluded. Not only did the trade of tent-making have no symbolic connotations, but the occupation of Paul, Aquila and Prisca must have been well known in Greece, Asia and Italy (Rom. 16: 3). Falsification would have brought ridicule. Silence is another matter. It is entirely possible that the Western text was motivated by a concern to raise Paul's social status by refusing to acknowledge the manual labour by which the Apostle himself claimed to earn his living (1 Thess. 2: 9; 2 Thess. 3: 8; 1 Cor. 4: 12).[45]

---

[42] See Ch. 7, 'The Founding of the Church'. [43] (1994b), 315 n. 3.
[44] So Haenchen (1971), 534, 539. [45] See Ch. 4, 'Learning a Trade'.

Silas and Timothy

The second redactional addition brings Silas and Timothy on the scene (18: 5). The consequence for the story-line is that Paul is presented as having ministered alone in Corinth for some time before being joined by his companions. From a historical point of view this most implausible. As I have argued above, the anxiety displayed in Paul's first letter to the Thessalonians guarantees that he remained in Athens until Timothy's return from Macedonia.[46] It is inconceivable that Paul would have made himself difficult to find by moving to another city, particularly one as large as Corinth. If anything he would have moved north to anticipate Timothy's arrival, as he later did in the case of Titus (2 Cor. 2: 12–13). Silvanus and Timothy were Paul's companions when he arrived in Corinth and participated in the evangelization of Corinth from the very beginning (2 Cor. 1: 19).

This conclusion calls into question two elements in the revised narrative, namely, the influx of funds which made it unnecessary for Paul to work, and so facilitated his move to a more up-market address (18: 5, 7). Timothy's mission to Macedonia had only one objective, namely, to reinforce the faith of the Thessalonians and to report back to Paul as quickly as possible (1 Thess. 3: 1–10). It was not the moment to solicit financial support, even if the Thessalonians had surplus funds, something which is far from certain (cf. Phil. 4: 15). While working in Thessalonica Paul had to be aided by the Philippians (Phil. 4: 16). Moreover, it is highly improbable that Timothy exceeded his mandate by making a visit to Philippi from Thessalonica.[47] The five-day journey each way, and a stay of several days, would have increased his already long absence from Athens by nearly three weeks. It is not impossible that some Philippians arrived in Thessalonica with a subsidy for Paul while Timothy happened to be there, but history has nothing to do with such wishful coincidences. Nothing suggests that, when the three missionaries arrived in Corinth, they had full pockets. They needed to find work, and quickly.

A Change of Location

As regards Paul's move (18: 7), the first redactor (in the Western text) says, 'Moving from Aquila's house he departed to the house of Justus', whereas his successor (in the Alexandrian text) notes, 'moving from there [the synagogue] he came to the house of a man called Titius Justus'. The former implies a change of residence, whereas the latter means only that he taught at a different location.

---

[46] See Ch. 5, 'Letter A'.
[47] As is assumed by Lüdemann (1984), 176.

The Western text is suspect because, from a literary point of view, it is a rather blatant symbol of a change in Paul's policy. Not only does he no longer preach to Jews but he does not even live among them!

In reality, it must be assumed that Prisca and Aquila were Christians, whom Paul would never have abandoned for the sake of a symbolic gesture. This couple were ex-slaves of Jewish origin, who had decided to leave Italy in AD 41, after the emperor Claudius had closed down a Roman synagogue as a result of continuous turmoil centring on the figure of Christ.[48] Jewish refugees who had lost everything through the machinations of Christian missionaries in Rome would hardly have given work and shelter in Corinth to the same sort of missionary, namely Paul.[49] Finally, it is certain that they were not converted by Paul in Greece (1 Cor. 16: 15).

If, as seems most likely, Prisca and Aquila were believers, the Alexandrian text can be seen as the correction of the Western text in the interests of historical probability.[50] As Paul's ministry expanded, particularly among God-fearers, the enmity of the Jews increased, and it became progressively impossible to preach in the synagogue; he had only to open his mouth to be shouted down.

The sort of little shops which artisans such as Prisca and Aquila occupied were scattered all over the city. They lined busy streets and were concentrated in specially built commercial developments. The Peribolos of Apollo just off the Lechaeum Road was the oldest such market in the city. Shortly before Paul arrived, the North Market was completed, and its arrangement is so typical as to serve as a valid illustration of the conditions under which Paul lived and worked in Corinth and later in Ephesus.[51]

The shops gave on to a wide, covered gallery running round all four sides of the square. They had a uniform height and depth of 4 m. (13 feet). The width varied from 2.8 m. (8 feet) to 4 m. (13 feet). There was no running water or toilet facilities. In one of the back corners, a series of steps in stone or brick was continued by a wooden ladder to a loft lit by an unglazed window centred above the shop entrance, which at night was closed by wooden shutters. Prisca and Aquila had their home in the loft, while Paul slept below amid the tool-strewn work-benches and the rolls of leather and canvas.

The workshop was perfect for initial contacts, particularly with women. While Paul worked on a cloak, or sandal, or belt, he had the opportunity for conversation which quickly became instruction (cf. 1 Thess. 2: 9), and further

[48] For details, see my (1992d).
[49] Haenchen (1971), 533 n. 4.
[50] There is thus no reason to think that Aquila and Justus were Paul's hosts on two different visits to Corinth. The duplication is understood by J. J. Taylor (1994b), 317, to imply that in Acts 18: 1–18 a redactor has fused the accounts of two visits of Paul to Corinth. He finds confirmation in the mention of two *archisynagogoi*, and dates the visits to AD 40 and 51 (325–6). In this he comes very close to the position of Lüdemann (1984), 171–3, against which I argued in (1982a).
[51] De Waele (1930); see Chapot (1899); Schneider (1932).

encounters were easily justified by the need for new pieces or other repairs. As his ministry expanded, however, something more suitable was required.[52] The space in the workshop was so limited that work had to stop if he addressed a group, and the assembly inevitably attracted the attention of passers-by. The lack of privacy precluded intimate discussions. Only the house of a relatively wealthy believer with its atrium and spacious rooms would provide the necessary space and seclusion. As we shall see shortly, Paul's first converts at Corinth were precisely such people. The fact that the owner of the house, (Titius) Justus, is not listed among them (cf. 1 Cor. 1: 14–16) might suggest that he came on the scene at a later stage.

Many Conversions

The Western text expands the very brief reference to converts in the original source ('brethren', Acts 18: 18) by an allusion to the conversion of a major figure associated with a synagogue, Crispus, together with his entire household, and a 'great crowd of Corinthians' (18: 8). The two events are simply juxtaposed. A causal relationship is excluded by the explicit mention that the latter were converted by the word of the Lord. The Alexandrian text, however, makes the former the cause of the latter. Most commentators accept the historicity of the conversion of Crispus, because he is mentioned in 1 Corinthians 1: 14.[53] It is undeniable that such an event must have had an impact on the God-fearers associated with the synagogue. The conclusion drawn by the Alexandrian text was perhaps inevitable. The conversion of an eminent Jewish personage can only have enhanced the credibility of a message which promised the purity of monotheism without the disadvantages of circumcision and dietary laws.

The source also mentions another synagogue personage, Sosthenes, who was beaten up after the failure of the Jewish appeal to Gallio (18: 17). The survival of this scrap of information probably implies that he subsequently became a Christian.[54] In opposition to Crispus, there is a certain hesitancy about identifying this individual with the Sosthenes who appears as co-sender of 1 Corinthians.[55] The fact that the name is widespread, however, has to be weighed against the only possible reason why Sosthenes should be cited by Paul as a co-author in this one instance, namely, intimate knowledge of the factionalization of the Corinthian community.[56]

The number of converts is also explained by the temporal precision. The 'many days' of the source is specified as 'one year and six months' (18: 11). The intrinsic plausibility of this figure is its strongest recommendation. Given the

[52] Haenchen (1971), 539.
[53] J. J. Taylor (1994*b*), 318. [54] Ibid. 324.
[55] Haenchen (1971), 536 n. 5, rejects it out of hand. Fee (1987), 30–1, is much more measured.
[56] See my (1993), 566–70.

limitations on travel, it would have been pointless to spend less time in such a great city. Paul would have arrived in Corinth in AD 50, when spring opened the roads to travellers, and left just before the end of the sailing season in AD 51.[57]

Having seen what can with some justification be accepted as historical in Acts 18: 1–18, we now have to integrate it with data from the letters to build up a picture of the evangelization of Corinth.

THE EVANGELIZATION OF CORINTH

If Paul foresaw employment in Corinth, as he recalled the demands of the Isthmian Games, he could not have imagined that he would find it with fellow-Christians. He and his companions were used to operating in virgin territory, and this custom eventually became a principle (Rom. 15: 20). How long he walked the streets of Corinth before finding Prisca and Aquila we can never know (Acts 18: 1–3).[58]

Perhaps they took him in, not because they needed help, but because they saw his craft as shared ground that would facilitate the conversion of a fellow-Jew. It is easy to imagine the stunned amazement when they realized that they were all followers of Christ. Paul must have been as disconcerted as Prisca and Aquila were overjoyed. They had had to struggle for the faith for nine years with only the rudimentary information acquired during their conversion process in Rome, and now they found themselves hosts, not only to an authorized emissary, but one who had been in Jerusalem!

The First Converts

Paul names the household of Stephanas as 'the first fruit of Achaia' (1 Cor. 16: 15), and lists that same household as one of the very few which he himself had baptized (1 Cor. 1: 16). If Paul made the first converts in Corinth, it means that Prisca and Aquila had not been successful as missionaries during their years in Corinth. One may wonder whether they had even tried. Their experience in Rome might have proved so traumatic that they felt inadequate to communicate their faith to others. They may have slid back into the life of the Jewish community, while still retaining the memory of Jesus whom they had accepted as the Messiah.

The arrival of Paul changed all that. The faith of Prisca and Aquila was given a new impetus, and they became two of the most committed members of his missionary team, even to the extent of risking their lives (Rom. 16: 4). They

[57] See Ch. 1, 'Dating the Stages of the Journey'.
[58] For a detailed description of the city, see my (1984) and Fig. 6.

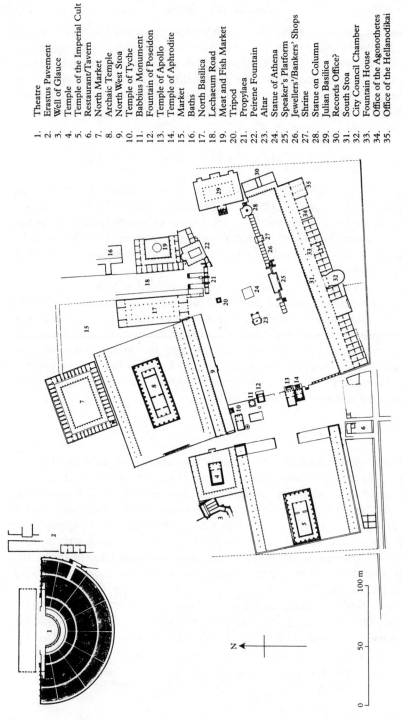

1. Theatre
2. Erastus Pavement
3. Well of Glauce
4. Temple
5. Temple of the Imperial Cult
6. Restaurant/Tavern
7. North Market
8. Archaic Temple
9. North West Stoa
10. Temple of Tyche
11. Babbius Monument
12. Fountain of Poseidon
13. Temple of Apollo
14. Temple of Aphrodite
15. Market
16. Baths
17. North Basilica
18. Lechaeum Road
19. Meat and Fish Market
20. Tripod
21. Propylaea
22. Peirene Fountain
23. Altar
24. Statue of Athena
25. Speaker's Platform
26. Jewellers'/Bankers' Shops
27. Shrine
28. Statue on Column
29. Julian Basilica
30. Records Office?
31. South Stoa
32. City Council Chamber
33. Fountain House
34. Office of the Agonothetes
35. Office of the Hellanodikai

FIG. 6 Central Corinth c. AD 50 (*Source*: C. K. Williams II, *The Corinthia in the Roman Period*, JRA Sup. 8, ed. T. E. Gregory (Ann Arbor, 1993))

prepared the way for Paul in Ephesus (1 Cor. 16: 19). and subsequently in Rome (Rom. 16: 3).

The fact that he was operating from a small workshop makes it most surprising that those whom Paul remembered baptizing were all from a superior class. The prejudice of such people against manual labourers is well documented.[59] The leadership role assumed by Stephanas and his family in the community (1 Cor. 16: 15) implies a degree of leisure difficult to associate with those who had to sweat for every morsel of food. If there is a deliberate contrast with the authority of charismatic gifts, as W. Meeks maintains, their contribution must have been in the form of benefactions.[60] Equally the freedom of Stephanas to take part in the delegation to Ephesus (1 Cor. 16: 17) means either that he was successfully self-employed or did not need to work.

Crispus (1 Cor. 1: 14) is identified by Acts 18: 8 as an *archisynagogos*. The title is not legally defined, and the common assumption that he was responsible for the public worship of the congregation[61] is not justified by the data.[62] All that one can say, in the light of Jewish inscriptional evidence, is that it was an honorific title awarded by a community in gratitude for a donation to their place of prayer (e.g. a whole building or parts thereof, a mosaic floor, a chancel screen, mural and ceiling paintings). Since sufficient superfluous wealth to become a patron was the only qualification the title could be given to a man or a woman.[63] Or even to a non-Jew. The title was also used in specifically pagan contexts,[64] with the meaning 'master of a guild or company'.[65] Absolutely speaking, therefore, Crispus may have been a God-fearer and not a Jew. In any case, he certainly was not a poor man.

Gaius (1 Cor. 1: 14) is also mentioned in Romans 16: 23 as 'host to me and to the whole church'. The adjective 'whole' is unnecessary if the Corinthian Christians met only as a single group (cf. 1 Cor. 14: 23).[66] Other sub-groups must have existed, and these can only have been house-churches on a smaller scale, 'the church in the home of X' (e.g. Rom. 16: 5; 1 Cor. 16: 19; Col. 4: 15; Philem. 2). An extra-large house was necessary to accommodate the entire community. Gaius, in consequence, must have been wealthier than the average believer.

It can hardly be coincidence that the few, whom Paul recalls having baptized at the very beginning of his ministry at Corinth, happened to be just the very

[59] See Ch. 4, 'Learning a Trade'.   [60] (1983), 58.
[61] So e.g. Schürer (1973–87), 2. 435.   [62] Rajak and Noy (1993), 82.
[63] Ibid. 87–8. This, of course, calls into question the thesis of Brooten (1982) that female *archisynagogoi* exercised a liturgical function in the synagogue.
[64] Six inscriptions are cited by Rajak and Noy (1993), 92–3.
[65] LSJ, s.v. 253.
[66] Theissen (1982), 89. Dunn (1988), 910, refutes the older view that Gaius was considered host to the universal church because of the hospitality he offered to travelling Christians. The allusion is to all believers in Corinth.

sort of people who could be of use to him. He had the practical sense to recognize that, while the gospel was offered to all, only those with initiative, leisure, and education could function as effective assistants in the spread of the gospel. Slaves might be model Christians, but they were not their own masters. They could not dispose of their time as they wished. Neither could Prisca and Aquila, who had a living to earn.

It would almost appear that Paul was deliberately following the recruiting policy of Jesus. His first disciples were Peter and Andrew, who were not 'uneducated, common men' (Acts 4: 13). Andrew had a purely Greek name, suggesting a bilingual family (cf. John 12: 20–2). The brothers worked in partnership (Luke 5: 7) with James and John, the sons of Zebedee (Luke 5: 10), who had employees (Mark 1: 20). One has the impression that they owned their own boats (Luke 5: 11). The brothers had moved from their hometown, Bethsaida (John 1: 44), in the territory of Philip, to Capernaum (Mark 1: 29) in the territory of Antipas, because they then paid less to have their fish processed for export at Magdala/Taricheae. They could afford to let others do the fishing in order to remain in Judaea as disciples of John the Baptist (John 1: 40–1). From our perspective they would be described as upper middle class, even though that terminology is anachronistic as regards the first century.

By the time he reached Corinth, therefore, it would appear that Paul had worked out a careful missionary strategy. In order to get anywhere quickly, he realized that the nascent church needed a solid nucleus of those who were in a position to furnish facilities, precisely the sort of person typified by Justus in the Lukan account (Acts 18: 7). A place of assembly was only one advantage. The house of Gaius was apparently the place where Tertius (the secretary of Gaius?) took down in shorthand Paul's dictation of the letter to the Romans; the host added his name when he heard others fill up the page with their greetings (Rom. 16: 23).

Paul's attitude to this nucleus of wealthy individuals in the Corinthian community was complex. While availing of the facilities they offered, he refused to permit himself to become dependent on them. He continued to support himself as best he could, and supplemented his income by gifts from Macedonia, in order to retain his independence (1 Cor. 9: 1– 18; 2 Cor. 11: 7–9). This willingness to take from Corinthian believers, but only in a very limited way, was to become a bone of contention at a later stage.[67]

Status Inconsistency

It is not known when another prominent member of the community became a believer. The epistle to the Romans also includes greetings from 'Erastus *oikon-*

[67] See Ch. 12, 'The Report of Titus'.

omos of the city' (16: 23). He is the only convert whose civil status is mentioned by Paul. This may be because it was exceptionally high, but it might have been evoked merely to distinguish this Erastus from others (2 Tim. 4: 20; Acts 19: 22).

Resolution of the problem is facilitated by the discovery of an AD 50–100 inscription, cut into the pavement between the North Market and the theatre at Corinth, which read '[——] Erastus in return for his aedileship paved (this area) at his own expense'.[68] Apart from Romans 16: 23 and the inscription, the name Erastus is not attested at Corinth and, unless one is prepared to accept the extraordinary coincidence that two individuals bearing the same unusual name both held public office in the same city at roughly the same time, their identity should be taken for granted.

To be an aedile, one of the four magistrates who governed the city, it was necessary to be a Roman citizen. The absence of any mention of the father of Erastus in the inscription suggests that he had once been a slave. Manifestly Erastus was one of those energetic freedmen, who flourished in the vigourously competitive atmosphere of Corinth, and who had the surplus funds which enabled him to undertake public office.[69] The normal translation of aedile in Greek, however, is *agoranomos*, not *oikonomos*, whence the well-founded suggestion that the latter was the equivalent of the Latin quaestor, an inferior financial position in the municipal hierarchy. Naturally, to accede to higher office one must have proved oneself in lower ones, and inscriptions reveal the office of quaestor to be a stepping-stone to that of aedile, as the latter was to that of duovir.[70]

Romans 16: 23, therefore, reveals Erastus at an early stage of his public career, which apparently was not affected in any way by his conversion to Christianity! We must assume that he somehow found a way to reconcile his new monotheistic belief with his duty to participate in the worship of the gods of the city; every meeting of the city council involved a pagan sacrifice. He may have seen it as a purely formal gesture without any real religious significance, as did his fellow-Christians who had no difficulty in eating meat offered to idols, because 'idols have no real existence' and 'there is no God but one' (1 Cor. 8: 4). Only later did participation in pagan rituals become a test of the faith.[71]

Whatever be the answer, the figure of Erastus gives us a privileged insight

---

[68] Kent (1966), 99. A photograph is given in Furnish (1984), plate vɪa.

[69] On servile origins and Roman citizenship, see Ch. 2, 'Roman Citizenship'.

[70] See the detailed discussion in Theissen (1982), 75–83, which has been accepted by Meeks (1983), 58–9.

[71] 'Among these I considered that I should dismiss any who denied that they were or ever had been Christians when they had repeated after me a formula of invocation to the gods and had made offerings of wine and incense to your [Trajan's] statue . . . and furthermore had reviled the name of Christ; none of which things, I understand, any genuine Christian can be induced to do' (Pliny, *Letters* 10. 96; trans. Radice).

into a section of the population of Corinth into which Christianity made inroads. However much Erastus may have achieved, he would never have felt fully at ease among the free-born. As with others of his class, the stigma of his servile origins blighted every pleasure. The fear of being patronized provoked an injudicious aggressiveness. The sense of insecurity of the successful freedman became a favourite topic in literature. Everyone knew instances of the affected culinary expertise of Nasidenus,[72] and of the pretentious learning of Trimalchio.[73] Corinth had its own celebrated example. A small circular structure on the agora carried the same inscription on both the pedestal and the band above the columns, 'Gnaeus Babbius Philinus, aedile and pontifex, had this monument erected at his own expense, and he approved it in his official capacity of duovir'.[74] He was not prepared to take the chance that his successor as chief magistrate might refuse the exercise in self-recommendation!

The root of such insecurity was the bitter awareness that one was not recognized for what one had achieved. Contemporaries, it was felt, imposed on the reality an unflattering portrait drawn from other sources. Erastus imagined that those who looked at him saw not the quaestor but merely an ex-slave. Freedmen, however, were not the only ones to feel the discomfort of ambiguous status. Phoebe of Cenchreae, though of sufficient independent wealth to be a patron to Paul and many others (Rom. 16: 2), would have been seen first and foremost as a woman, with the social and political disadvantages her sex implied. The pagan Justus experienced a dissonance in his own society which drove him to association with the synagogue (Acts 18: 7). Prisca and Aquila had the authority of needed experts, and Crispus and Sosthenes held responsible positions, but to their pagan neighbours they were above all Jews, who resided among them on sufferance.[75]

One factor that made the gospel attractive to such people was that it embodied the paradox they lived.[76] Its central thesis that the saviour of the world died under torture spoke to the contradictions of their existence. Though classed as weak they knew their own power, and thus could understand without difficulty the idea, revealed in the life of Christ, as in that of Paul, that 'power is made perfect in weakness' (2 Cor. 12: 9). To them Christianity made sense of

---

[72] 'There was a boar from Lucania, which our gracious host kept telling us was caught in a soft southerly breeze. . . . This lamprey was caught while she was pregnant; after spawning the flesh is inferior. The sauce has the following ingredients: Venafran oil (the *first* pressing of course), liguamen (from the guts of the *Spanish* Mackarel), wine that is five years old but grown in Italy' (Horace, *Satires* 2. 8. 6, 42–7; trans. Rudd).

[73] '"Tell me, my dear Agamemnon," continued Trimalchio, "do you remember the twelve labours of Hercules and the story of Ulysses—how the Cyclops tore out his eye with his thumb? I used to read about them in Homer, when I was a boy"'(Petronius, *Satyricon* 15. 48; trans. Sullivan). Homer wrote nothing about Hercules, and it was Ulysses who blinded the Cyclops.

[74] Kent (1966), 73.

[75] On status inconsistency, see Meeks (1983), 70, 73.

[76] Ibid. 191–2.

the ambiguity of their lives, and at the same time introduced them into a society committed to looking at them primarily as people, all equally valuable and valued. It gave them a space in which they could flourish in freedom.

The Composition of the Community

More is known about the composition of the Corinthian church than of any other except that of Rome. It is the one case in which Paul tells us something of the social structure of the group. The names of many individuals can be drawn from 1 Corinthians, Romans, and Acts.

'By the standards of the world, not many of you were wise, not many were powerful, not many were well born' (1 Cor. 1: 26). Despite subtle attempts to find in these words the description of specific philosophical groupings,[77] the sociological interpretation is demanded by common sense.[78] The 'wise' are the educated, and in particular those with a reputation for prudence and moderation, who exhibit sound judgement in politics or commerce. The 'powerful' are the influential, those whose opinions carry weight in civic life.[79] The 'well born' are those born into the aristocracy of wealth[80] created by the freedmen who were sent by Julius Caesar to found Corinth.[81] The overlap of the three terms needs no emphasis, and they carry the further connotation of freedom.[82] With admirable brevity Paul evokes a privileged élite, whose impact on Corinth was quite out of proportion to its numbers. It members were a minority in the city, just as were those from this class who became Christians; some of these have been named above. These latter no doubt played a dominant role in the affairs of the church; those who take their authority for granted do not need official positions to reinforce it.

The majority of believers were not so fortunate, but neither were they at the bottom of the the social scale. Among the unnamed members of the 'households' of Stephanas and Crispus, it is very probable that there were slaves (1 Cor. 7: 21). While legally disadvantaged, such house slaves often enjoyed a standard of living and education denied to those born free, and could look forward to exercising their trained talents in freedom.[83] Only then would they have to provide for themselves; a slave was guaranteed food and lodging.

[77] Munch (1959) 162–3; Horsley (1976), 282–3.
[78] Theissen (1982), 72.
[79] Sänger (1985).
[80] e.g. Lucius Castricius Regulus, see Kent (1966), 70.
[81] Strabo, *Geography* 8. 6. 23.
[82] 'It is impossible for anyone to be "noble" without being "well born" at the same time, or for one who is "well born" not to be free' (Dio Chrysostom, *Discourses* 15. 31; trans. Crosby).
[83] The lot of slaves who worked the fields of great estates or who laboured in the mines was completely different. The severity bordering on sadism of which they were the victims cannot be considered normal treatment for household slaves, though, of course, there were exceptions. 'To his quaking household he [Rutilus] is a monster, a mythical ogre, never so happy as when the torturer's

The secretarial ability of Tertius (Rom. 16: 22)[84] put him in the same class as Tiro, the famous secretary of Cicero. His skill made him valuable. If not a freedman already, he would certainly earn that status some day. Achaicus (1 Cor. 16: 17) is a nickname, 'the man from Achaia', and thus is more likely to have belonged to a slave than to a freeman. Evidently it was acquired outside Achaia,[85] and its Latin form would suggest somewhere to the west. In all probability he had been a slave in Italy, and returned to Corinth as a freedman. His participation in the delegation to Ephesus suggests at least that he was master of his own time. The same is true of Fortunatus (1 Cor. 16: 17). Nothing is known of Quartus (Rom. 16: 23); the name is common among slaves and freedmen.

There is some doubt whether Lucius, Jason, and Sosipater (Rom. 16: 21) were members of the church of Corinth or just happened to be there when the writing of Romans was coming to a conclusion.[86] The formulation separates them from Timothy, who is identified as Paul's co-worker. In consequence, they cannot be considered part of Paul's missionary team. A Sopater (possibly an abbreviated form of Sosipater) of Beroea appears in Acts 20: 4 as one of those accompanying Paul to Jerusalem with the money collected for the poor of Jerusalem (cf. 1 Cor. 16: 1–4). He, like Jason, is explicitly identified as a Jew. The latter may be the individual who hosted Paul at Thessalonica (Acts 17: 5–7). The possibility that these two were collection delegates of their respective churches has suggested that the same is true of Lucius. This individual is commonly identified as the Luke mentioned in Colossians 4: 14, Philemon 24, and 2 Timothy 4: 11, but why Paul should use the form 'Luke' in all these instances and then switch to 'Lucius' here is never explained. Moreover, were Luke in question he would have been grouped with Timothy rather than with the other two because of his long-standing association with Paul. The hypothesis that we have to do with three delegates cannot be excluded, but neither is there anything in its favour. In consequence, I prefer to consider them Corinthians.

It is unlikely that Chloe's people (1 Cor. 1: 11) were from Corinth.[87] The issues on which they report were not problems for the Corinthians themselves, e.g. divisions within the community and the way some men and women wore their hair at the liturgical assemblies. These matters, which were of crucial

---

there on the job, and some poor slave who's stolen a couple of towels is being branded with a red-hot iron' (Juvenal, *Satires* 14. 18–22; trans. Green). 'One victim has rods broken over his back, another bears bloody stripes from the whip, a third is lashed with a cat-o'-nine-tales. Some women pay their floggers an annual salary' (Juvenal, *Satires* 6. 479–80; trans. Green).

[84] Richards (1991), 171, rightly notes that the frequency of oratorical rhetoric in Rom. suggests that it was taken down at the speed of normal speech in shorthand.

[85] Meeks (1983), 56, appositely points out that 'It was not on Crete but in Toledo that Domenikos Theotokopoulos was named "El Greco"'.

[86] See Dunn (1988), 909.

[87] Against Theissen (1982), 92, and Meeks (1983), 59.

importance for Paul, were not mentioned in the letter (1 Cor. 7: 1) brought by the official delegation from Corinth (1 Cor. 16: 15–17). On the contrary, they are the sort of departures from the norm which have a great impact on visitors from another church, in this case Ephesus, and about which they would have been most eager to gossip when they returned home. Moreover, were Chloe's people Corinthians who had gone behind the community's back to run to Paul with tales, it would have been most insensitive of the latter to mention the source of his information; it could only make mischief.[88]

Of the members of the church at Corinth we know 16 individuals by name. Two of them (Prisca and Aquila) are married to each other. Two are women (Prisca and Phoebe). Six are explicitly of Jewish origin (Aquila, Crispus, Prisca, Sosthenes, Jason, Sosipater). Two are certainly Gentiles (Erastus, Justus). From these last two sets of figures it might appear that one could infer that Jewish believers predominated in the community. Any such extrapolation, however, is flatly excluded by the type of problems with which Paul has to deal in 1 Corinthians—appeal to pagan courts (6: 1–11); frequentation of prostitutes (6: 12–20); marriage and sex (7); participation in pagan temple meals (8–10)—and by his explicit statements that the majority of the community had at one time been idolators (6: 10–11; 8: 7; 12: 2).

The predominant group in the Corinthian church was made up of Gentiles of various grades of the middle of the social scale. Only the very top (great magnates) and the very bottom (field slaves) of that scale were lacking.[89] Jews were a minority, but two at least (Crispus and Sosthenes) stood out from the group.

A TURBULENT COMMUNITY

The potential for dissension within the community is evident. Most members had in common only their Christianity. They differed widely in educational attainment, financial resources, religious background, political skills, and above all in their expectations. A number were attracted to the church because it seemed to offer them a new field of opportunity, in which the talents whose expression society frustrated could be exploited to the full. They were energetic and ambitious people, and there was little agreement among their various hidden agendas. A certain competitive spirit was part of the ethos of the church from the beginning.

---

[88] Robertson and Plummer (1914), 10.
[89] Meeks (1983), 73.

The Arrival of Apollos

According to Acts 18: 24–8, Apollos came to Corinth from Ephesus some time after Paul's departure for Jerusalem in the late summer of AD 51. In saying 'I planted, Apollos watered' (1 Cor. 3: 6), Paul explicty confirms that Apollos exercised a ministry subsequent to his own in Corinth. Paul also tells us that Apollos was with him in Ephesus at the time of writing 1 Cor. 16: 12. Nothing similar appears in Acts which, however, tells us much more about this personage. How should it be evaluated?

According to Boismard and Lamouille there is nothing in the Western text of Acts 18: 24–8 which betrays the hand of a redactor. It is all of one piece, they assert, and its style points to Luke as the author.[90] The narrative, however, exhibits two serious internal contradictions, the first concerning Apollos' status as a Christian, the second his mission to Corinth. (1) Apollos taught accurately about Jesus (18: 25). None the less he needed further instruction (18: 26). In addition, how could he speak accurately about Jesus, when he knew only the baptism of John (18: 25)? (2) Apollos was invited to return with them by Corinthians resident at Ephesus. None the less the Ephesians wrote a letter requesting that he be received at Corinth (18: 27). He had no need of a letter of recommendation, however, since he was accompanied by Corinthians.

The Alexandrian text of this passage resolves the latter problem by making Apollos take the initiative in going to Achaia.[91] Why Apollos chose to go to Achaia is not explained. It might have been better had the Alexandrian text suppressed the mention of a letter of recommendation, because that would have made it obvious that a single hypothesis can explain the two contradictions.

The awkwardness of the clause, 'knowing only the baptism of John' (18: 25b) is manifest; it is simply juxtaposed to what precedes. Were a single author to have written the whole verse one would have expected something like, 'Although he knew only the baptism of John, he none the less taught accurately the things concerning Jesus.' The function of the allusion to John's baptism in the narrative is to justify further instruction, 'Aquila took him and explained the Way to him more accurately' (18: 26b); there had to be a defect to be remedied.[92] Apollos has been transformed from an unattached Christian of uncertain antecedents into a bearer of the Pauline gospel. The point is reinforced by having the Ephesians write a formal letter of recommendation for him. He has become an emissary of a Pauline church. Thoroughly domesticated, and integrated into an recognized channel of church development, he is now free to go his own way.

---

[90] (1990), 3. 237–8.　　　　　　　　　　　　　　　　　　　[91] Ibid. 2. 367.

[92] The choice of John's baptism as the defect may have been influenced by its presence in the next pericope (Acts 19: 1–7), on which see above Ch. 7, 'The Founding of the Church'.

When purged of the elements which impose a specific pattern of development on the growth of the church (cf. Acts 1: 8), the Apollos' story becomes perfectly coherent. The failure to rebaptize him in the name of Jesus (cf. 19: 5) underlines the fact that he was a fully qualified missionary.[93] No reason has been, or can be, given why a redactor should have created his racial origins, his place of birth and conversion, or his qualifications. The appearance of his name in the New Testament hapax form of Apollonius (corrected to the more normal New Testament form in the Alexandrian text) in precisely this context might suggest a source.[94]

Nothing is known about Christianity in Alexandria at this period, but if the faith had spread out of Palestine to Antioch in the north, and to Damascus in the north-east, nothing militates against it having also penetrated the great port city of Egypt. Jewish connections between Alexandria and Jerusalem had always been close,[95] and travel was facilitated by the great trade route, the Way of the Sea. Presumably merchants or returning pilgrims brought the faith to Alexandria, as they had done to Damascus.

Apollos is described as *logios* (Acts 18: 24). The adjective has two connotations, 'eloquent' and 'learned, cultured', and it can mean specifically one or the other.[96] In the present instance, however, no choice is necessary, for both are intended. Apollos spoke with inspirational enthusiasm (18: 25), and was well-versed in the Scriptures (18: 24). 1 Corinthians 1–4, where Paul has Apollos in view, confirms this interpretation, as Haenchen has very perceptively noted, 'Again and again Paul in these chapters comes back to two things which people missed in him but apparently detected in someone else, the gift of edifying speech, which was denied to Paul himself . . . and the gift of "wisdom".'[97] It is difficult to imagine that an Alexandrian Jew with precisely these qualifications, and with a mind so open that he eventually accepted Jesus, could have escaped the influence of Philo, the great intellectual leader of Alexandrian Jewry, particularly since the latter seems to have been especially concerned with education and preaching.[98] Philo's life-work was to give Hellenized Jews, such as Apollos, a perspective on the Law that would enable them to accept both it, and their ambient culture.

Acts gives the impression that Apollos arrived in Corinth before Paul returned to Ephesus towards the end of August AD 52 (Acts 19: 1). In this scenario Apollos would have been in Ephesus for almost a year, which is reasonable given the limitations imposed by the seasonal restrictions on travel.

---

[93] Haenchen (1971), 557.
[94] Hemer (1989), 233, notes, 'It is of unusual interest as it [the name Apollonius/ Apollos] seems to be almost unexampled outside Egypt, but is conspicuously common there.'
[95] See Smallwood (1981), 220–4.
[96] Excellent examples are assembled by Spicq (1978–82), I. 500–2.
[97] (1971), 555–6.
[98] Koester (1982), I. 175; Schürer (1973–87), 3. 818.

Regular sailings on the Ephesus–Corinth or the Troas–Neapolis routes came to an end in mid-September, and only in exceptional circumstances would ships have put to sea in October. The Corinthian merchants trading in Ephesus, who invited Apollos (the Western text of Acts 18: 27a), would have wanted to return home for the winter. It is intriguing to think that the two figures who were to dominate the immediate future of the church at Corinth might have missed each other by a couple of days.

This scenario very easily integrates the Previous Letter (1 Cor. 5: 9). It was provoked by news from Corinth, in response to which Paul told them 'not to associate with sexually immoral people' (1 Cor. 5: 9). The simplest hypothesis regarding the bearers of the bad news is that they were the Corinthian merchants who returned to Ephesus at the beginning of the new trading season in the spring of AD 53, or Apollos, who could not accept the way in which he was been set against Paul. Evidently nothing was said about divisions within the community, because Paul would certainly have dealt with the issue, and in 1 Corinthians would have reminded the Corinthians that they had not heeded his stress on the importance of unity for the local church. Divisiveness must have become a problem at Corinth only after the dispatch of the Previous Letter.

If this is correct, there must have been two phases in Apollos' activity at Corinth. In the first, as Acts 18: 28 says, he functioned as a missionary engaged in controversy with the Jews. When that proved unsuccessful, he turned his attention inward, and became as it were a theologian-in-residence of the Christian community.

Differences within the Community

Apollos quickly found a niche in the competitive world of the Corinthian church. Paul's preaching was anti–intellectual. He proclaimed a crucified Christ as the exemplar of authentic humanity (1 Cor. 2: 1–5), and saw no need for any speculative development. He was more concerned with evidence of the power of transforming grace in his life and that of others (2 Cor. 3: 2). He cut a poor figure by comparison with the orators who attracted followers by their eloquence. He also disappointed those believers who aspired to a real theology. Apollos met these needs. In addition to his oratorical gifts, he had the ability to connect things up, to establish relationships between different aspects of the faith. This was one of the fundamental aspects of rhetorical education. By using Philo's methods of interpretation, and his philosophical framework, Apollos provided intellectual fulfilment by building a rich synthesis of the elements which Paul had provided.

Human nature being what it is, the intellectual aspirations of those who clustered around Apollos certainly alienated others, perhaps the less well educated, who, in reaction, insisted on the importance of the bare minimum inculcated by the founder of the church; what mattered most was love of neighbour.

An Apollos group and a Paul group were inevitable once the former appeared on the scene. But apparently there was also a third group claiming allegiance to Cephas (1 Cor. 1: 12). Commentators debate whether Peter personally or merely his followers visited Corinth.[99] The answer is irrelevant, because in either case these scholars assume that he represented a Judaizing faction within the church, which among other things pushed for observance of Jewish dietary laws.[100] The most obvious candidates for membership in such a group are those Jewish converts who found difficulty in integrating into a predominantly Gentile community.[101] The secular style of Apollos might have contributed to their sense of isolation. From this perspective, 'Cephas' functions as a symbol for a type of Jewish Christian, who, for Paul, is exemplified, not by the consistent James, but by Peter, who surrendered his freedom under pressure. Time had not healed the bitterness of the memory of the incident at Antioch (Gal. 2: 11–14).

This conclusion is confirmed by the form of the slogans in 1 Corinthians 1: 12. L. Welborn reflected a wide consensus in asserting, 'A declaration of allegiance to a party so personal in organization could take no other form than that which is given in 1 Corinthians 1: 12—"I am of Paul!"'.[102] Unfortunately he produced no parallels to the pattern of a personal pronoun followed by 'to be' (explicit or understood) followed by the genitive of a proper name. M. Mitchell's study revealed that this pattern is virtually unattested as implying political affiliation or relationship to a teacher, but is most commonly used of the parent–child relationship and the master–slave relationship. The force of the formula, therefore, is to suggest that those who think of themselves as belonging to factions within the church are acting childishly (cf. 1 Cor. 3: 1–4) or slavishly (1 Cor. 7: 23). If such was the way they naturally would have been understood by the Corinthians, the slogans cannot be self-designations. They are a deliberate put-down on the part of Paul, who uses the rhetorical device of impersonation to suggest a strongly condemnatory judgement.[103]

The formation of these three groups[104] was facilitated by the fact that the church at Corinth was too numerous to be accommodated comfortably in a single house. Of the 16 known individuals, two were married (Prisca and

---

[99] Barrett (1963a).
[100] Vielhauer (1974); Barrett (1968), 44.   [101] Merklein (1992), 148.
[102] (1987), 91.   [103] (1991), 85–6.
[104] Mitchell's insight makes it possible to dispense with the enigmatic Christ party (83 n. 101). Paul was simply carried away by the rhetorical figure and embodied his own response in a formula identical with those he attributed to others. When taken in the sense just defined, it is an ideal description of all believers, "You are of Christ" (1 Cor. 3: 23; cf. 6:19–20). Paul, in writing to the Philippians not long before, had called himself a 'slave of Christ Jesus' (1: 1). For Paul, obviously, belonging to Christ did not exclude his relationship as apostle to the community, which is suggested as the only specific factor of the Christ party (against Sellin (1987), 3011–12).

Aquila), and one was single (Phoebe). One must presume that the remaining 13 were married. Two had been converted together with their households (Stephanas and Crispus). Hence, one should assume that there were at least 40 believers. There may have been many more. Even the house of a wealthy person, such as the villa at Anaploga,[105] could not accommodate such a number in the triclinium. Instead of reclining in comfort, some were forced to sit in the portico of the atrium.

This disparity of treatment at the liturgical meal highlights another division, which has already been suggested by the allusion to a privileged élite (1 Cor. 1: 26). Among the believers at Corinth there were 'haves' and 'have nots', and the former exhibited little or no concern for the latter (1 Cor. 11: 22). Despite virtually unlimited opportunity, economic inequality was a fact of life, and the believers did nothing to close the gap.

How the resultant class struggle related to the three other groupings is not clear. Welborn oversimplifies in considering it to be the fundamental division, and mistakenly dismisses all other differences as irrelevant.[106] It is tempting to identify the wealthy and better educated with the partisans of Apollos, and the lower class with those who preferred the elementary teaching of Paul.[107] One must keep in mind, however, the complexity of the human situation. Many people of low social status have a desperate urge to become educated, and many educated people desire simplicity in religion.

CONTACTS WITH CORINTH

When the sailing season opened in late April AD 54, a wealthy businesswoman of Ephesus, Chloe, sent some of her employees to Corinth, where shipments of new goods had arrived from the west. It may have been on their own initiative that they made contact with their fellow-Christians in the city, but it is most improbable that Paul would have failed to avail himself of the opportunity to obtain news of a church that he had not seen for three years. An important factor in his choice of Ephesus as his base had been the possibility of using precisely such travellers to carry his messages.

At this point in his career Paul had dealt with the teething and adolescent problems of a series of churches (Thessalonica, Galatia, Ephesus, Philippi, and Colossae). He had a very clear idea of the things that could go wrong, as a church struggled to define itself on the way to maturity. That he should have indicated to Chloe's people the areas of community life that might prove problematic is suggested not only by common sense, but by the fact that he does not know them individually. Evidently they were not important enough for him to

---

[105] See my (1992*e*), 163, and the reconstruction drawing in Pritchard (1987), 174–5.
[106] (1987), 89.   [107] Merklein (1992), 139.

retain their names, and there is a tendency to assume that members of the lower class need to have everything spelt out to them.[108] Sexual morality was certainly one of the sectors they were asked to assess, because Paul must have been concerned about the impact of the Previous Letter.

The report brought back by Chloe's people stunned Paul. No doubt they stressed the bizarre—an incestuous marriage (1 Cor. 5: 1–8), male homosexuals presiding at the liturgy (1 Cor. 11: 2–16), drunkenness at the eucharist (1 Cor. 11: 17–34)—but there were other observations, particularly regarding the factional divisions, which bought it home to Paul that the situation at Corinth was much more complex than anything he had dealt with hitherto.[109]

His immediate response was to send Timothy to investigate.[110] The excited gabble with which Chloe's people had poured out their experiences might have justified the hope that they were exaggerating. There was also the possibility that they might not have understood fully what was going on. After all, they were not experienced in church matters. It would have been unwise for Paul to react on the basis of what was essentially no more than gossip. Timothy, on the other hand, had already carried out a similar mission in Thessalonica (1 Thess. 3: 1–6); he was not only trustworthy, but experienced. Fortunately, he had not yet left for Philippi (Phil. 2: 19–24).

In view of the urgency of the matter, it is most improbable that Timothy took the land route to Corinth through northern Greece.[111] This hypothesis is the abortive fruit of an attempt to harmonize what Paul says with Acts 19: 22, which records a visit of Timothy and Erastus to Macedonia. This latter visit in fact took place, but after Timothy's return from Corinth. He replaced Paul on a planned visit to Macedonia (1 Cor. 16: 5) when the Apostle had to make a hasty and unforeseen visit to Corinth.

In all probability, therefore, Timothy took passage on one of the many ships plying between Ephesus and Corinth. The duration of the voyage is uncertain. It depended on so many factors, notably, fine weather, fair winds, and good omens. On official business and using naval equipment, which meant no cargo delays, it took Cicero from 6 to 22 July 51 BC to sail from Athens to Ephesus.[112] To allow Timothy two weeks in each direction is probably a minimum. He could hardly spend less than a fortnight in Corinth. Hence, we can safely

---

[108] On the social status of Chloe's people, see Theissen (1982), 92–4.

[109] Hurd (1965), 63, is wrong in reconstructing the report of Chloe's people by simply deducting from 1 Cor. those sections introduced by 'now concerning', which are assumed to belong to the letter from Corinth; see M. Mitchell (1989).

[110] While 1 Cor. 4: 17 could be read in such a way as to make Timothy the bearer of 1 Cor., this possibility is excluded by 1 Cor. 16: 10, which is better translated by the temporal 'when' (*RSV*) than the conditional 'if' (see Fee (1987), 821). Note also the absence of Timothy in the address of 1 Cor.

[111] *Pace* Barrett (1968), 390; Furnish (1984), 143; Lüdemann (1984), 93.

[112] *Letters to Att.*, 5. 11–13.

assume that he was away from Paul for six weeks, probably from the beginning of May to the middle of June.

During the early part of this period, a delegation, comprising Stephanas, Fortunatus, and Achaicus (1 Cor. 16: 15– 17), arrived from Ephesus bearing a letter, in which the church of Corinth asked for Paul's opinion on a number of issues (1 Cor. 7: 1). The presence of the delegation meant that all the information which Paul had expected to acquire through Timothy was now immediately available, and from a representative source. He had access to people who could speak authoritatively about the state of the community and its problems, and thus support or deny the gossip of Chloe's people. Paul, therefore, was in a position to formulate his response, not only to what the Corinthians saw as questions, but to aspects of their lives which he considered problematic. He did so in the letter we know as 1 Corinthians, which was written sometime before 2 June, the date of Pentecost (1 Cor. 16: 8) in AD 54.[113]

THE KALEIDOSCOPE LETTER

The organization of 1 Corinthians reflects the complexity of the situation at Corinth. Paul deals with a wide variety of issues both directly and indirectly, and a number of commentators have considered the arrangement so haphazard that they refuse to see in it a single letter. In their eyes it is no better than a collection of fragments thrown together without plan or design. This, as we have seen, is certainly an exaggeration.[114] Moreover, M. Mitchell has shown that 1 Corinthians is a consistent deliberative argument designed to show the Corinthians that the factionalism in which they indulged was not to their advantage; throughout Paul urges concord.[115] Finally, a close analysis of 1 Corinthians reveals that Paul was chiefly concerned with the attitude and activity of one group at Corinth. The spirit-people were at the root of the problems dealt with in 13 of the 16 chapters of 1 Corinthians.

The Influence of Apollos

If we look closely at 1 Corinthians 1–4, where Paul is most explictly concerned with divisions in the community, a group emerges whose members believed that their possession of 'wisdom' made them 'perfect' (2: 6). As possessors of 'the Spirit which is from God' (2: 12), they were 'spirit-people' (2: 15). They thought of themselves as 'filled (with divine blessings)', 'wealthy', 'kings', (4: 8), 'wise', 'strong', 'honoured' (4: 10). They looked down on others in the community who

---

[113] Jewett (1979), 48.
[114] See Ch. 11, '1 Corinthians'.
[115] (1991), 296–304.

had not attained their exalted spiritual status as 'children' capable of imbibing only 'milk' (3: 1), and as 'fools' who were 'weak' and 'dishonoured' (4: 10).[116]

While individual themes may be paralleled elsewhere in the Graeco-Roman world, R. A. Horsley has shown that there is a clear pattern within the wide selection of parallels provided by commentators.[117] The language used by the group at Corinth reflects Philo's distinction between the heavenly man and the earthly man. All the key elements just mentioned appear in two passages of a single work by the Alexandrian philosopher, *De Sobrietate*, 9–11 and 55–7. Hence, we are entitled to assume that other elements integral to Philo's understanding of the heavenly and earthly man also formed part of the religious outlook of the spirit-people, and that Paul has these latter in mind when he argues against such points.

The body was a fundamental point of disagreement between the heavenly and the earthly man. The wisdom possessed by the former revealed to him that 'the body is evil by nature and treacherous to the soul' (*Leg. All.* 3. 71), whereas the earthly man was 'a body lover' (*Leg. All.* 3. 74). If the body is 'a plotter against the soul, a corpse and always a dead thing' (*Leg. All.* 3. 69), it is natural to infer that the spirit-people were those who denied the resurrection (1 Cor. 15: 12). Death, from their perspective, was liberation from the weight and defilement of the body (*Som.* 148). To recover the body after death would have been meaningless. It is highly unlikely, therefore, that the spirit-people could have accepted Paul's preaching of Jesus as the Risen Lord in the sense that he intended. Perhaps they thought of him as a purely spiritual 'Lord of Glory' (1 Cor. 2: 8). In reality, they had no sense of Jesus; their attitude to him in effect said 'Anathema Jesus!' (1 Cor. 12: 3). In keeping with their sapiential orientation they were theists, and, in every instance where Paul confronts them, he has to remind them of the importance of Jesus Christ (1 Cor. 2: 16; 3: 23; 8: 6; 10: 16; 15: 3–5).

When Philo's disparagement of the body is associated with his dictum that 'only the wise man is free' (*Post.* 138), which means that 'he has the power to do anything and to live as he wishes' (*Prob.* 59), we see the basis for the Corinthian slogans 'all things are lawful to me' (1 Cor. 6: 12; 10: 23), and 'every sin which a man commits is outside the body' (1 Cor. 6: 18).[118] Their belief in the moral irrelevance of the body enabled the spirit-people to indulge their sexual appetites (1 Cor. 5: 1–8; 6: 12–20) and to eat what they wished (1 Cor. 8–10).

The importance that some Corinthians attached to glossolalia (1 Cor. 12–14) is drawn into this pattern,[119] when it is recognized that, for Philo, possession of the prophetic spirit expressed itself in ecstasy, madness, and inspired frenzy,

---

[116] On 'hubris' as the general category to describe this social attitude, see P. Marshall (1987), 182–218.

[117] (1976), (1977), (1978).   [118] See my (1978*a*).

[119] The 'adult–child' antithesis of 1 Cor. 3: 1 reappears again in 1 Cor. 13: 11.

since 'the mind is evicted at the arrival of the divine spirit' (*Heres* 264–5). When speaking of tongues Paul specifically mentions 'frenzy' (1 Cor. 14: 23) and the inactivity of the mind (1 Cor. 14: 14). Mysterious, unintelligible speech flattered the conviction of the spirit-people that they were superior.

Given that Jews were an alienated minority in the Corinthian church, the Diaspora synagogue is most unlikely to have been the source of Philonic influence at Corinth. The obvious channel by which Philo's philosophical framework entered the community was Apollos. What he said, however, and what his followers understood were not necessarily identical. If they mistook Paul's meaning so badly, it is improbable that they understood Apollos adequately. Certainly Paul's quarrel was with the practical implications of their interpretation of Apollos, rather than with the personality or teaching of the latter (1 Cor. 16: 12). Indeed Apollos may have left Corinth, and come to live with Paul at Ephesus, because he had become dismayed at the uses to which his teaching was being put.

An Unfortunate Strategy

What he heard about the situation at Corinth strengthened Paul's bias against speculative theology. It had revealed itself in its fruits. Not only was it unnecessary, it was pernicious. Hence, instead of a sincere effort to get to the root of the problem, and to understand the legitimate aspirations of the spirit-people, Paul's reaction was brutally dismissive. Confident that such people would always be a minority without popular support, Paul chose to play on the dark side of the majority by turning the spirit-people into figures of fun. Cruel laughter was the weapon he selected.

Perhaps if left to himself Paul would not have made what would prove to be a disastrous mistake. The suggestion may have come from his co-author, Sosthenes (1 Cor. 1: 1).[120] Even in that case, we must ask what motivated Paul to accept such an unchristian tactic. Personal factors were probably decisive. The influence acquired by Apollos was an implicit criticism of Paul's leadership. He might content himself with simple proclamation to those incapable of receiving anything higher, but the spirit-people expected much more. A latent note of challenge is easily detected (1 Cor. 2: 15; 4: 3). If Paul did not offer them 'wisdom', might it not be that he was incapable of the soaring religious speculation that they considered integral to true religious authority? Might not the absence of such a gift indicate that he was not fitted to lead the community? Such presumption, from Paul's perspective, merited a sharp put-down, and he provided it in 1 Corinthians 1–4, which sets the tone for his relationship with the spirit-people.

[120] See my (1993), 566–70.

The difficulties in the interpretation of 1 Corinthians 2: 6–16 are considered so severe that the passage has been dismissed as a post-Pauline interpolation.[121] The mistake of such commentators is to take Paul seriously. In fact he is playing a cruel intellectual game with his opponents. His whole purpose is to mystify them and thereby reduce them to confused silence amid the laughter of all the others who hear the letter read aloud. This he achieves by appropriating some of the most cherished terms in the lexicon of the spirit- people and giving them a meaning radically opposed to that intended by the latter. He agrees that they are 'spiritual' and possess 'wisdom', but their spirit is 'the spirit of the world' (2: 12) and their wisdom is 'the wisdom of this age' (2: 6). He consents to their self-designation of 'mature' only to redefine it as childishness (3: 1). Those who set themselves up as judges are revealed to be incompetent (2: 14). The unerring precision with which he goes to the heart of their beliefs is inexplicable without the assistance of Apollos, who was with him at the time of writing (1 Cor. 16: 12). The piercing tone of mockery pricks the bubble of their complacency.

The savagely sarcastic rhetorical questions with which this initial section terminates (3: 3–4) reappear in 4: 7, 'Who sees anything different in you? What have you that you did not receive?' The spirit-people have no qualities which would make them valuable as allies or supporters. They are recipients not creators. The contemptuous tone is heightened as Paul goes on to mock them by taking their spiritual language literally, thus transforming their legitimate religious aspirations into absurd social achievements. It takes little imagination to hear the rustle of laughter in the congregation and see the malicious sideward looks as the reader of the letter gave emphasis to the words, 'Without us you have become kings! And would that you did reign, so that we might share the rule with you!' (4: 8). This language has been interpreted as implying social status.[122] The point is debatable, but it does seem rather probable that the spirit-people were drawn predominantly from the wealthier, and better educated section of the church at Corinth. It is they who would have had the leisure and ability to indulge in religious speculation.

Paul's lack of charity in his treatment of the spirit-people must have diminished him in the eyes of sensitive souls in the community, who felt that he had gone too far. Thereafter his judgement became suspect, and unconditional support problematic. For the spirit-people it went much deeper. Profoundly wounded by the humiliation of public ridicule, they were completely alienated and became his implacable enemies. Since they could not attack him directly, they channelled their pain and anger into frustrating his ambitions for the community. Not long after receipt of the letter at Corinth in the summer of AD

[121] Widmann (1979). For a refutation, see my (1986a), 81–4.
[122] P. Marshall (1987), 214–17.

54, they gave hospitality to those whom Paul feared as the greatest threat to his ministry, the Judaizers from Antioch, who had troubled the churches in Galatia.

A Feast of New Insights

Despite his disparagement of the spirit-people, Paul knew that he had to deal with the issues they raised. He could not ignore their positions, while responding to those of others, without running the risk that his silence would be interpreted as approval. He thus found himself obliged to reflect formally on matters which previously had attracted his attention only fleetingly, or into which had he delved only superficially.

Paul probably undertook the task with some eagerness. Despite their faults, the Corinthians had accepted enthusiastically the challenge to translate the gospel into the realities of daily life. It was unfortunate that they made mistakes, but the fact that they at least made the effort put them far ahead of the Galatians, who were too timid even to try, and who sought the safety of authoritative rules and regulations. Whereas Paul ordered the Galatians to accept the burden of freedom, he entered into dialogue with the Corinthians.

The most noteworthy insights which were fostered by the confusion at Corinth concern the means of ministry, communal and personal freedom, the very nature of the Christian community, and the role of women in the church.

Transformed by Grace

His break with Antioch had forced Paul to reflect on the source of his authority as an apostle. If he had given further thought to the nature of ministry, it does not surface in any developed way in Philippians, Colossians, or Philemon. The opposition, which some at Corinth had set up between himself and Apollos, forced Paul to examine and evaluate the differences between them. Was it only a matter of style? Were there valid reasons why certain Corinthians opted for one or the other?

The basic insight is set out with exemplary simplicity: 'the kingdom of heaven does not consist in talk but in power' (1 Cor. 4: 20). The church is not a set of ideas which informs the mind, but a context of divine power which transforms the personality. Authentic ministers, in consequence, do not use their rhetorical training to develop persuasive arguments, but manifest the presence of the grace-giving Spirit. Faith is not the conclusion of a logical discourse, but is born of a vision of God at work here and now (1 Cor. 2: 1–5). Ministers assume the responsibility of being the place where the divine is active; their comportment must be such as to reveal the power of grace. They have to be able to say, 'Imitate me!' (1 Cor. 4: 17; 11: 1).

The need to define his relationship with Apollos prompts Paul to explain the description of a minister as 'God's co-worker', which he had used in a previous letter (1 Thess. 3: 2). God does not need any assistance in granting grace. But he has chosen humans as instruments through which he works to generate and sustain faith (1 Cor. 3: 5–9),[123] and their mutual relationship is decided by his will. Totally dedicated co-operation, therefore, is essential, and inappropriate behaviour empties the cross of Christ of its power (1 Cor. 1: 17). Inadequate ministers can block the passage of grace.

The importance of the witness value of the community in the proclamation of the gospel was apparent to Paul from his earliest letter (1 Thess. 1: 6–8; 4: 12; cf. Phil. 2: 14–16). The disparagement of the body by the spirit-people at Corinth made it imperative for him to emphasize this insight, which permeates the whole of 1 Corinthians, even though it surfaces explicitly only occasionally (10: 32; 14: 23–5). It is why, after dealing with the divisions within the community (chs. 1–4), Paul develops a tightly knit tripartite section, whose basic theme is the importance of the body as the sphere in which commitment to Christ becomes real (chs. 5–6). The key section is the central one devoted to lawsuits (6: 1–11). His point is not that the church should hide its dirty linen from the eyes of outsiders, but that they should grasp the opportunity to demonstrate the power of grace to non-believers by resolving such disputes themselves. Unless it is missionary, the church is untrue to itself. The world needs to see grace at work.

Personal and Communal Freedom

A feature of Paul's treatment of the incestuous marriage at Corinth is the stress he puts on his spiritual presence.

> As for me, absent in body but present in spirit, as one who is present, I have already judged the man who has done such a thing in the name of the Lord Jesus. When you are assembled, I being with you in spirit, and empowered by the Lord Jesus, such a person should be handed over to Satan.
>
> (1 Cor. 5: 3– 5)[124]

When dealing with a similar need for excommunication some years earlier Paul had simply ordered the community to refrain from all contact with the offender (2 Thess. 3: 14). In 1 Corinthians, on the contrary, we see a definite development in his sensitivity to the need of the local church for genuine autonomy if it is to develop normally. Those to whom all is dictated remain for ever immature. The lesson of Antioch had been thoroughly learned.[125]

Here, therefore, Paul begins by asserting the responsibility of the community for its own authenticity (5: 2). He thereby gives an unambiguous signal that he

---

[123] Translations of 1 Cor. 3: 9 such as 'we are God's servants, working together' (*NRSV*), 'God's we are, being fellow workers' (Fee (1987), 134), which put the emphasis exclusively on possession, fail to do justice to the productive activity in which God is involved. Ministers work with God.

[124] For the justification of this translation, see my (1977a).

[125] See Ch. 6, 'Pastoral Instruction'.

is not going to resolve the problem by fiat. It is the community which must decide. All he can do is to give himself a voice and vote in their council by claiming to be spiritually present. He spells out unequivocally what his position is, but does not pre-empt the decision of the church.

Precisely the same attitude emerges in his treatment of the eucharist (1 Cor. 11: 17–34). After analysing the situation in the light of his understanding of the eucharist, he informs the Corinthians that, even though they say the words of institution, their lack of charity positively excludes their meal being the Lord's Supper (11: 20). He does not tell them how to solve the problem. He simply lays out the need for self-examination on their part and specifies the criterion they must use, 'Anyone who eats and drinks without discerning the Body eats and drinks judgment on himself' (11: 29).[126] Interpersonal relations within the community are the crucial factor.

What was valid for the community as a whole was also true for individual members. Paul refuses to command their moral behaviour. Thus, in the case of eating meat offered to idols, he makes clear his disagreement with the position taken by some Corinthians, but his only conclusion is to tell them what he personally would do in the same circumstances, 'Therefore, if food is a cause of my brother's falling, I will never eat meat, lest I cause my brother to fall' (1 Cor. 8: 13).

It is unfortunate that this awareness of what the believing community needs from its leaders did not extend to the legitimate desires of the spirit-people. If it had, the situation in Corinth would have developed very differently.

The Body of Christ

In discussing Paul's response to the Galatians we saw that he attained a vague perception of the organic unity of Christians and their relation to Christ. They had 'put on Christ' and were 'one person in Christ Jesus' (Gal. 3: 27–8).[127] Col. 1: 18 and 2: 19 show that Paul continued to ruminate on the vistas opened by this perspective. It was available, and proved its value, when the factionalism of Corinth forced him to develop a definition of the community which would make the very existence of hostile groups appear self-contradictory. The need to exploit practical applications provoked a deepening and clarification of the fundamental insight.

In the first part of the letter, Paul four times introduces the name 'Christ' in contexts in which it cannot be understood of the individual Jesus Christ (1 Cor. 1: 13; 6: 15; 8: 12; 12: 12). It can only be a designation of the community.[128] But

---

[126] For the justification of giving Body a capital letter, see Fee (1987), 563–4, and Robinson (1952), 60 n. 1.

[127] See Ch. 8, 'The Living Christ'.

[128] For details, see my (1978*b*), 563–64; M. Mitchell (1991), 242 n. 321.

in what sense could the community be called 'Christ'?[129] Common sense excluded a static identity of being; the individual Jesus Christ was a completely different reality.[130] Only functionally could Christ and the community be considered one. Believers were the means by which the Risen Lord acted in the world. They were his ears, eyes, and hands. What he had done when physically present, they now do in his name and with his power. Thus to consider believers as agents of Christ, while true as far as it goes, does not exhaust the depth of Paul's meaning. The most fundamental activity of the church is an expression of its being.

'You are the Body of Christ and taken singly members of it' (1 Cor. 12: 27). The rather awkward formulation can only mean that collectively the Corinthians are the 'Body of Christ', whereas individually they are its members. In many instances the interpretation of this key verse is vitiated by the adoption of the holistic definition of 'body', which was first proposed by J. Weiss in his commentary on 1 Corinthians 6: 13 and then given the the status of common currency by R. Bultmann[131] and J. A. T. Robinson,[132] which prepared the way for the cosmic sacramental hypothesis of P. Benoit.[133] R. H. Gundry, however, has shown that no text demands the holistic interpretation, and that all passages become more intelligible when 'body' is permitted to retain its connotation of physicality or corporeity, and he perfectly grasps Paul's meaning in writing that 'a distinction between two bodies of Christ has to be drawn—an individual body, distinct from the believers, in which he arose, ascended, and lives on high, and an ecclesiastical Body consisting of believers, in which he dwells on earth through his Spirit'.[134] By calling the community the Body of Christ, therefore, Paul identifies it as the physical presence of Christ in the world. The mission of the church is a prolongation in time and space of the ministry of Christ by manifesting, as he did, the power and wisdom of God (1 Cor. 1: 24). Its role is to display God's intention for humanity and to enable those under the power of Sin to attain that ideal (Rom. 3: 9).

How the church exercises this ministry becomes clear when we note that for Paul the functional role of the Body is rooted in its very nature. What he once said in Galatians, 'you are all one person in Christ' (3: 28), is clarified by being reformulated as 'we who are many are one body' (1 Cor. 10: 17), which is later repeated as 'so we, many as we are, are one body in Christ' (Rom. 12: 5). With great insight Robinson has pointed out that in these texts 'the fact of unity, as

---

[129] The only commentator to attempt to answer this question is Wolff (1982), 107–8, who insists that 1 Cor. 12: 12b be read in the perspective of its bracketing verses, both of which evoke the activity of the Spirit. The community is 'Christ' in so far as it is the sphere where the saving power of the Spirit is at work.   [130] So rightly Barrett (1968), 287.
[131] (1965), I. 195, '*Man, his person as a whole*, can be denoted by *soma*.'
[132] (1952), 28–9, 'Indeed, *sôma* is the nearest equivalent to our word "personality". ... Frequently again, as in the case of *sarx*, *sôma* is simply a periphrasis for the personal pronoun.'
[133] (1956), 5–44.   [134] (1976), 228.

the basic datum, always stands for Paul in the main sentence; the multiplicity, on the other hand, is expressed by a subordinate phrase or clause with the sense of 'in spite of', and Robinson also draws the correct inference, 'the diversity is one that derives from the pre-existing nature of the unity as organic: it is not a diversity which has to discover or be made into a unity'.[135]

The church differs from all other human groupings in so far as its unity is not functional but organic. Its members are not merely united by a common purpose, but share a common existence. An autonomous Christian is as impossible as an independent arm or leg. Arms and legs exist only as parts. If they are given the status of independent wholes by amputation they are no longer an arm or a leg. For a while they may look as if they were, but corruption has begun, and they can neither grasp nor walk. The same is true of believers. Their existence is loving—'without love I am nothing' (1 Cor. 13: 2)[136]—which necessarily implies a relationship to another person. To love and be loved is of the essence of Christianity and is constitutive of the being of the believer. They are bound together by what makes them be what they are. Only now does it become clear what Paul tentatively envisaged when he said, 'It is no longer I who live, but Christ who lives in me' (Gal. 2: 20). The independent self, which the world takes for granted as normal, is absorbed into the authenticity of an organic community.

Paul's insight into the nature of the church as an organic unity inevitably conditions his understanding of individuation. Individuation by independence, e.g. 'I think, therefore I am', would be categorically refused by Paul, because it would destroy the unity which makes believers what they are. As arms and legs relative to the physical body, the members of the Body of Christ are differentiated by their various capacities for service. Each has a different spiritual gift which is necessary for the common good of the community (1 Cor. 12: 4–7). The authentically Christian use of the first-person singular must always be a variation of 'I exist to serve you'.

The most fundamental ministry of the church is to be the antithesis of a world which is characterized above all by divisions. Within the framework of hostile blocks (Gal. 3: 28), individuals are separated from one another by barriers of fear and suspicion (1 Cor. 5: 10–11; 6: 9–10).[137] The role of the church is to liberate the captives by revealing the opportunities of freedom in dependence on others.

---

[135] (1952), 60.

[136] The best commentary remains that of Spicq (1959), 71 n. 2, '*Outhen eimi* ... c'est presque l'équivalent du métaphysique *non-être* (*to mê on*, Platon, *Soph.* 238d; Aristote, *Métaph.* v, 2, 1026b 14). Chrétiennement, ce prophète ou gnostique sans charité n'existe pas (comparer *este en Christô*, 1 Cor 1, 30)'.

[137] The other Pauline vice-lists are Gal. 5: 19–21; Col. 3: 5, 8; 2 Cor. 12: 20–1; Rom. 1: 19–31; 13: 3). They exhibit an unusually high percentage of social vices which make real communication impossible; see my (1982*b*), 132–6.

When viewed against this background, the factions at Corinth appear as an abberration as radical as the misplaced ambitions of the leaders of house-churches at Philippi (Phil. 4: 2). The way in which he dealt with this latter problem shows that when Paul wrote that letter he did not yet have the conceptual tools which the Corinthian situation forced him to develop. He was aware that the rivalry was dangerous, but he did not then see it as contradictory of the very nature of the church.

Paul's vision of the church as an organic unity had far-reaching implications for his understanding of Christian morality. Thus he says to the litigants, 'To have lawsuits with yourselves is a total failure for you' (1 Cor. 6: 7). For one Christian to sue another is equivalent to bringing a case against oneself, because they are both members of one body.[138] Would it make sense for the arm to sue the leg? If the suit succeeded, would not both lose? In the same perspective, Paul will not accept that what is right in itself (e.g. the eating of meat offered to idols) is an adequate moral guideline for believers. The questions a believer must ask are: will the projected course of action empower or destroy others (1 Cor. 8: 9; 11: 29)? Will it build up the community (1 Cor. 14: 3–5)?

The Ministry of Women

Despite the ambitions of Euodia and Syntyche which troubled the church at Philippi, Paul took it entirely for granted that women were ministers of the church in precisely the same sense as men. He recognized their gifts as fruits of the Spirit, which he had neither the desire nor the authority to oppose. Given the androcentric world in which he lived, however, it would be surprising if there were not stirrings of opposition among those who failed to appreciate just how radical the gospel was.

We would never know how Paul might have dealt with such criticisms, were it not for an episode which took place during the visit of Chloe's people to Corinth. They participated in one of the liturgical assemblies, and were shocked at the leading role taken by a man, who was apparently homosexual, and a very strange woman.

Presumably Paul had emphasized to the Corinthians the difference between the church and the world in terms similar to those he penned to the Galatians, 'There is no more Jew nor Greek, no more slave nor free, no more male and female' (Gal. 3: 28).[139] No doubt he exhorted the Corinthians to work out a lifestyle which would incarnate the newness of the gospel and make them stand out in their environment. In a gesture typical of their infantile mentality (1 Cor. 3:

---

[138] So rightly Robertson and Plummer (1914), 116. Most versions and commentators translate *meth' heautôn* as if it were *met' allêlôn*.

[139] If the formula is not part of a baptismal liturgy (so Schlier (1962), 174–5: Betz (1979), 181–5; Longenecker (1990), 154–5), it certainly reflects a fixed pattern in Paul's preaching (cf. 1 Cor. 12: 13; Col. 3: 11; and the organization of 1 Cor. 7: 17–28).

2; 14: 20), the Corinthians decided to take the last phrase literally and set out to blur the distinction between the sexes! Unmasculine and unfeminine hairdos flew in the face of accepted convention, as did their approval of incest (1 Cor. 5: 1–2).

In his reaction (1 Cor. 11: 2–16) Paul develops three distinct arguments, but only the first is relevant here.[140] Its kernel is drawn from Genesis 2: 21–2, in which God is shown creating man and woman in different ways. From this Paul deduced that gender difference was part of God's plan for humanity, and so must be preserved as significant. Hence, a man should look like a man and a woman like a woman. For what this meant in practice, Paul, of course, was indebted to the fashions of his age. Men had short haircuts, and the long hair of women was plaited and wound round the top of the head to make a small hair cap.

Aware, however, that Genesis 2: 21–2 was used in Jewish circles to demonstrate the inferiority and subordination of women,[141] Paul immediately moved to ensure that nothing more than what he intended could be drawn from his premiss.[142] 1 Corinthians 11: 11–12 is the first and only explicit defence of the complete equality of women in the New Testament. Paul overturned the traditional argument from the chronological priority of the male in the creation narrative by pointing out that the chronological priority of woman in the birth of a male is just as much part of God's plan for the order of his creation (1 Cor. 11: 12). This elementary argument functions as proof for the principle, 'As Christians,[143] woman is not otherwise than man, and man is not otherwise than woman' (v. 11).[144] Equality is the issue here, not complementarity. The strength and clarity of this insight means that the directive that women must keep silent in church (1 Cor. 14: 34–5) cannot come from the pen of Paul.[145]

---

[140] For the details, and in particular for the references identifying the male homosexual, see my (1980), and (1988*b*).

[141] e.g. Josephus, *Against Apion* 2. 201; Philo, *Questions and Answers on Genesis* 1. 27.

[142] This danger is no less real today than it was in the first century; the majority of commentators interpret 1 Cor. 11: 8–9 as proving the inferiority of women.

[143] Bultmann (1965), 1. 329, has correctly noted that in contexts such as this 'in the Lord' merely fills 'the place of an adjective or adverb which the linguistic process had not yet developed: "Christian" or "as a Christian", "in a Christian manner"'.

[144] For the translation, see Kürzinger (1978).

[145] It was added by a later hand to bring it into line with the non-Pauline 1 Tim. 2: 11–14; see my (1986*a*), 90–2.

— 12 —

Corinth in Crisis

W HEN Paul finished writing 1 Corinthians and dispatched it with the returning delegation, he evidently considered that he had done all that was necessary for the moment. The care he had expended on the letter, he felt, justified a high degree of optimism as to its successful impact. A leisurely swing through the churches of Macedonia which, for all their problems, were as angels compared with the Corinthians, would refresh his spirit. It would be sufficient to reach Corinth at the end of the summer (of AD 54), and he could pass the winter there, if a long stay was indicated (1 Cor. 16: 5–7). These plans were completely disrupted by the need to make an unplanned visit to Corinth.

AN UNPLANNED VISIT

Whether or not Paul made a visit to Corinth between writing 1 Corinthians and 2 Corinthians—the so-called Intermediate Visit—depends on the interpretation of two difficult texts in 2 Corinthians.[1] The first is 2 Corinthians 12: 14 which can be translated in two ways: (a) 'Behold! This is the third time I am making preparations to come to you' and (b) 'Behold! I am ready to come to you for the third time.' The former means that Paul had planned to come to Corinth on several occasions but never succeed in accomplishing his purpose. The latter indicates that Paul had visited Corinth twice already. The same ambivalence plagues 2 Corinthians 13: 1a, which can be rendered: (a) 'this third time I am coming to you' (in opposition to two previous times when he tried but could not make it), or (b) 'this is the third time I am coming to you.'

Everything, therefore, hinges on the meaning of 13: 2, which may be translated very literally as 'I have said previously and I say beforehand, as being present the second time and being absent now, to those having sinned previously and to all the others that if I come again I will not spare.' The complexity of the sentence is due to the fact that Paul is thinking in two different time

---

[1] No such visit is mentioned in Acts.

frames. Once this is recognized, it is easy to separate and link the elements which go together:[2]

(a) 'I have said previously' = 'being present the second time'
(b) 'I say beforehand' = 'being absent now'

The prefix *pro-* in the verbs does not have the same point of reference in each case. In (a) it looks back to a time before the present, whereas in (b) it looks forward to a moment in the future. The latter—in the perspective of (a)—can only be a third visit. The ambiguity of 13: 1 is thereby removed; the only translation possible is: 'this is the third time I am coming to you', which in turn determines the sense of 12: 14. Paul's point is to repeat something—'If I come again I will not spare'[3]—which he once said in the presence of the Corinthians on the occasion of a second visit, and which he now repeats in anticipation of a third visit. The two statutory warnings having been given, he will be free to act as soon as he arrives.[4] The hint that the second visit was not a pleasant one is unmistakable. What happened?

The Mission of Timothy

We have no direct information as to what brought Paul to Corinth not long after he wrote 1 Corinthians. The data available, however, permit only one hypothesis. Timothy carried back news which galvanized Paul into action.[5] 1 Corinthians inadvertently betrays the latter's anxiety regarding how Timothy would be received at Corinth (4: 14–21). Paul's fear is not expressed in so many words, but the sudden shift in tone from the serene introduction of Timothy (4: 17) to the heated outburst, 'Some are arrogant, as though I were not coming to you' (4: 18), reveals Paul's emotional response to his sudden vivid image of what Timothy might be going through.

The possibility continued to worry him, and he made his concern explicit at the end of the letter. He did not want Timothy to be frightened or despised (1 Cor. 16: 10–11). Paul suspected that the spirit-people, who looked down on him for his lack of religious sophistication, would tend to adopt a similarly dismissive attitude towards his delegate. And Paul did not know how much support Timothy would find elsewhere in the community. Fear is often the concomitant of isolation. The agitated affection of Paul for his younger colleague culminated in a veiled threat, 'I am awaiting him with the brethren.' Commentators have made it clear that 'the brethren' cannot be either the three Corinthians mentioned in 16: 15 or other Corinthians supposed to come with Timothy. They must then be the members of the church at Ephesus. Timothy's

---

[2] Allo (1956*b*), 336.
[3] It is a direct quotation, so Plummer (1915), 374; Allo (1956*b*), 337; Furnish (1984), 570.
[4] Van Vliet (1958), 43–62.   [5] So rightly Fee (1987), 822.

fate would be a matter for the whole community—with possible repercussions for the business relations between the two churches.

The simplest explanation of Paul's surprise visit to Corinth is that it was motivated by Timothy's report.

What Happened at Corinth?

Much of what happened when Paul got to Corinth is shrouded in obscurity. He knew what had occurred and so did his readers. He did not have to rehearse the details. We have to try to work backwards from what is incidentally revealed of the aftermath in 2 Corinthians 2: 1–11 and 7: 6–16. The available clues must first be tabulated, and then assessed individually, if the reconstruction is to have any claim to objectivity.[6] The established facts concern both the offence and the response of the community: (1) a single Christian (2: 6; 7: 12) made a serious attack (2: 1, 3, 4) on Paul personally (2: 5, 10); (2) the members of the church did not manifest the personal loyalty and enthusiasm that Paul had expected (7: 12). They were sufficiently at fault to experience the need for repentance (7: 9). Yet they managed to convince Titus of their innocence in the matter (7: 11).

Barrett eases the palpable tension between the last two elements (7: 9 and 11) by suggesting that the offender, though a Christian, was only a visitor and not a member of the Corinthian community.[7] This inference is refused by Wolff on the grounds that the church could have disciplined only one of its own members (2: 6–7).[8] However, the most severe penalty that the community could inflict was to withdraw from all contact with an individual (cf. 1 Cor. 5: 11), and it was perfectly feasible for it to refuse the hospitality which the visitor had hitherto enjoyed. Hence, unless we are prepared to assume that Paul was telling a lie in order to exculpate the Corinthians (7: 11), we must conclude that the incident was provoked by an intruder.[9]

Why would an intruder challenge Paul's authority? This question is an invitation to unbridled speculation, and hypotheses have proliferated. It is much more profitable to ask: how could the intruder act in such a way that Paul would feel himself profoundly insulted and personally injured, but that the community would consider the issue to be none of its business? The answer, which builds on what we have already discovered, and which reveals the seeds of subsequent developments, is that the visitor was a spokesman of the

---

[6] This was done first by Allo (1956b), 55, whose observations and inferences were further refined by Barrett (1970a).

[7] Ibid. 155.

[8] (1989), 43. It is also refused by R. P. Martin (1986), 238, but without any justification.

[9] Thus we can exclude the classic views that the offender was the incestuous man of 1 Cor. 5: 1–5 or one of the litigants of 1 Cor. 6: 1–11; for details see Allo (1956b), 56– 60; Furnish (1984), 163–6; R. P. Martin (1986), 237.

delegation which Antioch had sent to exercise its rights over the churches that Paul had founded in its name.[10]

While still a prisoner in Ephesus, Paul had become aware of the ambitions of the Judaizers to follow him into Europe, and wrote to Philippi to alert the church there of its danger (Phil. 3: 2–4: 1).[11] Presumably, the reason for his planned visit to Macedonia (1 Cor. 16: 5) was to check whether the situation he feared had in fact developed. In these circumstances, the worst possible news that Timothy could have brought back to Paul was that the Judaizers were far ahead of him, and had already penetrated Corinth. Paul's sense of shock and outrage would have been exacerbated by the realization that they must have passed through Macedonia to get there. He would not have been caught so unawares had they passed through Ephesus. Philippi and Thessalonica, in consequence, were also at risk.

What was Paul to do? The first option was to continue with his original project, to go first to Macedonia and then to Corinth. This plan had little to recommend it. To follow the Judaizers around put him at a psychological disadvantage. They were making the running, and he was only playing catch-up. Moreover, Paul's relations with Thessalonica and Philippi were good. The problems of these churches with which he had to deal had not affected the affection in which he was held. At Corinth, on the contrary, he was not universally admired. The spirit-people, at least, were against him. And on reflection in this moment of crisis he may have realized that the strategy he had adopted in 1 Corinthians would not have made them any friendlier.

Recognition of this fact was decisive. Just because the Judaizers were opposed to Paul, they were more likely to get a favourable reception at Corinth than in Macedonia. Hence, it was imperative for Paul to go in person to Corinth. But something had to be done about Macedonia. Even though Philippi was aware of Paul's views on Judaizers (Phil. 3: 2 to 4: 1), a personal visit could only be beneficial. Even without the evidence of Acts 19: 22, it would have been natural to assume that the responsibility was entrusted to Paul's most valued co-operator, Timothy, particularly since his visit to Philippi had already been announced (Phil. 2: 19).

Timothy's heart must have been heavy as he headed north with Erastus (Acts 22: 22) to take ship from Troas. The impact of his news from Corinth on Paul had dismayed him, and he did not know what hostility he might face among the believers in Macedonia, where the Judaizers had had the time to establish a firm base. The burden he carried, however, could not be compared to that borne by Paul. As the ship swooped over the waves en route to Cenchreae, the eastern port of Corinth, it is easy to envisage the depressed state in which his imagina-

---

[10] See Ch. 8, 'Who were the Intruders?' Barrett (1970a), 156, with his usual fine perception, finds in the incident the seeds of the problems with which 2 Cor. 10–13 struggles.

[11] See Ch. 9, 'Letter of Warning'.

tion created one scenario worse than the other, while he mulled over different possible strategies. If he lost Corinth, his enemies would have completed their encirclement, and Ephesus at the centre could not long survive. The whole future of the Gentile church was at stake.

The confrontation at Corinth with the leader of the Judaizers was undoubtedly dramatic.[12] The line taken by the latter must have been very similar to the one he took in Galatia.[13] Paul, he asserted, was a dishonest representative of the church which had sent him out to proclaim the common faith. A traitor to his commission, Paul preached his own ideas, not the common gospel. The dismissive tone in which the slanders were pronounced added insult to injury. Paul leaves us in no doubt that he had been deeply humiliated, and his authority challenged in the most radical way possible.

The Neutrality of the Corinthians

Yet what perturbed him most was the attitude of the Corinthians. Their failure to come to his support cut him to the quick. From the perspective of an outsider this is not as disconcerting as Paul found it. The attitude of the church of Antioch was irrelevant as far as the Corinthians were concerned. They were confident of their own identity, and they had absorbed Paul's teaching on the autonomy of the local community. The idea of a claim based on a genetic connection with another church would have seemed rather unrealistic to them. The success of Jerusalem in imposing its ethos on Antioch was no concern of theirs. It might matter to Paul, but in that case he could fight it out with the intruder on a purely personal level.[14]

Naturally Paul did not see the episode in this light. Not only was there the matter of his wounded vanity—those whom he had engendered through the gospel (1 Cor. 4: 15) should have preferred him above all others—but the 'neutrality' of the Corinthians induced the suspicion that they were prepared to listen to the Judaizers. In fact the Corinthians might have replied, when Paul asked them to reject the intruder, that no one should be condemned without a fair hearing. This, of course, was precisely what Paul wanted to avoid. The report from Galatia had revealed to him the seductive character of the message of the Judaizers. One might reasonably suspect that the spirit-people were behind the refusal to let Paul lord it over their faith (2 Cor. 1: 23); the wounds caused by his treatment of them in 1 Corinthians were still open and bleeding.

The obstinacy of the Corinthians put Paul on the horns of a dilemma. On the one hand, he wanted to stay in Corinth to counter the arguments of the

---

[12] Barrett (1973), 7.

[13] See Ch. 8, 'Discrediting Paul'.

[14] This hypothesis not only integrates a large mass of data, but it has the added advantage of making unnecessary all theories of conspiracy or domination among the various factions at Corinth.

Judaizers, but on the other hand he believed himself to be necessary in Macedonia, where they were also active. He could not stay indefinitely in Corinth, and perhaps he recognized that his presence was only exacerbating the situation. The beneficial effects on everyone of a breathing space may have been a factor in his decision to leave. Where did he go?

A Journey to Macedonia

A number of commentators assume that Paul returned directly to Ephesus, on the grounds that 2 Corinthians 1: 15 plans a voyage with the following components: to visit Corinth first, from there to go to Macedonia and to return to Corinth, whence he would leave for Judaea.[15] In writing thus, however, Paul was attempting to justify not one but two changes of plan. The first he had announced to the Corinthians in 1 Corinthians 16: 5–6; he would come to Corinth via Macedonia. In fact, the circumstances discussed above forced him to go to Corinth first. It is this revised plan that he tries to present in the most attractive way possible by implying that he had changed his mind because of the unique importance of Corinth in his eyes; it would get two visits instead of one (2 Cor. 1: 15). At the time of writing he had in fact visited Corinth—the so-called intermediate visit—but subsequently changed the travel plans he had announced there, namely, to go to Macedonia and then return to Corinth (2 Cor. 2: 1). He went straight from Macedonia to Ephesus (2 Cor. 2: 12–13).[16]

This interpretation is reinforced by the inherent probabilities of the situation. If anything, Paul's reason for going to Macedonia (1 Cor. 16: 5–6) had been reinforced by his painful experience at Corinth. Since he himself had found it impossible to control the Judaizers, what chance would the less authoritative Timothy and Erastus have had? Even without 2 Corinthians 1: 15, we would be obliged to deduce that, from Corinth, Paul went to Macedonia in order to check on the situation there.

It is most improbable that Paul sailed north from Corinth. Not only was the voyage long and dangerous, but the Etesian winds began to blow strongly from the northern quadrant in mid-June and continued for three months, making northward navigation difficult if not impossible.[17] The overland route from

---

[15] e.g. Furnish (1984), 55, 143.

[16] So rightly Barrett (1973), 7; Fee (1978), 538; Lüdemann (1984), 94; R. P. Martin (1986), 24.

[17] Travel conditions would have been identical to those on the other side of the Aegean Sea as reported by Pliny the Younger to Trajan, 'I have arrived at Ephesus . . . and my intention now is to travel on to my province [Bithynia] partly by coastal boat and partly by carriage. The intense heat prevents my travelling entirely by road and the prevailing Etesian winds make it impossible to go all the way by sea' (*Letters*, 10. 15; trans. Radice), and a little later, 'I found the intense heat very trying when I went on to travel by road and developed a touch of fever which kept me at Pergamum. Then when I had resumed my journey by coastal boat, I was further delayed by contrary winds so that I did not reach Bithynia until 17 September' (*Letters*, 10. 17).

Corinth to Thessalonica is 580 km. (363 miles).[18] At an average of 32 km. (20 miles) per day the journey would have taken roughly three weeks. Whether Paul would have been able to maintain this average is another matter. He had to cross the great double plain of Thessaly, which in summer is one of the hottest places in Europe.[19] The mountains that ring it still contain the bears, wolves and wild boar, which are mentioned by Apuleius.[20] These, he feared, were awaiting him in the passes to the north, and apprehension must have intensified the exhaustion of the trek across the sun-seared plain.

There can be little doubt that Paul's physical state was an accurate reflection of his dispirited frame of mind as he walked slowly into Thessalonica sometime around mid-July AD 54. The trouble he anticipated, however, did not material-ize. There is no hint that he had to confront the Judaizers either there or in Philippi. On the contrary, the commitment of the Macedonians to his cherished project of the collection for the poor of Jerusalem reveals them to have been entirely on his side (2 Cor. 8: 1–4). The short shrift given to the Judaizers by the Macedonians explains how they reached Corinth so quickly. The contrast between their fidelity and the mocking neutrality of the Corinthians intensified the bitterness that Paul felt towards the latter.

Perhaps under the influence of Timothy, who is likely to have stayed on in Macedonia, Paul had the good sense to realize that it would be unproductive to return to Corinth with bile seething within him (2 Cor. 2: 1). It could only lead to another explosion and even greater damage. Hence, he decided to change his plans, even though this would give the Corinthians another stick with which to beat him. He found a ship sailing from Neapolis to Troas. From there it was a 350 km. (210 mile) walk to Ephesus. He could have been home in early August.

THE PAINFUL LETTER

Even though he had decided not to return to Corinth, Paul felt that he owed it to himself and to the Corinthians to explain how he felt about what had happened during his visit. It was not an easy letter to write. 'I wrote you out of much distress and anguish of heart and with many tears' (2 Cor. 2: 4). The letter has been lost and cannot be reconstructed in detail.[21]

From the way Paul speaks of the reception of the letter in 2 Corinthians 2 and 7, it is clear that the letter was in no way similar to the one he had written to the Galatians. He did not deal with the arguments of the Judaizers, but focused exclusively on his own relations with the Corinthians (2 Cor. 2: 9; 7: 12). His

[18] Rossiter (1981), 229, 499.
[19] Ibid. 415.
[20] *Metamorphoses* 4. 13; 7. 22–4; 10. 18.
[21] See Ch. 11, '2 Corinthians'.

strategy was to win their sympathy by revealing their treachery through the description of his hurt. The letter was designed to tug at the heartstrings, while at the same time administering a severe shock. The missive had to be strong enough to shake the Corinthians, but not so brutal as to alienate them. Effective reproof had to be blended with the assurance of his affection. The delicacy of the decisions made the writing an agonizing business. Even after it had been despatched, Paul fretted about the impact of the letter. It might do more harm than good. It might have stood a better chance of achieving his goal, had he said this rather than that. The uncertainty weighed upon him terribly.

The letter was entrusted to Titus (2 Cor. 2: 13; 7: 6). It would not have been tactful to send Timothy, who at the least had been the occasion of the blow-up between Paul and the Corinthians. No matter what other assistants may have been available, Titus had a special qualification. He had been with Paul at the Jerusalem Conference (Gal. 2: 1–3). What this meant, is well formulated by Barrett,

> There is thus the strong probability that Titus emerged from the Jerusalem meeting the uncircumcised Gentile he had always been, and that he would retain from this gathering a keen awareness of the peril of legalistic Judaism and of the activities of false brothers; also he would be aware of the quite different (even if not wholly satisfactory) attitude of the main Jerusalem apostles.[22]

Titus, in other words, made an admirable foil to the Painful Letter, in which Paul had poured out his anguished deception. He was in a position to report authoritatively on the agreement between Paul and the Mother Church and thereby to refute any claims, or highlight any distortions, of the Judaizers.[23]

DEPARTURE FOR TROAS

In 2 Corinthians 2: 13–14 Paul gives the impression that his departure from Ephesus was motivated by his affection for the Corinthians and his desire to have news of them.[24] In order to encounter Titus as soon as possible he moved north to Troas and then aborted a fruitful ministry there in order to cross over to Macedonia.

The first question raised by this scenario is: why did Paul settle down to minister in Troas? If he failed to find Titus there, would his concern not have driven him to sail to Neapolis, and then backtrack along the route he had taken only a month or so earlier?

---

[22] (1969), 5.
[23] Similarly Wolff (1989), 48.
[24] This is the standard interpretation; see Furnish (1984), 171; R. P. Martin (1986), 41.

The scenario also raises a second question: how did Paul know which way Titus would return to Ephesus? The obvious answer is that Paul had instructed him to return through Macedonia.[25] But this only generates another question: why would he have limited Titus' options in this way? If Paul were as anxious for news as he makes it appear, it would have been more sensible to let Titus make the decision. Under bad conditions a boat could make the transit from Corinth to Ephesus in two weeks. Under optimum conditions the land route of 1,082 km. (676 miles) could not be done in less than five weeks; the average time was probably a couple of weeks longer. It should have been left up to Titus to decide whether the risk of a late crossing was reasonable. Captains did not risk their ships and livelihood stupidly. And Paul was well aware that safety on the roads could not be taken for granted.[26] Either way there was danger, and only the person on the spot could weigh the pros and cons.

If Paul ordered Titus not to come by sea, it can only be because he suspected that he might not be in Ephesus when Titus arrived. Paul, in other words, was aware of some danger and had prepared a fall-back position.[27] If forced to leave Ephesus, he could be found at Troas, through which Titus would have to come on the overland route. Only when Titus had not appeared, and the sailing season was drawing to a definitive close around mid-September AD 54, did Paul take ship for Macedonia. If he missed the last boat, he would be separated from Titus for the winter, and he would have to wait for news until spring opened the seas to travellers.

Paul's sojourn in Ephesus had not been without its problems. He was opposed by certain members of the community (Phil. 1: 15–17). He had been imprisoned while under investigation, but whether that is what he meant by the reference to fighting with wild beasts (1 Cor. 15: 32) is an open question. The phrase has to be taken metaphorically,[28] but the precise form of the confrontation cannot be discerned. All of that, however, was in the past when he wrote 1 Corinthians, even though he still had enemies in the city (1 Cor. 16: 9).

Presumably it was these latter who were at the root of the 'affliction we experienced in Asia', a trial so grave that 'we despaired of life itself; indeed we felt that we had received the sentence of death' (2 Cor. 1: 8b–9). The formulation suggests less a juridical condemnation than Paul's conviction that his days were numbered. The introduction to this episode, 'we do not want you to be ignorant' (2 Cor. 1: 8a) indicates that it happened fairly recently and that the Corinthians are being made aware of it for the first time.[29] It is difficult to avoid the conclusion that it took place after Paul's return from the intermediate

---

[25] So rightly Furnish (1984), 172.
[26] See Ch. 4, 'Dangers on the Road'.
[27] So rightly Fee (1987), 821.
[28] See Osborne (1966); Malherbe (1968).
[29] Furnish (1984), 122.

visit.[30] The hints coalesce into a coherent picture. At the time when Titus was despatched to Corinth, opposition to Paul in Ephesus was growing. A sudden intensification of hostility forced Paul to leave the city. He moved north to Troas, where he began a new mission.

In Acts a violent episode—the disturbance instigated by Demetrius, the silversmith (Acts 19: 23–40)—is narrated just prior to Paul's departure (Acts 20: 1). If this narrative, which is all from the hand of Luke,[31] is assumed to be historical and subjected to close analysis, we find ourselves confronted with a whole series of unanswered questions and internal contradictions. The story can only be understood as a vehicle created by Luke to present, in a vivid scene, the rehabilitation of Paul by the authorities of the city, and the victory of Christianity over paganism.[32] Luke's care to anticipate Paul's departure by noting, in terms which appear to be based on Romans 15: 23–6, that it was planned before the riot, looks like a deliberate attempt to persuade the reader that Paul was not driven out of Ephesus.[33] Immediately one suspects that this is exactly what happened!

Two factors influenced Paul's choice of Troas as the area of his new apostolate. The first was personal; the city had to be on Titus' return route from Corinth. The second was strategic. The missionary expansion of the church of Ephesus had previously been limited to places within a week's walk of the city, the capital of Asia.[34] Now Paul decided to go further afield, and the two occasions on which he had already passed through Troas had shown him the value of a community there, which could serve as a link between the churches of Asia and those of Europe. Moreover, it would provide him with the large urban environment in which he worked most effectively.

At the time of Paul, Troas was 'one of the most notable cities of the world', a Roman colony founded by Augustus and encircled by a massive wall 8 km. (5 miles) long.[35] It resembled Corinth in its strategic location as a transit point for trade between Asia and Europe, and was very prosperous.[36] Its population has been estimated at between 30,000 and 40,000.[37] Paul's ministry there can hardly have lasted more than a month, but he hints that it was successful— 'despite the opportunity I took leave of them' (2 Cor. 2: 12–13)—and a Christian community at Troas is presumed by Acts 20: 7–12.

---

[30] Lüdemann (1984), 133 n. 174.
[31] So Boismard and Lamouille (1990), 2. 314–15; 3. 243.
[32] Haenchen (1971), 576–9.
[33] Ibid. 569.
[34] See Ch. 7, 'Missionary Expansion'.
[35] Strabo, *Geography* 13. 1. 26.
[36] For details, see Hemer (1975); Yamauchi (1992).
[37] Cook (1973), 383.

THE REPORT OF TITUS

The use of the first-person plural, 'when we came into Macedonia' indicates that Paul was not travelling alone. Certainly Timothy was with him (2 Cor. 1: 1). Once the Apostle reached Neapolis, one source of anxiety evaporated; he would not be cut off from Titus for the winter. But, as the days slowly passed, the stress of the delay began to wear him down. He mentions a period characterized by 'every kind of affliction, disputes without and fears within' (2 Cor. 7: 5), which was brought to an end only by the arrival of Titus. The strain of the uncertainty of how the Corinthians would react was exacerbated by his fears for the safety of Titus, now seriously overdue, and by squabbles of various sorts.[38] Paul by now was in a state of extreme tension where everything was an irritation. His emotional state inflated questions into accusations, and discussions into disputes.

It would have been impossible for Paul to have passed through Philippi without stopping to visit the believers. Were Titus not there, his stay would have been short. It is entirely possible that he had walked the 150 km. (90 miles) to Thessalonica before Titus appeared. The joyful reunion could have taken place anywhere. Now that winter was setting in, Paul and his companions settled down in the nearest community until good weather returned in the spring.

All we have of Titus' assessment of the situation at Corinth is the version reported by Paul to the Corinthians (2 Cor. 7: 7–16). The presentation is euphoric. Titus had been received in a way which justified the high report which Paul had given him of the Corinthians. The letter he bore, which had been written with such anguish, had achieved its purpose perfectly. The sincerity of the Corinthians' deep contrition for letting Paul down was underlined by the action they took against the intruder. Now they were totally on his side, and as far as Paul is concerned, 'I have every confidence in you' (7: 16).

In view of the shocking explosion which subsequently occasioned 2 Corinthians 10–13, it would be easy to accuse Titus of seeing what he wanted to see, and/or to indict Paul for an overly optimistic interpretation of what Titus told him. This, however, would be a little naïve and fails to do justice either to the intelligence of Titus or to the subtlety of Paul. In 2 Corinthians 7: 7–16, Paul says precisely what the Corinthians expected to hear. They had made an effort, and it was appropriate for Paul to recognize it in the most glowing language possible, particularly since he was going to introduce the topic of the collection for the poor of Jerusalem, with respect to which the Corinthians

---

[38] The support given by Furnish (1984), 394, to the suggestion of Windisch and Georgi that Paul had a confrontation with Judaizers is mistaken. The hypothesis is excluded by the speed with which Paul passed through Macedonia on his way from Corinth to Ephesus, and by his lavish praise of the believers there (2 Cor. 8).

had not been very energetic (2 Cor. 8–9). There are many hints earlier in 2 Corinthians 1–9 that Titus had been sharply observant, and appropriately critical of what was going on at Corinth,[39] and that he reported very accurately to Paul. We must now attempt to reconstruct the gist of what he said.

Judaizers and Spirit-People

Once Titus had assured Paul of the affection of the Corinthians, he produced evidence of their change of heart by detailing the action they had taken concerning the individual who had insulted Paul. The nature of the punishment is not specified, presumably because there was but one possibility, namely, complete ostracization. The believers simply refused to have anything to do with him. He was thrown on the mercies of a society which did not care whether he lived or died. Paul had sufficient imagination to feel the impact of such isolation. If he could be thin-skinned and prickly, he could also be generous, and his immediate concern was for the well-being of the offender. The penalty should be lifted and he should be taken back into the community (2 Cor. 2: 6–8).

It seems likely, however, that Paul's response was not entirely altruistic. It can be seen as an olive branch held out to a group that was still opposed to him, and about which Titus had brought disquieting information. We saw above that the individual who insulted him was in all probability an emissary from Antioch. This is confirmed by the nature of Paul's response in 2 Corinthians 1–9 to Titus' news; the intruders were Jewish Christians.[40] Only one church would send letters of recommendation to another (2 Cor. 3: 1), and they presented themselves as 'servants of the new covenant' (2 Cor. 3: 6). They chose this title in order to harmonize their belief that the eschaton had been inaugurated in Christ with their conviction, inspired by Jeremiah 31: 33, that the Law enjoyed enduring validity.[41] Their insistence on the role of the Law is highlighted by the abrupt introduction of 'on tablets of stone' (2 Cor. 3: 3) in place of the expected 'on parchments'.

The Judaizers, however, were not alone in opposition to Paul. There are also hints that point to the spirit-people.[42] The accusation that Paul's gospel was veiled (2 Cor. 4: 3) can only come from those who considered Paul to have an unimpressive personality and lack-lustre presentation (2 Cor. 10: 10), i.e. those who preferred the more speculative approach inculcated by Apollos on the basis of Philo. The sophisticated multilayered use of 'life' and 'death' in 2 Corinthians 2: 16 is adequately paralleled only in Philo (e.g. *Fuga*, 55), as is the language of 2 Corinthians 6: 14 to 7: 1. The attitude towards the body in 2 Corinthians 5: 6b is typically Philonic.

---

[39] So rightly Furnish (1984), 396.
[40] For details, see my (1986b) and (1987).
[41] See my (1989).   [42] See my (1988a) and (1988c).

The simplest and most natural explanation of the mixture of hints pointing to two very different groups is that Titus in his report had linked them as opponents to Paul's ministry. The mode of reference, moreover, suggests that they did not function as separate groups, but had formed an alliance against him. At first sight an alliance between free-thinking Hellenistic pseudo-philosophic believers, the spirit- people, and Law-observant Jewish Christians seems rather improbable. History, however, abounds in instances of minority groups with radically different aims uniting in order to overthrow a common enemy. When the Judaizers arrived in Corinth, their first act would have been to probe for weaknesses in the community which Paul had built up. In order to work from within, they had to be received by someone, and prudence would have indicated that they search out a group that was already at odds with Paul. In principle it should be more receptive than others to an alternative form of Christianity.[43]

The spirit-people had been brutally and publicly humiliated by 1 Corinthians. Naturally their pride sought revenge. Had Apollos remained in Corinth, they might have formed an alternative church, but he had left them to join Paul and apparently was not particularly interested in returning (1 Cor. 16: 12). Such betrayal, for which they might have blamed Paul, could only have intensified their bitterness. In this frame of mind, they would have been fair game for any of the Apostle's opponents. The alliance, in consequence, was one in which both parties gained something. The Judaizers found a welcome among the élite of the Corinthian community, and the spirit-people were given the means of damaging, if not destroying, Paul's achievement.

In addition to such negative common ground, both groups shared an interest in Moses. For the Judaizers he was the great Lawgiver, whose words had enduring value. For the spirit-people nourished on a form of Philonism, he was much more. Philo regularly presents Moses as the 'the perfect wise man' (*Leg. All.* 1. 395), who epitomized all Hellenistic virtues as 'king and lawgiver and high priest and prophet' (*Vita Mosis* 2. 3; cf. *Praem.* 53–6). Having alienated himself from the body (*Conf.* 82), Moses entered into the mysteries of God which, in consequence, he was able to reveal and teach (*Gig.* 54). In a word, Moses was everything that the spirit-people aspired to be.

It is easy to see how the Judaizers could have exploited this advantage in the interests of their mission. Philo insists on the honour in which all nations hold the Law of Moses (*Vita Mosis* 2. 17–24), and highlights the providential character of its availability in Greek (*Vita Mosis* 2. 25–44). The Law has a universal appeal because its statutes 'attain to the harmony of the universe and are in agreement with the principles of eternal nature' (*Vita Mosis* 2. 52), a perspective that is developed in *De Decalogo and De Specialibus Legibus*. Moses

---

[43] This obvious point has not always been recognized, but it has been evoked in passing by Ellis (1975), 287, and by Forbes (1986), 15.

himself was the living embodiment of the Law (*Vita Mosis* 1. 162); the Law-giver could not but act in accordance with the revelation he communicated. Others could reach the same heights of religious speculation by accepting the demands of the Law (*Mig.* 89–94). It is easy to see what attraction this approach would have had for the spirit-people. And once they were committed to Moses, the Judaizers were halfway home.

In order to enhance their appeal to the spirit-people, the Judaizers had to make some concessions. They would have become aware very quickly that the basis of the hostility to Paul among the spirit-people was rooted in his failure to meet their expectations concerning religious leadership. Thus the intruders were led to stress their superior qualifications. They proclaimed their credentials (2 Cor. 4: 5; 10: 12) by advertising their visions and revelations (2 Cor. 12: 1), and their miracles (2 Cor. 12: 12). If they did not know them already, they would have adopted conventions of Hellenistic rhetoric. Themes developed at some length and with a spice of mystery would have flattered the sensibilities of the spirit-people.[44]

The situation in Corinth, therefore, was anything but happy. The danger was much more insidious than at Galatia. The Judaizers had realized that their frontal attack on Paul had backfired. They were now consolidating their base among disaffected elements in the community. Once that had been achieved they would move into other sectors of the church, in which the ground had been prepared to some extent by Paul's own attitudes.

Travel Plans

Titus had also picked up criticism of Paul's inability to keep his word. Paul's vacillation regarding his travel plans was an easy terrain on which anyone with a grievance could score points. Paul had told them one thing (1 Cor. 16: 5–6), and he did not do it. He then promised them something else (2 Cor. 1: 15–16), and failed to do that. What finally he actually did had no resemblance to either; he merely wrote them a letter. The impact of such changes on the Corinthian church was minimal. Paul was not essential, either in theory or practice, to any aspect of its functioning. Why, then, did the changes become an issue?

The way in which Paul replies—'Do I make my plans like an opportunist, ready to say "Yes, yes" and "No, no"?' (2 Cor. 1: 17)—clearly indicates that he was being charged with the inconstancy of the flatterer, whose criterion of behaviour is the momentary pleasure of the listener. This is perfectly illustrated

---

[44] The 'Corinthianization' of the intruders was first recognized by Barrett (1970*b*), 251, and this is sufficient explanation of the Hellenistic traits which have led some commentators to identify Paul's opponents in 2 Cor. as Hellenistic–Jewish missionaries; e.g. Georgi (1986), 315; Furnish (1984), 53. They were in fact of Syro-Palestinian origin; so rightly Windisch (1924), 23–6, and Barrett (1973), 28–32.

by the self-description of a flatterer in Terence's play, *The Eunuch*, 'Whatever they say I praise; if again they say the opposite, I praise that too. If one says no, I say no; if one says yes, I say yes' (lines 251–3).[45] There were people at Corinth saying that Paul was entirely untrustworthy; his word could not be relied upon. In consequence, he could neither be a true friend nor, in this context, an authentic leader. The atmosphere in the community at Corinth was being deliberately poisoned, by continuous sniping at Paul. Titus must have warned him that his every word and gesture was liable to deliberate misconstruction (2 Cor. 1: 12–14). On the basis of what we know already, only the spirit-people had reason to justify such malice.

Financial Assistance

The same snide attitude surfaces in criticism of Paul's financial relations with the Corinthians. This had already been an issue at the time of writing of 1 Corinthians, and the spirit-people, who were also the ones most directly involved, now had even greater incentive to use it as a weapon against Paul. It gave them an opportunity to highlight a different aspect of Paul's inconstancy, and to elevate a hint of untrustworthiness into a charge that Paul did not practise what he preached. He had refused, and continued to refuse, a gesture of love.

The social cement which bound the inhabitants of the Graeco-Roman world together was the reciprocity of benefactions. Seneca in a work devoted to the topic, *De Beneficiis*, called it 'the chief bond of human society' (1. 4. 2). Mere possession of wealth was nothing. It was transmuted into status and power by being distributed.[46] A gift was a public gesture laying claim to superiority, and calling for honour from others. For the recipient, 'no duty is more imperative than that of *proving* one's gratitude.'[47] The gift had to be reciprocated. If the return was superior in value, the original recipient took the advantage. If of equal value, both remained level. If, however, the return was of less value, the recipient became a client, with an unrequited obligation to the giver. Refusal of a gift, though theoretically possible, was not a real option.

> Few were prepared to face the possible or likely hostilities inherent in a refusal. Rather, it was easier to accept an unwanted friendship and let the relationship take its unhappy course. The obligation to receive, then, was generally honoured, even though in many instances, carelessly, foolishly and begrudgingly.[48]

---

[45] See in particular P. Marshall (1987), 70–90, 319.

[46] 'The wealth that you esteem, that, as you think, makes you rich and powerful, is buried under an inglorious name so long as you keep it. It is but house, or slave or money. When you have given it away, it is a benefit.' (Seneca, *De Beneficiis* 6. 3. 4).

[47] Cicero, *De Officiis* 1. 47 (emphasis added). [48] P. Marshall (1987), 17–18.

Corinth in Crisis

At Corinth those capable of conferring benefits on Paul were the élite, among whom he had made his first converts.[49] Their bid to assist him could be justified, not only by the conventions of the period, but also by the fact that he had benefited them. In Corinth they were his oldest 'friends'.[50] The terminology used by Paul does not permit us to decide whether they offered a gift or a salary.[51] The distinction, however, is as irrelevant as any discussion of their motive, because Paul refused. Why did he take a step so much at variance with an honoured custom of his age?[52]

Paul tells us only that 'we endure anything rather than put an obstacle in the way of the gospel of Christ' (1 Cor. 9: 12b). This rather vague justification can be translated into specific reasons only on the basis of what we have seen to be the general principles on which Paul operated. His concern for existential witness guaranteed that he did not want to be compared to those philosophers and religious teachers who expected a return for their teaching. His refusal to conform to their comportment was intended to reinforce the difference in his message. His preoccupation with the unity of the community excluded any action whose result would be to make him a client of one segment of the community.

This latter point becomes clearer if Paul's refusal of Corinthian support is contrasted with the welcome he accorded subsidies from Philippi, both in Thessalonica (Phil. 4: 16), and at Corinth (2 Cor. 11: 9). The variation in practice has been commented on at length, but the essential point has not been highlighted. While one gift could be presumed to be communal, the other was necessarily individual. Distance made a crucial difference.

The Philippian gift represented a community effort. The church created a common fund to which all could contribute. The sum of money was brought by an official delegation, and presented in the name of the church. The implication, as far as Paul was concerned, was that all members of the church had participated, even though some may have given more than others. The individuality of each contribution was assumed into a whole, which symbolized the unity of the community. Thus the subsidy could be accepted by Paul as an offer of abiding friendship. His response was directed to the whole church (Phil. 4: 10–20).

At Corinth, on the contrary, because Paul lived there, all gifts were highly personal. Benefactions were necessarily particular. Not only because they were

---

[49] See Ch. 11, 'The First Converts'.

[50] 'Anyone who shows you some goodwill, or cultivates your society, or calls upon you regularly, is to be counted as a "friend"' (Cicero, *Handbook of Electioneering* 5. 16); cf. P. Marshall (1987), 24–5. [51] Ibid. 225.

[52] The various answers which have been given to this question are discussed by P. Marshall (1987), 233–58, who, unfortunately, does not take into account the fact that considerable water had flowed under the bridge between the writing of 1 Cor. 9 and of 2 Cor. 11–12.

handed over by specific individuals, but because they were in-kind. Lodging meant someone's house, a meal someone's kitchen. How was Paul to react to a multitude of individual gifts? According to the ethos of society, he would have had to portion out his time and energy in such a way that that those who had contributed the greatest amount received the most. The needy poor would have had little chance against the resources of the élite. Even with the best motives in the world, the wealthy would have monopolized Paul's attention to the detriment of the real needs of the community. Before he arrived in Corinth Paul must have seen that to accept a single gift would put him in an impossible situation. It is hardly surprising that he repudiated all.

Since the arrival of a delegation from a sister church could hardly be kept secret, there can be little doubt that Paul was forced to explain to the Corinthians why he refused their gifts while accepting that of the Philippians. In the light of 1 Corinthians 9: 1–12a it would appear that he prefaced his explanation with an assertion of his authority by drawing attention to his right to be subsidized. Such a paradoxical approach is unlikely to have enhanced the clarity of his presentation. To those of good will, the distinction between themselves and the Philippians would have made perfect sense. There may even have been some who thought of creating an organization similar to that in place at Philippi, whereby Paul could be helped and the identity of the donors blurred. Others among the élite, however, considered themselves slighted. Their quest for eminence in the community had been frustrated. It was easy for them to ignore the explanation, and to hammer at the facts. Paul refused them while taking from others. The discrepancy then became an opportunity for alternative explanations, none of which was favourable to Paul.

Titus must have been made aware that criticism of Paul had become habitual. His report to Paul, therefore, must have contained some mention of the way the latter's attitude towards support was being more and more seriously misrepresented.

The Collection for Jerusalem

The reason why Paul did not take up the issue of his personal finances in 2 Corinthians 1–9 was that the question of the collection for the poor of Jerusalem had come up during the visit of Titus to Corinth. The Corinthians had never been informed officially of the collection. They heard of it by accident, presumably from Chloe's people (1 Cor. 1: 11). Only this explains (a) how Paul could compliment them on taking the initiative as regards participation in the collection (2 Cor. 8: 10); and (b) why they had to request detailed organizational instructions, which Paul provided in 1 Corinthians 16: 1–4. Since Paul's one concern at the time of writing the Painful Letter was his relationship with the Corinthians, it is most unlikely that he instructed Titus to complicate an already tense situation by raising the question of the collection, particularly since Paul's attitude towards money was already under fire.[53]

In discussions with the Judaizers, however, the collection would have furnished a perfect *ad hominem* argument. The effort that Paul put into it demonstrated in the most practical way possible his love and concern for the Mother Church, which the Judaizers claimed to represent. They could not refuse the gift of the Pauline churches without endangering the survival of their compatriots in Jerusalem, and without putting themselves in precisely the position which the Corinthians found objectionable in Paul. Once Titus was convinced that the Corinthians had accepted the reprimand of the Painful Letter, it would have been natural for him to remind them gently of their commitment to the collection, a gesture which the Judaizers could only second (2 Cor. 8: 6)! In that instant at least, Paul, the Corinthians, and the Judaizers would have been at one, and an enthusiastic note in Titus' report becomes more understandable.

WINTER IN MACEDONIA

Paul was fortunate that winter had begun by the time Titus returned from Corinth. Since there was no question of the latter going back there immediately, the possibility of a hasty reaction to the situation at Corinth, similar to 1 Corinthians, was excluded. Paul had time to write and tear up many drafts, before a messenger could get through to Corinth the following spring. At the earliest, the letter was sent in March or April AD 55. Climatic conditions, therefore, forced on Paul a period of reflection on the best strategy to deal with a very complex situation. This time he had the additional advantage of having at his side Timothy, who proved to be a much better co-author than Sosthenes had been for 1 Corinthians.

Co-authorship

The contribution of Sosthenes to 1 Corinthians was limited to 1: 18–31 and 2: 6–16. He appears to have been one of those individuals who are briskly insightful in conversation, but who prove to be complicated and overly subtle in formulating a text. Paul gave him two chances, and then in irritation abandoned him; appended to the co-operative sections is Paul's own frank formulation of what he was trying to get across (2: 1–5; 3: 1–4).[54] Timothy was much closer to Paul, and thus had greater influence on him. Manifestly he played a much more significant role in the composition of 2 Corinthians 1–9.

[53] Against Barrett (1969), 10, who is followed by Furnish (1984), 397. F. Watson (1984), 335, is one of the few to disassociate the collection from the Painful Letter.
[54] See my (1993), 566–70.

Whereas 1 Corinthians is a first-person singular letter in which it is necessary to explain the irruption of the first-person plural, 2 Corinthians 1–9 is precisely the opposite; 74 per cent of the letter is expressed in the first-person plural, and only 26 per cent in the first-person singular. All the latter passages deal with situations in which Timothy was not involved, namely, the consequences of the intermediate visit (2 Cor. 1: 15–17; 1: 23 to 2: 13; 7: 3–12), and the issue of the collection at Corinth (2 Cor. 8: 8–15; 9: 1–15). The precise reference of the first-person plural can vary,[55] but in no case is it necessary to exclude Timothy. He and Paul worked consistently and well together, notably in the major section on the apostolate (2 Cor. 2: 14 to 7: 2), but the nature of some of the material obliged Paul to be highly personal. Such interventions tended to run on a little too long. Each time, however, Timothy was able to get Paul back to the co-operative task, which eventually produced the most extraordinary letter of the New Testament.

Weaning the Spirit-People from the Judaizers

In opposition to 1 Corinthians, where Paul jumps straight into the most diffi-cult problem after a rather perfunctory thanksgiving (1: 4–9), which shows his tongue to have been firmly in his cheek,[56] 2 Corinthians 1–9 begins very cautiously and with a subtlety which sets the tone of the letter. The extremely careful craftsmanship betrays the refinement that is the product of long thought and numerous drafts.

The introductory paragraph begins with a blessing (1: 3) and the idea of thanksgiving appears only at the very end of the paragraph (1: 11). The hint that Paul has deliberately diverged from his customary practice is borne out by a close analysis,

> Ordinarily, Paul is the subject of the verb action, here it is the Corinthians; ordinarily, the addressees are referred to in the adverbial phrases, here it is Paul (*hyper hêmôn* — twice; *eis hêmas*); ordinarily the principal *eucharistô*—clause is followed by a final clause, here *eucharistô* is the verb of the final clause; ordinarily, the *eucharistô*—clause forms the beginning of the proemium, here it forms the conclusion; ordinarily, the verb is used in the active, here it is used in the passive.[57]

The Corinthians can hardly have been unaware of the systematic way in which Paul inverted his usual pattern. Many in the community flattered themselves on their intelligence, and they had one if not two letters with

---

[55] The most detailed analysis is that of Carrez (1980).
[56] The Corinthians are complimented on speech and knowledge, which do not rate very high on Paul's scale of values (cf. 1 Cor. 8: 1 and 13: 1 to 14: 40). The Thessalonians had been praised sincerely for their faith, hope, and charity (1 Thess. 1: 3; 2 Thess. 1: 3).
[57] Schubert (1939), 50.

thanksgivings in their possession for comparison, namely, the Previous Letter (cf. 1 Cor. 5: 9) and 1 Corinthians It would have been difficult to avoid the (correct) inference that Paul was sending them a subtle message. First, there is a suggestion that he cannot be unequivocally grateful for the state of the Corinthian church. A breach has been repaired (2 Cor. 7: 5–16), but difficulties remain, and Paul's subversion of the normal thanksgiving prepares for similar sleight of hand with Philonic terminology in the body of the letter. Secondly, Paul's unusual focus on his own experience prefigures the major theological theme of the letter, namely, that suffering and weakness, not power and eloquence, are the distinctive signs of the true apostle.

Paul's two-pronged approach was designed, not only to re-establish his authority, but to drive a wedge between the spirit-people and the Judaizers. If he could rob the latter of their base at Corinth, they would be rendered impotent. Thus, he had to wean the spirit-people away from their guests. To this end he offers a critique of the Mosaic dispensation in terms to which the spirit-people would be particularly sensitive, while at the same time presenting the Christian dispensation in a light which they should find attractive.[58]

In 2 Corinthians 3: 7–18 Paul focuses on the figure of Moses, which was the lever used by the Judaizers to pry their way into the favour of the spirit-people. The polemic edge of his exposition of 'When Moses had finished speaking with them he put a veil on his face' (Exod. 34: 34–5) becomes explicit in the contrast he establishes between his own behaviour and that of Moses, 'we act with confident boldness, not like Moses who put a veil over his face' (3: 12–13). The implications are well brought out in a passage from Philo, for whom the theme had special importance:

> Let men who do injurious things be put to shame, and seeking hidden places and recesses in the earth, and deep darkness, hide themselves, *veiling* their lawless iniquity from sight so that no one may behold them. But to those who do such things as are for common advantage, let there be *confident openness*, and let them go by day through through the middle of the market-place where they will meet with the most numerous crowds, to display their own manner of life in the pure sun. (*Spec. Leg.* 1. 321; trans. Yonge adapted; emphasis added)

By presenting himself simply as he was, without any pretentions, Paul implicitly claimed a whole range of other qualities, which Moses must have lacked because he dissimulated by veiling himself.

Having thus sown the seeds of doubt in the minds of the spirit-people concerning the stature of Moses, Paul goes on to associate Moses' achievement with intellectual blindness (3: 14–15). The Judaizers had played into his hands by introducing the idea of a *new* covenant, for this enabled him to stigmatize the Law as the *old* covenant, thereby making it supremely unattractive to those

[58] For greater detail see my (1986*b*), 50–5.

who thought of themselves as in the forefront of religious thought. By using 'Moses' alone in 3: 15, instead of 'the book of Moses' (Esd. 23 [Neh. 13]: 1; 2 Chr. 35: 12), Paul cleverly attaches the pejorative connotation of 'old' to the figure of Moses. Paul then goes on to reinforce this point by a subtle adaptation of Exodus 34: 34, which is nothing more than a simple and effective *ad hominem* argument. The presence of Jews in the Corinthian community showed that they had found something lacking in their previous mode of life based on the Law. They had been blind and now they see. Why then, Paul implies, would the spirit-people want to commit themselves to the darkness of intellectual sclerosis, when they could have the light of authentic glory in the gospel?

In 3: 17 Paul shifts from indirect criticism to seduction by appropriating two key terms in the lexicon of the spirit-people, namely, 'spirit' and 'freedom'. If in 3: 14–15 Paul associated the Law with intellectual blindness, the vice that the spirit-people most despised, here he identifies the gospel with the values they most esteemed. 'Spirit' evoked the Philonic heavenly man, and for Philo 'freedom' carried the connotations of virtue, perfection, and wisdom.[59]

This brief summary only hints at the intricacy of Paul's argumentation. But it is enough to illustrate the change from the brutal tactics of I Corinthians. Paul was capable of learning from his mistakes. With the assistance of Timothy, and possibly also of Apollos, he thought his way into the religious world of the spirit-people, and chipped away delicately at their convictions. His subtle denigration of Moses diminished the common ground on which the Judaizers had relied. His reformulation of the gospel was carefully calculated both to harmonize with, and gently but firmly refashion, the Philonic perspective which the spirit-people had received from Apollos. These latter, as Paul presumably realized, would have been constitutionally opposed to the restrictions imposed by the Law. It needed but little to tip the balance against the Judaizers. Paul's discreet, indirect approach obviated the danger of a perverse reaction such as had been the outcome of I Corinthians.

Manifesting the Life of Jesus

The attitude of the spirit-people to the historical Jesus is summarized by Paul in the shocking phrase 'Anathema Jesus!' (1 Cor. 12: 3).[60] They found the idea of a crucified saviour repugnant and preferred to think in terms of 'the Lord of Glory' (1 Cor. 2: 8), a super-human saviour from above. Paul could not accept this separation of Jesus and Christ, because, as one of Paul's oldest commentators most perceptively put it, Jesus is the truth of Christ (Eph. 4: 21).[61] Only a being of flesh and blood, anchored in space and time, can demonstrate the real

---

[59] *Quod omnis Probus Liber sit*, 41–2, 47, 113, 117, 131.
[60] See Ch. 11, 'The Influence of Apollos'.
[61] De la Potterie (1963).

possibility of the restored humanity proclaimed in the gospel. Unless the ideal is lived, it remains a purely theoretical possibility, beautiful to contemplate, but without any guarantee that achievement is feasible. Paul, therefore, had to insist that Jesus exhibited love, as opposed to merely talking about it.

Even though the gospels narrated how Christ died on the cross, the preaching tradition of the early church spoke only of the death of Jesus.[62] For Paul, this made its reality too easy to ignore, and, in consequence, he consistently insisted that Jesus died in a particularly horrible way, even though he recognized that a crucified Christ was 'a stumbling block to Jews and folly to Gentiles' (1 Cor. 1: 23).[63] The spirit-people preferred to avert their thoughts from this dimension; it cannot be integrated into any philosophical approach to religion. No doubt the Judaizers co-operated. They could assert, with perfect justification, that Paul's stress on the manner of Christ's death was exceptional. Moreover, their adaptation to what the spirit-people expected of religious leaders meant a life-style more compatible with that of the Lord of Glory than with that of a tortured criminal.

These attitudes obliged Paul to defend both his ministry and the historicity of Jesus. An integrated approach was indispensable, and the quest forced Paul's thought into a new dimension. It was in reflecting on the conditions of Jesus' ministry that Paul saw its relevance to his own situation. In the process he gave new depth to the understanding of Christ's ministry reflected in the gospel tradition.

The manner of Christ's ministry was determined by God, 'For our sake he made him to be sin who knew no sin' (2 Cor. 5: 21). In other words, God willed Christ to be subject to the consequences of sin. Jesus was so integrated into humanity-needing-salvation that he endured the penalties inherent in its fallen state. Jesus saved humanity from within by accepting its condition and transforming it. He became as other human beings were in order to reveal to them what they had the potential to become. Thus he suffered as others suffer, and died as others die, even though he in no way merited such affliction.

If 2 Corinthians 5: 21 highlights the divine plan, other texts emphasize the freedom of Christ's co-operation, 'he became poor for your sake' (2 Cor. 8: 9), and the reason for his choice, 'one died for all' (2 Cor. 5: 14). His life and death were a deliberate sacrifice of self in order that others might benefit. The fundamental lesson of the self-oblation of Christ is that 'those who live might live no longer for themselves' (2 Cor. 5: 15). Prior to Christ it was taken for granted that the primary goals of human existence should be survival, comfort, and

[62] The texts commonly cited are Rom. 1: 3–4; 4: 25; 8: 34; 1 Cor. 15: 2–7; Gal. 1: 3–4; 1 Thess. 1: 10.

[63] Even though the Philippian hymn was based on Paul's teaching, it spoke only of the death of Jesus. Paul had to add 'death on a cross' (Phil. 2: 8). Similarly it was Paul who added 'by the blood of his cross' to the Colossian hymn (Col. 1: 20).

success. In the light of Christ's radical altruism, such a life-style can only be perceived as the 'death' of selfishness. It is the antithesis of genuine 'life', which is totally concerned with benefiting the other.

The presentation of Christ as 'the image of God' (2 Cor. 4: 4) reveals the essence of authenticity to be empowerment, the ability to reach out to enable others.[64] In the chapter of Genesis in which this formula appears (Gen. 1: 26–7), God is presented exclusively as the Creator. In consequence, creativity remains the primary referent in determining the meaning of the phrase. Humans resemble God in so far as they are creative. Christ is, like Adam before the Fall, 'the image and glory of God' (1 Cor. 11: 7) in the sense that he gives glory to God precisely by being what the Creator intended.

The creative power which made Christ the New Adam (cf. 1 Cor. 15: 45) was exercised in and through poverty and ignominy. His whole existence was a 'dying' (2 Cor. 4: 10), but he brought into being 'a new creation' (2 Cor. 5: 17). Once Paul had been led to this insight, it was easy for him to see it as the archetype of his own situation. He was conscious of his 'weakness' (1 Cor. 9: 22), yet he disposed of a 'power' (2 Cor. 4: 7), which created new communities of transformed individuals (2 Cor. 3: 2–3). The basis of Paul's identification with Jesus, which is the distinctive feature of his understanding of ministry in 2 Corinthians, was their shared experience of suffering.

Hitherto Paul had accepted suffering as integral to the human condition. His experiences would not have set him apart in the ancient world. Life was harsh and survival very much a matter of luck. None of Paul's acquaintances would have dissented from Homer's insight, 'The sorrowless gods have so spun the thread that wretched mortals live in pain' (*Iliad* 24. 525). Now Paul saw an opportunity to give meaning to suffering. Even though he thought in terms of his own ministry, his insight is valid for all believers. Suffering can be revelatory when the unchangeable is accepted with grace. If the achievement is disproportionate to the means, the power of God becomes visible.

Paul perceived himself as one of the prisoners of war destined for execution at the climax of a Roman victory parade (2 Cor. 2: 14).[65] His first insight is to see his suffering as a prolongation of the sacrifice of Christ. He is 'the aroma of Christ' (2 Cor. 2: 15). As smoke wafting across the city from the altar conveyed the fact of sacrifice to those who were not present in the temple, so Paul in his wanderings proclaimed Jesus to the world, not merely in words, but more fundamentally in his comportment. He speaks of himself as 'always carrying in the body the dying[66] of Jesus, so that the life of Jesus may be manifest in our bodies. For while we live, we are always being given up to death for Jesus' sake,

[64] On this point, see in particular Macquarrie (1966), 212 (§35).
[65] Hafemann (1986), 12–39.
[66] *Necrôsis* is not attested prior to Paul's use here. If in Rom. 4: 19 it means 'a death-like state', the context here clearly indicates a process; hence it should be translated 'dying' rather than 'death'.

so that the life of Jesus may be made visible in our mortal flesh' (2 Cor. 4: 10–11).

This extraordinary statement is the summit of 2 Corinthians, and the most profound insight ever articulated as to the meaning of suffering and the nature of authentic ministry. Death shadowed Paul's every step; he could die at any moment. As one headed towards a fate which seemed inevitable, he saw his life as a 'dying', which he identified with that of Jesus, who had also foreseen his death (e.g. Mark 8: 31). Paul's acceptance of his sufferings created a transparency, in which the authentic humanity of Jesus became visible. By grace Paul is what Jesus was.

Paul, however, did not put himself on the same level as Jesus; what he achieved would not have been possible without Jesus. None the less he recognized that, were Jesus to have been the only one to demonstrate the type of humanity desired by the Creator, its revelation could have been dismissed as irrelevant, a unique case without meaning for the rest of humanity. Hence, his acceptance of the responsibility of being Jesus for his converts. The explicitness of this presentation of the minister as an *alter Christus* is unique in the New Testament. It was forced upon Paul by the spirit-people/Judaizers' denial of the reality of Jesus' terrestrial existence and their disparagement of Paul's ministry.

PLANS FOR THE COLLECTION

The initial enthusiasm of the Corinthians for the collection for the poor of Jerusalem had evaporated in the heated atmosphere of the factional disputes within the community. Deeply offended by the way they had been pilloried in 1 Corinthians, the spirit-people, who were potentially the major donors, retaliated by refusing to take part in a project so dear to Paul's heart. Titus, however, had won the consent of their allies, the Judaizers, by a clever *ad hominem* argument, and Paul decided to exploit the opening.

2 Corinthians 8–9 reveals Paul at his best in terms of religious leadership. His consummate skill in the art of persuasion underlines how much he has matured in a single year. Even though he has to stretch the truth to do so, he praises what can be praised—the willingness of the Corinthians (although it was now a year old; 9: 2)—and sedulously avoids even a hint of criticism. He explicitly states that he is not ordering them to contribute (8: 8a), but merely expressing his opinion (8: 10). The example of the Macedonians is introduced in such a way as to permit the Corinthians' self-respect to function as an internal incentive. In order to assuage any possible anxiety on their part as to the sum expected, he is at pains to emphasize that their attitude is more important than the value of the gift (8: 12). Near the end, however, a hint of the old Paul surfaces in the way he highlights the possibility that he and the Corinthians might be humiliated by

the much poorer Macedonian church (9: 4). Fortunately, he immediately excludes the hint of moral blackmail, by denying that he wants to extort money from them (9: 5).

Once before, however, the Corinthians had given their assent and then done nothing. This time Paul was not prepared to rely on words alone, and decided to send emissaries to Corinth, whose presence would be a continuous reminder of his invitation. Even such discreet pressure, however, might be resented by the Corinthians as interference in the internal affairs of a local Church. Paul's nervousness is palpable in his presentation of Titus. He emphasizes that he is not really sending Titus, as 8: 6 might imply. The latter had volunteered to return to Corinth in response to Paul's appeal (8: 17)! This little vignette tells us something about Paul's treatment of his associates. He does not order a subordinate, but requests 'a partner and co-worker' (8: 23). Naturally Titus was the bearer of the letter which recommended him so highly.

With Titus will go a brother selected by the churches of Macedonia to act as their delegate in the actual assembling of the money for Jerusalem (8: 19). It is curious that, while his qualifications are given prominence, his name is never mentioned. Many explanations have been suggested,[67] but, in the light of the contacts between the Corinthian and Macedonian churches (1 Thess. 1: 7–9; 2 Cor. 11: 9), the simplest hypothesis is that he was a Corinthian Christian, who had gone to aid the spread of the church in Macedonia, and who there had established himself as an exceptional preacher of the gospel. When the Corinthians recognized him, and heard Paul's eulogy, they would have been both flattered and relieved. Their contribution to a sister church was publically praised, and Paul's emissary was not a critical Macedonian (9: 4), but one of their own. His specific role was to guarantee the integrity of the collection (8: 20–1).

The third member of the party (8: 22) is also unnamed. The way he is described suggests that he was a long-time associate of Paul, who had some relationship to the Corinthians. He may have been with Paul on the intermediate visit, or he may have accompanied Titus when the latter carried the Painful Letter. It was Paul's practice to travel with others, and it is most unlikely that he permitted Titus to go to Corinth alone. A travelling companion was indispensable, not merely to present a stronger front to robbers, but to guard whatever property they had, while the other went to the bath or elsewhere.

---

[67] See e.g. Furnish (1984), 435–6; R. P. Martin (1986), 275.

ONCE AGAIN A MISSIONARY

According to 2 Corinthians 9: 4, Paul planned to go to Corinth in the near future, i.e. during the summer of AD 55, in order to finalize the collection, on which he had now been working for four years. It would have been clear to him, however, that he could not just breeze in, make contact with the Corinthian delegation, and leave for Jerusalem. Despite his optimistic words in 2 Corinthians 7: 5–16, he was fully aware that the re-establishment of relations with the church left a number of serious problems unresolved. An extended stay was imperative. Exactly how long would depend on circumstances, but he could not risk spoiling the process of reconciliation by fixing a premature departure date. The more he reflected, the clearer it became that he would have to spend the winter of AD 55–56 in Corinth.

The Macedonians, however, might not want to delay. It would be natural to want to be rid as soon as possible of the heavy responsibility represented by the money collected for the poor. Only in summer could they travel to Jerusalem, and the round trip took several months. Any delay now would mean postponing the trip for a year. Hence, the note of hesitation, '*if* some of the Macedonians come with me' (2 Cor. 9: 4); the matter had not been decided when 2 Corinthians 1–9 was sent.

The more Paul thought about his plans for the future, however, the more reasons he found not to hasten to Corinth. 2 Corinthians 1–9 demanded time for the subtlety of its message to be assimilated adequately. It could only be to Paul's advantage to have his arguments discussed at length. He could be sure that Titus would nudge reflection in the right direction, and such delicate manipulation should not be hurried. Paul was prepared to find such reasons convincing, because for four years he personally had done little real missionary work. His agents had founded churches in Asia, and he had begun a new community at Troas. This latter episode was brief, and for the most part his energy had been focused on maintaining existing communities. Crisis after crisis in one church or another had demanded his attention. Now all were tranquil. A free summer was a golden opportunity to again seek virgin territory, and to be what he was divinely chosen to be, a founder of churches, who preached Christ where he had not yet been named (Rom. 15: 20). The prospect must have been irresistible. In any case Paul did not restrain himself. He went to Illyricum (Rom. 15: 19).[68]

While there might be a slight theoretical doubt as to what precisely Paul meant by this term,[69] there is little real uncertainty as to where he was.

[68] The unjustified scepticism of Fitzmyer (1990), 43, is as mistaken as efforts to interpret this verse purely symbolically, e.g. Geyser (1959), on which, see Knox (1964), 8 n. 10.
[69] See Appian, *Roman History* Book 10, 'The Illyrian Wars'.

Wherever Paul and his associates had passed the winter in Macedonia, Paul would certainly have accompanied Titus and his two companions as far as Thessalonica, if they had planned to go south by ship, or to near Pella, if they preferred the land route. To reach virgin territory all Paul had to do was to continue along the Via Egnatia to the west, and in ten days or so he would have been in Illyricum.[70] The area is described by Strabo, but in the opposite direction to that travelled by Paul (see Fig. 7):

> Of this Adriatic coast, then, the first parts are those about Epidamnus and Apollonia. ... From Apollonia to Macedonia one travels the Egnatian Road, towards the east. It has been measured by Roman miles and marked by pillars as far as Cypsela and the Hebrus River—a distance of 535 miles. ... And it so happens that travellers setting out from Apollonia and Epidamnus meet at an equal distance from the two places on the same road. Now although the road as a whole is called the Egnatian Road, the first part of it is called the Road to Candavia—an Illyrian mountain—and passes through Lychnidus, a city, and Pylon, a place on the road which marks the boundary between the Illyrian country and Macedonia. From Pylon the road runs to Barnus through Heracleia, and the country of the Lyncestae and that of the Eordi into Edessa and Pella and as far as Thessalonica. And the length of this road, according to Polybius, is 267 miles. (*Geography* 7. 7. 4; trans. Jones)[71]

On modern roads, only part of which coincides with the Via Egnatia, the distance between Thessalonica and what Strabo considered the fringes of Illyrian territory is approximately 320 km. (200 miles).[72] If Paul set off in mid-April, he would have been among the Illyrians by the end of the month and could look forward to at least three months of intense missionary work before having to head south in August in order to reach Corinth before the onset of winter.

BAD NEWS FROM CORINTH

How much Paul had invested in his plans for the summer of AD 55 can be gauged from the depth of his frustration when news from Corinth forced him to change them.

2 Corinthians 1–9, it will be remembered, had two objectives: to drive a wedge between the Judaizers and the spirit-people, and to win the latter to Paul's side. How well this latter goal was achieved is an open question, but it appears that he did succeed in isolating the Judaizers. Having lost what they hoped would be a firm base at Corinth, the Judaizers could only redouble their

---

[70] Sanday and Headlam (1902), 407–8.

[71] The modern equivalents of the place-names are: Epidamnus = Dyrrachium = Durrës in Albania; Cypsela = Ipsala in Turkey; Lychnidus = Ochrida in Serbia; Heracleia = Bitola in Serbia; Edessa = Edhessa in Greece. A detailed map of the road is given in Papazoglou (1979), 304.

[72] Rossiter (1981), 504, 526, 541.

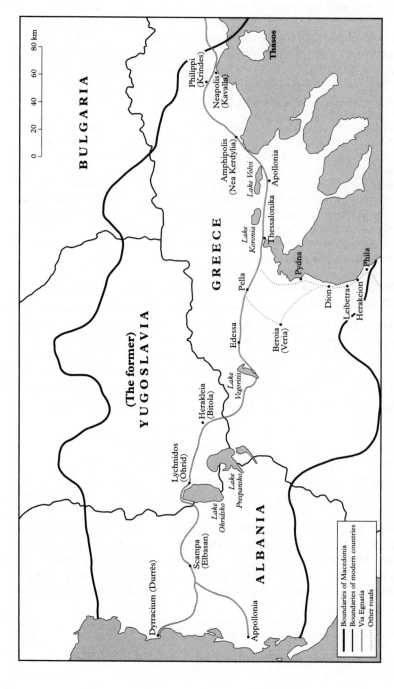

FIG. 7 The Roman Province of Macedonia and the Via Egnatia (*Source:* F. Papazoglou, ANRW II, 7/1 (1980))

attacks on Paul's person and authority. If there was now little chance of converting the spirit-people into Law-observant Christians, there was always the possibility that they might still be receptive to criticism of Paul.

Titus, or someone sent by him, found Paul in Illyricum and informed him that the old criticism of his unimpressive presence and uninspired preaching (2 Cor. 10: 10) had been revived in a more vicious form. The Judaizers had managed to convince a number that their spiritual gifts raised them far above Paul (2 Cor. 11: 5). The latter's failure to take strong action during the intermediate visit, they suggested, perhaps indicated that he did not have the authority. Certainly his flight, and failure to return, could only be interpreted as cowardice.

The importance which Paul attached to the collection for the poor of Jerusalem gave the Judaizers the opportunity to highlight his suspiciously ambiguous attitude towards money. He apparently refused money for himself, but solicited it for the poor. Would it all really go to Jerusalem? All the Judaizers had to do, when questioned by the Corinthians about the poverty of the Jerusalem church, was to shrug their shoulders. They did not have to deny the need for the collection. All they had to do was to insinuate that the questioners were a little naïve in taking Paul's statements at face value. By harping on the fact that Paul had taken money from Philippi (2 Cor. 11: 9), they could make a case that Paul did not love the Corinthians whose generosity he had refused.

Speaking as a Fool

Paul could only take such criticisms as a malicious distortion of his motives and actions. His bitter anger was intensified by the awareness that, if he was discredited, his version of the gospel was at risk. Another gospel might take its place. In a mood of desperate anxiety for the future of the Corinthian community, he dashed off 2 Corinthians 10–13. The reasonable tone and subtle arguments of 2 Corinthians 1–9 are replaced by a wild outburst, in which Paul gives his capacity for sarcasm and irony free rein.

The language in which he excoriates the gullibility of the Corinthians is a perfect illustration of the character of 2 Corinthians 10–13, 'You gladly bear with fools, being wise yourselves! You put up with it when someone makes slaves of you, or eats you out of house and home, or swindles you, or walks all over you, or smacks your face. To my shame, I must say, we were too weak for that!' (2 Cor. 11: 19–21). The 'wisdom' of the Corinthians is to be so lacking in self-respect that they eagerly accept their own exploitation!

What is said in this text has been interpreted literally, metaphorically, and rhetorically.[73] A choice between these different options is less important than an appreciation of the quality of the writing.

---

[73] So respectively Allo (1956*b*), 290; Windisch (1924), 347; and Betz (1972), 116–17.

The style has many impressive features, such as *enumeration*, as five verbs are listed in succession, and in a climactic way, following the 'law of increasing members'[74] with each verb adding extra weight to Paul's exposé. The emphasis of a string of verbs, with two *hapaxes*, found only here in the NT, is enhanced by the anaphorical repetition of [*ei*] *tis* five times, 'the one who,' and epiphoric assonance (hence the sonorous *-oi, -ei, -ei, - ai, -ei*).[75]

The quality of the writing is matched by the authority of the strategy. His opponents have forced him to compare himself with them, and what he does is to display his contempt for their pretensions by turning rhetorical convention upside down. After noting his breeding (2 Cor. 11: 22–3), he goes on to parody the self-display of the Judaizers by highlighting what should be hidden, and minimizing what should be accentuated (2 Cor. 11: 23–30).[76] Churches and converts are only hinted at; the spotlight is on situations in which he has been degraded. With great dramatic flair he concludes his list of 'accomplishments' with a graphic account of his humiliating escape from Damascus, lowered down the wall like a helpless baby in a basket (2 Cor. 11: 32–3)! He is the anti-thesis of the winner of the well-known 'wall crown',[77] which, according to Aulus Gelius, 'is that which is awarded by a commander to the man who is first to mount the wall and force his way into an enemy town; therefore it is orna-mented with representations of the battlements of a wall' (*Attic Nights*, 5. 6. 16; trans. Rolfe).

Parody is not the only weapon in Paul's rhetorical armoury. He deflates his opponents' claim to visions and revelations by speaking of his own experience in the third person (2 Cor. 12: 2–4). The technique distances him from the episode, and thereby underlines its irrelevance for his ministry.[78] It did not change him in any way, and did not provide him with any information he could use. The criticism of his opponents is all the more effective for being unstated. If their experience was the same as Paul's, it contributed nothing to their minis-try. If it was something about which they could talk, it was less ineffable than his!

2 Corinthians 10–13 is extraordinarily revelatory of a Paul rarely apparent elsewhere. Here the rigid control he normally imposed on his passionate nature dissolves in the heat of his anger. He gives full rein to his emotions, and in so doing betrays the quality of his education, which he usually denied (cf. 1 Cor. 2: 1–5). The fluid creativity of his thought is matched by the masterful facility and freedom with which he employs a number of the techniques of rhetoric. The assurance of his adept use of rhetorical devices can only be the fruit of long

---

[74] Zmijewski (1978), 207.

[76] Forbes (1986).

[75] R. P. Martin (1986), 364.

[77] Judge (1968), 47.

[78] The precision of the date, 'fourteen years ago', betrays how vivid was the memory of the experience. The date would be about AD 41, a period of Paul's life about which we know nothing.

study and practice.[79] There can be little doubt that Paul was brought up in a socially privileged class, which he was formed to adorn.[80]

The Thorn in the Flesh

Paul concludes his defence with a rhetorical tour de force, a humble admission which leads into a paradox, 'And to keep me from being too elated by the abundance of revelations, a thorn was given me, a messenger of Satan, to buffet me, to keep me from being too conceited . . . when I am weak then I am strong' (2 Cor. 12: 7–10). The nature of the thorn in the flesh has intrigued commentators from the early patristic period to the present day, and the wide variety of interpretations bears witness to the inexhaustible creativity of the human spirit.[81]

The vast majority of scholars consider that Paul had a physical ailment or a psychic problem. The suggestions—and they cannot be considered anything more—regarding the latter betray a very refined imagination: a real demon, who accompanied Paul on his heavenly journey, agony at the refusal of the Jews to respond to the gospel, sexual temptations, hysteria, depression. Somatic illnesses appear to have a better foundation: epilepsy (Paul fell to the ground during his conversion, Acts 9: 4), poor eyesight (he desired the eyes of the Galatians, Gal. 4: 15), a speech defect (he made a bad first impression, Gal. 4: 13 ff., and spoke badly, 2 Cor. 10: 10; 11: 6), recurring malarial fever, headache or earache. It will be obvious that the majority of these proposals depend on gratuitous and/or forced interpretations of texts, which are in no way related to physical ailments, be they those of Paul or of anyone else. Moreover, in order to have achieved all that he did, Paul must have been blessed with robust health and a strong constitution.

The only hypothesis for which a serious case can be made is that by the thorn in his flesh Paul meant opposition to his ministry.[82] His mention of 'a messenger of Satan' implies an external, personal source of affliction, and previously he had identified as 'servants of Satan' (2 Cor. 11: 14–15) his adversaries at Corinth. In the Old Testament, 'thorns' are a metaphor for Israel's enemies, both within (Num. 33: 35) and without (Ezek. 28: 24). This latter reference,

---

[79] Paul also gives his instincts free rein in Rom., and experts recognize the quality of his education. Fitzmyer (1993), 92, approvingly cites the judgement of Sanday and Headlam (1902), p. lv, 'the rush of words is always well under control. Still there is a rush of words, rising repeatedly to passages of splendid eloquence; but the eloquence is spontaneous, the outcome of strongly moved feeling; there is nothing about it of labored oratory. The language is rapid, terse, incisive; the argument is conducted by a quick cut and thrust of dialectic.' The unschooled could never exhibit this combination of qualities.

[80] See in particular Forbes (1986), 22–4, and P. Marshall (1987), 400.

[81] Good surveys are provided by Allo (1956*b*), 313–23; Furnish (1984), 547–9; R. P. Martin (1986), 412–16.

[82] Mullins (1957); Barré (1980).

when coupled with Paul's use of 'to buffet' in 1 Corinthians 4: 11, has been taken to mean that Paul had persecution in mind. I doubt, however, that Paul would have prayed to be delivered from persecution.[83] He saw such sufferings as the means whereby he was assimilated to Jesus (2 Cor. 4: 10–11).

What was a continuous source of pain to Paul was the fact that none of his churches measured up to his expectations. There was always someone, in every community he founded, who caused him grief—the idlers at Thessalonica, Euodia and Syntyche at Philippi, those paralysed by prudence in Galatia, the resentful at Ephesus, the mystics at Colossae, the spirit-people at Corinth. There was no grouping of his converts on which he could look with complacent pride. Any tendency to conceit, or even satisfaction, was immediately countered by evidence of some sort of dissent. Such divisions, however, were opposed to the plan of God, for whom the church should exhibit the organic unity of a living body. Hence, Paul could pray legitimately that they would come to an end.

The wry humour of his self-assessment continues into the presentation of the response to his prayer in the form of a divine oracle, 'It suffices for you my grace, for this power in weakness is perfected' (2 Cor. 12: 9a).[84] The paradox is as extreme as the meaning is profound. The thorn reminds him that he has none of the qualities which the world considered essential prerequisites for the success of his mission. Yet he serves as a channel of divine grace expressed in the power of Christ (2 Cor. 12: 9b), whose 'life' he exhibits (2 Cor. 4: 10–11), and the world is changed.

---

[83] So rightly R. P. Martin (1986), 415.
[84] See esp. O'Collins (1971).

— 13 —

Looking Westward

THE messengers who had brought the bad news from Corinth returned there with 2 Corinthians 10–13. In it Paul promised a visit in the near future (2 Cor. 12: 14; 13: 1–2). Anxiety for the Corinthians raged in his heart, but he was not free to leave Illyricum immediately. His experience had taught him that he could not simply abandon new converts to the care of the Holy Spirit. God acted through human agents (1 Cor. 3: 5–9), and it was Paul's responsibility to set the infant community on a secure foundation. It is easy to imagine the re-doubled fervour with which he worked, attempting to pack in as much as possible before the onset of winter obliged him to start the long journey to the south. His need to capitalize on the shock effect of 2 Corinthians 10–13 makes it unthinkable that he should have postponed his visit to Corinth until the following spring. It would have been out of character for him to leave one of his co-workers to direct the nascent church. Certainly Timothy travelled with him to Corinth (Rom. 16: 21).

Of Paul's reception at Corinth we know nothing, but it would seem that 2 Corinthians 10–13 had a salutary effect. If the Corinthians in fact contributed to the collection for the poor of Jerusalem (Rom. 15: 26), it is unlikely that the community as a whole, or even a majority, was alienated from Paul. Moreover, during the winter of AD 55–56, he had the leisure to compose his most developed theological argument, which was part of his preparations for the future. After accompanying the collection to Jerusalem, he planned to go to Rome, first, and then to Spain (Rom. 15: 24, 28). One has the impression that he had no intention of returning to Corinth. He had given what he could, and the future of the believers there was in their hands and God's. Since Paul never took a boat going west, his natural route from Palestine to the capital of the Empire ran through Asia Minor to Troas, and then along the Via Egnatia. Before he left the north he may even have promised the Illyrians that he would return as soon as possible.

THE TEXTUAL PROBLEMS OF ROMANS

Our knowledge of the next steps in Paul's career come from the epistle to the Romans but, before it can be used as evidence, note must be taken of a number of textual problems. The manuscript tradition attests eight different forms of Romans. In essence these are created by the presence or absence of chs. 15 and 16, and the positioning of the doxology (Rom. 16: 25– 7).[1] In addition, a number of manuscripts lack the italicized words in 1: 7, 'To all God's beloved *in Rome* who are called to be saints', and in 1: 15, 'I am eager to preach the gospel to you also who are *in Rome*.' Without the specific address, and the highly personal list of greetings in ch. 16, the rest of the letter is so generic that it could have been addressed to any and every church.

There is now virtual unanimity among scholars that the letter which Paul wrote contained 1: 1 to 16: 23. There is some doubt about 16: 24,[2] but none about the concluding doxology (16: 25–7). Considerations of content, style, and epistolary practice conspire to make it unlikely that Paul was the author.[3] He does not end his letters in this way. The absence of any mention of Christ in 'the revelation of the mystery concealed for long ages, but now made manifest through the prophetic scriptures' (16: 25b–26a) betrays its un-Pauline character, even though the language evokes Colossians 1: 26–7.

As regards the varying lengths of the letter in different manuscripts, it is most probable that the elements which related the letter to a specific community were deliberately edited out in order to give the letter greater universality. It is known that the particularity of the Pauline letters created certain difficulties in the second century,[4] and the easiest solution to this type of problem is to eliminate the causes. Thus the words 'in Rome' were excised from 1: 7 and 15. Were Romans the only case, this hypothesis would lose much of its appeal. But Ephesians is certainly a generic letter, and Dahl has plausibly suggested that the two different positions for the phrase 'which is in Corinth' (1 Cor. 1: 2) in the manuscript tradition is due to the phrase having been excised in certain manuscripts to generalize the letter, and then restored but in the wrong place.[5] Similarly certain scribes considered that the personal details about Paul which ch. 15 contained threatened the universality of the letter, whereas others did not. The highly specific greetings of ch. 16 were a much graver cause for concern.[6] And they have remained so for modern critical scholars.

---

[1] See Gamble (1977), Aland (1979), 284–301, and Fitzmyer (1993), 44–54.

[2] It is rejected by Dunn (1988), 901, and Fitzmyer (1993), 751, but the argument of Gamble (1977), 130, in favour of its authenticity is far from negligible.

[3] See in particular Elliott (1981); Dunn (1988), 913–16; Fitzmyer (1993), 753.

[4] Dahl (1962).

[5] Ibid. 266–7.

[6] Gamble (1977), 128.

Doubts that 16: 1–23 belonged to a Pauline letter addressed to Rome arose in the eighteenth century, and eventually evolved into the hypothesis that these verses originally constituted a separate letter destined for Ephesus. Since then the hypothesis has both won influential support and encountered vigourous criticism.[7] The publication in 1937 of P[46], which is the only manuscript attesting a 15–chapter version of Romans, was greeted with delight by those who had detached ch. 16 from Romans on purely literary grounds. It appeared to provide objective verification of their theory. It was the first manuscript confirmation of a literary analysis. Regretably P[46] does not contain the original text of Romans. In consequence, it is not surprising that its publication failed to put an end to the controversy, which continues unabated.[8]

In my view 16: 1–23 cannot be an independent letter, and must be an integral part of Romans. The chapter opens with the words 'now I commend to you'. The particle *de* indicates that 16: 1 is not a beginning but a continuation. Moreover, without ch. 16 Romans would contain only one of the three elements with which Paul regularly concludes his letters, namely, the peace wish (15: 33).[9] The greeting, particularly that expressed by the kiss (16: 16), and the final blessing (16: 20b) would be lacking. If ch. 16 has the characteristic features of a conclusion, it cannot have been an independent letter, and must always have been associated with chs. 1–15, which otherwise would be incomplete. The only possible objections, namely, that Paul arbitrarily departed from his consistent practice, or that ch. 16 was the conclusion to another now lost letter, are unworthy of serious consideration.

Admittedly the recommendation of Phoebe (16: 1–2) is an unusual feature in a Pauline conclusion. The only comparable element in the conclusions to Paul's other letters is the praise of Tychicus and Onesimus in Colossians 4: 7–9. The formal differences in expression diminish the force of the parallel, but the uniqueness of 16: 1–2 should not be exaggerated. Cicero occasionally slipped a note of commendation into the conclusion of his letters, sometimes in reference to the bearer, but not always.[10] In consequence, this element cannot be used to prove that ch. 16 was originally independent.[11]

Two of the groups greeted in Romans 16 were almost certainly domiciled at Rome. 'Those among the [slaves] of Aristobulus'[12] worked for the grandson of

---

[7] Without attempting to be exhaustive Kümmel (1975), 318– 20 lists 21 eminent scholars in favour and a similar number against. [8] e.g. Refoulé (1990).

[9] Gamble (1977), 89. While pointing out minor flaws, Fitzmyer (1993), 63, agrees that the argument is not only sound, but crucial.

[10] *Fam.* 3. 1; 8. 8; 12. 24; 16. 21; *Att.* 1. 19; *Q.Fr.* 2. 14; see Gamble (1977), 85–6.

[11] Lampe (1989), 131–5, has argued that the language of ch. 16 is closer to that of Rom. 1–15 than to that of any other Pauline letter. In itself this proves that they both were written about the same time, not that they were part of the same letter. His observations acquire significance only from the preceding arguments.

[12] BDF §162(5); cf. Phil. 4: 22. Paul is definitely not speaking of a house-church within a family (as the *NRSV* suggests); in that case his formula would be *hē kat' oikon ekklēsia* (cf. Rom. 16: 5; 1

Herod the Great who died in Rome in the latter part of the 40s AD.[13] By his will they would have been incorporated into another great household while retaining their distinctive name. In the imperial household we know, for example, of *Maecenatiani* and *Germaniciani*, who had been the slaves of Gaius Maecenas (d. 8 BC) and Germanicus Julius Caesar (d. AD 19), respectively. In this perspective Paul's formula would translate *Aristobuliani*.[14] This interpretation is made virtually certain by the name of the next individual singled out. 'Herodion' unambiguously suggests a connection with the family of Herod since freed slaves took the name of their patron.

In the light of the foregoing, 'those among the [slaves] of Narcissus' can be identified plausibly as the slaves of the influential freedman of Claudius, who was killed shortly after his patron died in AD 54.[15] They, together with his vast wealth, passed into the household of Nero as *Narcissiani*, for which there is inscriptional evidence.[16]

If these identifications are correct—and they are widely accepted[17]—both of these groups can be sought only in Rome. One cannot conceive such 'legacies' being moved en bloc outside the Eternal City, let alone to Ephesus.

Paradoxically the extensive list of persons greeted in 16: 1–23 proves that Paul was writing to a community in which he had not lived and worked. It was Paul's normal practice not to name the recipients of greetings, presumably because it would have been invidious to single out individuals in a group all of whose members were known to him. Romans 16, therefore, is a radical departure from Paul's normal practice, and unambiguously indicates that he was writing to a church in which he was not known personally. This conclusion is confirmed by Colossians, in which Paul singles out 'Nympha and the church in her house' (4: 15). Paul was not the founder of the churches in the Lycus valley and had never visited them. Individuals are also named in 2 Timothy where Paul salutes 'Prisca and Aquila, and the household of Onesiphorus' (4: 19), but this is a letter to a private individual, and the couple were among Paul's oldest friends and closest collaborators.

Cor. 16: 19; Col. 4: 15; Philem. 2). Had he the whole household in mind, the formulation would be *hoi Aristoboulou*.

[13] *Josephus, JW* 2. 221; *AJ* 20. 13. Lampe (1989), 136, notes that the name Aristobulus is extremely rare in Rome, and suggests that Christianity began there by Aristobulus bringing Christian slaves with him from the east. Little or nothing is known about the career of Aristobulus, but there is no indication that he spent any time in Judaea. If anything it is more likely that one channel of penetration was through those who carried messages between himself and his brothers in the east, Agrippa I in Judaea and Herod in Chalcis. It is equally possible, however, that either or both these latter sent him slaves of Jewish origin in order to maintain the character of his household, but who happened to be Christians.

[14] Lightfoot (1908), 175.

[15] Tacitus, *Annals* 13. 1; Dio Cassius, *History* 60. 34; Juvenal, *Satires* 14. 329–31.

[16] *CIL* 3. 3973; 6. 15640.

[17] e.g. most recently Dunn (1988), 896. Fitzmyer (1993), 740–1, is less certain.

The identifying notes attached to some individuals also militate against the Ephesian hypothesis. The description of Epaenetus as 'the first convert to Christ in Asia' makes more sense if he were now outside the province.[18] Surely the fact would be as well known to the church of Ephesus, as Timothy's status as Paul's 'co-worker' (16: 21). The need for such clarification implies non-acquaintance.

In the body of the letter, Paul praises the Romans for their obedience 'to the form of teaching to which you were committed' (6: 17).[19] In 16: 17 he evokes 'the teaching which you learned'. In both cases *didachê* connotes the common body of Christian 'teaching'. The unusual verb form in 6: 17 suggests that Paul did not know who had instructed the Romans. The same is true in 16: 17. Had Paul taught the recipients, as the Ephesian hypothesis demands, he would have used the first-person singular (e.g. 1 Cor. 15: 1; Gal. 1: 8).

Finally we return to the manuscript tradition associating ch. 16 with chs. 1–15. This simple fact is a serious objection to the Ephesian hypothesis, but the ingenuity of Manson has been equal to the challenge.[20] After Paul had completed Romans 1–15, he suggests, and was preparing to send the letter to Rome, he realized the contents would also be of interest to the Ephesians. While in residence there, as we know from Galatians, he had begun to deal with the relationship between Judaism and Christianity, and now thought it appropriate that they should have his mature reflections. Thus he had a copy made for Ephesus, and attached to it greetings to friends, and a recommendation for Phoebe, who was presumably the bearer.

Clever as this hypothesis may be, it fails to answer the obvious question: if the 16–chapter version is the Ephesian version, why is that church not mentioned in the address rather than Rome? The only way out of this difficulty is to maintain that the association of ch. 16 and chs. 1–15 is entirely accidental, and came about when the collection of Paul's letters was being assembled at Ephesus. For reasons of piety and local pride, we are told, the Ephesians wanted to include their letter, namely, ch. 16, but it looked so unimpressive beside the other letters that they feared it might get lost or be neglected. To preserve it they decided to tack it on to Romans.[21] This hypothesis is attractive in its naïve romanticism, but it embodies so many highly speculative components that no discussion is possible. It suffices to note that nowhere else in the Pauline corpus are letters to different churches combined.

---

[18] 1 Cor. 16: 15 is not an objection. If Paul there notes that the household of Stephanas were 'the first converts in Achaia' in a letter to Corinth, it is in order to underline their length of service to the community.

[19] The view of Bultmann (1967), 283–4, that this verse is an interpolation is rendered unnecessary by better exegesis; see Borse (1968).

[20] (1962), 225–41.

[21] So, among many, Goodspeed (1937), 85–6.

Although certitude is not possible, it is far more probable that ch. 16 was the original conclusion to Paul's letter to the Romans.[22]

PLANNING FOR THE FUTURE

In ch. 16 Paul greets 26 individuals, 24 of whom are named. In addition mention is made of three house-churches (vv. 5, 14, 15), and two groupings of (ex-) slaves (vv. 10, 11), which may also have been house-churches.[23] Some of these he certainly knew personally. Prisca and Aquila were with him at Corinth, and later at Ephesus (1 Cor. 16: 19). The qualification of Epaenetus, Ampilatus, and Stachys as 'my dear friend' (*NJB*) cannot be an empty formula. If the mother of Rufus had also 'mothered' Paul,[24] he must have known both her and her son. An element of doubt clouds the case of Andronicus and Junia because 'my fellow prisoners' could mean only that they had suffered imprisonment as Paul had done. In Colossians 4: 10 and Philemon 23 the context indicates that Aristarchus and Epaphras were imprisoned with Paul, but such is not the case here. There is nothing in the least surprising that Paul, during his ministry in Greece and Asia Minor should have known a minimum of seven and a maximum of nine Christians who had ended up in Rome.[25]

What is significant is that Paul knew where these individuals were and had learned the names and praiseworthy achievements of many others in the Roman church.[26] The conclusion that he had contacts with Rome prior to writing the letter is inescapable. Were these accidental or intentional? The presence of Prisca and Aquila in Rome argues for careful advance planning. This couple, as we have seen, had provided Paul with a base in Corinth,[27] and had prepared the ground for his ministry in Ephesus.[28] If they now appear in Rome, the obvious inference is that they had been sent by Paul in anticipation of his arrival there.[29]

---

[22] So rightly Dunn (1988), 884; Fitzmyer (1993), 57–64, against Georgi (1992), 110–12.

[23] Gerlemann (1989), 86–95, treats the names in Rom. 16 as a hidden code which when deciphered reveals the following list of those who are really greeted: Phoebus Apollo, Legions and Senate, Aeneas, Ulysses, Andromache, Plato, Livy, Vergil, Ovid, Aristotle, Herodotos, Pericles, Sybaris, Xenophon, Sallust, Cato, Cicero, Caesar, Augustus, Claudius, and Nero!

[24] The view of Baslez that she was actually Paul's mother has been discussed above, Ch. 2, 'Paul's Relatives'.

[25] Fitzmyer (1993), 734, also includes Apelles, Urbanus, and Persis, but on inadequate grounds.

[26] On the ministerial implications of 'labouring', see von Harnack (1928).

[27] Ch. 11, 'The First Converts'.

[28] Ch. 7, 'The Founding of the Church'.

[29] So F. Watson (1986), 105; Lampe (1991), 220. The objection of Dunn (1988), 892, is based on the unfounded assumption that they returned to Rome to assume control of the business which they had confided to trusted slaves when they were expelled by Claudius.

A Mission to Spain

Rome, however, was not the sort of virgin missionary territory in which an advance guard had proved its value. It had a well-established church, many of whose members had been formed in the cultural ethos of the eastern Mediterranean. For such people, hospitality to travellers was second nature.[30] They did not need to be placated or warned. Paul could rely confidently on their generosity. Why then did he prepare the ground so carefully by sending close collaborators, and assiduously collecting information on members of the Roman church? He must have had in mind something much more important than a friendly visit.

What Paul was thinking of emerges, if close attention is paid to his exact wording, 'I hope to see you as I pass through [Rome] and to be helped on my way there [Spain] by you' (Rom. 15: 24), Rome is not his goal; it is merely a staging-point (Rom. 15: 28) en route to Spain, which was only four days away by sea.[31] But he wants something from the Roman believers. The way the verb *propempô* is used in the New Testament[32] makes it 'almost a technical term for the provision made by a church for missionary support'.[33] Commentators generally think exclusively in terms of of a material contribution, and this aspect cannot be excluded, as we shall see. It is unlikely, however, to have been the only consideration. If all Paul needed was financial support, it would have been much more prudent to have collected funds from the communities he had founded in Asia Minor and Greece.[34] These churches would have been on his route west from Jerusalem and they owed him everything. Philippi at least had helped him before. Why should Paul wait until the last stage of his journey, and risk everything on the problematic generosity of a community, which did not know him?

Even though M. Prior makes it clear that he is thinking in logistical terms, he formulates Paul's desire in a way which permits a broader and more complete interpretation, 'Paul hoped that the Roman community would "own" the mission to Spain, in the way the Antioch community did his earlier ones.'[35] Antioch, however, had commissioned Paul (Acts 13: 1– 3), and this was the reason why it supported him. After the incident with Peter (Gal. 2: 11–14), Paul felt that he could no longer represent a church which took such an erroneous

---

[30] See in particular Koenig (1985).

[31] Speaking of flax from which sails were made, Pliny says, 'what is more marvellous than the fact that there is a plant . . . that brings Cadiz within seven days' sail from the Straits of Gibraltar to Ostia, and Hither Spain within four days' (*NH* 19. 4).

[32] Acts 15: 3; 20: 38; 21: 5; 1 Cor. 16: 6, 11; 2 Cor. 1: 16; Titus 3: 13; 3 John 6.

[33] Dunn (1988), 872.

[34] Once the collection for the poor had been deposited in Jerusalem, he would be free to appeal for funds for his own support.

[35] (1989), 135.

stance with respect to its Gentile members. Ever since, he had been without a
legitimizing home base, a point on which his opponents at Corinth and Galatia
had capitalized. Paul, they had insisted, was an unrepresentative maverick. His
retort was to draw out the implications of his conversion, and to insist that he
had been commissioned by God through Christ (Gal. 1: 1). While theologically
valid, this claim could be verified only indirectly and with hindsight (1 Cor. 9:
1–2; 2 Cor. 3: 2). He had no credentials to produce. Similar difficulties in the
future could be avoided if he were 'sent' to Spain by the church of Rome. Once
again he would be integrated into the Christian movement.

Why did Paul choose Spain? While no definite answer can be given, it has
been suggested that he was inspired by the eschatological vision of Isaiah, 'I am
coming to gather all nations and tongues, and they shall come and see my glory,
and I will set a sign among them. From them I will send survivors to the nations,
to Tarshish, Put, and Lud—which draw the bow—to Tubal and Javan, to the
coastlands far away that have not heard of my fame or seen my glory' (66: 18–
19).[36] Although no certain location is ever given for Tarshish, the hints of the
Old Testament point to the western end of the Mediterranean (Jonah 1: 3; 4:
2).[37] It would be perfectly in keeping with Paul's Messianic understanding of his
vocation to understand this prophecy as outlining his responsibility to extend
the kingdom of God to the limits of the known world.[38] In Galatians 1: 15 Paul
makes it clear that he considered himself to be prefigured in the Servant Song
(Isa. 49: 1), which contains the words, 'I will make you a light of the nations to
bear my salvation to the end of the earth' (Isa. 49: 6).[39] For Paul's contemporaries, Cape St Vincent on the west coast of Spain was the end of the
world.[40]

Some such powerful theological motive must be postulated, because in
practical terms Spain had little to recommend it. In the eastern Mediterranean
Paul moved in a world whose language he spoke, and which had a network of
Jewish institutions of which he could avail himself. In Spain both these
advantages were lacking, and an entirely new missionary strategy would have
to be developed.[41] The Jewish Diaspora did not extend westward beyond Italy.[42]
Hence, there were no God-fearers whose minds had been prepared for the
gospel by the reading of the Scriptures. Nor were there many who spoke Greek.
The language survived in the few old Greek colonies along the east coast, but

[36] First suggested by Spicq (1969), 132, this hypothesis has been fully worked out by Aus
(1979b), but in a detail that is not aways convincing, see Prior (1989), 129.
[37] Baker (1992), 332.
[38] See my (1964), 106–14.
[39] Note that Isa. 49: 8 is cited in 2 Cor. 6: 2.
[40] 'Let me describe Iberia in detail, beginning with the Sacred Cape [= Cape St Vincent] which is
the most westerly point, not only of Europe, but of the whole inhabited world' (Strabo, *Geography*
3. 1. 4).
[41] Jewett (1988).
[42] See Bowers (1975), with the map in Prichard (1987), 170–1.

the hinterland was dominated by a bewildering number of Iberian dialects. Latin was the language of the Roman administration, but not of any significant portion of the population.[43] To what extent Paul was aware of these difficulties is impossible to say with any precision, but he must have known that the western end of the Mediterranean would be dissimilar to what he had previously experienced. Paul realized that he needed the expertise of those closer to that strange land.

A Winter in Greece

When did Paul determine to go to Spain? It may have been when he anticipated being forced out of Ephesus in the late summer of AD 54. He had prepared a fall-back ministry in Troas and might have advised Prisca and Aquila to return to Rome, with a view to linking up with them later. At that particular time, however, his relationship with Corinth was so tense that it is unlikely that he looked very far into the future. One question monopolized his attention: how would the Corinthians respond to the Painful Letter?

The summer of AD 55 is a more realistic candidate for the decision to go to Spain. As far as Paul was concerned, he had made his peace with the Corinthians in 2 Corinthians 1–9, and at last was free to undertake a new mission. When he reached Illyricum, Rome was just over the horizon. The pull of an ambition long frustrated (Rom. 1: 13; 15: 22) must have been very strong. True he had to go to Corinth to pick up the collection and then to Jerusalem, but that was a formality. Then he could look forward to a voyage to the other end of the Mediterranean. It is in this brief moment of euphoria that the formulation of the plan to send Prisca and Aquila to Rome is most plausible.

Then came the bombshell. Far from improving, the situation at Corinth had deteriorated. Bearers of bad news shattered Paul's Illyrian idyll. In frustration his resolve hardened. He would spend the winter in Corinth as planned, but after that would devote no more time to the childishness (1 Cor. 3: 1; 14: 20) of what should have been the most brilliant of his churches. It was time to devote his energies to something more profitable. Rome, as a springboard to Spain, was no longer a vague future hope; it became the next item on Paul's agenda. The visit to Rome was firmly fixed for the following summer (AD 56). He would head west as soon as he had deposited the collection in Jerusalem. This made it all the more urgent that Prisca and Aquila should go to Rome immediately. It may have been at this moment that Paul, or one of his entourage, conceived the idea of writing a letter to the Romans.

Obviously Prisca and Aquila could speak of him, and detail his qualities, but would not a personal letter in which he revealed something of himself be so

[43] See esp. the linguistic map in García y Bellido (1972), 476–7.

much effective as an introduction? Paul, of course, had never written such a letter. All his previous writings had been in response to particular problems in communities whose members he knew. What would he say to members of a strange church in the capital of the empire? Moreover, right now he had much more urgent matters on his mind. Illyricum and Corinth demanded all his attention. There was no time to compose such a letter. Hence, it could not be sent before the following spring. The realization that he had the winter in Corinth to work out what to say may have been a factor in Paul's acceptance of the idea. The breathing space also meant that he had time to acquire some information about the membership of the Roman church. The sensible course was to ask Prisca and Aquila to send him a report as soon as possible,[44] certainly no later than the spring of AD 56, when he had to leave Corinth for Jerusalem with the collection for the poor.

THE WRITING OF THE LETTER TO THE ROMANS

The assumption that Romans had a single purpose has given rise to a long and inconclusive debate.[45] Recent commentators have drawn the correct inference from the conflicting observations; the letter was written to achieve a number of different goals.[46] Paul's basic problem was to find a topic which would be of interest to the Romans, and at the same time serve as an introduction to his person and his gospel. Thus he had to know something of the composition of the community. The minimum he needed to know in order to write Romans 1–12 was that the members of the community were predominantly God-fearers.[47] This he could have learnt from Prisca and Aquila prior to their departure for the Eternal City. Their recollections of the church they had left some fifteen years earlier may have been brought up to date by travellers passing through Corinth or Ephesus. Paul could also have extrapolated from his own experience of other European churches. The writing of Romans 13–16, however, demanded detailed knowledge of the contemporary Roman ecclesiastical scene. Hence I suspect that Paul worked on the material in Romans 1–12 during the winter of AD 55–56, and added the rest of the letter when information arrived from Prisca and Aquila in the spring of AD 56.

[44] Dunn (1988), 909, speculates that Tertius (Rom. 16: 22) may have spent some time in Rome. While not impossible, it is less probable.

[45] Schmithals (1975), 10–94, remains the most detailed survey of the history of research.

[46] Dunn (1988), p. lv; Fitzmyer (1993), 79.

[47] The popularity of this hypothesis is recorded by Schmithals (1975), 58–63, and most recently it has been adopted by Lampe (1989), 54; Dunn (1988), p. xlviii; and Fitzmyer (1993), 64.

Jews and Gentiles in Rome

The origins of the church in Rome, like those of Damascus, are shrouded in obscurity. Despite the lack of any evidence putting Rome in a special relationship with Jerusalem,[48] it would seem none the less that Rome had been evangelized by Christians of Jewish origin. The earliest mention of missionary activity is the remark of Suetonius, 'He expelled from Rome the Jews constantly making disturbances at the instigation of Chrestus' (*Claudius* 25. 4). This imperial action has become the key element in a reconstruction of Roman Christianity which has profoundly influenced the interpretation of the letter in recent years.[49]

The essence of the scenario is that Christianity in Rome grew out of the synagogue. Missionaries converted a number of Jews and a greater number of God-fearers, who had attached themselves to the synagogue. Already attracted by the ethos of Judaism, these latter would have had little difficulty in adopting Jewish practices, circumcision excepted. Pre-Claudian Christianity, therefore, was very much a Judaizing version of the faith, and one which proved increasingly intolerable to the various Roman synagogues in which it sought to find a home. Opposition became progressively more violent, and the Roman authorities had to step in. The lack of any centralized Jewish leadership left Claudius no option but to expel all Jews. Believers of Gentile origins, in consequence, ended up as the sole representatives of Christianity in Rome. Inevitably they became gradually less Jewish, particularly as they attracted new converts. Their institutional focus was no longer the synagogue but the house-church. With the accession of Nero, the ban was lifted. Jews and Christians of Jewish origins were permitted to return to the city, but not to assemble. The latter discovered a version of Christianity which they hardly recognized. Distributed among the various house-churches, inevitably they resented their minority position and their second-class status.

As I argued above, however, there is little chance that this scenario corresponds to reality.[50] In all probability the action of Claudius affected only one synagogue, which is why the expulsion order is not noted in any Jewish source as a disaster for the Jews of Rome. Only a tiny proportion of the 20,000[51] or 40,000 to 50,000[52] Jewish inhabitants of the city were involved. The Jewish vacuum, which is essential to the theory that the content of Romans was determined by a unique feature of Christianity in the Eternal City, is a myth.

---

[48] Against Brown and Meier (1983), 103–4; Brown (1990), 107; Fitzmyer (1993), 33.

[49] First proposed by Marxen (1968), 92–109, the hypothesis has been espoused by Dunn (1988), p. liii; Fitzmyer (1993), 77–8. [50] Ch. 1, 'The Edict of Claudius'.

[51] The figure given by Penna (1982), 328; repeated in his (1992), 1075.

[52] The figure given by Leon (1960), 135–6, which is accepted by Brown and Meier (1983), 94, and Dunn (1988), p. xlvi.

Paul's focus on the relationship between Judaism and Christianity is more likely to have been directly inspired by the problems he encountered with the Judaizers during the winter in Corinth, and by his concern as to how the collection would be accepted in Jerusalem (Rom. 15: 30–1).[53] The Jewish and Christian issue was forced upon him by circumstances. Why should he sacrifice time to the development of another subject in his introductory letter to Rome, when a central problem for the future of Christianity claimed his attention?

A salient feature of the style of Romans 1–11 confirms that Paul had only a generic understanding of the church in Rome. He addresses an interlocutor,[54] deals with objections and false conclusions,[55] and expresses himself in a dialogical exchange complete with an example.[56] This combination of censure and persuasion is typical of the diatribe, a teaching technique whose setting is the classroom of a philosophical school. The choice of the technique reveals another facet of the quality of Paul's education, and the sophisticated fashion in which he exploits it betrays the give and take of many discussions.[57] A further inference is that the questions and objections do not articulate the specific problems of those to whom he is writing, but synthesize the experiences of many new converts. Paul had had to deal with many Jews and Gentiles, as they struggled to understand their past and present. Paul would not have had to use imaginary, typical, interlocutors if he were aware of what was actually being said and done in Rome. The absence of the diatribe technique in other letters is due to his detailed knowledge of the local scene.[58] The way, for example, in which Paul handles the concrete and highly specific slogans of the Corinthians has nothing in common with the stylized objections in Romans.

Sin, Law, and Death

The fundamental thrust of Romans is that God, not only desires the salvation of all, but has put a plan into effect whereby grace can reach each and every individual (3: 29). The general *propositio*, which commands the whole development of the letter is: 'The gospel is the power of God for salvation to everyone who has faith, to the Jew first but also the Gentile' (1: 16). This insight had been the guiding principle of Paul's life since the moment of his conversion, and in this sense Romans crystallizes his gospel. How the power of the gospel produced its effect, however, was much more adequately presented in his earlier letters. Their insights must be kept in the forefront of the mind, if the full scope

[53] This does not mean that I believe that Rom. is a version of a speech which Paul planned to give in Jerusalem, *pace* Jervell (1991).
[54] Rom. 2: 1–5, 17–24; 9: 19–21; 11: 17–24.
[55] Rom. 3: 1–9, 31; 6: 1, 15; 7: 7, 13; 9: 14, 19; 11: 1, 11, 19.
[56] Rom. 3: 27 to 4: 25.
[57] See in particular Stowers (1981), and Schmeller (1987).
[58] Aune (1987), 201.

of the Pauline gospel as hinted at in Romans is to be properly understood. What the earlier letters did not spell out adequately was Paul's perception of humanity's need for salvation. The originality of Romans is not its teaching on the how of salvation (save in its insight into the ultimate salvation of the Jews), but its explanation of the why of salvation.

One of the most distinctive features of Romans is its use of *hamartia* in an unusual sense. It first appears in Romans 3: 9, 'all, both Jews and Gentiles, are under sin'. Clearly Paul is not thinking in terms of the personal sins of individuals, and in order to underline the difference 'Sin' should be capitalized. The same usage is found in a whole series of texts, which can be classified under three heads:

Sin
'Sin came into the world through one man' (5: 12a).
'That we might no longer be enslaved to Sin' (6: 6).
'Do not present your members to Sin' (6: 13).
'You were slaves of Sin' (6: 16, 17, 20).
'Having been set free from Sin' (6: 18).
'Sold under Sin' (7: 14).
'Sin dwells within me' (7: 20).

Sin and Law
'Sin will have no dominion over you, since you are not under Law, but under grace' (6: 14).
'Apart from the Law Sin lies dead' (7: 8).
'I am captive to the Law of Sin' (7: 23, 25).
'Has set me free from the Law of Sin and Death' (8: 2).

Sin and Death
'Death (came into the world) through Sin' (5: 12b).
'Sin reigned in Death' (5: 21).
'The wages of Sin is Death' (6: 23).

Manifestly Sin functions as a myth or symbol. What is the meaning of the myth? What is hidden behind the symbol? When taken out of context Sin could easily appear to be but another name for Satan, of whom Paul sometimes speaks.[59] Not only has this hypothesis nothing to recommend it, but it is positively excluded by the way Paul uses the two names. Satan is invariably mentioned in connection with those who are already believers, whereas Sin is exclusively associated with unbelievers.

Paul first used the term in Galatians 3: 22 ('The Scripture shut up all under Sin'). In Romans 3: 9–18 he clarifies this enigmatic statement by citing a catena of Old Testament passages,[60] in which humanity appears as unrighteous,

---

[59] Rom. 16: 20; 1 Cor. 5: 5; 7: 5; 2 Cor. 2: 11; 11: 14; 12: 7; 1 Thess. 2: 18; 2 Thess. 2: 9.
[60] Eccles. 7: 20; Ps. 13: 2–3; 5: 10; 139: 4; 9: 28; Isa. 59: 7–8; Ps. 35: 2.

ignorant of God, evil-working, deceitful, murderous (cf. Rom. 1: 29–31). In opposition to his Jewish forebearers, however, Paul refuses to see this situation as one in which human responsibility is engaged. The failure of individuals is not their personal responsibility. It is ascribed to the power of Sin. The human race is twisted and distorted by a power greater than any of its members.[61] The basis of this insight was Paul's own experience. As a traveller he found himself forced to be other than he wished to be. His commitment was to be another Christ, totally dedicated to the service of others. But if he was to survive on the road, he had to look after his own interests first. The conditions under which he lived obliged him to be selfish, to mistrust others instead of loving them.[62]

Paul chose the word Sin to crystalize his vision of society as the victim of a massive disorientation, because its origins were to be traced back to the sin of one person (5: 12a; cf. 5: 19). The point of Genesis 3 is that, at some point in the history of humanity, a false decision was made. From then on, according to Genesis 4–11, evil developed exponentially (Gen 6: 5). Wickedness became endemic, as sinners interacted with each other. All those born into a warped society inherit its defects. They have no choice but to internalize its values, and to pass them on reinforced to the next generation. They are enslaved to Sin, which dwells within them. Sin, for Paul, was not an extra-terrestrial force, but a reality within humanity, the accumulated power of lived assent to a false value system. 'God has imprisoned all human beings in their own disobedience' (Rom. 11: 32).[63]

As far as Paul was concerned, one of the false values which Jews inherited was a particular attitude towards the Law, which distorted its true purpose (Rom. 7: 10).[64] The fundamental component of the theological system of all Jews of Paul's time was belief in their election by a gratuitous divine act. God's giving of the Law established the covenant. Membership in the covenant was necessary for salvation, and involved obedience to its regulations as expressed in the precepts of the Law. The tricky point for Jewish theologians was the precise relationship of divine initiative and human response. How were unmotivated mercy and the demands of the Law reconciled? The solution proposed by E. P. Sanders—'*Obedience to the commandments* was not thought of as earning salvation, which came rather by God's grace, but was nevertheless required as *a condition of remaining in the covenant*; and not obeying the commandments would damn'[65]—is certainly justified by documents contemporary with Paul,

---

[61] See in particular Dunn (1988), 148.
[62] See Ch. 4, 'Dangers on the Road'.   [63] See esp. Dunn (1988), 696.
[64] The definition of Law implicit in what follows in this section has been well formulated by Westerholm (1988), 220, 'Crucial to an appreciation of Paul's understanding of the law is the realization that he normally means by *nomos* the divine commandments imparted to Israel on Mount Sinai with their accompanying sanctions. The "works of the law" are concrete deeds which this legislation manifestly requires.'
[65] (1977), 320—his italics.

but its subtlety highlights the practical problem. The human mind instinctively simplifies. If disobedience to the commands of the Law caused damnation, then it seemed logical that obedience to such precepts won salvation. Thus, while lip-service was paid to the fundamental concept of gratuitous grace in election, in practice all attention was concentrated on observance of the commandments. A religion of grace which expresses itself in covenant form quickly becomes a religion of meritorious achievement, certainly in the popular mind, if not in the dissertations of theologians. What concerned Paul, however, was less the objectionable idea of buying salvation, than the inversion of values consequent on the importance attached to obedience to the Law.

This distinction between what is true in principle and what is real in fact is nowhere more graphically illustrated than in the the extraordinary inversion of the positions of God and the Law in the rabbinic writings. According to the rabbis, 'There are twelve hours in the day; during the first three the Holy One, blessed be He, occupies Himself with the Torah.'[66] As a student it is not surprising that God should take his place with other scholars, 'Now they were disputing in the Heavenly Academy ... the Holy One, blessed be He, ruled, "He is clean"; whilst the entire Heavenly Academy maintained, "He is unclean". Who shall decide it? said they.—Rabbah b. Nahmani; for he said, "I am pre-eminent in the laws of leprosy and tents".'[67] Divine authority gives way before rabbinic expertise. Elsewhere when questioned God can only reply, 'My son Abiathar says So-and-so, and my son Jonathan says So-and-so. Said R. Abiathar: "Can there be uncertainty in the mind of the Heavenly One?"'[68] God takes a position on the problem of the Red Heifer by citing a ruling of R. Eliezer.[69]

On another occasion, R. Eliezer is supported by direct divine intervention,

> A Heavenly Voice cried out: 'Why do you dispute with R. Eliezer, seeing that in all matters the halachah agrees with him!' But R. Joshua arose and exclaimed: '*It is not in heaven* (Deut. 30: 12).' What did he mean by this?—Said R. Jeremiah: 'That the Torah had already been given at Mount Sinai: we pay no attention to a Heavenly Voice, because Thou hast long since written in the Torah at Mount Sinai; *After the majority must one incline* (Exod. 23: 2).' R. Nathan met [the prophet] Elijah and asked him: 'What did the Holy One, blessed be He, do in that hour?—He laughed [with joy], he replied, saying, 'My sons have defeated Me, My sons have defeated Me.'[70]

A good loser, God recognizes that he has been side-lined. He had failed to realize that, once he had given the Law to the Jewish people, it was out of his hands.

---

[66] *b. Aboda Zara* 3b; trans. Epstein (1935), 4. 9. Similarly, the Jerusalem Targum on Deut. 32: 4, trans. Le Deaut (1980), 264–5, 'For three hours does He occupy Himself with the Torah.'
[67] *b. Baba Mezia* 86a; trans. Epstein (1935), 1. 495.
[68] *b. Gittin* 6b; trans. Epstein (1936), 4. 21.
[69] Urbach (1979), 307.
[70] *b. Baba Mezia* 59b; trans. Epstein (1935), 1. 353.

Now only the voice of the rabbis counted. God himself is bound by their decisions!

Whatever their date, and even when given their most benign interpretation as assertions of the freedom of human reason, these quotations unambiguously illustrate what happens when intangible grace is confronted with the concrete specificity of the Law. Jews debated points of Law, not the mystery of grace; manipulation of the controllable supplanted contemplation of the ineffable. Gratitude for election could not be expressed merely in psalms of praise; performance of the works of the Law was necessary.

Dunn denies that Paul criticized the concept of earning salvation through obedience, and instead asserts that 'Paul's negative thrust against the law is against the law taken over too completely by Israel, the law misunderstood by a misplaced emphasis on boundary-marking ritual, the law become a tool of sin in its too close identification with matters of the flesh, the law side tracked into a focus for nationalistic zeal'.[71] Were this correct,[72] Paul would have the same objection. However it was conceived, be it as a means of guaranteeing salvation or as a nationalistic imperative, the Law absorbed everybody and everything in its orbit. It left no real space for God or grace or faith. It had room only for obedience.

In order to ensure that the gracious gift of God in Jesus Christ would retain its primacy in practice Paul had to insist that the Law was completely irrelevant for all believers, both Jews and Gentiles. His fundamental objection to the Law was that, once admitted into a community, it inevitably created an attitude which monopolized the religious perspective. To focus on the Law was necessarily to ignore Christ. There could not be two ways of salvation.[73] The authentic response to God's grace is revealed in the self-sacrifice of Christ, which was in no way anticipated in the Law. If anything, Christ is the New Law (Gal. 6: 2). Understandably, therefore, Paul insists that Christ had written *finis* to the Law as far as humanity is concerned (Rom. 10: 4).[74] Once the goal (the probable sense of *telos* here) of the Law had been achieved definitively in and through Christ, the means thereto (the Law) no longer had any raison d'être.

The condition of Gentiles, distorted by the egocentric values of the society

---

[71] (1988), p. lxxii.

[72] The most serious objection is the conclusion which flows from Dunn's hypothesis. He (1988), p. lxxi, tells us that Paul's purpose in criticizing the Law was 'to free both promise and law for a wider range of recipients'. In other words, Paul's concern was to increase the sphere of grace to include Gentiles, who then would maintain their place in the new covenant by doing those works of the Law that were not Jewish identity-markers. This, however, cannot be what Paul means. To adopt covenantal nomism as the pattern of Christianity would be to re-create the conditions which produced the absorption of Jews in the Law. See Hooker (1982), and my (1989).

[73] This point is rightly emphasized by Sanders (1977), 482, 497, 506.

[74] Interpretations of this crucial text are regularly coloured by the conscious or unconscious desire of commentators to find in Paul legitimation of the moral code of the Old Testament. See in particular Badenas (1986).

into which they were born, and that of the Jews, made 'captive to the Law of Sin' (7: 23), is summed up by Paul in one and the same word, Death. This vision of the human condition is derived from Paul's conviction—already hinted at in Galatians—that the criterion of authentic humanity is the self-sacrificing love revealed in Jesus who 'did not please himself' (Rom. 15: 3) but suffered on behalf of all human beings (8: 17), to the point of dying for the godless (5: 6).[75] Lacking this creative outreach, a life turned inward on itself by society or the Law can only be imaged as the existence of a corpse. The unloving are the walking dead.

This brief synthesis of what Paul meant by Sin, Law, and Death reveal how deeply he had reflected on the circumstances of his ministry. An analysis of the human condition was essential, if he was to give a precise focus to his preaching. His recognition that his hearers had to be freed from Sin before they could respond to the gospel, and thus be raised from Death to Life, flowered into a commitment to be for others the Christ who was the power of God and the wisdom of God (1 Cor. 1: 24; 2 Cor. 4: 10–11).

The Salvation of the Jews

Although the tentacles of Sin reached out into every section of humanity, Paul considered his own people to be its greatest victim. He, and others, had freed multitudes of Gentiles from the power of Sin, but nothing like as many Jews (although there were some) had accepted Christ as the Messiah. Why those who had been the most privileged by God in terms of preparation for the advent of the Messiah should have been the most adamant in their refusal of Jesus was a mystery with which Paul struggled during all his apostolic life. What was going to happen ultimately? The urgency of this question was enhanced by his apprehension regarding his reception in Jerusalem (Rom. 15: 31),[76] with its implicit recognition of the increasing hostility of Jews to the Jesus movement. It seemed most improbable that Christian missionaries to Jews would be more successful in the future. The winter in Corinth gave him the leisure to bring together the partial insights which had occurred to him over the years, and to formulate a comprehensive answer.

The intricate argumentation of Romans 9–11 is impressive evidence of the depth of Paul's knowledge of the Jewish scriptures. The sophistication of his interpretation once again betrays the strength of his intellectual formation. The quality of the writing is also remarkable, and the hymn, in which he sings out his adoration with extraordinary eloquence (Rom. 11: 33–6), is arguably his greatest literary achievement.[77]

---

[75] See Ch. 8, 'The Living Christ'.   [76] See Ch. 14, 'A Sense of Foreboding'

[77] Dunn (1988), 698, draws attention to 'the number of echoes of language which was evidently much in Paul's mind in dealing with the church at Corinth'.

The outpouring of gratitude is a fitting conclusion to the summation of the argument in Romans 11: 25–32, in which Paul reveals his solution to the problem of the salvation of Israel. He had never wavered in his conviction that God could not deny himself, and abandon those whom he had chosen and gifted (Rom. 11: 1–2). Paul recognized the truth of this in his own ministry. The book of Isaiah had always played a key role in his understanding of his apostolate to the Gentiles. He saw himself as part of the faithful remnant, which proclaimed salvation to the nations, thereby fulfilling the eschatological obligation laid upon Israel.[78] Not surprisingly, it was in reading Isaiah that he realized the means whereby the Jews would be saved. In Romans 11: 26 in order to support his thesis that 'all Israel will be saved', he quotes 'From Sion will come the Redeemer, he will banish ungodliness from Jacob, and this will be my covenant with them' (Isa. 59: 20–1); 'when I take away their sins' (Isa. 27: 9). The allusion is to the Parousia of Christ.[79] The Jews, in other words, will be saved in exactly the same way as Paul was.[80] His commitment to the Law had not only blinded him to the true role of Christ, but it had engendered bitter hostility. That attitude was changed by a completely unexpected encounter on the road to Damascus, where Christ took the initiative. So will it be for all Israel, at the Parousia when Christ appears in glory. Then the Jews will no more be capable of rejecting him than Paul had been.

[78] Niebuhr (1992), 170.
[79] Sanday and Headlam (1902), 336; Dunn (1988), 682.
[80] Hofius (1986), 320.

— 14 —

The Last Years

T HE depth of Paul's concern about his reception in Jerusalem can be gauged from the fact that he mentions it in his letter to the Romans, a group whose support he needed if he was to preach in Spain. One might assume that he was thinking primarily of those members of the church whom he knew personally, but to even hint at difficulties with the mother church could only have damaged his image in the eyes of those he hoped would be allies. The slip betrays the intensity of his anxiety.[1]

AN UNCERTAIN FUTURE

Paul asked the Romans to pray for three things '(1) that I may be preserved from the disobedient in Judaea, (2) and that my service for Jerusalem may be acceptable to the saints, (3) in order that I may come to you with joy by God's will' (Rom. 15: 31–2).

'The disobedient in Judaea' is an unusual formulation, but what Paul means is clear from Romans 11: 30–2; he is thinking of Jews who have not accepted Christ. The verb he uses connotes a vivid sense of danger. When speaking of his two previous visits to Jerusalem Paul never suggests that he felt physically threatened (Gal. 1: 18–19; 2: 1–10). There is no basis for the view that he delayed three years in Damascus because he felt that a visit to Jerusalem would be too much of a risk.[2] It is possible to detect an element of secrecy in Paul's first visit as a Christian—he saw no one but Peter and James—but that may have been due to embarassment; he did not want to confront those whom he had persecuted not many years earlier. Nothing similar is in evidence during his second visit. The series of meetings must have been known to the whole community, but other Jews apparently exhibited no interest. Were they aware of the

---

[1] Brown & Meier (1983), 110, and Fitzmyer (1993), 726, suggest that Paul expected the Roman church to intercede on his behalf with Jerusalem. Nothing in Rom. 15: 30–1 supports this interpretation, and with eminent common sense Bruce (1991), 192, has pointed out that there was no way that emissaries from Rome could have reached Jerusalem before Paul.
[2] *Pace* Longenecker (1990), 38.

decision not to circumcize Gentile believers, one might assume that they welcomed the refusal of Christians to make pagans nominally Jews.

Why should Paul now take it for granted that Palestinian Jews would be hostile to him? One might speculate that he somehow learnt that his name was on a death-list circulating in Palestine. To make this extreme scenario plausible demands a complex series of assumptions, e.g. a Diaspora synagogue noted Paul's radically antinomian stance and, taking advantage of the sending of the Temple tax to Jerusalem, informed the authorities there, who in turn decided that it was equivalent to apostasy and thus worthy of death. Such pyramiding of possibilities does not enhance their probability. Moreover, such interest on the part of the Temple authorities in the opinions of a single Diaspora Jew seems rather implausible, especially given the number of Jews in the dispersion (several million) and the problems in Jerusalem.

Alternatively one might assume that Paul's fear was entirely subjective. In this minimalist scenario we should envisage Paul the Pharisee assessing and finding guilty Paul the antinomian. Reflection on the way he had once acted towards followers of Christ, whose deviations from Judaism were minor, and on the way he would have wanted to act against those who repudiated the Law as irrelevant, stimulated Paul's imagination to the point where the threat from observant Jews became very real. He postulated what he feared.

While there is uncertainty regarding the attitude to Paul of Palestinian Jews in general, there is none regarding that of the Jerusalem church. Emissaries from Antioch, as we have seen, had been on Paul's heels for four years, challenging his attitude towards the Law. In Galatia they had the opportunity to read Paul's letter to the churches there, and it is far from impossible that they sent a copy to Antioch. In any case, it is unlikely that they stayed away for several years without reporting back to their home base. At least an oral report of Paul's radical antinomianism reached Antioch. Regular contacts between Antioch and Jerusalem can be safely assumed. The practice attested in the Acts of the Apostles would have become even more necessary as anti-Semitic pressures intensified in the eastern Mediterranean. Paul could be quite sure, therefore, that James and his cohorts knew perfectly well that Paul's position had hardened into a stance which was the antithesis of theirs. It was perfectly reasonable, in consequence, for Paul to wonder if James would accept a gift with which he was so intimately associated.

In principle Jews had no compunction about accepting gifts from Gentiles. Not only had the Temple been graced by the donations of foreigners,[3] but the implication of Leviticus 22: 25 that Gentiles could offer sacrifices in the Temple is well documented by Josephus. It may be just a legend that Alexander

---

[3] *JW* 2. 412. The only coinage accepted in the Temple was that of pagan Tyre (*m. Bekorot* 8. 7). 'All the money spoken of in the Torah is Tyrian money' (*t. Ketubot* 13. 3).

the Great offered sacrifice there,[4] but there seems to be no doubt that Ptolemy III of Egypt,[5] Antiochus VII of Syria (during a truce when he was beseiging Jerusalem!),[6] and the Romans Marcus Agrippa,[7] and Lucius Vitellius sent offerings to Jerusalem.[8] As relations with Rome deteriorated, however, such pagan participation in the Jewish cult became progressively less acceptable to the more extreme elements. The climax came in AD 66. 'Eleazar, the son of Ananias the high priest, a very bold youth, who at that time was governor of the Temple persuaded those who officiated in the divine service to receive no gift or sacrifice for any foreigner. This was the true beginning of our war with the Romans' (*JW* 2. 409; trans. Whiston and Margoliouth).

It is only in this political context that Paul's apprehension regarding the reception of the collection becomes understandable. He had experienced James' nationalistic attitude both positively (Gal. 2: 3) and negatively (Gal. 2: 12),[9] and was well aware that a gesture which could be understood as forging a bond with Gentiles might meet with a rebuff. Paul could not be sure, however, because he did not know how much the Jerusalem community needed the money.

Paul could have decided not to return to Jerusalem. His participation in the delegation was not imperative. The delegates of the contributing churches were with him,[10] and he could have given back the money and opted out. Or they could have gone ahead without him. The only injury would have been to his pride. His decision to persevere, despite mortal danger and the possible futility of the gesture, underlines how deeply he felt about the relationship between the Jewish and Gentile churches. No one was more conscious of the profundity of the widening gap between those for whom Christ was central and those for whom he was not. Yet it was desperately important to fling across the abyss a fragile bridge of charity. He would risk all in the attempt.

A FAREWELL CIRCUIT OF THE AEGEAN SEA

Luke provides a long account of Paul's journey to Jerusalem with the collection (Acts 20: 3 to 21: 17). He is aware of so many details because he derives virtually his whole narrative from the Travel Journal, one of the oldest sources of Acts, which was the work of an eyewitness.[11] With the exception of a short discourse at Miletus, it is simply a list of times and places, which correspond very closely

---

[4] *AJ* 11. 329–30.
[5] *Against Apion* 2. 48.
[6] *AJ* 13. 242–3.
[7] *AJ* 16. 14.
[8] *AJ* 18. 122
[9] See Ch. 6, 'Why did James Agree with Paul on Circumcision?'
[10] This is not absolutely certain from the letters (1 Cor. 16: 3; 2 Cor. 8: 19; 9: 4), but it is the common interpretation of the list in Acts 20: 4; see Haenchen (1971), 581; Georgi (1992), 122; Hemer (1989), 188.
[11] Boismard and Lamouille (1990), 2. 221–7.

with the theory and practice of coastal navigation in antiquity.[12] Starting from Philippi, more accurately from Neapolis, each day the boat made what distance the wind permitted, as it worked south along the coast of Asia Minor, and sought a harbour for the night. Only in Myra did they find a boat sailing directly for Tyre.

The plausibility of this scenario cannot dispense us from raising certain questions regarding the substance of the story. Nothing in the reconstruction of the Travel Journal hints that Paul and his companions were burdened by the responsibility of conveying a considerable sum of money to Jerusalem. Moreover, the departure point is not the expected one of Cenchreae (cf. Acts 18: 18), the eastern port of Corinth, but Philippi in far away Macedonia. Does the Travel Journal really describe the journey to Jerusalem which took place after the letter to Rome had been dispatched? Or does it narrate a different journey, which Luke adapted to his purpose because it happened to end in Jerusalem?

Two features of the Travel Journal suggest that the first of these two options is to be preferred. The discourse of Paul to the elders of Ephesus at Miletus looks back to his first stay in Ephesus (Acts 20: 18). Hence the voyage must be dated after AD 54. Between AD 54 and 56, however, Paul's travels were limited to a sweep out to the west and south. From Ephesus he reached Corinth via Troas, Macedonia, and Illyricum. Unless one is prepared to consider the voyage as taking place sometime later than the collection voyage, one is forced to identify it with the latter. Further, the note of apprehension in 'And now, as a captive to the Spirit, I am on my way to Jerusalem, not knowing what will happen to me there' (Acts 20: 22) echoes the feeling of Romans 15: 31.

How then is Paul's appearance in Macedonia to be explained? According to Boismard and Lamouille's reconstruction of the Travel Journal, 'He was about to sail for Syria when a plot was made against him by the Jews, and so he decided to return through Macedonia' (Acts 20: 3). This verse, however, could equally well be a Lukan redactional insertion. Luke knew that the natural course would have been for Paul to take a ship going east from Cenchreae (Acts 18: 18), but his source gave Philippi as the departure point. Hence, he had to postulate a scenario to explain why Paul did not do the normal thing, and so Luke opted for his standard device of a Jewish plot in order to move Paul to Macedonia.[13] He had done this sort of thing earlier to explain Paul's departure from Damascus (Acts 9: 24), where we know that the real reason was quite different (2 Cor. 11: 32–3).

[12] Casson (1971).
[13] Taking the statement at face value, Georgi (1992), 124, suggests that 'the plot was to be executed on the high sea. This, in turn, tells us that there must have been a considerable number of Jews on the ship; that is to say, it must have been a shipload of pilgrims'! This interpretation goes back to Ramsay; see Haenchen (1971), 581 n. 3.

Transporting the Collection

If the Jewish piracy hypothesis is not acceptable, what reason could Paul have for going to Macedonia at this point? The possibility that he had to go north from Corinth in order to pick up the contributions to the collection from Thessalonica and Philippi is excluded by Romans 15: 26 (cf. 2 Cor. 9: 4), which gives the impression that the cash was already in hand. As a preliminary to an exploration of this issue, it is important to have as clear an idea as possible of what was involved in the transportation of the collection.

The model that would have spontaneously occurred to people of the Apostle's background was the procedure for transmitting the annual half-shekel Temple tax from the Diaspora to Jerusalem. The money collected from the various communities was reduced to the smallest volume by being exchanged for metal of the highest value, namely, gold.[14] In 59 BC Lucius Valerius Flaccus, sometime governor of Asia, was brought to court for (among other charges) refusing to permit Jews to export 100 pounds of gold from Apamea, 20 pounds from Laodicea, 100 pounds from Adramyttium, and a lesser amount from Pergamum.[15] The precautions taken by the Jews to protect their sacred funds in transit is perhaps exaggerated by Josephus,[16] but security was no doubt a major preoccupation.

Exactly how much money Paul collected can never be known. It is certain, however, that it was considerable. The symbolic value of the gesture would have been negated were the sum derisory (1 Cor. 16: 2); it would have been seen by the Jerusalemites as an expression of contempt.[17] Unless an impressive amount of cash had been assembled, it is most probable that Paul would have considered the exercise a failure, and would have returned the contributions to the communities, accompanied, no doubt, by a bitter comment on their lack of generosity.

Given the conditions of travel in the first century,[18] Paul's major preoccupation had to be the security of the funds entrusted to him. He was hardly in a position to hire armed guards. Hence we can exclude the use of pack animals to transport coffers of specie or sacks full of clinking coins. Imagine the effect on the bystanders, when the containers were unloaded at the first inn, and how quickly word would spread among the local underworld types! Paul's best

---

[14] 'Shekels may be changed into darics because of [lightening] a journey's load' (*m. Shekalim* 2. 1). A daric was a Persian gold coin worth about sixteen shekels (Danby (1967), 153 n. 9), and was one of the very few internationally accepted currencies (Casson (1979), 75).

[15] Cicero, *Pro Flacco* 66–9. The right of Jews of Asia to send money to Jerusalem was reaffirmed by an edict of Augustus in AD 2–3 (Josephus, *AJ* 16. 163); see Saulnier (1981), 185–8.

[16] From Babylon, he says, 'many myriads' accompanied the transfer of the Temple tax to Jerusalem (AJ 18. 312–13).

[17] On the cost of living in Jerusalem, see Jeremias (1969), 120–3.

[18] See Ch, 4, 'Dangers on the Road'.

protection was absolute secrecy. Each member of the party carried their personal funds for the journey in the usual moneybelt or little bag suspended from a cord around the neck,[19] but in addition each had a number of gold coins sewn into his or her garments in such a way that they would not chink. Since gold is heavy for its volume, the danger of distorting the shape of the garment would have limited the number of coins that any one individual could carry. In consequence, the number of Paul's companions was conditioned by the amount of money that had to be transported.

Even with such precautions, Paul and his companions were more at risk on land than at sea. The crew and passengers on a ship were a fixed and known quantity whose movements could be monitored. A thief had a chance of escaping only if the theft went undiscovered until the ship arrived in port. Casual acquaintances on the road, or fellow-guests at an inn, were constantly changing. Paul could never know what they had noticed, or what plans were being made to attack his group. On the road, therefore, the anxiety level was consistently high, whereas at sea there were long periods of relaxation.

A Sense of Foreboding

Paul's decision to go to Macedonia committed him to the land route, because to sail north against the Etesian winds would have meant interminable delays.[20] In order to justify the greater risk of losing the funds for Jerusalem, there must have been a proportionately serious reason. The need to pick up further contributions to the collection having already been excluded, only one serious possibility remains. Despite Paul's optimistic plans for the future as revealed in Romans 15–16, the pessimistic side of his nature subsequently gained the upper hand, and he became convinced that he would never return from Jerusalem. He took the risk of going north to Macedonia to say farewell to the communities he had founded there. This hypothesis is confirmed by what the Travel Journal tells of Paul's conduct in Asia. Despite his haste to reach Jerusalem, he devoted time to his two foundations.

Paul stayed in Troas, even though a ship sailing as far as Assos was ready to weigh anchor. He told his companions to take it, after arranging to join them there by cutting across the base of the great headland on foot (Acts 20: 13). Speculation on the reason for this decision has given rise to the rather far-fetched suggestions that he wanted to be alone with God, or that he feared a rough sea voyage.[21] It is preferable to assume that he was retained by unfinished business in Troas. The community there was one to which Paul had not

---

[19] Casson (1979), 176.
[20] See Ch. 12, 'A Journey to Macedonia'.
[21] Haenchen (1971), 587.

been able to devote much time (2 Cor. 2: 12–13).[22] Presumably, when he crossed to Macedonia to meet Titus, he had promised to return. Now when he did, he told them that it was unlikely that he would ever see them again. Not unnaturally they clung to him until the last possible moment.

Only in this perspective does it become possible to explain why Paul decided to avoid a visit to Ephesus, and instead summoned the Elders to Miletus. He did not want to repeat the same wrenching experience of the farewells at Troas. The reason given by the Travel Journal—that Paul was in a hurry to get to Jerusalem (Acts 20: 16)—does not resist examination.[23] In antiquity—before the silting up of the Gulf of Latmus—Miletus was 80 km. (50 miles) from Ephesus.[24] The journey involved going round the head of the Gulf of Latmus, crossing the river Meander and two mountain ranges. A messenger would have thought in terms of a three-day journey, and one can be sure that the Elders would have strolled along at a pace appropriate to their dignity. Driven by a sense of urgency, Paul could have made the journey in two days, provided he increased his usual daily average by 8 km. (5 miles). In other words, if Paul's real concern was to save time, it would have been much quicker to go to Ephesus himself.

For the majority of exegetes Paul's discourse at Miletus was inserted by Luke into the Travel Journal. Boismard and Lamouille, however, have shown that Luke merely amplified a brief discourse which already formed part of the Travel Journal.[25] In that discourse Paul does not explicitly bid farewell to the Elders, but the terminal character of the visit is clearly signalled by his apprehension as to what will happen in Jerusalem (Acts 20: 22), and by his provision for the future by alerting the Elders to the fact that henceforward they would have sole responsibility for the church of Ephesus (Acts 20: 28).

PAUL'S RECEPTION IN JERUSALEM

This section of the Travel Journal culminates with Paul's arrival in Jerusalem (Acts 21: 17); it continues with his departure for Rome (Acts 27: 1). For all that happened in between we are entirely dependent on Luke, and his other sources.

It is rare for Luke merely to report an event to his readers; he permits them to see it happening. He creates each episode out of broad strokes of bright colour, laid on at a pace which leaves his readers with a vivid impression of the central point but no clear memory of the details. But if those details are given any importance as historically reliable information, the narrative falls apart,

[22] See Ch. 12, 'Departure for Troas'.
[23] The author of the Travel Document, it will be recalled, was not a witness of the heart-rending farewells at Troas (Acts 20: 13). [24] McDonagh (1989), 297.
[25] (1990), 3. 247–51. The original discourse is reconstructed in their (1990), 2. 222.

because 'then the persons act very strangely in an improbable and incomprehensible manner'.[26] There is a growing consensus that Luke had only a very rudimentary idea of what actually happened to Paul on his last visit to Jerusalem.

This basic framework probably comprised no more than the following elements: Roman intervention saved Paul from a Jewish mob. The tribune in Jerusalem, Claudius Lysias, transferred the responsibility to his superior in Caesarea, Felix, who had not disposed of the case when he was replaced by Festus. Eventually Paul claimed his right as a Roman citizen to be tried by the emperor and was sent to Rome. With marvellous literary skill, Luke put flesh on this skeleton. His concern, however, was not to document the details of what really happened on the basis of reliable sources, but to elaborate and manipulate those events with a view to getting across his own convictions regarding Paul's role in the plan of salvation. It is a logical fallacy to infer from the precision of Luke's legal knowledge that events in fact took place exactly as he describes them.[27]

Rather than trying to establish what might be historical in Luke's presentation, which would mean writing a commentary on the Acts of the Apostles, it is preferable to take up from his perspective the two issues which preoccupied Paul in Romans 15: 31–2, namely, his apprehension regarding the reception of the collection, and his fear of assassination. Were his fears justified?

Only one verse in Acts can possibly be construed as an allusion to the collection, namely, 'After many years I have come here with alms for my people and offerings' (24: 14), but none of Luke's readers would have understood it in this sense.[28] Did Luke intend a subtle reference to the collection, which followers of Paul might recognize? Not necessarily. It was customary for devout pilgrims to bring alms for distribution in the Holy City, as well as money to pay for sacrifices.[29] The point of the reference is to show that Paul believed 'everything that is written in the Law and the prophets' (24: 14). In other words, Acts 24: 17 is precisely what Luke would have postulated, even if he had no knowledge of what Paul had actually done.[30]

But is it really possible that Luke was ignorant of the collection? If he was a companion of Paul, as some have suggested,[31] or if he knew the Pauline letters, as others have maintained,[32] he must have been aware of the importance which the Apostle attached to the gesture. In the case that neither of these possibilities

[26] Haenchen (1971), 639.
[27] Against Tajra (1989). Many entirely imaginative detective stories are set in a framework of meticulously exact police procedures.
[28] So rightly Haenchen (1971), 655, against Hemer (1989), 189, and Légasse (1991), 202.
[29] Jeremias (1969), 129–30.
[30] Luke does not necessarily rely on a source here, *pace* Lüdemann (1984), 24.
[31] Trocmé (1957), 138, for whom Luke was the author of the Travel Document.
[32] Boismard and Lamouille (1990), I. 39.

is admitted, we cannot say what Luke might or might not have known. It is more profitable to ask what James and others in the church of Jerusalem knew. Even if Cephas and John were no longer in the Holy City, James must certainly have remembered that it was he who had requested financial aid for his church from Paul. Not only was that request no more than five years old, but Paul's response must have become a subject of continual discussion.

At the time of the Jerusalem Conference in the autumn of AD 51, a financial contribution from Gentile believers seemed like a reasonable *quid pro quo* for Jerusalem's concession on circumcision, and no doubt would have been proclaimed as such to the church by the three Pillars. But as Paul's radically antinomian stance became known in an ever more nationalistic Jerusalem church, there must have been those who insisted that they would accept nothing from hands so soiled. The charge that Paul taught 'all the Jews who live among the Gentiles to forsake Moses, telling them that they should neither circumcise their children nor live according to the customs' (Acts 21: 21) was fully justified.[33] Others, more pragmatically, would have asserted that money has no smell, that it was necessary, and that it could be used to good ends without accepting Paul's interpretation that it constituted a bond between the Jewish and Gentile churches. The final decision was up to James, but why should he endanger his authority by taking sides on a purely hypothetical problem? If and when the money arrived, which was not at all guaranteed, would be time enough to make up his mind.

What happened when Paul finally put in an appearance? The number of possibilities is limited: (1) the collection was accepted;[34] (2) the collection was refused;[35] (3) some Jerusalemites accepted the collection over the objections of others;[36] (4) the handing over of the collection was impeded.[37] In all cases, the one argument invoked is the silence of Luke! (1) Luke simply did not know of this happy ending. (2) Luke did not mention the refusal in order to preserve the image of a unified church. (3) The grudging 'unofficial' acceptance was omitted by Luke as insulting to the Pauline churches. (4) Since nothing happened, Luke had no need to mention it.

What differentiates the fourth possibility from the other three is that it takes into account what Luke does say. In order to maintain faith with a project which he had initiated, James' first reaction would have been to look for a way which would make it possible for him to accept the collection. His basic concern, then, would have been to satisfy himself and his right-wing constituency that Paul remained a practising Jew, and that his antinomian reputation was

[33] See Ch. 8, 'Faith and Law'; and Ch. 13, 'Sin, Law, and Death'.
[34] Koester (1982), 2. 144.
[35] Lüdemann (1989), 61; Légasse (1991), 203.
[36] Georgi (1992), 125–6.
[37] Morton Smith in a letter cited by Lüdemann (1989), 250 n. 115; Becker (1989), 485.

unjustified. A profession of faith alone would not have sufficed. Paul had to make a public gesture demonstrating his Jewishness.

The simplest act, and the minimum which James could have accepted, would have been the purification required of all Jews coming from pagan territory, and who wished to enter the Temple. It was assumed that they had incurred corpse-uncleanness, and this levitical impurity had to be removed by having a priest sprinkle them with the water of atonement on the third and seventh days.[38] Only by going through this ritual would Paul be in a position to relate to the strictly observant members of the Jerusalem community, who, in addition, would have been gratified by the implicit condemnation of Gentiles.

Even if we did not have the witness of Luke, this basic commitment on the part of Paul would have to be postulated in order to do justice to what is known of James, and of the increasing pressures under which his community lived. Luke's lack of clarity, however, in confusing the purification which Paul had to undergo, and the ceremony celebrating the termination of the Nazarite vow, suggests that he was drawing on a source.[39] The intrinsic probability of one element in this source (Paul's purification) enhances the credibility of the other (Paul's involvement with the Nazarites).

It was not cheap to acquit oneself of a Nazarite vow. The obligatory offering consisted of, 'an unblemished male yearling lamb as a burnt offering, an unblemished yearling ewe lamb as a sacrifice for sin, an unblemished ram as a peace offering, and a basket of unleavened loaves made of fine flour mixed with oil, and of unleavened wafers spread with oil, with the cereal offerings and libations appropriate to them' (Num. 6: 14–15). For a poor person the financial burden was considerable, and it might take an excessively long time to assemble the money. Hence, it was considered meritorious of the Jewish community to help out.[40] Given what we know of Paul's personal financial situation from his letters,[41] it is most improbable that he had the wherewithal to pay for four Nazarites.[42] It would have been necessary to draw on the collection money. The gesture was one which the Jerusalem church could hardly refuse from a visitor who had been ritually purified. Moreover, it relieved the church of a financial burden. Acceptance, however, meant that the church had already profited by the collection! Refusal, on the other hand, would mean recognizing Paul by reimbursing him.

---

[38] Num. 19: 11–16; Acts 21: 27; see Billerbeck (1922–8), 2. 759.

[39] Haenchen (1971), 612; Lüdemann (1989), 56–7.

[40] This is the implication of *Genesis Rabbah*, 91, where Shimon ben Schatach found funds for 150 Nazarites and petitioned Alexander Jannaeus for support for 150 more (Billerbeck (1922–8), 2. 755), and of Josephus, *AJ* 19. 294, where Agrippa I, on assuming the throne, paid for the sacrifices of many Nazarites.

[41] See Ch. 12, 'Financial Assistance'.

[42] Little historical value can be accorded to Luke's statements that Felix assumed that Paul was capable of paying a large bribe (Acts 24: 26), and that Paul rented an apartment in Rome for two years at his own expense (Acts 28: 30); see Hemer (1989), 192.

Luke gives the credit for this strategy to the leaders of the Jerusalem community (Acts 21: 24). What his source said we have no way of knowing, but it is far from impossible that Paul himself was responsible. The anxiety exhibited in Romans 15: 31–2 ensured that his mind continued to worry at the problem of his reception in Jerusalem. He had plenty of time on the long journey from Greece to work out a plan that would both placate the Jerusalem believers, and confront them with a *fait accompli*.

Tragically, the plan was initiated but never terminated. Before the seven days of his purification were completed (Acts 21: 27), and thus before he could do anything for the Nazarites, the second danger that Paul anticipated became a reality. Non-Christian Jews attempted to lynch him. The garrison in the Antonia intervened to save him. Thereafter he was in Roman custody. What his companions did with the collection will never be known. Once it was out of Paul's hands, Luke (like biographers of Paul!) loses interest.

A ROMAN PRISONER

A feature of Luke's narrative subsequent to Paul's arrest is the number of discourses attributed to the Apostle. In Jerusalem he addresses the populace (Acts 22: 1–21), and the Sanhedrin (Acts 23: 1–10). In Caesarea the Roman governors Felix (Acts 24: 1–21), and Festus and the Jewish king Agrippa II (Acts 26: 1–23), hear him speak. Paul's message reaches every section of the population of Palestine, from the city mob to the royal court, from the religious leadership to the secular authority. The improbability of a prisoner under investigation being offered the opportunity to disseminate the heresy/treason of which he is accused needs no emphasis. The voice may be that of Paul, but the words are those of Luke. Since much of the intervening narrative sections are designed to set the stage for the discourses, their historical value is also severely compromised.

A Hazardous Journey to Rome

Problems multiply when we come to the narrative of the journey to Rome (Acts 27: 1 to 28: 14). Not only do the Western and Alexandrian texts offer significantly different versions,[43] but both manifest traces of sources and editorial contributions. There can be no question here of an exhaustive analysis; a few examples of the lack of literary unity within the Western text must suffice.

1. Acts 27: 8–10 cannot belong to the same narrative as 27: 11–12. The former concerns a dangerous voyage on the open sea starting from Fair Havens

---

[43] A synoptic presentation in Greek is to be found in Boismard and Lamouille (1984), 1. 219–24, and in French in their (1990), 1. 168–74.

in Crete, whereas the latter deals with a similar voyage, whose success is not at all assured, and whose goal is Phoenix in Crete. When these two incompatible texts are juxtaposed, the result is a routine coastal voyage of some 100 km. along the south coast of Crete.[44]

2. Acts 27: 13–17 recounts a stormy passage to the Isle of Clauda where the ship takes shelter, but in 27: 18 the ship is still at sea.[45]

3. The unusual Homeric phrase *epekilan tên naun*, meaning to land a boat on an open beach (which has nothing to do with shipwreck), appears in both 27: 29 and 27: 41. Such a doublet betrays the redactional technique of the *Wiederaufnahme*.[46]

Only Boismard and Lamouille have made a serious effort to offer a detailed reconstruction of the sources employed by Luke.[47]

Travel Journal

(**27**: 1) When it was decided that we were to sail for Italy, (3) we came to Sidon. (7b) Putting out to sea from there, we sailed under the lee of Crete, (8) and came to Fair Havens near a city. (9) Since we had lost much time, and sailing was now dangerous because even the Fast had already passed, Paul advised them (10) saying, 'Men, I can see that the voyage will be with danger and much heavy loss, not only of the cargo and the ship, but also of our lives.' (12 TA) But the majority was in favour of putting to sea. (13) When a moderate south wind began to blow we sailed along the coast of Crete. (14) A storm struck from the south-east. (18) The next day, since the boat was being pounded so violently, (19) we jettisoned the cargo. (20) When the storm had raged for many days, when neither sun nor stars appeared, we

Act I

(**27**: 1b) The following day having called a centurion called Julius, he delivered to him Paul and some other prisoners. (6) The centurion found a ship of Alexandria sailing for Italy and put (them) on board. (11–12a) The captain and the owner decided to put to sea on the chance that somehow they could reach Phoenix, a harbor of Crete. (13) Supposing that they had obtained their purpose they weighed anchor and sailed (to) Crete. (14) But soon a tempest struck. (15) When the ship was caught and could not face the wind, they took measures to undergird the ship. Then fearing they should run on the Syrtis, they lowered the gear, and so were driven. (18b) They began next day to throw the cargo overboard. (19) And the third day they cast out with their own hands the tackle of the ship (38) and the ship was lightened. (27) When the

[44] Ibid. 3. 300.
[45] Boismard and Lamouille (1990), 3. 306. The name of the island in the Western text is Klauda, but in the Alexandrian text Kauda. Boismard and Lamouille (1990), 3. 307, suggest that the reference should be to Gaudos, the modern Gozo near Malta (Strabo, *Geography* 6. 2. 11).
[46] Boismard and Lamouille (1990), 3. 310–11.
[47] Ibid. 2. 225–6, 260.

finally lost all hope of surviving. (37) We were about 70 people. (21) The food having completely run out, Paul stood up and said, 'Men, you should have listened to me and not set sail from Crete and thereby avoided this danger and loss. (22a) I urge you now to keep up your courage, for there will be no loss of life among us. (26) We must strike an island.' (15) The ship was pushed by the wind, and we arrived at an island called [Gaudos]. (17b) There we rested. (28: 2) The natives welcomed us (10a) and showed us unusual honours. (11) We sailed on an Alexandrian ship which had wintered at the island. (14b) And so we came to Rome. (15a) The believers, when they heard, came to meet us.

fourteenth night had come, about midnight, the sailors suspected that they were nearing land. (41) They ran the vessel aground. (28: 1) Having landed, they recognized the region which was called Malta. (2a) The natives showed them unusual kindness, (10b) putting on board whatever we needed. (16) When we came to Rome, Paul was permitted to stay by himself with the soldier who guarded him.

The similarity between the two narratives is manifest. From the eastern littoral of the Mediterranean a ship sailed to Crete as autumn was edging into winter. The risk was justified by a troublefree voyage. On the next stage to Italy, the travellers were not so fortunate. Caught in a storm, the crew were forced to jettison the cargo. They survived to find refuge on an island, where they were hospitably received by the natives, and eventually sent forward on their way to Rome. So many points in common unambiguously indicates that we are confronted by two versions of the same event.[48] In addition, the two voyages must be dated at approximately the same time. The journey described in the Travel Journal must be dated after Paul's collection visit to Jerusalem, because the immediately preceding journey to Jerusalem in the Travel Journal is that evoked by Romans 15: 25. According to Act I, the process leading to the voyage began with Paul's arrest in Jerusalem.

There is, however, a major difference between the two texts, namely, the status of Paul. According to Act I, he is a prisoner, but in the Travel Journal he is free, as he is in the other sections of that document. Does this mean that we have to dismiss the similarities as mere coincidence and to postulate that Paul made two journeys to Rome, one of his own volition and the other under coercion? To state the problem in this way is to forget that, according to Act I, the decision to go to Rome was made by Paul (Acts 25: 11–12). Festus could not

[48] For Boismard and Lamouille (1990), I. 23, Act I depends on the Travel Document.

have refused his appeal to the emperor.[49] It was perfectly feasible, therefore, to tell the story of the journey in a way which ignored the fact that Paul was technically a prisoner.

Moreover, there are hints in the Travel Journal version which suggest that Paul was not master of his own destiny. From his letters we know that it was his normal practice when travelling west to go overland through Asia Minor. Why should he now change his custom, particularly at the season when sailing became particularly dangerous? Secondly, why did Paul not heed his own warning that it was too risky to continue (Acts 27: 11), and leave the ship at Fair Havens? One might answer to both these questions that Paul was in a great hurry to get to Rome. But there is no hint of haste in this section of the Travel Journal, and speed is mentioned explicitly when it was a factor at an earlier stage (Acts 20: 16). In the light of Act I, the correct answer to both questions is that Paul had no choice. The decisions were made by the centurion, in whose custody he travelled.

Incarceration in the Eternal City

The conditions of Paul's imprisonment in Rome are described by Luke. He lived in private lodgings under the supervision of a single soldier (Acts 28: 16, 30). There is nothing intrinsically implausible in this form of detention.[50] The alternative was to be held in jail.[51] In a prison, however, it would be difficult to receive the visitors and make the speech that Luke attributes to Paul. The implausiblity both of the address to the Jews, and their reaction to it, reveals the agenda of Luke,[52] and once again compromises his credibility as a historian. If the narrative is his creation, might he not have also created the setting, which permits Paul to carry on his missionary activity 'quite openly and unhindered' (Acts 28: 31)? A large question mark, therefore, hangs over the type of detention to which Paul was subject in Rome. He may well have been extremely uncomfortable.[53]

It is unlikely that Paul was down-hearted, at least at the beginning. He had been in many prisons (2 Cor. 11: 23),[54] but had always been released. He had

---

[49] See in particular Tajra (1989), 144–51.

[50] Mommsen (1955), 317 n. 5; Humbert (1899b).

[51] Juvenal regrets 'the good old days of kings and tribunes, when Rome made do with one prison only' (*Satires* 3. 312–14; trans. Green). He is thinking of the Tullianum, today known as the Mamertine prison; where the others were located is unknown, perhaps in military barracks (Juvenal, *Satires* 6. 561; Tacitus, *Annals* 1. 21).   [52] Haenchen (1971), 727.

[53] 'In the prison, when you have gone up a little way towards the left, there is a place called the Tullianum, about twelve feet below the surface of the ground. It is enclosed on all sides by walls, and above it is a chamber with a vaulted roof of stone. Neglect, darkness, and stench make it hideous and fearsome to behold' (Sallust, *War against Jugurtha* 55. 3–4; trans. Rolfe).

[54] According to Clement of Rome, Paul was imprisoned seven times (1 Clement 5. 6). The perfect number excites suspicion, but history is not immune to coincidence.

been investigated on several occasions, but nothing had been found against him. His experience provided no basis for pessimism regarding the outcome of the present enquiry. This fact has led some exegetes to interpret Luke's highly ambiguous ending of Acts positively. Paul, we are invited to infer, was liberated. An effort has been made to sustain this interpretation by asserting that the function of Luke's mention of the duration of Paul's detention (two years; Acts 28: 30) was to draw attention to a rule of Roman law whereby defendants were released if their accusers failed to appear within a fixed time.[55] It has been shown, however, that this is a misreading of the evidence, which in fact assigns penalties for plaintiffs who failed to appear, but is silent regarding the liberation of the defendant.[56] There was no discharge by default in first-century Rome.

Against this background, Haenchen has argued that Luke has provided clear hints that the outcome of Paul's trial was not a happy one. The evangelist makes the Apostle say 'all you among whom I have gone about preaching the kingdom will see my face no more' (Acts 20: 25; cf. 20: 38), and consistently emphasizes that Paul deserved neither imprisonment nor death (Acts 28: 18). Luke, Haenchen tells us, knew that Paul had died under Nero, but decided to remain silent in order not to exacerbate the relationship of Christianity with the empire. In Luke's view, Rome, if handled carefully, could repudiate the unfortunate decision of a single emperor. The apparatus of state is discreetly invited to see Nero's act as an unfortunate aberration. Imperial officials at Philippi, Corinth, and Caesarea, had held a different opinion.[57]

The weakness of this ingenious hypothesis is that it rests on the improbable assumption that Luke really believed that highly placed Romans would devote sufficient attention to his story of Peter and Paul to note all its subtle implications. If Luke's strategy was as Haenchen claims, the evangelist would have made certain that it was understood, by giving Paul a brilliant speech from the dock after sentence had been passed. It is entirely possible that Luke did not know what had happened to Paul—or to Peter—and simply terminated his account when his data ran out.

From the perspective of a Roman official, Paul was not a particularly important prisoner and, as Sherwin-White has pointed out, he may have benefited from a purely casual release, 'if, for example, a show of clemency were thought desirable at some moment, or simply to shorten the court list by dropping the arrears'.[58]

[55] Cadbury (1933).
[56] Sherwin-White (1963), 113; Tajra (1989), 192–6.
[57] Haenchen (1971), 731–2.
[58] (1963), 119.

THE PASTORAL LETTERS

If, in the last analysis, nothing can be deduced from Acts regarding subsequent developments in Paul's career, is there any other source from which information can be drawn? An immediate affirmative answer comes from those who maintain the authenticity of the three Pastoral Letters (1 & 2 Timothy, Titus). Recognizing that the events mentioned in, or supposed by, these three letters cannot be fitted into the framework of Paul's life as revealed by his acknowledged letters and the Acts of the Apostles, they assign the Pastorals to the period between the Roman imprisonment mentioned by Acts, and the traditional date of the death of Paul in AD 67.[59]

The Problem of Authenticity

Just to read through the Pastorals, however, makes it clear that they are not identical with the other letters of the Pauline corpus. Those who attempt to quantify impressions draw attention to differences in (1) language, (2) literary style, (3) theological perspective, (4) church organization, and (5) the nature of the opposition the church has to face. The dominant view among critical scholars, in consequence, is that the bulk at least of these three letters was not written by Paul.[60] In this hypothesis the writer was concerned to invoke the authority of Paul in an attempt to deal with the problems of the church about AD 100. Thus he created three fictitious letters. How did they win acceptance? The word was circulated that they had remained outside the developing Pauline corpus because of their private character. Now, however, the needs of the church demanded that they be made public.

This hypothesis is perfectly possible in itself. Most scholars will admit that Ephesians was not composed by Paul himself, and to what extent he explicitly covered it with his authority is dubious. There are, in addition, a number of pseudepigraphic letters, e.g. the extant *Letter to the Laodiceans,* and *3 Corinthians* (both attributed to Paul), and the *Letter of Titus,* in addition to the supposed letters of Paul to the Macedonians and to the Alexandrians, which are known only through vague and late allusions.[61] Compared to the apocryphal gospels, however, the letter-form was not a popular genre.[62] It cannot be claimed, therefore that there was a climate of acceptance, which would make it easy for the forged Pastorals to enter the mainstream of church life. Moreover,

[59] Spicq (1969), 121–46; Kelly (1963), 34–6. An exception is de Lestapis (1976), who dates them during Paul's last journey to Jerusalem.
[60] It is articulated in most detail by Brox (1969), 22–60.
[61] Hennecke and Schneemelcher (1965), 2. 90–1.
[62] Ibid. 2. 90.

the Pastorals, addressed to individuals and replete with personal details, are not at all like the extant pseudepigraphic letters, which are addressed to churches and contain little or no personal news.

Equally, it should not be assumed that Christians of the first two centuries were characterized by credulity and naivety. Paul himself warned the Thessalonians of the possibility of forged letters purporting to come from him (2 Thess. 2: 2), and insisted on their checking the authenticity of his signature (2 Thess. 3: 17). The author of Revelation menaces those who might be tempted to capitalize on his authority, 'I warn everyone who hears the words of the prophecy of this book; if anyone adds to them God will add to him the plagues described in this book' (22: 18). The *Muratorian Canon* qualifies the forged *Letter to the Laodiceans* and the *Letter to the Alexandrians* as gall compared to the honey of the genuine Pauline letters. Hence, if the Pastoral letters did in fact win acceptance,[63] there must have been a very solid link with Pauline circles.

The Authenticity of Second Timothy

Realistically, the only scenario capable of explaining the acceptance of the Pastorals, is the authenticity of one of the three letters. Were one to have been long known and recognized, then the delayed 'discovery' of two others with the same general pattern could be explained in a variety of convincing ways. An approach along these lines has been pioneered by M. Prior.[64]

Even though commentators have insisted that the three Pastorals form a totally homogeneous block,[65] they have always recognized the special affinity of 1 Timothy and Titus.[66] Once the generalizations of vocabulary, style, and doctrine are reduced to specific details, however, it appears impossible that 2 Timothy should be the product of the author who composed 1 Timothy and Titus. If one reads closely what the three letters have to say regarding the status of the sender, the recipient, Christology, ministry, the gospel, the attitude towards women, and false teaching, it is possible to tabulate 30 points where something in 2 Timothy is missing in both 1 Timothy and Titus, or where something shared by the two latter epistles is lacking in 2 Timothy.[67]

Prior, therefore, is entirely justified in his insistence that the authenticity of 2 Timothy should be judged without any reference to 1 Timothy and Titus.[68] The closest one can get to a direct proof of the authenticity of any literary work is a statement of the author backed up by the evidence of witnesses who saw him

---

[63] Barrett (1963b), 4, notes that 'No one in antiquity appears to have doubted the Pauline origin of the letters. They appear to have been accepted by all, whether orthodox or heretical, who knew them.' [64] (1989), 61–90.

[65] Spicq (1969), 31; Kelly (1963), 28, 38; Koester (1982), 2. 297; Kümmel (1975), 367.

[66] Spicq (1969), 139; Kümmel (1975), 367.

[67] See my (1991a). [68] (1989), 67.

pen the document. But even then he might merely be putting on paper what he had memorized from the work of another writer! In practice one is forced to trust the claim implicit in the superscription. In other words, there must be a presumption of authenticity, which stands unless it is overturned by convincing arguments that the author claiming responsibility could not possibly have written the document in question.

One has only to read a detailed treatment of the problem to realize that the overwhelming number of the arguments against the authenticity of 2 Timothy are drawn, not from the letter itself as common sense demands, but from 1 Timothy and Titus.[69] In other words, what is true for 1 Timothy and/or Titus is assumed to be true of 2 Timothy, despite the lack of evidence. Incredible in its methodological assumption, this approach is condemned by its practitioners' recognition that 2 Timothy is different from 1 Timothy and Titus, and that, in consequence, the three letters are not the homogeneous block they are claimed to be. If the authenticity of 2 Timothy is maintained, such differences would be the clearest proof of the inauthenticity of 1 Timothy and Titus.

Even though 2 Timothy differs from the other Pauline letters in that it does not have a co-author and is addressed, not to a church, but to an individual,[70] a recent sophisticated stylometric study concluded that '2 Timothy, one of the commonly rejected Pastoral Epistles, is as near the centre of the [Pauline] constellation as 2 Corinthians, which belongs to the group most widely accepted as authentic'.[71] The extent to which 2 Timothy is at home in the Pauline corpus is graphically illustrated by the great number of parallels with the authentic letters.[72] To treat these parallels as meaningless coincidences,[73] or as betraying literary indebtedness on the part of a post-Pauline editor,[74] is methodologically unjustified. Equally unacceptable as a criterion is the absence of such common Pauline terms as 'body (of Christ)', 'cross', 'freedom', 'covenant'.[75] No commentator can dictate what Paul should or should not say.

Previous attempts to draw biographical material from 2 Timothy have been made within a framework created by 1 Timothy and Titus, and by the Captivity Epistles (Phil., Col., Philem.). J. N. D. Kelly is far from atypical in writing, '[Paul] wants Timothy, who is probably still at Ephesus to come to him, picking up Mark on the way. We might be tempted to infer that this is the Roman captivity of Acts xxviii, were it not that Timothy and Mark were then in Rome

---

[69] e.g. Brox (1969), 22–60.

[70] Prior (1989), 37–59, has insisted on these points in order to exclude stylistic comparisons with the rest of the Pauline corpus.

[71] Kenny (1986), 100. See, however, Neumann (1990), 213, which must be read in the light of his warning on p. 218.

[72] These are well laid out in Harrison (1921), 167–75; see also Barnett (1941), 262–71, 275–6.

[73] The opinion of Lock (1936), p. xxiv.

[74] The thesis of Barnett (1941), 251–77.

[75] So Wild (1990), 6.

with the Apostle (Col. i. 1; iv. 10; Phm. 24)'.[76] We have already seen, however, that the Captivity Epistles are much more likely to have been written during Paul's sojourn in Ephesus than during his detention in Rome, and so reflect the Apostle's situation in AD 52–54.[77] In consequence, the personal notes they contain cannot be used to interpret 2 Timothy. The influence of 1 Timothy and Titus has been no less pernicious. The suggestion of 2 Timothy 4: 12 (cf. 1: 18) that Timothy is not in Ephesus is never followed up, because of the explicit statement of 1 Timothy 1: 3 that he was left in charge of the church of Ephesus. The personal data furnished by 2 Timothy needs to be studied in and for itself.

ROME AND SPAIN

When he wrote 2 Timothy, Paul was chained (2 Tim. 1: 8, 16) as an 'evildoer' (2 Tim. 2: 9). Among his contemporaries, the term was used of hardened criminals, e.g. the two thieves who died on crosses with Jesus (Luke 23: 32). The precise force Paul attached to it is difficult to determine because he uses it nowhere else. He may mean simply that he was chained 'as if he were a criminal' or he may intend to indicate that this imprisonment was somehow more severe than those which he had experienced previously. This latter possibility, however, which has led some to think in terms of the persecution of Christians under Nero,[78] is excluded by the fact that Paul enjoys the same facilities to receive visitors and send letters as in his previous periods of detention.

The visitor in question was Onesiphorus, who 'often refreshed me. He was not ashamed of my chains, but when he arrived in Rome he diligently searched and found me' (1: 16–17).[79] These simple words carry a weight of information, of which the most important is the place of the Apostle's imprisonment. Paul's delight at the courage of Onesiphorus is matched by his appreciation of the tenacity displayed by his visitor in the latter's quest for him. Rome had over a million inhabitants, but no street names, and no house numbers. Onesiphorus in fact had two problems. He had to find out where Paul was, and then work out how to get there, both tasks made horrendously difficult if he did not know his way around the city, and so could not follow directions. A faint hint of his difficulties may be gauged from a passage of Terence (195–159 BC) describing the same problem for someone who knew much smaller Athens:

> *Demea:* Tell me the place then. *Syrus:* Do you know the colonnade by the meat-market down that way? *Demea:* Of course I do. *Syrus:* Go that way

[76] (1963), 7.
[77] See Ch. 7, 'Imprisonment'.
[78] So Kelly (1963), 177.
[79] Harrison (1921), 129, appositely evokes the commendation of another prison visitor, Epaphroditus (Phil. 2: 25–30).

straight up the street. When you get there the Slope is right down in front of you: down it you go. At the end there is a chapel on this side. Just by the side of it there's an alley. *Demea:* Which? *Syrus:* That where the great wild fig-tree is. *Demea:* I know it. *Syrus:* Take that way. *Demea:* That's a blind alley. *Syrus:* So it is, by Jove. Tut, tut, you must think me a fool. I made a mistake. Come back to the colonnade: yes, yes, there's a much nearer way, and much less chance of missing it. Do you know Cratinus's house, the millionaire man there? *Demea:* Yes. *Syrus:* When you are past it turn to your left, go straight along the street and when you come to the temple of Diana turn to the right. Before you come to the town-gate, close by the pool there's a baker's shop and opposite it is a workshop. That's where he is. (*The Brothers* 571–85; trans. Sargeaunt)

The hint that Onesiphorus was his only visitor, and had little help in finding him, is confirmed by Paul's feeling of isolation (2 Tim. 4: 11). Evidently there was little warmth in whatever contact Paul had with the Roman church. In the conclusion of 2 Timothy, Paul transmits greetings from Prudens, Linus, Claudia, and all the believers (4: 21). All these are Latin names, and Linus is identified as Peter's successor as head of the church of Rome.[80] There can be no serious doubt that they were members of the local church. It is equally clear, however, that they distanced themselves from the Apostle; 'at my first defense no one took my part; all deserted me' (2 Tim. 4: 16). When balanced against Paul's manifest desire to have loyal friends around him (2 Tim. 4: 11), it would appear that Paul's reason for mentioning a few names was to give Timothy the impression that his relations with the Roman church were better than they really were.

This situation is difficult to reconcile with the conditions of Paul's first imprisonment in Rome, however much it might have been prolonged.[81] At that stage, as we know from Romans 16, he not only had old friends in the church there (e.g. the mother of Rufus), but two of his most trusted collaborators, Prisca and Aquila, were responsible for one of the house-churches. When 2 Timothy was written, these latter are no longer in Rome (2 Tim. 4: 19). While it is possible that Prisca and Aquila might have moved because of the delay occasioned by Paul's detention in Caesarea, it seems more likely that their loyalty to him would lead them to carry out his directives until told to cease. Hence we are forced to envisage a second Roman imprisonment some time after the first.[82]

This, of course, means that Paul was freed after his first imprisonment, and we must assume that he continued to engage in missionary activity. Where did

---

[80] Eusebius, *History of the Church* 3. 2. Where, one might ask, was Peter?

[81] Against de Lestapis (1976), 83; Prior (1989), 89.

[82] So Spicq (1969),140–1; Kelly (1963), 36. According to Eusebius, 'There is evidence that, having been brought to trial, the apostle again set out on the ministry of preaching, and having appeared a second time in the same city found fulfilment in his martyrdom. In the course of this imprisonment he composed the second Epistle to Timothy' (*History of the Church* 2. 22; trans. Williamson).

he preach? His plan was to go to Spain (Rom. 15: 24), and Clement of Rome, writing about AD 95, asserts that this is precisely what he did:[83]

> Paul because of jealousy and contention has become the very type of endurance rewarded. He was in bonds seven times, he was exiled, he was stoned. He preached in the East and in the West winning a noble reputation for his faith. He taught righteousness to all the world; and after reaching the boundary of the setting [of the sun], and bearing his testimony before kings and rulers, he passed out of this world and was received into the holy places.
>
> (5. 5–7; trans. Staniforth adapted)

For those who deny any historicity to the Pastorals, this is nothing but a projection based on Romans 15: 25. Clement, it is claimed, thought it would enhance Paul's reputation to say that his plans had come to fruition. The gratuity of this assumption highlights the uncritical bias of such an approach. Were the Pauline letters Clement's sole source of information, it is improbable that he would have mentioned an 'exile' of which Paul never speaks,[84] and that he would not have used Paul's term 'Spain' for the land of the Apostle's ambition. From the perspective of someone writing from Rome, 'the boundary of the setting (of the sun)' can only mean Spain.[85] But why should an author, whose exclusive concern was to have a prophecy fulfilled, use a term which, absolutely speaking, was susceptible of other interpretations depending on the geographical location of his readers? From Rome the West was Spain, but to someone in Palestine it would more naturally be Greece or Rome.

Travel to Spain did not involve exceptional exertion. Under optimum conditions, the Iberian peninsula was only a seven-day sail from Ostia, the port of Rome.[86] There is not a shred of evidence as to where Paul went in Spain.[87] How long he spent there is somewhat easier to answer, because he was executed under Nero, who died in AD 68, and other activities have to be fitted into that period in addition to the Spanish mission, as we shall see below.

[83] So does the Muratorian fragment, 'Paul's departure from the City, setting out for Spain' (lines 38–9), but this document is at least 100 years later and possibly more; see Gamble (1985), 32.

[84] Inevitably it has been suggested that Paul was exiled to Spain; see Pherigo (1951), 278; Bruce (1977), 445.

[85] 'In the third book of his Geography Eratosthenes, in establishing the map of the inhabited world, divides it into two parts by a line drawn from west to east, parallel to the equatorial line; and as the ends of this line he takes, on the west, the Pillars of Hercules, and on the east, the capes and most remote peaks of the mountain-chain that forms the northern boundary of India' (Strabo, *Geography* 2. 1. 1; trans. Jones). '[Cape St Vincent] is the most westerly point, not only of Europe, but of the whole inhabited world' (Strabo, *Geography* 3. 1. 4). 'Now in regard to the Pillars, which they say Hercules fixed in the ground as limits of the earth. ... The extremes of Europe and Libya border a strait 60 stadia wide through which the ocean is admitted into the inner sea' (Philostratus, *Life of Apollonius of Tyana* 5. 1; trans. Conybeare).

[86] 'There is a plant [flax from which linen sails were made] which brings Cadiz within seven days' sail from the Straits of Gibraltar to Ostia, and Hither Spain within four days' (Pliny, NH 19. 1. 4; trans. Rackham); cf. Plutarch, *Galba* 7.

[87] See Meinardus (1978).

When discussing Paul's project of a mission in Spain, I argued that Paul hoped that the Roman church would appoint him its delegate, as the church of Antioch had done at the beginning of his missionary career.[88] The miserable relations between the imprisoned Paul and the Christians in Rome is perhaps best explained by the hypothesis that such sponsorship had not been forthcoming. The community might have been prepared to give him its blessing, but he needed much more. Spain was a new world in which he would be the complete alien. He needed, not only financial aid, but above all linguistic assistance, if the mission was to have any chance of success. As Paul's ever more desperate appeals for volunteers fell on deaf ears, his relations with the community deteriorated. His total commitment stigmatized their detachment, and guilt bred resentment.

Given what we know of Paul's impetuous character, it is easy to visualize him deciding to attempt the mission to Spain aided only by those with whom he had worked in the East. This may have transformed the resentful passivity of the local church into active hostility. The Iberian peninsula was the most romanized of all the territories under Roman control. Writing of the southern part, Strabo could say, 'The Turdetanians, however, and particularly those that live about the Baetis [= the Guadalquivir river] have completely changed over to the Roman way of life, not even remembering their own language any more. And most of them have become Latins, and they have received Romans as colonists, so that they are not far from being all Romans' (*Geography* 3. 2. 15; trans. Jones). Just as Romans went to Spain, so Spaniards flocked to Rome, and, while many remained on the level of the gypsy dancers,[89] others achieved positions of eminence, e.g. Pomponius Mela, Seneca, Lucan, Martial, Quintilian, Columella,[90] and probably Juvenal.[91]

In the perspective of this close relationship, it is not improbable that Roman believers considered Spain to belong to their sphere of influence and argued that it was their responsibility to carry the gospel there. They were certainly the best equipped, and it was up to them to choose the most opportune moment. In their eyes it would have been most presumptuous of Paul to go ahead without their approval. Not only would he be flouting the dignity of the church of the capital of the empire, but his inevitable failure might make it difficult for them to mount their own missionary expeditions at a later date.

The speculative nature of this hypothesis needs no emphasis, but if under these conditions, Paul attempted the conversion of Spain and failed, the hostility of the Roman church becomes explicable, as does Paul's silence.[92] It would

[88] See Ch. 13, 'A Mission to Spain'.
[89] Juvenal, *Satires* 11. 162.   [90] Spicq (1969), 133 n. 2.
[91] Green in his translation of Juvenal, *The Sixteen Satires* (London: Penguin, 1984), 21.
[92] The discretion of the apocryphal literature, so notorious for its imaginative details, is highly significant. The *Acts of Peter* twice speaks of Paul's departure for Spain (1. 1; 2. 6), but offers no information as to where he went or what he did.

have been as ignominious an episode as his abortive attempt to convert the Nabataeans (Gal. 1: 17).[93] In terms of chronology it should not be assigned more than a single summer. While there are other possibilities—witness Paul's failures in Arabia (Gal. 1: 17) and in Athens (Gal. 3: 1)—the fact that Greek was not widely spoken remains the most probable explanation of the failure of the Spanish mission. Such being the case, the southern littoral of Gaul and northern Italy would not have been more fertile soil. Where was Paul to go?

ONCE MORE THE AEGEAN

Returning defeated from Spain, it would be unlike Paul to flaunt his failure before the Roman church. A stop in Rome on his way back to the East is most improbable. Unless the ship docked at a port to the south of Ostia, one should rather imagine him swinging around the south of the city to pick up the Appian Way, which cut across the shin of Italy to the port of Brindisi on the Adriatic Sea. A detailed description of conditions on that famous road is given by Horace, who travelled it in the suite of Maecenas in the spring of 37 BC.[94] At a leisurely pace, the 544 km. (340 miles) journey took just under two weeks.[95] Provided that it was not too late in the year, it would have been easy to find a boat crossing to Dyrrachium, the eastern terminal of the Via Egnatia, and the doorway to Illyricum (see Fig. 7).

It should not be forgotten that Illyricum was the only region in the East in which Paul had unfinished business. His mission there had been hastily terminated by the crisis in Corinth, which provoked 2 Corinthians 10–13.[96] From long experience, Paul knew that his foundations needed to be nurtured, but what had happened to him subsequently—imprisonment in Caesarea and Rome, the visit to Spain—had made it difficult, if not impossible, for Illyrian believers to have had any contact with him. He owed them the follow-up visit which he had granted all his other communities. It is in this concern that we find the motivation for Paul's return to the East.

If he followed his customary practice in Macedonia, Paul spent at least a year labouring among the Illyrians. Once satisfied that the church was solidly established, he knew that his continued presence would be an impediment to the normal development of the community. As long as he remained, he would automatically be the authority figure in all aspects of its life. Only when freed of the weight of his prestige could the charisma of other members develop naturally. Where did Paul then go?

---

[93] See Ch. 4, 'The Situation when Paul Arrived'. [94] *Satires* 1. 5.
[95] See Strabo, *Geography* 6. 3. 7. Cato is supposed to have done it in five days (Plutarch, *Cato Major* 14), and Helius Caesarianus reached Greece in seven days (Dio Cassius, *History* 63. 19).
[96] See Ch. 12, 'Once Again a Missionary'.

2 Timothy provides a series of geographical references, but only five are relevant to Paul's itinerary. Two allude to what Timothy already knows; the other three are new information. The phrases 'you are aware' (1: 15) and 'you well know' (1: 18) bracket two types of behaviour which should have an educational value for Timothy. Paul contrasts the comportment of those 'in Asia who turned away from me' with the dedication of Onesiphorus as expressed in 'the contributions he made at Ephesus'. It is clear that Timothy has direct knowledge of these two episodes, and the latter would place him at Ephesus with Paul at the time of their occurrence. Indirect confirmation that the capital of Asia was Timothy's base is furnished by the fact that he was ignorant of what had happened in two port cities, one north of Ephesus, the other to the south. 'When you come, bring the cloak that I left with Carpus at Troas, and the books, and above all the parchment note-books' (2 Tim. 4: 13); 'Trophimus I left ill at Miletus' (2 Tim. 4: 20). The fifth reference stands in sharp contrast to the last mentioned, 'Erastus remained in Corinth' (4: 20).

This latter allusion permits a choice between the options open to Paul after leaving Illyricum. It excludes a journey south to Corinth, because then he would have been able to tell Timothy of Erastus' decision when they subsequently met in Ephesus. Hence, Paul continued east along the Via Egnatia and presumably visited Thessalonica and Philippi before crossing over to Troas. The simplest explanation of why Paul left his cloak, books, and notebooks there was they would have been an impossible burden to carry as he tramped the 350 km. (210 miles) to Ephesus in the heat of summer.[97] How long he spent in the capital of Asia, and whether he visited the churches of the Lycus valley, we do not know. Eventually he moved to Miletus, whence he took a ship to Corinth. The purpose of this pastoral circuit of the Aegean can only have been the edification, encouragement and consolation (1 Cor. 14: 3) of the churches he had founded.

THE PROBLEM OF TIMOTHY

Where was Timothy when Paul wrote 2 Timothy? For those who accept the authenticity of the Pastorals, the question is answered by 'Tychicus I have sent to Ephesus' (4: 12).[98] A little reflection, however, shows that this in fact implies that Timothy must have been elsewhere. If Tychicus had been dispatched before the letter was sent, why does Paul tell Timothy in Ephesus something he must have known already? If Tychicus was the bearer of the letter, why should

[97] Pliny wrote to the emperor Trajan at the very end of August or the beginning of September, 'I kept in excellent health throughout my voyage to Ephesus, but I found the intense heat very trying when I went on to travel by road and developed a touch of fever which kept me at Pergamum' (*Letters* 10. 17; trans. Radice).
[98] So Kelly (1963), 214.

Paul stress the obvious? Nothing in the formulation suggests emphasis on the fact that it was Paul, and no other, who had sent Tychicus to Ephesus.

If Timothy had been in Ephesus with Paul (1: 15–18), but was there no longer, and Paul none the less knew where to send the letter, it seems clear that Paul had sent him to another community. Since Timothy was ignorant of what had happened north and south of Ephesus (4: 13, 20), and was in a position to pass through Troas on his way to Rome, then his new mission can only have been to the east, among the churches of the Lycus valley or those in Galatia.

There are clear hints in 2 Timothy that Paul was not pleased with his disciple's performance. The repeated injunction, 'Take your share of suffering' (1: 8; 2: 3), when read in the perspective of 'rekindle the gift of God that is within you . . . for God did not give us a spirit of timidity but a spirit of power and love and self control' (1: 6), reveals a certain lack of commitment on Timothy's part. To some extent at least, he was not doing the work of an evangelist; he was not fulfilling his ministry (4: 5).[99] The most obvious source of Paul's knowledge of such failure is his first-hand experience of how Timothy conducted himself in Ephesus. Paul's initial attempt to remedy the situation was to devise a mission that would take Timothy out of the city. With Paul himself taking charge in Ephesus, the move could be made to look like an extension of Timothy's responsibilities. Whether Timothy was deceived was another matter; his tears are mentioned in 2 Timothy 1: 4.

Paul's sojourn in Ephesus did not improve the situation. In fact, 'all in Asia turned away from me' (2 Tim. 1: 15) suggests that he exacerbated whatever tensions wracked the community. There had always been opposition to Paul in Ephesus (Phil. 1: 14–18),[100] and it is very likely that the church had developed in ways alien to the Pauline pattern during the seven or more years (from AD 54) when Paul had been out of touch. The false teaching which Paul exhorts Timothy to avoid probably reflects his personal experience at Ephesus on his last visit. In opposition to the specifically Jewish teaching, which is combatted in 1 Timothy and Titus, the principal preoccupation of 2 Timothy is futile debate, 'fighting with words' (2: 14), 'profane and empty talk' (2: 16), 'foolish and inexpert research' (2: 23). Such meaningless discussions, Paul insists, lead only to 'the demoralization of the hearers' (2: 14), to 'greater ungodliness' (2: 16), and to 'quarrels' (2: 23). An example of such debate is the argument that the resurrection has already taken place (2: 18), which could be presented as a perfectly reasonable interpretation of Paul's own preaching, 'you also were raised with him' (Col. 2: 12; cf. 3: 1). To all appearances there was a group at Ephesus,

---

[99] Lemonnyer's comment on 2: 4 is quoted by Spicq (1969), 742, 'It seems that Timothy, modelling himself on his Master, but probably also out of timidity, dedicated himself to earning his own living in order not to burden the community. . . . If we are to judge by the context in which Paul places these exhortations, perhaps Timothy was not unhappy to find a plausible reason to escape the heavy duties of his charge.'

[100] See Ch. 9, 'Opposition at Ephesus'.

whose intellectual aspirations paralleled those of the spirit-people in Corinth.[101] Paul's mentality was such that he simply could not understand or appreciate either one or the other.

Paul's move to Miletus (2 Tim. 4: 20) may have been prompted by the realization that he was incapable of dealing with the problems of the community at Ephesus. It demanded qualities which he lacked (2 Tim. 2: 24–5). To abandon the believers completely, however, was impossible, and he made a great sacrifice in an attempt to guarantee the stability of the community. He turned over responsibility to the founders of the church, Prisca and Aquila. The hint that they were at Ephesus comes from their association with the household of Onesiphorus (2 Tim. 4: 19; cf. 1: 18). In addition, the city would have been on Timothy's route from the Lycus valley to Rome. In this perspective, the sending of Tychicus, whom Luke identifies as an Asian (Acts 20: 4), would have been designed to reinforce the Pauline party within the Ephesian church.

On reflection, Paul realized that it was not wise to permit Timothy to remain in the region. He had neither the skill nor the authority to combat the theological verbalism which had invaded Ephesus, and which could very easily spread to the hinterland. Evidently Timothy was one of those people who make wonderful assistants but poor leaders. Hence, Paul summons him to Rome on the pretext that it was for Paul's benefit (1: 4; 4: 9, 21). The latter's affection was in no way diminished by the failure of his disciple to live up to his expectations.

WHAT DID THE FUTURE HOLD?

'Do your best to come to me before winter' (2 Tim. 4: 21). This laconic directive yields two important items of information. The letter must have been sent from Rome some time in the late spring or early summer, and Paul fully expected to be in Rome to receive Timothy at least four months later.[102] Moreover, he also expected to have a use for his winter cloak (2 Tim. 4: 13).

The further command to Timothy to bring Mark with him is justified by a phrase (2 Tim. 4: 11), which can be read in two different ways, depending on the sense given to *diakonia*. 'For he is very useful in serving me' (*RSV*) or 'for he is useful in my ministry' (*NRSV*). In one case Paul wants Mark as a personal servant, while in the other he is concerned to recruit a member for a new missionary team. Although it is preferred by some scholars,[103] the first option appears highly improbable. In two other letters Mark is identified as 'a co-

---

[101] See Ch. 11, 'The Influence of Apollos'.

[102] The land route from Rome to Ephesus, using the Via Appia and the Via Egnatia, was 1,944 km. (1,200 miles).

[103] Kelly (1963), 214.

worker for the kingdom of God' (Col. 4: 10–11; Philem. 24), and it is inconceiv-able that he should be demoted to the status of a valet. Paul certainly had a ministerial function in view.[104] In either case, it is clear that Paul is looking for-ward to an active future.

Even though no one supported him during his first hearing (2 Tim. 4: 16), the decision had not gone against him, despite the fact that he did not hide who he was or what he did, 'The Lord stood by me and gave me strength to proclaim the word fully, that all the Gentiles might hear it. So I was rescued from the lion's mouth' (2 Tim. 4: 17). He cannot envisage what new evidence might be produced against him, and so looks confidently to a favourable outcome, 'The Lord will rescue me from every evil' (2 Tim. 4: 18).

This optimistic perspective poses serious problems for the understanding of 2 Timothy 4: 6–8.[105] The normal translation of these verses—'I am already on the point of being sacrificed; the time of my departure has come. I have fought the good fight, I have finished the race, I have kept the faith. Henceforth there is laid up for me a crown of righteousness, which the Lord, the righteous judge, will award to me on that Day' (*RSV*)—conveys a note of finality, which is generally interpreted as meaning Paul's 'consciousness that his own martyr-dom cannot be long delayed'.[106] In order to integrate this interpretation with the succeeding verses, those who accept the unity and authenticity of 2 Timothy have to make two assumptions: (1) the second examination will go against Paul, and (2) subsequently his execution will be deferred for several months. Neither can be considered plausible. Paul was surely a better judge of his situation than any modern commentator, and once sentence was passed, it was not in the interest of the authorities to delay in carrying it out.

The only serious effort to reconcile 2 Timothy 4: 6–8 with 2 Timothy 4: 9–21 is that of M. Prior, who offers a new interpretation of *egô gar spendomai kai ho kairos tês analyseôs ephestêken* (4: 6). He denies that *spendô* refers to the spilling of blood, and that *analysis* means 'death'. On the contrary, he maintains, *spendô* 'reflects Paul's sense of his own total dedication to his task', while the connota-tion of *analysis* is 'release'.[107] Hence, Prior translates, 'For my part I am already spent, and the time for my release is at hand.'[108] Paul's need for Timothy and Mark is rooted in the conviction that he will soon get out of prison and will be free to undertake missionary work.

While Prior's critique of the current consensus regarding the meaning of 4:

---

[104] Prior (1989), 146–9.
[105] So rightly Prior (1989), 104–7, who has little difficulty in manifesting the utter subjectivity (*a*) of those who attribute 4: 6–8 and 4: 9–21 to fragments of different letters, and (*b*) of those who deny authenticity and see the personal references as window-dressing so meaningless that incon-sistencies were irrelevant.
[106] Kelly (1963), 205.
[107] (1989), 111.
[108] Ibid. 108.

6–8 is both apposite and convincing, his suggestion as to how these verses should be construed is not quite as satisfying. Paul never uses the rare term *analysis* anywhere else, but he does employ the cognate verb *analyô* in another letter also written when he was a prisoner, 'my desire is to depart and be with Christ' (Phil. 1: 17). The allusion here is to his death. Even though in this text the ambiguity of 'to depart' is removed by 'to be with Christ', the point is that Paul, not only could, but did think of his death in terms of a 'departure'. In the same letter Paul also uses *spendomai*, 'even if I am to be poured out as a libation on the sacrificial offering of your faith' (Phil. 2: 17). The conditional mode excludes the interpretation of the verb as a reference to Paul's apostolic activity; the allusion is to a future event outside his control.

Perhaps the most serious flaw in the exegesis of both Prior and his opponents is their neglect of the tenses of the verbs in 2 Timothy 4: 6–7. All are in the perfect tense with the exception of *spendô*, which is brought into line by the addition of 'already'. Paul, in other words, is not looking forward, but backward! The sense of completion could not be more emphatic. However, it is not the anxious finality experienced by a prisoner on death row, but the complacent recognition of a life well spent. Paul, who by now was close to 70, realized that his best years were behind him. In terms of the normal life-span he was living on borrowed time, particularly for one who for so many years had borne in his body the dying of Jesus (2 Cor. 4: 10). Each day was a grace, and he intended to make the best possible use of every moment. He might not live long after his release from prison, but that did not exempt him from the obligations of his ministry. He could plan for the future and, if he was taken, then Timothy and Mark could carry on.

MARTYRDOM

Even though no precise calculation is possible, it seems certain that Paul's abortive visit to Spain and his pastoral circle of the Aegean sea absorbed enough time to ensure that his return to Rome must have occurred after the great fire which raged in Rome for 9 days (19–28 July) in AD 64, and destroyed 10 of the 14 quarters of the city. Despite Nero's concern for the homeless, his replanning of the city in order to ensure that the tragedy would not be repeated, and his propitiation of every god who might have felt slighted, the idea circulated that the emperor had resorted to a rather drastic type of urban renewal.

> But neither human resources, not imperial munificence, nor appeasement of the gods, eliminated sinister suspicions that the fire had been instigated. To suppress this rumour, Nero fabricated scapegoats, and punished with every refinement the notoriously depraved Christians (as they were popularly called). ... Their deaths were made farcical. Dressed in wild animals' skins, they were

torn to pieces by dogs, or crucified, or made into torches to be ignited after dark as substitutes for daylight. Nero provided his Gardens for the spectacle, and exhibited displays in the Circus, at which he mingled with the crowd, or stood in a chariot, dressed as a charioteer.

(Tacitus, *Annals* 15. 44; trans. Grant)

The implied sequence of events suggests that the persecution of Christians was not an immediate consequence of the fire. Some time had to elapse for the failure of the imperial propiatory gestures to become manifest, and for the rumours to build to a climax. Even though they differ by a year, the Armenian version and Jerome's translation of the *Chronicle* of Eusebius assign a four-year interval between the fire and the persecution.[109] The reliability of this assessment is open to doubt. A spontaneous whispering campaign could not have lasted that long, particularly if the rebuilding of the destroyed quarters revealed a significant improvement in urban life. No more than a year should be allowed. The persecution, therefore, should probably be dated to the spring of AD 65.[110] It is doubtful that it lasted longer than was necessary to give the population something else to think about. To prolong the hunt excessively risked directing attention to the motivation of the emperor. Moreover, that same spring Nero had to deal with the Pisonian conspiracy to assassinate him. The unfortunate long-term consequence of the episode was the creation of a sinister precedent that the guilt of Christians could be presumed.

Those who accept the reality of a second Roman imprisonment for Paul tend to assume that it parallelled the first imprisonment, namely, that Paul was arrested in the East and was brought to Rome in chains.[111] While this hypothesis is certainly not impossible, its basis is extremely fragile, and an alternative possibility deserves consideration, even though it is equally incapable of proof.

The news of Nero's bestial ferocity spread like wildfire through the empire. At least by the autumn of AD 65, it had reached the Pauline churches in Greece and Asia. The believers were agast at the thought of the horrible deaths their fellow-believers had suffered. Seeing the impact such frightfulness made on his own converts, Paul had little difficulty in estimating the consequences for the few Christians in Rome, who by chance had escaped the emperor's dragnet. Brutal reality challenged their idealistic acceptance of martyrdom as a remote future possibility. The vision of an extremely painful, prolonged death was unlikely to have had a favourable effect on morale. If the Roman church was to survive, other communities had to come to its aid. Here we have a motive which adequately explains both Paul's return to Rome, and the decision of Erastus to remain at Corinth (2 Tim. 4: 20). The latter thought the risk too great. Paul

---

[109] Schoene (1875), 2. 154–7.
[110] Légasse (1991), 244.
[111] Spicq (1969), 141, 814; Kelly (1963), 215; Baslez (1991), 282–4.

knew he was taking his life into his hands, but the need was imperative. Did he recall what he had once written? 'The love of Christ constrains us, because we are convinced that one has died for all; therefore all have died. And he died that those who live might live no longer for themselves, but for him who for their sake died and was raised' (2 Cor. 5: 14–15).

If Paul returned to Rome under such circumstances, it is improbable that he slipped secretly into the city, and thereafter slid cautiously from one refuge to another. Unobtrusiveness was a first step towards apostasy. The restoration of the community demanded a high-profile presence. In order to communicate courage and hope, he had to assume an overt leadership stance. Witness had to be public. How long was he able to maintain this role?

The upper limit is the suicide of Nero on 9 June 68. The hint of 1 Clement 6: 1 that Peter and Paul were martyred in the Neronian persecution is made explicit by Eusebius:

> It is recorded that in Nero's reign Paul was beheaded in Rome itself, and that Peter likewise was crucified and the record is confirmed by the fact the the cemeteries there are still called by the names of Peter and Paul . . . . That they were both martyred at the same time Bishop Dionysius of Corinth [c. AD 170] informs us in a letter written to the Romans: 'In this way by your impressive admonition you have bound together all that has grown from the seed which Peter and Paul sowed in Romans and Corinthians alike. For both of them sowed in our Corinth, and taught us jointly; in Italy too they taught jointly in the same city, and were martyred at the same time'.
>
> (*History of the Church* 2. 25; cf. 3. 1; trans. Williamson)

This translation admirably respects the vagueness of the formula 'at the same time', which does not necessarily mean 'on the same day'. Paul certainly never worked with Peter in Corinth, and it is unlikely that he did so in Rome. The martyrdoms of the two apostles are presented in such a way in the the second-century *Acts of Peter* and *Acts of Paul* that the natural inference is that they died at different times. The late fourth-century *Peristephanon* (12. 5) of Prudentius Aurelius Clemens claims that they died on the same day, but a year apart!

The year of Paul's death, which is lacking in the *Church History* of Eusebius, is supplied by the latter's *Chronicle*. Unfortunately the the versions do not agree. According to the Armenian version, the persecution unleashed by Nero took the lives of Peter and Paul in the emperor's thirteenth year (13 Oct. 66–12 Oct. 67), whereas Jerome's translation dates it to the fourteenth year (13 Oct. 67–9 June 68).[112] Elsewhere the latter wavers between the two dates. In his note on Paul in *Famous Men*, Jerome says 'he was beheaded in the 14th year of Nero on the same day as Peter died', whereas the entry on Seneca in the same work says, 'Seneca was executed by Nero two years before the martyrdom of Peter

---

[112] Schoene (1875), 2. 156–7.

and Paul.' Since Seneca died in April AD 65, the eleventh year of Nero, this would date the death of the apostles to the 13th year of Nero. The discrepancy may be insignificant, because those accustomed to counting in years beginning in January—as had been the case for the Romans since 153 BC[113]—could easily be confused regarding the status of the last quarter of the year. In theory it was the beginning of the fourteenth year of the Emperor's reign, but it was the conclusion of a calendar year most of which belonged to the thirteenth year of Nero's rule.

The error of Eusebius concerning the beginning of the persecution of Christians becomes explicable if we assume that his sources contained only two items of information: (1) a persecution of Christians instigated by Nero after the fire of AD 64; (2) the death of Peter and Paul at the very end of the reign of Nero. In combining the two, the former was inevitably attracted into the time-frame of the latter. As regards the personal involvement of Nero, it should be remembered, that he was absent in Greece from autumn 66 to the end of 67 or the beginning of 68.[114]

The manner of Paul's death, beheading, is understood to imply that he was condemned by a regularly constituted court. Where the execution took place, and where he was buried are unknown. No public liturgical cult of the martyrs in Rome is attested before the middle of the third century. Hence no reliable local tradition can be assumed. It has been persuasively argued that the veneration of Peter and Paul at *Ad Catacumbas* on the Via Appia originated in a private revelation of suspect origins, and that the Roman church dealt with the situation very astutely by claiming that the bodies had been secretly transferred to other locations, that of Peter to the Vatican hill, and that of Paul to the Via Ostiense.[115]

---

[113] OCCL 109. January was explicitly retained by Nero, against the wishes of the senate which decreed that the year should begin in December, the month of his birth (Tacitus, *Annals* 13. 10).

[114] Spicq (1969), 146 n. 3.

[115] Chadwick (1957).

Bibliography

Classical Authors

AELIUS ARISTIDES, *The Complete Works*, ii. *Orations XVII–LIII*, trans. by C. A. Behr (Leiden: Brill, 1981).

APPIAN, *History*, trans. by H. White (LCL; Cambridge, Mass.: Harvard, 1913).

APULEIUS, *Metamorphoses or The Golden Ass*, trans. by R. Graves (Penguin Classics; London: Penguin, 1950).

ATHENAEUS, *The Deipnosophists*, trans. by C. G. Gulick (LCL; Cambridge, Mass.: Harvard, 1951).

AULUS GELLIUS, *Attic Nights*, trans. by J. C. Rolfe (LCL; Cambridge, Mass.: Harvard, 1927–8).

CICERO, *De Officiis*, trans. by W. Miller (LCL; London: Heinemann, 1921).

—— *De Oratore*, trans. by E. W. Sutton and H. Rackman (LCL; London: Heinemann, 1942).

—— *Handbook of Electioneering*, trans. by M. Henderson (LCL; Cambridge, Mass.: Harvard, 1958).

CLEMENT OF ROME, *The First Letter to the Corinthians*, in *Early Christian Treaties: The Apostolic Fathers*, trans. by M. Staniforth (Penguin Classics; London: Penguin, 1968).

DIDORUS SICULUS, *The Library of History*, trans. by C. H. Oldfather (LCL; Cambridge, Mass.: Harvard, 1939).

DIO CASSIUS, *Roman History*, trans. by E. Cary (LCL; Cambridge, Mass.: Harvard, 1924–5).

—— *The Roman History*, trans. by. I. Scott-Kilvert (Penguin Classics; London: Penguin, 1987).

DIO CHRYSOSTOM, *Discourses*, trans. by J. W. Cohoon and H. L. Crosby (LCL; Cambridge, Mass.: Harvard, 1949–51).

EUSEBIUS, *The History of the Church from Christ to Constantine*, trans. by G. A. Williamson (Penguin Classics; London: Penguin, 1989).

FRONTINUS, *Strategems and Aqueducts*, trans. by C. E. Bennett and M. B. McElwain (LCL; Cambridge, Mass.: Harvard, 1925).

Greek Anthology, trans. by W. R. Paton (LCL; Cambridge, Mass.: Harvard, 1939).

HERODOTUS, *Histories*, trans. by A. D. Godley (LCL; Cambridge: Harvard, 1920–5).

HOMER, *The Iliad*, trans. by E. V. Rieu (Penguin Classics; London: Penguin, 1950).

HORACE, *Satires and Epistles*, trans. by N. Rudd (Penguin Classics; London: Penguin, 1979).

—— *The Complete Odes and Epodes with the Centennial Hymn*, trans. by W. G. Shepherd (Penguin Classics; London: Penguin 1983).

Itinerarium Burgdigalense, ed. by P. Geyer and O. Kuntz (CCSL 175; Turnholt: Brepols, 1965).

JOSEPHUS, *The Works of Flavius Josephus*, trans. by W. Whiston and D. S. Margoliouth (London: Routledge, 1906).

JUVENAL, *The Sixteen Satires*, trans. by P. Green (Penguin: Classics; London: Penguin, 1974).

— *The Satires*, trans. by N. Rudd (World's Classics; Oxford: OUP, 1991).

LIVY, *Rome and the Mediterranean: Books XXI–XLV of The History of Rome from its Foundations*, trans. by H. Betenson (London: Penguin,1976).

OROSIUS, *Historiae adversus paganos*, trans. by M.-P. Arnaud-Lindet (Collection Budé, Paris: Belles Lettres, 1990–1).

PAUSANIAS, *Guide to Greece*, trans. by P. Levi (Penguin Classics; London: Penguin, 1979).

PETRONIUS, *The Satyricon*, trans. by J. P. Sullivan (London: Penguin, 1986).

PHILO, *The Works of Philo*, trans. by C. D. Yonge (Peabody: Hendrickson, 1993).

PHILOSTRATUS, *The Life of Apollonius of Tyana*, trans. by F. C. Conybeare (LCL; Cambridge, Mass.: Harvard, 1912).

PLINY THE ELDER, *Natural History*, Bks. 1–19, 33–7, trans. by H. Rackham (LCL; Cambridge, Mass.: Harvard, 1944–52).

— *Natural History*, Bks. 20–32, trans. by W. H. S. Jones (LCL; Cambridge, Mass.: Harvard, 1951–63).

PLINY THE YOUNGER, *Letters and Panegyricus*, trans. by B. Radice (LCL; Cambridge, Mass.: Harvard, 1969).

QUINTILIAN, *Institutio Oratoria*, trans. by H. E. Butler (LCL; Cambridge, Mass.: Harvard, 1920–2).

SALLUST, *War against Jugurtha*, trans. by J. C. Rolfe (LCL; Cambridge, Mass.: Harvard, 1935).

SENECA, *Ad Lucilium epistulae morales*, trans. by R. M. Gummere (LCL; Cambridge, Mass.: Harvard, 1953).

— *De Beneficis*, trans. by J. W. Basore (LCL; Cambridge, Mass.: Harvard, 1935).

STRABO, *Geography*, trans. by H. L. Jones (LCL; Cambridge, Mass.: Harvard, 1931–2).

SUETONIUS, *The Twelve Caesars*, trans. by R. Graves, revised by M. Grant (Penguin Classics; London: Penguin, 1979).

— *Lives of the Caesars*, trans. by J. C. Rolfe (LCL; Cambridge, Mass.: Harvard, 1914–15).

TACITUS, *The Annals of Imperial Rome*, trans. by M. Grant (Penguin Classics; London: Penguin, 1977).

— *The Annals*, trans. by J. Jackson (LCL; Cambridge, Mass.: Harvard, 1931–7).

— *The Histories*, trans. by C. H. Moore (LCL; Cambridge, Mass.: Harvard, 1925–31).

TERENCE, *The Brothers*, trans. by J. Sargeaunt (LCL; Cambridge, Mass.: Harvard. 1912).

XENOPHON, *Anabasis*, trans. by C. L. Brownson (LCL; Cambridge: 1922).

Modern Authors

ABBOTT, T. K. (1897), *A Critical and Exegetical Commentary on the Epistles to the Ephesians and to the Colossians* (ICC; Edinburgh: Clark).

ABEL, F.-M. (1938), *Géographie de la Palestine* (ÉBib; Paris: Gabalda).

ADINOLFI, M. (1969), 'Giscala e San Paolo', in his *Questioni bibliche di storia e storiografia* (Brescia: Paideia).

ALAND, K. (1979), *Neutestamentliche Entwürfe* (Munich: Kaiser).

ALBERTZ, M. (1922), 'Zur Formgeschichte der Auferstehungsberichte', *ZNW* 21: 259–69.

ALETTI, J.-N. (1993), *Saint Paul: Épître aux Colossiens* (ÉBib NS 20; Paris: Gabalda).

ALLISON, D. C. (1982), 'The Pauline Epistles and the Synoptic Gospels: The Pattern of the Parallels', *NTS* 28: 1–32.

ALLO, E.-B. (1956a), *Saint Paul: Première Épître aux Corinthiens* (ÉBib; Paris: Gabalda).

—— (1956b), *Saint Paul: Seconde Épître aux Corinthiens* (ÉBib; Paris: Gabalda).

ALZINGER, W. (1970), 'Ephesos', PWSup 12. 1588–1704.

AMBELAIN, R. (1972), *La Vie secrète de saint Paul* (Paris: Laffont).

ANDRÉ, J.-M., AND BASLEZ, M.-F., (1993), *Voyager dans l'Antiquité* (Paris: Fayard).

APPLEBAUM, S. (1976), 'The Social and Economic Status of the Jews in the Diaspora', in *The Jewish People in the First Century: Historical Geography, Political History, Social, Cultural, and Religious Life and Institutions*, ed. S. Safrai *et al.* (Compendia Rerum Iudaicarum ad Novum Testamentum, sect. 1, vol. 2; Philadelphia: Fortress), 701–27.

ARNOLD, C. E. (1992), 'Colossae', *ABD* 1. 1089–90.

AUNE, D. (1987), *The New Testament in its Literary Environment* (Philadelphia: Westminster).

AUS, R. D. (1979a), 'Three Pillars and Three Patriarchs: A Proposal Concerning Gal 2:9', *ZNW* 70: 252–61.

—— (1979b), 'Paul's Travel Plans to Spain and the "Full Number of the Gentiles" of Rom XI, 25', *NovT* 29: 232–62.

BADENAS, R. (1986), *Christ the End of the Law: Romans 10. 4 in Pauline Perspective* (JSNTSup 10; Sheffield: JSOT Press).

BAILEY, J. A. (1978), 'Who Wrote II Thessalonians?', *NTS* 25: 131–45.

BAKER, D. W. (1992), 'Tarshish (Place)', *ABD* 6. 331–3.

BANKS, R. (1980), *Paul's Idea of Community* (Exeter: Paternoster Press).

BARCLAY, J. M. G. (1987), 'Mirror-Reading a Polemical Letter: Galatians as a Test Case', *JSNT* 31: 73–93.

BARNETT, A. E. (1941), *Paul becomes a Literary Influence* (Chicago: University of Chicago Press).

BARRÉ, M. (1980), 'Qumran and the Weakness of Paul', *CBQ* 42: 216–27.

BARRETT, C. K. (1963a), 'Cephas and Corinth', in *Abraham unser Vater: Festschrift für Otto Michel*, ed. O. Betz *et al.* (AGJU 5; Leiden: Brill), 1–12.

—— (1963b), *The Pastoral Epistles in the New English Bible* (NCB; Oxford: Clarendon).

—— (1968), *A Commentary on the First Epistle to the Corinthians* (HNTC; New York: Harper & Row).

—— (1969), 'Titus', in *Neotestamentica et Semitica: Studies in Honour of Matthew Black*, ed. E. E. Ellis and M. Wilcox (Edinburgh: Clark), 1–14.

—— (1970a), 'Ho adikesas (2. Cor. 7, 12)', in *Verborum Veritas: Festschrift für Gustav Stählin zum 70. Geburtstag*, ed. O. Bocher and K. Haacker (Wuppertal, Brockhaus), 149–57.

—— (1970b), 'Paul's Opponents in II Corinthians', *NTS* 17: 233–54.

—— (1973), *A Commentary on the Second Epistle to the Corinthians* (HNTC; New York: Harper & Row).

—— (1985), *Freedom and Obligation: A Study of the Epistle to the Galatians* (London: SPCK).

—— (1987), *The New Testament Background: Selected Documents* (2nd. edn.; London: SPCK).

—— (1994), *Paul: An Introduction to his Thought* (Outstanding Christian Thinkers Series; London: Chapman).

BARTCHY, S. S (1992), 'Slavery (Greco-Roman)', *ABD* 6. 65–73.

BASLEZ, M.-F. (1991), *Saint Paul* (Paris: Fayard).

BATES, W. H. (1965), 'The Integrity of II Corinthians', *NTS* 12: 56–69.

BAUCKHAM, R. (1970), 'Barnabas in Galatians', *JSNT* 2: 61–70.

BEARE, F. W. (1969), *A Commentary on the Epistle to the Philippians* (2nd. edn.; BNTC; London: Black).

BECKER, J. (1989), *Paulus Apostel der Völker* (Tübingen: Mohr).

BENOIT, P. (1952), 'Prétoire, Lithostroton, et Gabbatha', *RB* 59: 530–50.

—— (1956), 'Corps, tête et plérôme dans les épîtres de la captivité', *RB* 63: 5–44.

—— (1959), 'Le Deuxième Visite de saint Paul à Jérusalem', *Biblica*, 40: 778–92.

—— (1975), 'L'Hymne christologique de Col i, 15–20: Jugement critique sur l'état des recherches', in *Christianity, Judaism, and Other Greco-Roman Cults Studies for Morton Smith at Sixty*, ed. J. Neusner (SJLA 12; Leiden: Brill), 1. 226–63.

BERCOVITZ, J. P. (1985), '*Kalein* in Gal 1:15', *Proceedings of the Eastern Great Lakes and Midwest Biblical Societies*, 5: 28–37.

BEST, E. (1979), *A Commentary on the First and Second Epistles to the Thessalonians* (BNTC; London: Black).

BETZ, H. D. (1972), *Der Apostel Paulus und die sokratische Tradition: Eine exegetische Untersuchung zu seiner 'Apologie' 2 Korinther 10–13* (BHT 45; Tübingen: Mohr).

—— (1979), *Galatians: A Commentary on Paul's Letter to the Churches in Galatia* (Hermeneia; Philadelphia: Fortress).

—— (1985), *2 Corinthians 8 and 9: A Commentary on Two Administrative Letters of the Apostle Paul* (Hermeneia: Philadelphia: Fortress).

—— (1992) 'Paul', *ABD* 5. 186–201.

BEYSCHLAG, W. (1867), 'Das geschichtliche Problem des Römerbriefs', *TSK* 40: 627–50.

BIETENHARD, H. (1977), 'Die syrische Dekapolis von Pompeius bis Trajan', *ANRW* II, 8. 220–61.

BILLERBECK, P. (1922–8), *Kommentar zum Neuen Testament aus Talmud und Midrasch* (München: Beck).

BLINZLER, J. (1963), 'Lexikalisches zu dem Terminus *stoicheia tou kosmou* bei Paulus', in *Studiorum Paulinorum Congresus Internationalis Catholicus*, ii (AnBib 18; Rome: Pontifical Biblical Institute), 429–42.

BLOOMQUIST, L. G. (1993), *The Function of Suffering in Philippians* (JSNTSup 78; Sheffield: JSOT Press).

BOËTIUS, A. (1934), 'Remarks on the Development of Domestic Architecture in Rome', *AJA* 38: 158–70.

BOISMARD, M.-E., AND LAMOUILLE, A. (1977), *L'Évangile de Jean: Commentaire* (Synopse des quatre évangiles en Français, 3; Paris: Cerf).

—— (1984), *Le Texte occidental des Actes des Apôtres: reconstitution et réhabilition* (Synthèse 17; Paris: Éditions Recherche sur les Civilisations).

—— (1990), *Les Actes des deux apôtres*, i–iii (ÉBib; Paris: Gabalda).

BONSIRVEN, J. (1939), *Exégèse rabbinique et exégèse paulinienne* (Paris: Beauchesne).

BORGEN, P. (1961), '"At the Age of Twenty" (1QSa)', *RQ* 3: 267–77.

BORNKAMM, G. (1961), 'The History of the Origin of the So-Called Second Letter to the Corinthians', *NTS* 8: 258–64.

—— (1971), *Paul* (New York: Harper & Row).

BORSE, U. (1968), '"Abbild der Lehre" (Rom 6:17) im Kontext', *BZ* 12: 95–103.

—— (1972), *Der Standort des Galaterbriefes* (BBB 41; Köln-Bonn: Hanstein).

—— (1984), *Der Brief an die Galater* (RNT; Regensburg: Pustet).

BOURGUET, E. (1905), *De rebus Delphicis imperatoriae aetatis capita duo* (Montpellier).

BOWEN, C. R. (1924), 'The Original Form of Paul's Letter to the Colossians', *JBL* 43: 177–206.

BOWERS, W. P. (1975), 'Jewish Communities in Spain in the Time of Paul the Apostle', *JTS* 26: 395–402.

—— (1979), 'Paul's Route through Mysia: A Note on Acts xvi. 8', *JTS* 30: 507–11.

BRASSAC, A. (1913), 'Une inscription de Delphes et la chronologie de saint Paul', *RB* 10: 36–53, 207–17.

BRONEER, O. (1962), 'The Isthmian Victory Games', *AJA* 66: 259–63.

BROOTEN, B. (1982), *Women Leaders in the Ancient Synagogue* (Chico: Scholars Press).

BROSHI, M. (1979), 'The Population of Western Palestine in the Roman-Byzantine Period', *BASOR* 236: 1–10.

BROUGHTON, T. R. S. (1937), 'Three Notes on Saint Paul's Journeys in Asia Minor', in *Quantulacumque: Studies Presented to Kirsopp Lake*, ed. R P. Casey *et al.* (London: Christophers), 131–8.

BROWN, R. E. (1990), 'Further Reflections on the Origins of the Church of Rome', in *The Conversation Continues: Studies in Paul and John in Honor of J. Louis Martyn*, ed. R. T. Fortna and B. R. Gaventa (Nashville: Abingdon), 98–115.

—— and Meier, J. P. (1983), *Antioch and Rome: New Testament Cradles of Catholic Christianity* (New York: Paulist).

BROX, N. (1969), *Die Pastoralbriefe* (RNT 7/2; Regensburg: Pustet).

BRUCE, F. F. (1977), *Paul: Apostle of the Free Spirit* (Exeter: Paternoster).

—— (1982*a*), *1 & 2 Thessalonians* (WBC 45; Waco, Tex.: Word Books).

—— (1982*b*), *The Epistle of Paul to the Galatians: A Commentary on the Greek Text* (NIGTC; Exeter: Paternoster).

—— (1983), *Philippians* (Good News Commentary; San Francisco: Harper & Row).

—— (1985), 'Chronological Questions in the Acts of the Apostles', *BJRL* 68: 273–95.

—— (1991), 'The Romans Debate—Continued', in *The Romans Debate: Revised and Expanded Edition*, ed. K. Donfried (Edinburgh: Clark), 175–94.

—— (1992*a*), 'Hierapolis', *ABD* 3. 195–6.

—— (1992*b*), 'Laodicea', *ABD* 4. 229–31.

BUJARD, W. (1973), *Stilanalystische Untersuchungen zum Kolosserbrief als Beitrag zur Methodik von Sprachvergleichen* (SUNT 11; Göttingen: Vandenhoeck & Ruprecht).

BULTMANN, R. (1963), *The History of the Synoptic Tradition* (Oxford: Blackwell).

—— (1964), 'Paul', in *Existence and Faith: Shorter Writings of R. Bultmann*, ed. S. Ogden (London: Collins), 130–72.

—— (1965), *Theology of the New Testament*, i–ii (London: SCM).

—— (1967), 'Glossen im Römerbrief', in his *Exegetica*, ed. E. Dinkler (Tübingen: Mohr), 278–84.

BURCHARD, C. (1970), *Der dreizehnte Zeuge* (FRLANT 103; Göttingen: Vandenhoeck & Ruprecht).

BURTON, E. DE WITT (1921), *A Critical and Exegetical Commentary on the Epistle to the Galatians* (ICC; Edinburgh: Clark).

BURTON, G. P. (1975), 'Proconsuls, Assizes and the Administration of Justice under the Empire', *JRS* 65: 92–106.

CADBURY, H. J. (1933), 'Roman Law and the Trial of Paul', in *The Beginnings of Christianity*, Part 1. *The Acts of the Apostles*, ed. F. Jackson and K. Lake (London: Macmillan), 5. 326–38.

CAIRD, G. (1976), *Paul's Letters from Prison* (NCB; Oxford: OUP).

CALLANDER, T. (1904), 'The Tarsian Orations of Dio Chrysostom', *JHS* 24: 64–5.

CARCOPINO, J. (1981), *Daily Life in Ancient Rome* (London: Penguin).

CARREZ, M. (1980), 'Le "nous" en 2 Corinthiens: Paul parle-t-il au nom de toute la communauté, du group apostolique, de l'équipe ministérielle ou en son nom personnel? Contribution à l'étude de l'apostolicité dans 2 Corinthiens', *NTS* 26: 474–86.

CARRINGTON, R. C. (1933), 'The Ancient Italian Town-House', *Antiquity*, 7: 133–52.

CARROLL, S. T. (1992), 'Mysia', *ABD* 4. 940–1.

CASSON, L. (1971), *Ships and Seamanship in the Ancient World* (Princeton: Princeton University Press).

—— (1979), *Travel in the Ancient World* (London: Allen & Unwin).

CAVALLIN, H. C. C. (1974), *Life after Death: Paul's Argument for the Resurrection of the Dead in 1 Cor 15*, Part I. *An Enquiry into the Jewish Background* (ConBNT 7/1; Lund: Gleerup).

CHADWICK, H. (1950), '1 Thess 3:3, *sainesthai*', *JTS* 1: 156–8.

—— (1957), 'St. Peter and St. Paul in Rome: The Problem of the Memoria Apostolorum ad Catcumbas', *JTS* 8: 31–52.

—— (1965), 'St. Paul and Philo of Alexandria', *BJRL* 48: 286–307.

CHANTRAINE, P. (1968), *Dictionnaire Etymologique de la langue Grecque* (Paris: Klincksieck).

CHAPOT, F. (1899), 'Taberna', in *Dictionaire des Antiquités grecques et romains*, ed. C. Daremberg and E. Saglio (Paris: Hachette), 5. 8–11.

CHARLESWORTH, M. P. (1950), 'Tiberius', *CAH* 10: 648–52.

CHEVALLIER, R. (1972), *Les Voies romaines* (Paris: Colin).

COHEN, S. (1986), 'Was Timothy Jewish (Acts 16: 1–3)? Patristic Exegesis, Rabbinic Law, and Matrilineal Descent', *JBL* 105: 251–68.

COLLINS, J. J. (1985), 'A Symbol of Otherness: Circumcision and Salvation in the First Century', in *'To See Ourselves as Others See Us': Christians, Jews and 'Others' in Late Antiquity*, ed. J. Neusner and E. Frerichs (Chico: Scholars Press), 163–86.

CONZELMANN, H. (1975), *1 Corinthians: A Commentary on the First Epistle to the Corinthians* (Hermeneia; Philadelphia: Fortress).

COOK, J. M. (1973), *The Troad: An Archaeological and Topographical Study* (Oxford: Clarendon).

COX, P. (1983), *Biography in Late Antiquity: A Quest for the Holy Man* (Berkeley: University of California Press).

CRANFIELD, C. (1979), *A Critical and Exegetical Commentary on the Epistle to the Romans* (ICC; Edinburgh: Clark).

CROSS, F. M. (1961), *The Ancient Library of Qumran and Modern Biblical Studies* (rev. edn.; Garden City, NJ: Doubleday).

CROUCH, J. E. (1972), *The Origin and Intention of the Colossian Haustafel* (FRLANT 109; Göttingen: Vandenhoeck & Ruprecht).

CULLMANN, O. (1953), *Peter: Disciple, Apostle, Martyr* (London: SCM).

CUNTZ, O. (1929) (ed.), *Itineraria Romana,* i. *Itineraria Antonini Augusti et Burdigalense* (Leipzig: Teubner).

CURRAN, J. (1945), 'Tradition and the Roman Origin of the Captivity Letters', *TS* 5: 163–205.

DAHL, N. A. (1962), 'The Particularity of the Pauline Epistles as a Problem in the Ancient Church', in *Neotestamentica et Patristica* (NovTSup 6; FS O. Cullmann; Leiden: Brill), 261–71.

DANBY, H. (1967) (trans.), *The Mishnah* (Oxford: OUP).

DAUBE, D. (1956), *The New Testament and Rabbinic Judaism* (London: Athlone Press).

DAVIES, P. R. (1983), *The Damascus Covenant: An Interpretation of the 'Damascus Document'* (JSOTSup 25; Sheffield; JSOT Press).

DAVIES, W. D. (1962), *Paul and Rabbinic Judaism: Some Rabbinic Elements in Pauline Theology* (2nd. edn.; London: SPCK).

—— (1977), 'Paul and the People of Israel', *NTS* 24: 4–39.

DE LA POTTERIE, I. (1963), 'Jésus et la vérité d'après Eph 4, 21', in *Studiorum Paulinorum Congressus Internationalis Catholicus 1961* (AnBib 18; Rome: Pontifical Biblical Institute), 2. 45–57.

DENIS, A.M. (1957), 'L'Investiture de la fonction apostolique par "apocalypse". Étude thématique de Gal., 1:16', *RB* 64: 335–62, 492–515.

—— (1958), 'La Fonction apostolique et la liturgie nouvelle en esprit: Étude thématique des métaphores pauliniennes du culte nouveau', *RSPT* 42: 401–36, 617–56.

DINKLER, E. (1953), 'Der Brief an die Galater', *Verkündigung und Forschung* 7: 182–3.

DODD, C. H. (1953), '*Ennomos Christou*', in *Studia Paulina in honorem J. de Zwaan*, ed. W. C. van Unnik and J. N. Sevenster (Haarlem: Bohn), 96–110.

DONFRIED, K. P. (1984), '1 Thessalonians 2: 13–16 as a Test Case', *Int* 38: 242–53.

—— (1985), 'The Cults of Thessalonica and the Thessalonian Correspondence', *NTS* 31: 336–56.

—— (1990), '1 Thessalonians and the Early Paul', in *The Thessalonian Correspondence* ed. R. F. Collins (BETL 87; Leuven: University Press), 3–26.

DOWNEY, G. (1961), *A History of Antioch in Syria: From Seleucus to the Arab Conquest* (Princeton: Princeton University Press).

DUFF, A. M. (1928), *Freedmen in the Early Roman Empire* (Oxford: Clarendon).

DUNCAN, G. S. (1929), *St. Paul's Ephesian Ministry: A Reconstruction with Special Reference to the Ephesian Origin of the Imprisonment Epistles* (London: Hodder & Stoughton).

DUNGAN, D. G. (1971), *The Sayings of Jesus in the Churches of Paul* (Philadelphia: Fortress).

DUNN, J. D. G. (1980), *Christology in the Making: A New Testament Inquiry into the Origins of the Doctrine of the Incarnation* (London: SCM).

—— (1983), 'The Incident at Antioch (Gal 2: 11–18)', *JSNT* 18: 3–57.

—— (1985), 'Once More—Gal 1: 18 *historêsai Kêphan:* In Reply to Otfried Hofius', *ZNW* 76: 138–9.

—— (1987), '"A Light to the Gentiles": The Significance of the Damascus Road Christophany for Paul', in *The Glory of Christ in the New Testament: Studies in Christology in Memory of George Bradford Caird*, ed. L. D. Hurst and N. T. Wright (Oxford: Clarendon), 251–66.

—— (1988), *Romans* (WBC 38; Dallas: Word Books).

—— (1989), 'Paul's Knowledge of the Jesus' Tradition: The Evidence of Romans', in *Christus Bezeugen*, ed. K. Kertelge *et al.* (FS W. Trilling; (Leipzig: Benno-Verlag), 193–207.

—— (1992), 'Jesus, Table-fellowship and Qumran', in *Jesus and the Dead Sea Scrolls*, ed. J. H. Charlesworth (ABRL; New York: Doubleday), 254–72.

—— (1993), *The Epistle to the Galatians* (BNTC; Peabody MA: Hendrickson).

EDLERKIN, G. W., STILLWELL, R., WAAGE, F. O., WAAGE, D. B., and LASSUS, J. (1934–72), *Antioch-on-the-Orontes*, i–v (Princeton: Princeton University Press).

EDSON, C. (1948), 'Cults of Thessalonica', *HTR* 41: 153–205.

EISENMAN, R. (1983), *Maccabees, Zadokites, Christians and Qumran* (SPB 34; Leiden: Brill).

ELLIGER, W. (1978), *Paulus in Griechenland: Philippi, Thessaloniki, Athen, Korinth* (SBS 92–3; Stuttgart: Katholisches Bibelwerk).

ELLIOTT, J. K. (1981), 'The Language and Style of the Concluding Doxology at the End of Romans', *ZNW* 72: 124–30.

ELLIS, E. E. (1975), 'Paul and His Opponents: Trends in Research', in *Christianity, Judaism and Other Graeco-Roman Cults*, ed. J. Neusner (FS Morton Smith; Leiden: Brill, 1975), I. 264–98.

—— (1981), *Paul's Use of the Old Testament* (2nd edn., Grand Rapids, Mich.: Eerdemans).

ENGELS, D. (1990), *Roman Corinth: An Alternative Model for the Classical City* (Chicago: University of Chicago Press).

EPSTEIN, I. (1935), *The Babylonian Talmud: Seder Nazikin* (London: Soncino Press).

—— (1936), *The Babylonian Talmud: Seder Nashim* (London: Soncino Press).

ERDEMGIL, S. (1988), *Les Maisons du flanc à Ephese* (Istanbul: Hitit Color).

—— (1989), *Ephese* (Istanbul: Net Turistic Yayinlar).

EVANS, E. C. (1935), 'Roman Descriptions of Personal Appearance in History and Biography', *HSCP* 46: 43–84.

—— (1941), 'The Study of Physiognomy in the 2nd Century AD', *TPAPA* 72: 96–108.

FASCHER, E. (1929), 'Zur Witwerschaft des Paulus und der Auslegung von I Kor 7', *ZNW* 28: 62–9.

FEE, G. D. (1978), '*Charis* in II Corinthians I.15: Apostolic Parousia and Paul–Corinth Chronology', *NTS* 24: 533–8.

—— (1987), *The First Epistle to the Corinthians* (NICNT; Grand Rapids, Mich.: Eerdmans).

FEINE, P. (1916), *Die Abfassung des Philipperbriefes in Ephesus* (Gütersloh: Mohn).

FESTUGIÈRE, A. (1959), *Antioche païenne et chrétienne* (Paris: Boccard).

FINLEY, M. I. (1977), *Atlas of Classical Archaeology* (London: Chatto and Windus).

FITZMYER, J. A. (1989), 'Another Look at *kephale* in I Corinthians II. 3', *NTS* 35: 503–11.

FITZMYER, J. A. (1990), 'Paul', *NJBC* §79.

—— (1993), *Romans* (AB 33; New York: Doubleday).

FJÄRSTEDT, B. (1974), *Synoptic Tradition in 1 Corinthians: Themes and Clusters of Theme Words in 1 Corinthians 1–4 and 9* (Uppsala: Theologiska Institutionen).

FORBES, C. (1986), 'Comparison, Self-Praise and Irony: Paul's Boasting and the Conventions of Hellenistic Rhetoric', *NTS* 32: 1–30.

FORTNA, R. T. (1990), 'Philippians: Paul's Most Egocentric Letter', in *The Conversation Continues: Studies in Paul and John in Honor of J. Louis Martyn*, ed. R. T. Fortna and B. R. Gaventa; Nashville: Abingdon), 220–34.

FRAME, J. E. (1912), *A Critical and Exegetical Commentary on the Epistles of St. Paul to the Thessalonians* (ICC; Edinburgh: Clark).

FRASER, J. W. (1970), 'Paul's Knowledge of Jesus: 2 Cor 5: 16 Once More', *NTS* 17: 293–313.

FRENCH, D. H. (1980), 'The Roman Road System of Asia Minor', *ANRW* II, 7/2. 698–729.

FREYNE, S. (1980), *Galilee from Alexander the Great to Hadrian 323 BCE to 135 CE* (Wilmington, Del.: Glazier).

FRIER, B. R. (1980), *Landlords and Tenants in Imperial Rome* (Princeton: Princeton University Press).

FUCHS, H. (1950), 'Tacitus über die Christen', *VC* 4: 65–93.

FURNISH, V. P. (1963), 'The Place and Purpose of Philippians III', *NTS* 10: 80–8.

—— (1968), *Theology and Ethics in Paul* (Nashville: Abingdon).

—— (1984), *II Corinthians* (AB 32A; Garden City, NJ: Doubleday).

—— (1992), 'Colossians, Epistle to the', *ABD* 1. 1090–6.

GAGER, J. G. (1981), 'Some Notes on Paul's Conversion', *NTS* 27: 697–704.

GAMBLE, H. (1977), *The Textual History of the Letter to the Romans* (SD 42; Grand Rapids, Mich.: Eerdmans).

—— (1985), *The New Testament Canon* (Philadelphia: Fortress).

GARCÍA Y BELLIDO, A. (1972), 'Die Latinisierung Hispaniens', *ANRW* I, 1. 462–91.

GARDINER, J. F. (1986), 'Proofs of Status in the Roman World', *JCS* 33: 1–14.

GARLAND, D. E. (1985), 'The Composition and Unity of Philippians: Some Neglected Literary Factors', *NovT* 27: 141–73.

GARNSEY, P. (1966), 'The Lex Julia and Appeal under the Empire', *JRS* 56: 167–89.

GEAGAN, G. (1979), 'Roman Athens: Some Aspects of Life and Culture, I. 86 BC–AD 267', *ANRW* II, 7/1. 371–437.

GEORGI, D. (1986), *The Opponents of Paul in Second Corinthians* (Philadelphia: Fortress).

—— (1991), *Theocracy in Paul's Practice and Theology* (Minneapolis: Fortress).

—— (1992), *Remembering the Poor: The History of Paul's Collection for Jerusalem* (Nashville: Abingdon).

GERLEMANN, G. (1989), *Der Heidenapostel: Ketzerische Erwägungen zur Predigt des Paulus zugleich ein Streifzug in der Griechischen Mythologie* (Scripta Minora Regiae Societatis Humaniorum Litterarum Lundensis; Stockholm: Almquist and Wiksell).

GEYSER, A. (1959), 'Un essai d'explication de Rom. xv. 19', *NTS* 6: 156–9.

GIBLIN, C. H. (1990), 'The Second Letter to the Thessalonians', *NJBC* §53.

GIET, G. (1953), 'Les Trois Premiers Voyages de saint Paul à Jérusalem', *RSR* 41: 321–47.

GILLARD, F. (1989), 'The Problem of the AntiSemitic Commas between 1 Thessalonians 2.14 and 15', *NTS* 35: 481–502.

GILLMAN, J. (1992), 'Titus', *ABD* 6. 581–2.

GNILKA, J. (1968), *Der Philipperbrief* (HTKNT; Freiburg: Herder).

—— (1980), *Der Kolosserbrief* (HTKNT; Freiburg: Herder).

GOLDSTEIN, J. (1981), 'Jewish Acceptance and Rejection of Hellenism', in *Jewish and Christian Self-Definition*, ii. *Aspects of Judaism in the Graeco-Roman Period*, ed. E. P. Sanders, A. I. Baumgarten, A. Mendelson (London: SCM), 64–87.

GOODBLATT, D. (1989), 'The Place of the Pharisees in First-Century Judaism: The State of the Debate', *JSJ* 20: 12–30.

GOODSPEED, E. J. (1937), *An Introduction to the New Testament* (Chicago: University of Chicago Press).

—— (1951), 'Phoebe's Letter of Introduction', *HTR* 44: 55–7.

GRAF, D. F. (1992), 'Nabateans', *ABD* 4. 970–3.

GRANT, R. M. (1982), 'The Description of Paul in the Acts of Paul and Thecla', *VC* 36:1–4.

GUNDRY, R. H. (1976), *Sōma in Biblical Theology with Emphasis on Pauline Anthropology* (SNTSMS 29; Cambridge: CUP).

—— (1987), 'The Hellenization of Dominical Tradition and the Christianization of Jewish Tradition in the Eschatology of 1–2 Thessalonians', *NTS* 33: 161–78.

GUTHRIE, D. (1966), *New Testament Introduction: The Pauline Epistles* (London: Tyndale Press).

GUTSCHMID, A. VON (1885), 'Verzeichniss der nabatäischen Könige', in *Nabatäische Inschriften aus Arabien*, ed. J. Euting (Berlin: Reimer).

HAENCHEN, E. (1966), 'The Book of Acts as Source Material for the History of Early Christianity', in *Studies in Luke–Acts: Essays Presented in Honour of Paul Schubert*, ed. L. E. Keck and J. L. Martyn (New York/Nashville: Abingdon), 268–9.

—— (1971), *The Acts of the Apostles: A Commentary* (Philadelphia: Westminster).

HAFEMANN, S. J. (1986), *Suffering and the Spirit: An Exegetical Study of II Cor. 2:14–3:3 within the Context of the Corinthian Correspondence* (WUNT 2/29; Tübingen: Mohr).

HALL, R. G. (1988), 'Epispasm and the Dating of Ancient Jewish Writings', *JSP* 2: 71–86.

HAMEL, G. (1990), *Poverty and Charity in Roman Palestine: First Three Centuries C.E.* (University of California Publications, Near Eastern Studies, 23; Berkeley: University of California Press).

HAMMOND, N. G. L. (1974), 'The Western Part of the Via Egnatia', *JRS* 64: 185–94.

HARDING, M. (1993), 'On the Historicity of Acts: Comparing Acts 9: 23–25 with 2 Corinthians 11: 32–33', *NTS* 39: 518–38.

HARNACK, A. VON (1912), 'Chronologische Berechnung des Tages von Damaskus', *Sitzungsberichte der preussischen Akademie der Wissenschaft zu Berlin*, 37: 675–6.

—— (1928), '*Kopos* (*kopian, hoi kipiôntes*) im frühchristlichen Sprachgebrauch', *ZNW* 27: 1–10.

HARRER, G. A. (1940), 'Saul Who is also Called Paul', *HTR* 33: 19–33.

HARRISON, P. N. (1921), *The Problem of the Pastorals* (Oxford: OUP).

—— (1964), *The Problem of the Pastoral Epistles* (Oxford: OUP).

HAUSRATH, A. (1870), *Der Vier-Capitel-Brief des Paulus an die Korinther* (Heidelberg: Bassermann).

HAWTHORN, G. F. (1983), *Philippians* (WBC 43; Waco: Word Books).

HAYS, R. B. (1989), *Echoes of Scripture in the Letters of Paul* (New Haven/London: Yale University Press).

HEDRICK, C. W. (1981), 'Paul's Conversion/Call: A Comparative Analysis of the Three Reports in Acts', *JBL* 100: 415–32.

HEMBERG, B. (1950), *Die Kabieren* (Uppsala: Almquist).

HEMER, C. J. (1975), 'Alexandria Troas', *TynB* 26: 79–112.

—— (1985), 'The Name of Paul', *TynB* 36: 179–83.

—— (1989), *The Book of Acts in the Setting of Hellenistic History*, ed. C. H. Gempf (WUNT 49; Tübingen: Mohr).

HENDRIX, H. (1992a), 'Philippi', *ABD* 5. 313–18.

—— (1992b), 'Thessalonica', *ABD* 6. 523–7.

HENGEL, M. (1974), *Judaism and Hellenism: Studies in their Encounter in Palestine in the Early Hellenistic Period* (Philadelphia: Fortress).

—— (1985a), 'Jakobus der Herrenbruder—der erste "Papst"?' in *Glaube und Eschatologie* (FS W. G. Kümmel; Tübingen: Mohr), 71–104.

—— (1985b), *Studies in the Gospel of Mark* (London: SCM).

—— (1989), *The 'Hellenization' of Judaea in the First Century after Christ* (London: SCM).

—— with R. DEINES (1991), *The Pre-Christian Paul* (London: SCM).

HENNECKE, E., AND SCHNEEMELCHER, W. (1965), *New Testament Apocrypha* (London: Lutterworth).

HÉRING, J. (1959), *La Première Épitre de saint Paul aux Corinthiens* (CNT 7; Neuchâtel-Paris: Delachaux et Niestlé).

HILL, A. E. (1980), 'The Temple of Asclepius: An Alternative Source for Paul's Body Theology?', *JBL* 99: 437–9.

HOCK, R. F. (1978), 'Paul's Tentmaking and the Problem of his Social Class', *JBL* 97: 555–64.

—— (1980), *The Social Context of Paul's Ministry: Tentmaking and Apostleship* (Philadelphia: Fortress).

HOEHNER, H. (1972), *Herod Antipas* (SNTSMS 17; Cambridge: CUP).

HOFFMANN, R. J. (1984), *Marcion: On the Restitution of Christianity: An Essay on the Development of Radical Paulinist Theology in the Second Century* (AARAS 46; Chico: Scholars Press).

HOFIUS, O. (1986), 'Das Evangelium und Israel: Erwägungen zu Römer 9–11', *ZTK* 83: 297–324.

HOLLAND, G. (1990), '"A Letter Supposedly from Us": A Contribution to the Discussion about the Authorship of 2 Thessalonians', in *The Thessalonian Correspondence* (BETL 87; ed. R. Collins; Leuven: University Press), 394–402.

HOLMBERG, B. (1978), *Paul and Power: The Structure of Authority in the Primitive Church as Reflected in the Pauline Epistles* (Lund: Gleerup).

HOOKER, M. D. (1982), 'Paul and "Covenantal Nomism"', in *Paul and Paulinism: Essays in Honour of C. K. Barrett*, ed. M. D. Hooker and S. G. Wilson (London: SPCK), 47–56.

HORSLEY, R. (1976), 'Pneumatikos vs Psychikos: Distinction of Status among the Corinthians', *HTR* 69: 269–88.

—— (1977), 'Wisdom of Words and Words of Wisdom in Corinth', *CBQ* 39: 224–39.

—— (1978), '"How can some of you say that there is no resurrection from the dead?" Spiritual Elitism in Corinth', *NovT* 20: 203–31.

HOWARD, G. (1990), *Paul: Crisis in Galatia* (2nd edn.; SNTSMS 35, Cambridge: CUP).

HUGHES, F. W. (1990), 'The Rhetoric of 1 Thessalonians', in *The Thessalonian Correspondence* (BETL 80; ed. R. F. Collins; Leuven: University Press), 94–116.

HULTGREN, A. J. (1976), 'Paul's Pre-Christian Persecutions of the Church: Their Purpose, Locale, and Nature', *JBL* 96: 105–7.

HUMBERT, G. (1899*a*), 'Carcer', *Dictionnaire des antiquités grecques et romaines*, ed. C. Daremberg and E. Saglio (Paris: Hachette), I/2. 916–19.

—— (1899*b*), 'Custodia', *Dictionnaire des antiquités grecques et romaines*, ed. C. Daremberg and E. Saglio (Paris: Hachette), I/2. 1672–3.

—— and LÉCRIVAN, C. (1899), 'Insula', *Dictionnaire des antiquités grecques et romaines*, ed. C. Daremberg and E. Saglio; Paris: Hachette), III/1. 546.

HURD, J. C. (1965), *The Origin of 1 Corinthians* (London: SPCK).

HYLDAHL, N. (1973), 'Die Frage nach der literarischen Einheit des zweiten Korintherbriefes', *ZNW* 64: 289–306.

JAUBERT, A. (1958), 'Le Pays de Damas', *RB* 65: 214–48.

JEREMIAS, J. (1926), 'War Paulus Witwer?', *ZNW* 25: 310–12.

—— (1929), 'Nochmals: War Paulus Witwer?', *ZNW* 28: 321–2.

—— (1953), 'Zur Gedankenführung in den paulinischen Briefen', in *Studia Paulina in honorem J. de Zwaan*, ed. J. N. Sevenster and W. C. van Unnik (Haarlem: Bonn), 152–4.

—— (1961), *Verzeichnis der Schriftgelehrten: Geographisches Register*, Vol. 6 of P. Billerbeck, *Kommentar zum Neuen Testament aus Talmud und Midrash* (München: Beck).

—— (1963), 'Zu Phil. 2: 7', *NovT* 6: 186–7.

—— (1969), *Jerusalem in the Time of Jesus: An Investigation into Economic and Social Conditions during the New Testament Period* (London: SCM).

—— (1971), *New Testament Theology*, i. *The Proclamation of Jesus* (NT Library; London: SCM).

JERVELL, J. (1991), 'The Letter to Jerusalem', in *The Romans Debate* (rev. edn.; ed. K. Donfried; Edinburgh: Clark), 53–64.

JEWETT, R. (1970), 'The Agitators and the Galatian Congregation', *NTS* 17: 196–212.

—— (1979), *Dating Paul's Life* (London: SCM).

—— (1986), *The Thessalonian Correspondence: Pauline Rhetoric and Millenarian Piety* (Foundations and Facets; Philadelphia: Fortress).

—— (1988), 'Paul, Phoebe, and the Spanish Mission', in *Essays in Tribute to Howard Clark Kee: The Social World of Formative Christianity and Judaism*, ed. J. Neusner *et al.* (Philadelphia: Fortress), 142–61.

—— (1993), 'Tenement Churches and Communal Meals in the Early Church: The Implications of a Form-Critical Analysis of 2 Thessalonians 3:10', *BR* 38: 23–43.

JOHNSON, L. (1956), 'The Pauline Letters from Caesarea', *ExpTim* 68: 24–6.

JONES, A. H. M. (1971), *The Cities of the Eastern Roman Provinces* (2nd edn.; Oxford: Clarendon).

JUDGE, E. A. (1968), 'Paul's Boasting in Relation to Contemporary Professional Practice', *AusBR* 16: 37–50.

KÄSEMANN, E. (1964), *Essays on New Testament Themes* (SBT 41; London: SCM).

KELLY, J. N. D. (1963), *The Pastoral Epistles* (BNTC; London: Black).

—— (1975), *Jerome: His Life, Writings and Controversies* (London: Duckworth).

KENNEDY, G. A. (1963), *The Art of Persuasion in Greece* (Princeton: Princeton University Press).

—— (1972), *The Art of Rhetoric in the Roman World 300 BC–AD 300* (Princeton: Princeton University Press).

—— (1984), *New Testament Interpretation through Rhetorical Criticism* (Chapel Hill/London: University of North Carolina Press).

KENNY, A. (1986), *A Stylometric Study of the New Testament* (Oxford: Clarendon).

KENT, J. H. (1966), *Corinth VIII/3: The Inscriptions 1926–1950* (Princeton: American School of Classical Studies at Athens).

KILPATRICK, G. D. (1959), 'Galatians 1: 18 *historesai Kêphan*', in *New Testament Essays*, ed. A. J. B. Higgins (FS Manson; ed. A. J. B. Higgins Manchester: University Press), 144–9.

—— (1968), '*Blepete*, Philippians 3: 2', in *In Memoriam Paul Kahle*, ed. M. Black and G. Fohrer (BZAW 103; Berlin: Töpelmann), 146–8.

KIM, CHAN-HIE (1972), *Form and Structure of the Familiar Greek Letter of Recommendation* (SBLDS 4; Missoula, Mont.: Scholars Press).

KIM, S. (1984), *The Origin of Paul's Gospel* (2nd edn.; WUNT 2, 4; Tübingen: Mohr).

KLAUCK, H. J. (1984), *1 Korintherbrief* (Neue Echter Bibel; Würzburg: Echter).

—— (1986), *2 Korintherbrief* (Neue Echter Bibel; Würzburg: Echter).

KLEIN, G. (1960), 'Galater 2: 6–9 und die Geschichte der Jerusalemer Urgemeinde', *ZTK* 57: 275–95.

KLINZING, G. (1971), *Die Umdeutung des Kultus in der Qumrangemeinde und im Neuen Testament* (SUNT 7; Göttingen: Vandenhoeck and Ruprecht).

KNAUF, E. A. (1983), 'Zum Ethnarchen des Aretas 2 Kor 11, 32', *ZNW* 74: 145–7.

KNIBBE, D., AND ALZINGER, W. (1980), 'Ephesos vom Beginn der römischen Herrschaft in Kleinasien biz zum Ende der Principatszeit', *ANRW* II, 7/2. 748–830.

KNIBBE, D. (1970), 'Ephesos', PWSup 12. 248–97.

KNOX, J. (1950), *Chapters in a Life of Paul* (New York/Nashville: Abingdon-Cokesbury).

—— (1962), *The Ethic of Jesus in the Teaching of the Church* (London: SPCK).

—— (1964), 'Romans 15: 14–33 and Paul's Conception of his Apostolic Mission', *JBL* 83: 1–11.

—— (1987), 'The Meaning of Gal 1: 15', *JBL* 106: 301–4.

—— (1990), 'On the Pauline Chronology: Buck-Taylor-Hurd Revisited' in *The Conversation Continues: Studies in Paul and John in Honor of J. Louis Martyn*, ed. R. T. Fortna and B. R. Gaventa (Nashville: Abingdon), 258–74.

KOCH, D. A. (1986), *Die Schrift als Zeuge des Evangeliums: Untersuchungen zur Verwendung und zum Verständnis der Schrift bei Paulus* (BHT 69; Tübingen: Mohr).

KOENIG, J. (1985), *New Testament Hospitality: Partnership with Strangers as Promise and Mission* (Philadelphia: Fortress).

KOESTER, H. (1982), *Introduction to the New Testament*, i–ii (Philadelphia: Fortress).

KOLB, F. (1974), 'Zur Geschichte der Stadt Hierapolis in Phrygien', *ZPE* 15: 255–70.

KRAABEL, T. (1979), 'The Diaspora Synagogue: Archaeological and Epigraphic Evidence since Sukenik', *ANRW* II, 19/1. 477–510.

KRAELING, C. H. (1932), 'The Jewish Community at Antioch', *JBL* 51: 130–60.

KREMER, J, (1956), *Was an den Leiden Christi nochmangelt: Eine interpretationsgeschichtliche und exegetische Untersuchung zu Kol. 1, 24b* (BBB 12; Bonn: Hanstein).

KÜMMEL, W. G. (1975), *Introduction to the New Testament* (rev. edn.; London: SCM).

KUNTZ, O. (1929) (ed.),*Itineraria Romana*, i. *Itineraria Antonini Augusti et Burdigalense* (Leipzig: Teubner).

KÜRZINGER, J. (1978), 'Frau und Mann nach 1 Kor 11, 11f.', *BZ* 22: 270–5.

LAGRANGE, M. J. (1925), *Saint Paul: Épitre aux Galates* (ÉBib; Paris: Gabalda).

LAMBRECHT, J. (1990), 'Thanksgivings in 1 Thess 1–3', in *The Thessalonian Correspondence*, ed. R. F. Collins (BETL 80; Leuven: University Press), 183–205.

LAMPE, P. (1985), 'Keine "Sklavenflucht" des Onesimus', *ZNW* 76: 135–7.

—— (1987), 'Paulus—Zeltmacher', *BZ* 31: 256–61.

—— (1989), *Die stadtrömischen Christen in den ersten beiden Jahrhunderten: Untersuchungen zur Socialgeschichte* (2nd edn.; WUNT 2/18; Tübingen: Mohr).

—— (1991), 'The Roman Christians of Romans 16', in *The Romans Debate*, ed. K. Donfried (rev. edn.; Edinburgh: Clark), 216–30.

LANOWSKI, J. (1965), 'Weltwunder', *PWSup* 10: 1020–30.

LAPIDE, P. (1993), *Paulus—zwischen Damaskus und Qumran: Fehldeutungen und Übersetzungsfehler* (Gütersloh: Mohn).

LASSUS, J. (1977), 'La Ville d'Antioche à l'époque romaine d'après l'archéologie', *ANRW* II/8. 54–102.

LEARY, T. J. (1992), 'Paul's Improper Name', *NTS* 38: 467–9.

LE DÉAUT, R. (1980), *Targum du Pentateuque IV. Deutéronome* (SC 271; Paris: Cerf).

LÉGASSE, S. (1991), *Paul Apôtre: Essai de biographie critique* (Paris: Cerf/Fides).

LENTZ, J. C., Jr. (1993), *Luke's Portrait of Paul* (SNTSMS 77; Cambridge: CUP).

LEON, H. J. (1960), *The Jews of Ancient Rome* (Philadelphia: Jewish Publication Society).

LESTAPIS, S. DE (1976), *L'Énigma des Pastorales de saint Paul* (Paris: Gabalda).

LIEBSCHUTZ, J. (1972), *Antioch: City and Imperial Administration in the Later Roman Empire* (Oxford: Clarendon).

LIFSHITZ, B. (1967), *Donateurs et fondateurs dans les synagogues juives* (CRB 7; Paris: Gabalda).

LIGHTFOOT, J. B. (1904), *Saint Paul's Epistles to the Colossians and to Philemon* (London: Macmillan).

—— (1908), *Saint Paul's Epistle to the Philippians* (London: Macmillan).

—— (1910), *Saint Paul's Epistle to the Galatians* (London: Macmillan).

LITHGOW, W. (1974), *The Rare Adventures and Painful Peregrinations of William Lithgow*, ed. and with an introduction by G. Phelps (London: Folio Society).

LOCK, W. (1936), *A Critical and Exegetical Commentary on the Pastoral Epistles* (ICC; Edinburgh: Clark).

LOHMEYER, E. (1974), *Der Brief an die Philipper* (MeyerK; Göttingen: Vandenhoeck & Ruprecht).

LOHSE, E. (1968), *Die Briefe an die Kolosser und an Philemon* (MeyerK; Göttingen: Vandenhoeck and Ruprecht).

LOISY, A. (1920), *Les Actes des Apôtres* (Paris: Nourry).

LONGENECKER, R. (1990), *Galatians* (WBC 41; Dallas: Word Books).

386 *Bibliography*

LOUTH, A. (1989), Introduction to *Eusebius, The History of the Church from Christ to Constantine* (trans. G. A. Williamson; London: Penguin).

LÜDEMANN, G. (1984), *Paul Apostle to the Gentiles: Studies in Chronology* (London: SCM).

—— (1987), *Das frühe Christentum nach den Traditionen der Apostelgeschichte: Ein Kommentar* (Göttingen: Vandenhoeck and Ruprecht).

—— (1989), *Opposition to Paul in Jewish Christianity* (Minneapolis: Fortress).

—— (1991), 'Das Judenedikt des Claudius (Apg 18,2)', in *Der Treue Gottes trauen. Beiträge zum Werk des Lukas. Festschrift für Gerhard Schneider*, ed. C. Bussmann and W. Radl (Freiburg/Basel/Wien: Herder) 289–98.

LYONS, G. (1985), *Pauline Autobiography: Toward a New Understanding* (SBLDS 73; Atlanta: Scholars Press).

MACAULAY, D. (1974), *City: A Story of Roman Planning and Construction* (Boston: Houghton, Mifflin).

MCDONAGH, B. (1989), *Turkey: The Aegean and Mediterranean Coasts* (Blue Guide: London: Black/New York: Norton).

MCELENEY, N. (1973), 'Conversion, Circumcision and the Law', *NTS* 20: 319–41.

MACMULLEN, R. (1966), *Enemies of the Roman Order: Treason, Unrest and Alienation in the Empire* (Cambridge, Mass.: Harvard University Press).

—— (1976), *Roman Social Relations 50 BC to AD 284* (New Haven: Yale University Press).

MACQUARRIE, J. (1966), *Principles of Christian Theology* (London: SCM).

MALHERBE, A. J. (1968), 'The Beasts at Ephesus', *JBL* 87: 71–80.

—— (1986), 'A Physical Description of Paul', in *Christians among Jews and Gentiles*, ed. G. W. E. Nickelsburg and G. W. MacRea (FS K. Stendahl; Philadelphia: Fortress), 170–5.

—— (1987), *Paul and the Thessalonians. The Philosophic Tradition of Pastoral Care* (Philadelphia: Fortress).

—— (1990), 'Did the Thessalonians Write to Paul?', in *The Conversation Continues: Studies in Paul and John in Honor of J. Louis Martyn*, ed. R. T. Fortna and B. R. Gaventa (Nashville: Abingdon), 246–57.

MANSON, T. W. (1952), 'Paul in Greece: The Letters to the Thessalonians', *BJRL* 35: 428–47.

—— (1962), 'St. Paul's Letter to the Romans—and Others', in *Studies in the Gospels and Epistles*, ed. M. Black (Manchester: Manchester University Press), 225–41.

MARROU, H.-I. (1948), *Histoire de l'éducation dans l'antiquité* (Paris: Seuil).

MARSHALL, I. H. (1980), *The Acts of the Apostles: An Introduction and Commentary* (TynNTC; Leicester: Inter-Varsity Press).

MARSHALL, P. (1987), *Enmity in Corinth: Social Conventions in Paul's Relations with the Corinthians* (WUNT 2/23; Tübingen: Mohr).

MARTIN, R. P. (1976), *Philippians* (NCB; London: Oliphants).

—— (1983), *Carmen Christi: Philippians 2: 5–11 in Recent Interpretation and the Setting of Early Christian Worship* (rev. edn.; Grand Rapids, Mich.: Eerdmans).

—— (1986), *2 Corinthians* (WBC 40; Waco, Tex.: Word Books).

MARTYN, J. L. (1985), 'A Law-Observant Mission to Gentiles: The Background of Galatians', *SJT* 38: 307–24 = *MQR* 22: 221–36.

MARXEN, W. (1968), *Introduction to the New Testament: An Approach to its Problems* (Oxford: Blackwell).

MASSON, C. (1962), 'A propos de Actes 9, 19b–25: Note sur l'utilisation de Gal. et de 2 Cor. par l'auteur des Actes', *TZ* 18: 161–6.

MATERA, F. J. (1992), *Galatians* (Sacra Pagina, 9; Collegeville, Minn.: Liturgical Press).

MEARNS, C. L. (1981), 'Early Eschatological Development in Paul: The Evidence of I and II Thessalonians', *NTS* 27: 137–57.

MEEKS, W. (1983), *The First Urban Christians: The Social World of the Apostle Paul* (New Haven/London: Yale University Press).

—— and Wilken, R. (1978), *Jews and Christians in Antioch in the First Four Centuries of the Common Era* (Missoula, Mont.: Scholars Press).

MEIER, J. P. (1990), 'Jesus in Josephus: A Modest Proposal', *CBQ* 52: 76–103.

MEINARDUS, O. (1978), 'Paul's Missionary Journey to Spain: Tradition and Folklore', *BA* 41: 61–3.

MENDELSON, A. (1982), *Secular Education in Philo of Alexandria* (Monographs of the Hebrew Union College, 7; Cincinnati: Hebrew Union College).

MENGEL, B. (1982), *Studien zum Philipperbrief* (WUNT 2/8; Tübingen: Mohr).

MERKLEIN, H. (1984), 'Die Einheitlichkeit des ersten Korintherbriefes', *ZNW* 75: 153–83.

—— (1992), *Der erste Brief an die Korinther: Kapitel 1–4* (Ökumenischer Taschenbuchkommentar zum Neuen Testament, 7/1; Gütersloh: Mohn/Würzburg: Echter).

MESHORER, Y. (1975), *Nabataean Coins* (Qedem, 3; Jerusalem: Hebrew University Press).

MESSADIÉ, G. (1991), *L'Incendiaire: Vie de Saul, apôtre* (Paris: Laffont)

METZGER, B. (1971), *A Textual Commentary on the Greek New Testament* (London/New York: United Bible Societies).

MICHEL, A., AND LeMOYNE, J. (1965), 'Pharisiens', *DBSup* 7: 1022–1115.

MICHEL, O. (1929), *Paulus und seine Bibel* (Gütersloh: Bertelsmann).

MILLAR, F. (1981), 'The World of the *Golden Ass,*' *JRS* 71: 63–75.

MITCHELL, M. (1989), 'Concerning *peri de* in 1 Corinthians', *NovT* 31: 229–56.

—— (1991), *Paul and the Rhetoric of Reconciliation: An Exegetical Investigation of the Language and Composition of 1 Corinthians* (HUZT 28; Tübingen: Mohr).

MITCHELL, S. (1980), 'Population and Land in Roman Galatia', *ANRW* II, 7/2. 1053–81.

—— (1992), 'Galatia', *ABD* 2. 870–2.

—— (1993), *Anatolia. Land, Men and Gods in Asia Minor,* i. *The Celts in Anatolia and the Impact of Roman Rule* (Oxford: Clarendon).

MOMMSEN, T. (1955), *Römisches Strafrecht* (Graz: Akademische Druck- und Verlagsanstalt).

MORRIS, L. (1956), '*Kai hapax kai dis*', *NovT* 1: 205–8.

MOULE, C. F. D. (1957), *The Epistles to the Colossians and to Philemon* (CGTC; Cambridge, Mass.: CUP).

—— (1965), 'The Problem of the Pastoral Epistles: A Reappraisal', *BJRL* 47: 430–52.

MOWINCKEL, S. (1959), *He that Cometh* (Oxford: Blackwell).

MULLINS, T. Y. (1957), 'Paul's Thorn in the Flesh', *JBL* 76: 299–303.

MUNCH, J. (1959), *Paul and the Salvation of Mankind* (London: SPCK).

MUNRO, J. A. R., AND ANTHONY, H. M. (1897), 'Explorations in Mysia', *Geographical Journal,* 9: 150–69, 256–76.

MURPHY-O'CONNOR, J. (1964), *Paul on Preaching* (London/New York: Sheed and Ward).

Murphy-O'Connor, J. (1965), 'Philippiens (Épitre aux)', *DBSup* 7: 1211–33.

—— (1974*a*), 'The Essenes and their History', *RB* 81: 215–44.

—— (1974*b*), *L'Existence chrétienne selon saint Paul* (LD 80; Paris: Cerf).

—— (1976), 'Christological Anthropology in Phil 2: 6–11', *RB* 83: 25–50.

—— (1977*a*), 'I Corinthians. V. 3–5', *RB* 84: 239–45.

—— (1977*b*), 'Works without Faith in 1 Cor 7: 14', *RB* 84: 349–61.

—— (1978*a*), 'Corinthian Slogans in 1 Cor 6: 12–20', *CBQ* 40: 391–6.

—— (1978*b*), 'Freedom or the Ghetto (1 Cor. 8: 1–13; 10:23–11: 1', *RB* 85: 543–74.

—— (1979), 'Food and Spiritual Gifts in 1 Cor 8: 8', *CBQ* 41: 292–8.

—— (1980), 'Sex and Logic in 1 Corinthians 11: 2–6', *CBQ* 42: 482–500.

—— (1981*a*), 'The Divorced Woman in 1 Cor 7: 10–11', *JBL* 100: 901–6.

—— (1981*b*), 'Tradition and Redaction in 1 Cor 15: 3–7', *CBQ* 43: 582–9.

—— (1982*a*), 'Pauline Missions before the Jerusalem Conference', *RB* 89: 71–91.

—— (1982*b*), *Becoming Human Together: The Pastoral Anthropology of St. Paul* (2nd edn.; GNS 2; Wilmington, DE Glazier).

—— (1984), 'The Corinth that Saint Paul Saw', *BA* 47: 147–59.

—— (1985), 'The *Damascus Document* Revisited', *RB* 92: 239–41.

—— (1986*a*), 'Interpolations in 1 Corinthians', *CBQ* 48: 81–94.

—— (1986*b*), '*Pneumatikoi* and Judaizers in 2 Cor 2:14–4:6', *AusBR* 34: 42–58.

—— (1987), 'A Ministry beyond the Letter (2 Cor 3: 1–6)', in *Paolo: Ministro del nuovo testamento (2 Co 2,14–4,6)*, ed. L. De Lorenzi Serie Monografica di 'Benedictina' 9; Roma: Benedictina Editrice), 105–29.

—— (1988*a*), 'Philo and 2 Cor 6:14–7: 1', *RB* 95: 55–69.

—— (1988*b*), '1 Corinthians 11: 2–26 Once Again', *CBQ* 50: 265–74.

—— (1988*c*), '*Pneumatikoi* in 2 Corinthians', *PIBA* 11: 59–66.

—— (1989), 'The New Covenant in the Letters of Paul and the Essene Documents', in *To Touch the Text: Biblical and Related Studies in Honor of Joseph A. Fitzmyer, S.J.*, ed. M. P. Horgan and P. J. Kobelski (New York: Crossroad), 194–204.

—— (1990), 'Another Jesus (2 Cor 11: 4)', *RB* 97: 238–51.

—— (1991*a*), '2 Timothy contrasted with 1 Timothy and Titus', *RB* 98: 403–18.

—— (1991*b*), 'The Date of 2 Corinthians 10–13', *AusBR* 39: 31–43.

—— (1992*a*), 'A First-Century Jewish Mission to Gentiles?', *Pacifica* 5: 32–42.

—— (1992*b*), 'Corinth', *ABD* 1. 1134–9.

—— (1992*c*), 'Lots of God-fearers: *Theosebeis* in the Aphrodisias Inscription', *RB* 99: 418–24.

—— (1992*d*), 'Prisca and Aquila: Travelling Tentmakers and Church Builders', *BR* 8/6: 40–51.

—— (1992*e*), *St. Paul's Corinth. Texts and Archaeology* (expanded edn., GNS 6; Collegeville, Mass.: Liturgical Press).

—— (1993), 'Co-authorship in the Corinthian Correspondence', *RB* 100: 562–79.

—— (1995), *St Paul the Letter-Writer: His World, His Options, His Skills* (GNS 41; Collegeville: Liturgical Press).

Negev, A. (1977), 'The Nabataeans and the Provincia Arabia', *ANRW* II/8. 549–635.

Neirynck, F. (1986), 'Paul and the Sayings of Jesus', in *L'Apôtre Paul: Personalité, style et conception du ministère*, ed. A. Vanhoye (BETL 73; Leuven: Leuven University Press), 265–321.

NEUMANN, K. J. (1990), *The Authenticity of the Pauline Epistles in the Light of Stylostatistical Analysis* (SBLDS 120; Atlanta: Scholars Press).

NEUSNER, J. (1971), *The Rabbinic Traditions concerning the Pharisees before 70*, i–iii (Leiden: Brill).

—— (1972), 'Josephus's Pharisees', in *Ex Orbe Religionum*, i (FS G. Widengren; Studies in the History of Religions, 21; Leiden: Brill), 224–44.

NIEBUHR, K.-W. (1992), *Heidenapostel aus Israel: Die jüdische Identität des Paulus nach ihrer DArstellung in seinen Briefen* (WUNT 62; Tübingen: Mohr).

NOLLAND, J. (1981), 'Uncircumcised Proselytes?', *JSJ* 12: 173–94.

NORRIS, F. W. (1992), 'Antioch of Syria', *ABD* 1. 265–9.

O'BRIEN, P. (1982) *Colossians, Philemon* (WBC 44; Waco: Word Books).

O'COLLINS, G. (1971), 'Power made Perfect in Weakness: 2 Cor 12: 9–10', *CBQ* 33: 528–37.

O'SULLIVAN, F. (1972), *The Egnatian Way* (Newton Abbot: David and Charles).

OBERHUMMER, E. (1905), 'Egnatia, via', *PW* 5. 1988–93.

—— (1936), 'Thessalonike', *PW* zweite Reihe 6/A1: 143–63.

OLIVER, J. H. (1970), 'The Epistle of Claudius Which Mentions the Proconsul Junius Gallio', *Hesperia*, 40: 239–40.

OSBORNE, R. E. (1966), 'Paul and the Wild Beasts', *JBL* 85: 225–30.

OSTER, R. E. (1992), 'Ephesus', *ABD* 2. 542–49.

PACKER, J. E. (1967), 'Housing and Population in Imperial Ostia and Rome', *JRS* 57: 80–95.

PAPAZOGLOU, F. (1979), 'Quelques aspects de l'histoire de la province de Macédoine', *ANRW* II, 7/1. 302–69.

—— (1982), 'Le Territoire de la colonie de Philippes', *BCH* 106: 86–106.

PEARSON, B. (1971), '1 Thessalonians 2: 13–16: A Deutero-Pauline Interpolation', *HTR* 64: 79–94.

PELLETIER, A. (1970), 'Les Apparitions du Ressuscité en termes de la Septante', *Biblica* 51: 76–9.

PENNA, R. (1982), 'Les Juifs à Rome au temps de l'apôtre Paul', *NTS* 28: 321–47.

—— (1992), 'Judaism (Rome)', *ABD* 3. 1073–6.

PERRIN, N. (1976), *Rediscovering the Teaching of Jesus* (New York: Harper and Row).

PFITZNER, V. (1976), *Paul and the Agon Motif: Traditional Athletic Imagery in the Pauline Literature* (NovTSup 17; Leiden: Brill).

PHERIGO, L. P. (1951), 'Paul's Life after the Close of Acts', *JBL* 70: 277–84.

PHILLIPS, J. B. (1955), *Letters to Young Churches: A Translation of the New Testament Epistles* (London: Fontana).

PLASSART, A. (1967), 'L'Inscription de Delphes mentionnant le proconsul Galion', *REG* 80: 372–8.

—— (1970), *Les Inscriptions du temple du IV siècle* (Fouilles de Delphes, III/4; Paris: École Française d'Athénes).

PLEVNIK, J. (1984), 'The Taking up of the Faithful and the Resurrection of the Dead in 1 Thessalonians 4: 13–18', *CBQ* 46: 274–83.

PLUMMER, A. (1915), *A Critical and Exegetical Commentary on the Second Epistle of St Paul to the Corinthians* (ICC; Edinburgh: Clark).

POLENZ, M. (1946), 'Paulus und die Stoa', *ZNW* 42: 69–104.

POLLACK, A. (1899), 'Carcer', PW 3. 1576–82.

PORTEFAIX, L. (1988), _Sisters Rejoice: Paul's Letter to the Philippians and Luke-Acts as Seen by First-Century Philippian Women_ (ConBNT 20; Stockholm: Almquist and Wiksell).

PREMERSTEIN, G. VON (1917), 'Ius Italicum', PW 10. 1238–53.

PREUSCHEN, E. (1901), 'Paulus als Antichrist', _ZNW_ 2: 169–201.

PRIOR, M. (1989), _Paul the Letter-Writer and the Second Letter to Timothy_ (JSNTSup 23; Sheffield: JSOT Press).

PRITCHARD, J. B. (1987) (ed.), _The Times Atlas of the Bible_ (London: Times Books).

PROBST, H. (1991), _Paulus und der Brief: Die Rhetorik des antiken Briefes als Form der paulinischen Korinther-korrespondenz (1 Kor 8–10)_ (WUNT 2/45; Tübingen: Mohr).

RADKE, G. (1973), 'Viae publicae Romanae', PWSup 13. 1418–1686.

RAJAK, T., AND NOY, D. (1993), '_Archisynagogoi_: Office, Title and Social Status in the Greco-Jewish Synagogue', _JRS_ 83: 75–93.

RAMSAY, W. M. (1897), _St. Paul the Traveller and Roman Citizen_ (London: Hodder and Stoughton).

—— (1899), 'The Philippians and their Magistrates, 2. On the Greek Form of the Name Philippians', _JTS_ 1: 116.

—— (1900), _A Historical Commentary on St. Paul's Epistle to the Galatians_ (2nd edn.; London: Hodder and Stoughton).

—— (1904), 'The Tarsian Citizenship of St Paul', _ExpTim_ 16: 18–21.

—— (1907), _The Cities of St. Paul and their Influence on his Life and Thought: The Cities of Eastern Asia Minor_ (London: Hodder and Stoughton).

REFOULÉ, F. (1990), 'A contre-courant Rom 16: 3–16', _RHPR_ 70: 409–20.

RICHARDS, E. R. (1991), _The Secretary in the Letters of Paul_ (WUNT 2/42; Tübingen: Mohr).

RIESENFELD, H. (1960), 'Le Langage parabolique dans les épitres de saint Paul', in _Littérature et théologie pauliniennes_, ed. A. Descamps (RechBib 5; ed. A. Descamps; Bruges: Desclée de Brouwer), 47–59.

RIGAUX, B. (1956), _Saint Paul: Les épitres aux Thessaloniciens_ (ÉBib; Paris: Gabalda).

RITTERLING, E. (1927), 'Military Forces in the Senatorial Provinces', _JRS_ 17: 28–32.

ROBERTSON, A., AND PLUMMER, A. (1914), _A Critical and Exegetical Commentary on the First Epistle of St Paul to the Corinthians_ (ICC; Edinburgh: Clark).

ROBINSON, J. A. T. (1952), _The Body: A Study in Pauline Theology_ (SBT 5; London: SCM).

—— (1976), _Redating the New Testament_ (London: SCM).

ROLLAND, P. (1992), 'Discussions sur la chronologie paulinienne', _NRT_ 114: 870–89.

ROSSITER, S. (1981), _Greece_ (Blue Guides; London: Benn).

SABUGAL, S. (1976), _Análisis exegético sobre la conversión de san Pablo: El problema teológico e histórico_ (Barcelona: Herder).

SAFRAI, S. (1974), 'Relations between the Disapora and the Land of Israel', in _The Jewish People in the First Century: Historical Geography, Political History, Social, Cultural, and Religious Life and Institutions_, ed. S. Safrai _et al._ (Compendia Rerum Iudaicarum ad Novum Testamentum, section 1, vol. 1; Assen: Van Gorcum), 184–215.

—— (1976), 'Education and the Study of the Torah', in _The Jewish People in the First Century: Historical Geography, Political History, Social, Cultural, and Religious Life_ *and Institutions*, ed. S. Safrai *et al.* (Compendia Rerum Iudaicarum ad Novum Testamentum, section 1, vol. 2; Philadelphia: Fortress), 945–70.

SALDARINI, A. J. (1988), *Pharisees, Scribes and Sadducees in Palestinian Society* (Edinburgh: Clark).

SALMON, J. B. (1984), *Wealthy Corinth: A History of the City to 338 BC* (Oxford: Clarendon).

SANDAY, W., AND HEADLAM, A. C. (1902), *A Critical and Exegetical Commentary on the Epistle to the Romans* (ICC; Edinburgh: Clark).

SANDERS, E. P. (1977), *Paul and Palestinian Judaism: A Comparison of Patterns of Religion* (London: SCM).

—— (1990), 'Jewish Association with Gentiles and Galatians 2: 11–14', in *The Conversation Continues: Studies in Paul and John in Honor of J. Louis Martyn*, ed. R. T. Fortna and B. R. Gaventa (Nashville: Abingdon), 170–88.

SANDERS, J. T. (1962), 'The Transition from Opening Epistolary Thanksgiving to Body in the Letters of the Pauline Corpus', *JBL* 81: 348–62.

SANDNES, K. O. (1991), *Paul—One of the Prophets? A Contribution to the Apostle's Self-Understanding* (WUNT 2, 43; Tübingen: Mohr).

SÄNGER, D. (1985), 'Die *dynatoi* in 1 Kor 1:26', *ZNW* 76: 285–91.

SAPPINGTON, T. J. (1991), *Revelation and Redemption at Colossae* (JSNTSup 53; Sheffield: JSOT Press).

SAULNIER, C. (1981), 'Lois romaines sur les Juifs selon Flavius Josèphe', *RB* 88: 161–98.

—— (1984), 'Hérode Antipas et Jean le Baptiste: Quelques remarques sur les confusions chronologiques de Flavius Josèphe', *RB* 91: 365–71.

SAUNDERS, E. W. (1967), 'The Colossian Heresy and Qumran Theology', in *Studies in the History and Text of the New Testament in Honor of K. W. Clark*, ed. B. L. Daniels and M. J. Suggs (SD 29; Salt Lake City: University of Utah Press), 133–45.

SCHENKE, W. (1969), 'Der 1. Korintherbrief als Briefsammlung', *ZNW* 60: 219–43.

SCHLIER, H. (1962), *Der Brief an die Galater* (MeyerK; Göttingen: Vandenhoeck & Ruprecht).

SCHMELLER, T. (1987), *Paulus und die 'Diatribe': Eine vergleichende Stilinterpretation* (NTAbh NF 19; Münster: Aschendorff).

SCHMIDT, A. (1992), 'Das Missionsdekret in Galater 2. 7–8 als Vereinbarung vom ersten Besuch Pauli in Jerusalem', *NTS* 38: 149–52.

SCHMIDT, J. (1938), 'Philippoi', *PW* 19/2. 2206–44.

SCHMITHALS, W. (1964), 'Die Thessalonicherbriefe als Briefkompositionen', in *Zeit und Geschichte: Dankesgabe an Rudolf Bultmann zum 80. Geburtstag*, ed. E. Dinkler (Tübingen: Mohr), 295–315.

—— (1973), 'Die Korintherbriefe als Briefsammlung', *ZNW* 64: 263–88.

—— (1975), *Der Römerbrief als historisches Problem* (StNT 9; Gütersloh: Mohn, 1975).

SCHNACKENBURG, R. (1965), *Das Johannesevangelium. I. Teil. Einleitung und Kommentar zu Kap. 1–4* (HTKNT; Freiburg: Herder).

SCHNEIDER, K. (1932), 'Taberna', in *PW Zweite Reihe* 4. 1863–72.

SCHOENE, A. (1875) (ed.), *Eusebi Chronicorum libri duo* (Berlin: Weidmann).

SCHOTT, H. A. (1830), *Isagoge historico-critica in libros Novi Foederis sacros* (Jena: Beck).

SCHUBERT, P. (1939), *Form and Function of the Pauline Thanksgivings* (BZNW 20; Berlin: Töpelmann).

SCHULZ, F. (1942–3), 'Roman Registers of Birth and Birth Certificates', *JRS* 32: 78–91; 33: 55–64.

SCHÜRER, E., VERMES, G., AND MILLAR, F. (1973–87), *The History of the Jewish People in the Age of Jesus Christ (175 B.C.–A.D. 135)*, i–iii (Edinburgh: Clark).

SCHWANK, B. (1983), 'Neue Funde in Nabatäerstädten und ihre Bedeutung für die neutestamentliche Exegese', *NTS* 29: 434–5.

SCHWARTZ, D. R. (1990), *Agrippa I: The Last King of Judaea* (TSAJ 23; Tübingen: Mohr).

SCULLARD, H. H. (1982), *From the Gracchi to Nero: A History of Rome from 133 B.C. to A.D. 68* (5th edn.; London/New York: Methuen).

SEGAL, A. F. (1990), *Paul the Convert: The Apostolate and Apostasy of Saul the Pharisee* (New Haven/London: Yale University Press).

SELLIN, G. (1987), 'Hauptprobleme des ersten Korintherbriefes', ANRW II, 25/4. 2941–3044.

SENFT, C. (1979), *La Première Épitre de saint Paul aux Corinthiens* (CNT 7 NS; Neuchâtel-Paris: Delachaux et Niestlé).

SHERK, R. K. (1980), 'Roman Galatia: The Governors from 25 BC to AD 114', *ANRW* II, 7/2. 954–1052.

SHERWIN-WHITE, A. N. (1963), *Roman Society and Roman Law in the New Testament* (Oxford: Clarendon).

—— (1972), 'The Roman Citizenship: A Survey of its Development into a World Franchise', *ANRW* I, 2. 23–58.

—— (1973), *The Roman Citizenship* (2nd edn.; Oxford: Clarendon).

SLINGERLAND, D. (1988), 'Suetonius *Claudius* 25.4 and the Account in Cassius Dio', *JQR* 79: 305–22.

—— (1989), 'Chrestus: Christus?', in *New Perspectives on Ancient Judaism*, iv. *The Literature of Early Rabbinic Judaism*, ed. A. J. Avery-Peck (Lanham: University Press of America), 133–44.

—— (1990), 'Acts 18: 1–17 and Luedemann's Pauline Chronology', *JBL* 109: 686–90.

SMALLWOOD, E. M. (1981), *The Jews under Roman Rule from Pompey to Diocletian: A Study in Political Relations* (SJLA 20; Leiden: Brill).

SMITH, D, E. (1992), 'Table Fellowship', *ABD* 6. 302–4.

SPICQ, C. (1937), 'L'Image sportive de 2 Cor 4: 7–9', *ETL* 13: 209–29.

—— (1959), *Agapè dans le Nouveau Testament: Analyse des textes*, ii (ÉBib; Paris: Gabalda).

—— (1969), *Saint Paul: Les Épitres Pastorales* (2nd edn.; ÉBib; Paris: Gabalda).

—— (1978–82), *Notes de lexicographie néo-testamentaire*, i–iii (OBO 22/1–3; Fribourg: Editions Universitaires).

STANLEY, D. M. (1953), 'Why Three Accounts?', *CBQ* 15: 315–38.

STARCKY, J. (1966), 'Pétra et la Nabatène', *DBSup* 7: 886–1017.

STEGEMANN, H. (1994), *Die Essener, Qumran, Johannes der Täufer und Jesus: Ein Sachbuch* (4th revised edn.; Freiburg/Basel/Wien: Herder).

STEGEMANN, W. (1987), 'War der Apostel Paulus ein römischer Bürger?', *ZNW* 76: 200–29.

STENDAHL, K. (1963), 'The Apostle Paul and the Introspective Conscience of the West', *HTR* 56: 199–215.

—— (1977), *Paul among Jews and Gentiles* (Philadelphia, Fortress).

STOWERS, S. K. (1981), *The Diatribe and Paul's Letter to the Romans* (SBLDS 57; Chico: Scholars Press).

—— (1986), *Letter Writing in Greco-Roman Antiquity* (Philadelphia: Westminster).

—— (1987), Review of Betz (1985) in *JBL* 106: 727–30.

STROBEL, A. (1965), 'Der Begriff des "Hauses" im griechischen und römischen Privatrecht', *ZNW* 51: 91–100.

—— (1969), 'Schreiben des Lukas? Zum sprachlichen Problem der Pastoralbriefe', *NTS* 15: 191–210.

SUHL, A. (1975), *Paulus und seine Briefe: Ein Beitrag zur paulinischen Chronologie* (Gütersloh: Mohn).

TAJRA, H. W. (1989), *A Juridical Exegesis of the Second Half of the Acts of the Apostles* (WUNT 2/35; Tübingen: Mohr, 1989).

TAYLOR, J. J. (1992), 'The Ethnarch of King Aretas at Damascus: A Note on 2 Cor 11: 32–33', *RB* 99: 719–28.

—— (1994a), 'Why were the Disciples first called "Christians" at Antioch?', *RB* 101: 75–94.

—— (1994b), *Les Actes des deux apôtres*, v. *Commentaire historique (Act 9, 1–18, 22)* (ÉBib; Paris: Gabalda).

—— (forthcoming), 'Why did Paul Persecute the Church?'.

TAYLOR, N. (1993), *Paul, Antioch and Jerusalem: A Study in Relationships and Authority in Earliest Christianity* (JSNTSup 66; Sheffield: JSOT Press).

TCHERIKOVER, V. (1959), *Hellenistic Civilization and the Jews* (Philadelphia: Jewish Publication Society).

THEISSEN, G. (1982), *The Social Setting of Pauline Christianity: Essays on Corinth* (Philadelphia: Fortress).

THORNTON, T. C. G. (1972), 'Jewish Batchelors in New Testament Times', *JTS* 23: 444–5.

TRILLING, W. (1972), *Untersuchungen zum zweiten Thessalonicherbrief* (Leipzig: St Benno).

TROBISCH, D. (1989), *Die Entstehung der Paulusbriefsammlung. Studien zu den Anfängen christlicher Publizistik* (NTOA 10; Freiburg: Universitätsverlag/Göttingen: Vandenhoeck and Ruprecht).

TROCMÉ, E. (1957), *Le Livre des Actes et l'histoire* (Paris: Presses Universitaires de France).

TUCKETT, C. M. (1990), 'Synoptic Tradition in 1 Thessalonians', in *The Thessalonian Correspondence* (BETL 87; ed. R. F. Collins; Leuven: Leuven University Press), 160–82.

UNNIK, W. C. VAN (1943), 'Aramaisms in Paul', *Vox Theologica*, 23: 117–26.

—— (1962), *Tarsus or Jerusalem: The City of Paul's Youth* (London: Epworth, 1962).

URBACH, E. E. (1979), *The Sages: Their Concepts and Beliefs* (2nd edn.; Jerusalem: Magnes Press).

VAUX, R. DE (1973), *Archaeology and the Dead Sea Scrolls* (London: Oxford University Press).

VIELHAUER, P. (1974), 'Paulus und die Kephaspartei in Korinth', *NTS* 21: 341–52.

VIVIANO, B. T. (1978), *Study as Worship: Aboth and the New Testament* (SJLA 26; Leiden: Brill).

VLIET, H. VAN (1958), *No Single Testimony: A Study on the Adoption of the Law of Deut. 19: 15 Par. into the New Testament* (STRT 4; Utrecht: Kemink and Zoon).

WACHOLDER, B. Z. (1962), *Nicolaus of Damascus* (UC Publications in History 75; Berkeley and Los Angeles: University of California Press).

WAELE, F. J. DE (1930), 'The Roman Market North of the Temple at Corinth', *AJA* 34: 432–54.

WALKER, W. O. (1992), 'Why Paul Went to Jerusalem: The Interpretation of Galatians 2: 1–5', *CBQ* 54: 503–10.

WATSON, D. F. (1988), 'A Rhetorical Analysis of Philippians and Its Implications for the Unity Question', *NovT* 30: 57–88.

WATSON, F. (1984), '2 Cor x–xiii and Paul's Painful Letter to the Corinthians', *JTS* 35: 324–46.

—— (1986), *Paul, Judaism and Gentiles: A Sociological Approach* (SNTSMS 56; Cambridge, CUP).

WEDDERBURN, A. J. M. (1988), *The Reasons for Romans* (Edinburgh: Clark).

WELBORN, L. (1987), 'On the Discord in Corinth: 1 Corinthians 1–4 and Ancient Politics', *JBL* 106: 85–111.

WESTERHOLM, S. (1988), *Israel's Law and the Church's Faith: Paul and his Recent Interpreters* (Grand Rapids, Mich.: Eerdmans).

WHITE, J. L. (1986), *Light from Ancient Letters* (Foundations & Facets: NT; Philadelphia: Fortress).

WIDMANN, W. (1979), '1. Kor 2: 6–16: Ein Einspruch gegen Paulus', *ZNW* 70: 44–53.

WIEFEL, W. (1990), 'The Jewish Community in Ancient Rome and the Origins of Roman Christianity', in *The Conversation Continues: Studies in Paul and John in Honor of J. Louis Martyn*, ed. R. T. Fortna and B. R. Gaventa (Nashville: Abingdon), 85–101.

WILD, R. A. (1990), 'The Pastoral Letters', *NJBC* §56.

WILKINSON, J. (1974), 'L'Apport de saint Jérome à la topographie', *RB* 81: 245–57.

—— (1977), *Jerusalem Pilgrims before the Crusades* (Jerusalem: Ariel).

WILL, E., AND ORRIEUX, C. (1992), *Prosélytisme juif? Histoire d'une erreur* (Paris: Belles Lettres).

WILLIAMS, S. K. (1987), 'Again *pistis Christou*', *CBQ* 49: 431–47.

—— (1989), 'The Hearing of Faith: *Akoê pisteôs* in Galatians 3', *NTS* 35: 82–93.

WINDISCH, H. (1924), *Der zweite Korintherbrief* (MeyerK; Göttingen: Vandenhoeck and Ruprecht).

—— (1934), *Paulus und Christus: Ein biblisch-religionsgeschichtlicher Vergleich* (UNT 24; Leipzig: Hinrich).

WISEMAN, J. (1978), *The Land of the Ancient Corinthians* (Göteburg: Aström).

WOLFF, C. (1982), *Der erste Brief des Paulus an die Korinther: Zweiter Teil, Auslegung der Kapital 8–16* (THKNT 7/2; Berlin: Evangelische Verlagsanstalt).

—— (1989), *Der zweite brief des Paulus an die Korinther* (THKNT 8; Berlin: Evangelische Verlagsanstalt).

WOLTER, M. (1987), 'Apollos und die ephesinischen Johannesjünger (Act 18:24–19:7)', *ZNW* 78: 49–73.

WREDE, W. (1903), *Die Echtheit des zweiten Thessalonicherbriefs untersucht* (TU 24/NF 9; Leipzig: Hinrichs).

WRIGHT, N. T. (1990), 'Poetry and Theology in Colossians 1, 15–20', *NTS* 36: 444–68.

WUELLNER, W. (1990), 'The Argumentative Structure of 1 Thessalonians as Paradoxical Encomium', in *The Thessalonian Correspondence*, ed. R. F. Collins (BETL 80; Leuven: Leuven University Press), 117–36.

YAMAUCHI, E. M. (1992), 'Troas', *ABD* 6. 666–7.

ZAHN, T. (1890), *Geschichte des Neutestamentlichen Kanons* (Erlangen/Leipzig: Deichert).

—— (1900), *Einleitung in das Neue Testament* (2nd edn.; Leipzig: Deichert).

ZERWICK, M. (1953), *Analysis Philologica Novi Testamenti Graeci* (Rome: Pontifical Biblical Institute).

ZIMMER, G. (1985), 'Romische Handwerker', *ANRW* II, 12/3. 205–28.

ZMIJEWSKI, J. (1978), *Der Stil der paulinischen 'Narrenrede': Analyse der Sprachgestaltung in 2 Kor 11,1–12,10 als Beitrag zur Methodik von Stiluntersuchungen neutestamentlicher Texte* (BBB 52; Köln-Bonn: Hanstein).

Index of Passages Cited

I. OLD TESTAMENT

III. JEWISH WRITINGS

IV. CHRISTIAN WRITINGS

Acts of John
61–2 99

Acts of Paul
3. 1 44

Acts of Peter
1. 1 362
2. 6 362

Ascension of James
frag. 63

Clement of Alexandria
Stromata
3. 6. 52 64

Clement of Rome
Epistle to the Corinthians
5. 5–7 361
5. 6 354

Epiphanius
Panarion
30. 16. 9 63

Epistle of Peter to James
2: 3–5 134

Epistula Apostolorum
33 53

Eusebius
History of the Church
2. 15 233
2. 22 360
3. 2 360
3. 30. 1 64
3. 36–9 233

Muratortian Fragment
38–9 361

Jerome
Comm. in Ep. ad Galatas
2. 3 189
Comm. in Ep. ad Philemon
vv. 23–4 37, 38

Epistola 108
13. 5 38
De Viris Illustribus
5 37, 38

Pseudo-Clement
Recognitions
1. 42. 1 135
17. 19. 4 79

V. CLASSICAL AUTHORS

Aelius Aristides
Discourses
46. 24 109
48 22

Aelius Gellius
Attic Nights
1. 8. 3–4 109
1. 8. 4 258

Appian
Civil Wars
4. 105–6 212
4. 106 211
History
5. 1. 7 34
10 316

Apuleius
Metamorphoses
1. 7 98
1. 24 99
2. 18 98, 100
3. 3 100
3. 28 99, 100
3. 29 98
4. 3 99
4. 4–18 99
4. 10 99

4. 12 99
4. 13 98, 297
7. 1 99
7. 7 97, 100
7. 11–12 100
7. 22–4 297
7. 23 99
7. 24 98
8. 4 98
8. 15 98
8. 15–23 99
8. 16 98
8. 17 99
9. 35 100
10. 18 297

Arrianus
Dissertationes
2. 5. 24–7 245
2. 9. 20 138
2. 10. 3–4 245

Athenaeus
Deipnosophistae
4. 151e–152b 186
4. 154b 186
6. 246b 187
13. 603a 187

Bordeaux Pilgrim
Itinerarium 99, 103

Cicero
Ad Atticum
1. 19 325
3. 14. 2 114
5. 11–13 279
5. 15 233
Contra Verrem
2. 5. 30 87
2. 5. 80 87
2. 5. 139 39
2. 5. 149–51 39
Ad Familiares
3. 1 325
3. 5 233
8. 8 325
12. 24 325
16. 21 325
De Officiis
1. 47 305
150–1 40, 89
De Oratore
1. 11. 47 51
Pro Flacco
66–9 345
68 232

6. 34 33
7. 22 179

Pliny

Natural History
2. 122 20
2. 276 62
4. 9–11 258
4. 36 115
5. 74 81
5. 105 232
5. 115 166
5. 147 159
11. 131 45
11. 275–6 44
19. 4 329
19. 22 87
19. 23–4 87
29. 33 191
31. 62 19
31. 84 191

Pliny the Younger

Letters
2. 14 213
6. 25 97
9. 21 177
9. 24 177
10. 15 296
10. 17 296, 364
10. 96 68, 269

Plutarch
Anthony
25–8 34
36. 2 81

Cato Major
14 363

Quaestiones conviviales
5. 3. 1–3 (675D–677B) 259

Quintilian

Institutio Oratoria
2. 13. 1–7 254
4. 2. 41 202
4. 2. 83–4 94, 132
4. 2. 101 202

Rhetorica ad Herrenium
3. 16 254

Sallust
War against Jugurtha
55. 3–4 354

Seneca
De Beneficiis
1. 4. 2 305
6. 3. 4 305
Epistulae morales
6. 5 121
104. 1 19

Strabo

Geography
2. 1. 1 361
2. 5. 32 81
3. 1. 4 330, 361
3. 2. 15 362
4. 1. 13 159, 185
4. 4. 2 190
4. 4. 5 190
6. 2. 11 352
6. 3. 7 363
7 frag. 10 104
7 frag. 21 114, 115
7 frag. 24 114
7. 7. 14 102
8. 6. 20 109, 217, 258
8. 6. 23 271
9. 1. 1–16 257
9. 1. 4 257
9. 1. 15–16 108
10. 3. 12 192
12. 5. 1 186
12. 5. 3 189, 192
12. 5. 4 189
12. 6. 1 190, 191
12. 7. 2 191
12. 8. 8 163
13. 1. 26 300
13. 4. 14 232
14. 1. 20 166
14. 1. 22–3 167
14. 1. 24 166
14. 2. 5 167
14. 2. 16 167
14. 2. 29 165, 231
14. 5. 11 49
14. 5. 13 35, 52, 108
14. 5. 13–14 49
14. 5. 14 35, 51
16. 1. 5 167
16. 2. 20 81
16. 4. 21 81
16. 4. 26 81
17. 1. 21 81
17. 1. 33 167

Suetonius
Augustus
32. 1 97
48 82
Caesar
39. 4 87
Claudius
15 20
18 20
25 140, 147
25. 4 9, 11
28 218
42 21
Nero
16 147
Tiberius
8 97, 233
36. 1 139
37. 1 97

Tacitus
Annals
1. 21 354
1. 76. 4 115
1. 80. 1 115
2. 57 7
2. 85. 5 139
6. 31–7 84
12. 54 23
13. 1 326
13. 10 371
14. 27. 1 233
14. 44. 2 147
15. 44 369
16. 23 166

Terence
The Brothers
571–85 360

The Eunuch
251–3 305

Vegetius
Epitoma rei militaris
4. 39 20

Vitruvius
De Architectura
1. 2. 7 246
8. 3. 10 232

Xenophon
Anabasis
1. 2. 6 232
1. 2. 23 33

General Index

view of slavery 249
visit to Spain 361–3
visits to Jerusalem 94, 130
wilful tantrum 223–4
workshop 117, 263, 267
Pergamum 175
persecution of church:
 Luke's version 65
 Paul's version 67
Pessinus 162, 164, 186, 189, 191, 192
Pharisees:
 in the Diaspora 57
 in Galilee 56, 58
 knowledge of Jesus 73
 origins and ethos 54–6, 150
 studies 59
Philadelphia 175
Philemon, location of house church 176
Philemon, written at Ephesus 178
Philinus, Gnaeus Babbius 270
Philippi:
 benefactions 217
 deities 213
 description 211–12
 founding of church 213
 judicial process 214–15
 place of prayer 213
 threat of Judaizers 228
 visits of Paul 178, 222, 364
 women leaders 224
Philippians:
 integrity 215–20
 written at Ephesus 178
Philo, influence on Apollos 275, 282
Phoebe 270, 325
Phrygia:
 judged by Greeks 189
 supreme deity 192
 territory 162
physiognomy 44
politeuma 222
postscript 254
praitorion 220, 222
precepts:
 destroy freedom 155–6
 Jesus 154
Previous Letter 252, 276, 279
Prisca and Aquila:
 converted in Rome 263
 employed Paul in Corinth 234, 261
 founded church in Ephesus 171–2
 placed in Ephesus 131, 171
 sent to Rome 329, 331
proverbs about Corinth 258

relatives of Paul 45–6

rhetorical techniques 320
returnees, advantages as missionaries 235
robbers 97
Roman:
 citizenship 39–41
 law, illustrated by *Acts* 179
 name 41–3
Romans:
 diatribe 334
 occasion 332
 originality 335
 textual problems 324
Rome:
 composition of church 333
 expulsion of Jews 333
 fire of 368
 journey from Jerusalem 351
 relations with Spain 362
 Paul's first imprisonment 354, 360
 Paul's second imprisonment 359
 plan to visit 323
 reason for Paul's return 369
 refused to commission Paul 362
 visitor's problem 359
runaway slaves 177

sailing, contrary winds 145, 165, 183, 256,
  296, 346
salvation of Jews 339–40
Sangarios, river 162, 189, 190, 192
Sardis 175
Saul 42
Sceironian rocks 257
Seneca 19, 21, 362
seven ages 1
seven wonders of the world 167
sin 100, 335–9
slavery 249, 271
slogans at Corinth 277
Smyrna 175
societal pressure 100, 208, 210, 227, 336, 338
society, characteristics of 288
Sosthenes, co-author 264, 308
South Galatia hypothesis 159
Spain:
 linguistic problems 362
 mission to be sponsored by Rome 329–30
 Paul's abortive visit 361–3
 reasons for choice 330
spirit-people:
 alliance with Judaizers 294, 302, 309
 humiliated by *1 Corinthians* 282, 294, 295,
  302, 304
 influenced by Philo 280–2
spiritual:
 gifts 288